Proactive Purchasing in
the Supply Chain

Proactive Purchasing in the Supply Chain

The Key to World-Class Procurement

David N. Burt

Sheila D. Petcavage

Richard L. Pinkerton

McGraw
Hill

New York   Chicago   San Francisco
Lisbon   London   Madrid   Mexico City
Milan   New Delhi   San Juan
Seoul   Singapore   Sydney   Toronto

1 2 3 4 5 6 7 8 9 0   DOC/DOC   1 7 6 5 4 3 2 1

ISBN 978-0-07-177061-3
MHID 0-07-177061-5

Sponsoring Editor	**Copy Editor**
Judy Bass	Susan Fox-Greenberg
Editing Supervisor	**Proofreader**
Stephen M. Smith	Ashish Janmeja
Production Supervisor	**Indexer**
Pamela A. Pelton	Robert Swanson
Acquisitions Coordinator	**Art Director, Cover**
Bridget L. Thoreson	Jeff Weeks
Project Manager	**Composition**
Vastavikta Sharma, Cenveo Publisher Services	Cenveo Publisher Services

Printed and bound by RR Donnelley.

McGraw-Hill books are available at special quantity discounts to use as premiums and sales promotions, or for use in corporate training programs. To contact a representative, please e-mail us at bulksales@ mcgraw-hill.com.

This book is printed on acid-free paper.

To Robert Porter Lynch, friend, colleague, and mentor for the past 20 years. Robert is the thought leader in collaboration, alliances, value network management, and trust. I am honored to call him a friend.

David Burt

This book is in part the result of a long friendship and working partnership with my colleague and mentor, Dr. Kenneth Killen. Ken has been instrumental in the development and direction of my career as I transitioned from industry to academics.

Sheila Petcavage

Dick Pinkerton dedicates this book to his long-time partner, Shirley Chalfont Temby, the world's greatest expediter for Ryan Aeronautical and Ling-Temco-Vought, whose loving hands chased parts for the Lunar Module.

About the Authors

David N. Burt is Professor Emeritus of Supply Management at the University of San Diego (USD). In 2008, he completed 50 years in the fields of procurement, supply management, and value network management. He was recognized by the National Association of Purchasing Management as a NAPM Professor in 1992.

Dr. Burt created USD's undergraduate and graduate programs in procurement during the 1980s. His greatest joy has been in finding meaningful employment for several hundred of his students. He is Founder and Director Emeritus of USD's Supply Chain Management Institute and of the Strategic Supply Management Forum, an annual meeting of innovative supply management professionals from North America, Europe, and Australia. Under Dr. Burt's leadership, the University initiated a graduate hybrid resident/Internet program in 2002. This program grants a Master of Science in Supply Chain Management (MS SCM) on successful completion of the two-year program. It transfers cutting-edge knowledge and practices to working professionals while minimizing time away from the participant's workplace, develops leaders in supply chain management, and provides immediate payback to sponsoring firms. USD's MS SCM is the only such program endorsed by the Institute for Supply Management.

Dr. Burt's publications include seven books and numerous articles. His articles have appeared in *The Journal of Purchasing, Harvard Business Review, Thexis,* MIT's *Sloan Management Review, The Journal of Marketing Research,* and *The California Management Review.*

He has been a buyer, negotiator, and CPO (Chief Procurement Officer) of three organizations. Dr. Burt began his career as an Air Force Procurement Officer during the 1950s at a time when the Air Force was one of the few organizations in the world to recognize the critical importance of procurement. As a result, at the tender age of 22, he gained responsibilities and experience which few purchasing people in their 30s and 40s enjoyed. His most productive years in the Air Force were as an Associate Professor of Logistics at the Air Force Institute of Technology's Graduate School of Logistics. During this assignment, he was an in-service consultant to the Air Force Director of Procurement and the program directors of the F-15 and B-1 programs. His final assignment in the Air Force was as Chief of Military Sales based in Canberra, Australia. Dr. Burt then began his civilian career as an Associate Professor of Acquisition Management at the Naval Postgraduate School in Monterey, California.

He studied under the great Lamar Lee, Jr., at Stanford University and subsequently became co-author with the late Admiral Lee and Dr. Donald W. Dobler of early editions of the textbook *Supply Management* (then known as *Purchasing and Materials Management: Text and Cases*).

In addition to his hands-on and academic experience, Dr. Burt has consulted with small, medium, and large businesses including IBM, Motorola, Lockheed, Avery Dennison, and Southern California Edison, with the objective of upgrading their procurement operations to world-class status. Dr. Burt received his B.A. in Economics from the University of Colorado, M.S.I.A. from the University of Michigan, and Ph.D. in Logistics from Stanford University.

Sheila D. Petcavage is Assistant Professor and Coordinator of Business, Math, and Technology Associate Degree Programs at Cuyahoga Community College Western Campus in Cleveland, Ohio. She received a B.A. in Business Administration with honors and an M.B.A. from Baldwin Wallace College in Cleveland, Ohio.

Ms. Petcavage has over 15 years of purchasing and operations management experience in both manufacturing and distribution, including Corporate Purchasing Manager at Premier Industrial

Corporation; Senior Buyer at Johnson & Johnson Corporation, Technicare Division; and Material Planner/Buyer at Tappan Corporation, Air Conditioning Division.

She is an active member of the Institute for Supply Management (ISM) and the Purchasing Management Association of Cleveland, Ohio, where she has served on the Board of Directors. She is a board member of the ISM Management Materials Group and serves as Editor of its news publication. She is a Certified Purchasing Manager (CPM) with a lifetime designation. In 2003, Ms. Petcavage received the Ted R. Thompson Award for outstanding leadership in purchasing education from the Columbus, Ohio, Purchasing Management Association.

Richard L. Pinkerton is Professor Emeritus of Marketing and Logistics at California State University, Fresno. He received his B.A. in Economics from the University of Michigan in Ann Arbor; his M.B.A. from Case Western Reserve in Cleveland, Ohio; and a Ph.D. in Marketing and Curriculum Studies from the University of Wisconsin, Madison. His dissertation, "A Curriculum for Purchasing" (1969), was sponsored by the National Association of Purchasing Management (NAPM) and is regarded as one of the key benchmark studies in the field of purchasing. A past Dean of the Graduate School of Administration at Capital University in Columbus, Ohio, he also served as Chair and Business Center Director of the Craig School of Business, California State University, Fresno. He is a Certified Purchasing Manager (CPM) with a lifetime designation.

Dr. Pinkerton's industrial experience includes Market Research Analyst for the Harris Corporation of Melbourne, Florida, and Manager of Sales Development at the Webb-Triax Company of Cleveland, Ohio. His consulting experience includes Pepsi-Cola, Quaker Oats, Perkin Elmer Laser Lamps, Con Edison of New York City, PG&E of San Francisco, IBM, and the Oracle Corporation. For almost 40 years he has served the Institute for Supply Management (ISM; formerly NAPM) in various leadership roles at both the national and local levels, including Chair of the Academic Planning Committee. He was a long-time member of the *Purchasing Today* (now called *Inside Supply Management*) and *Info-Edge* Editorial Review Boards, Chair of the 1992 NAPM Research Symposium, and the NAPM representative at the 1992 convention of the Chile Procurement Society in Santiago, Chile. At the local level, Dr. Pinkerton was very active in the Columbus, Ohio; Fresno, California; and Silicon Valley Purchasing Management Associations and is a life member of the Northern California Purchasing Management Association.

His research on European Union procurement issues, competitive intelligence, and supply chain management is quoted in eight leading supply chain books. He has taught and consulted in Singapore, the United Kingdom, Germany, Slovakia, Chile, Poland, and Saudi Arabia.

He is co-author with David N. Burt of *A Purchasing Manager's Guide to Strategic Proactive Procurement*, published by the American Management Association. In 2002, he was given the Ted R. Thompson Award for outstanding leadership in purchasing education by the Columbus, Ohio, Purchasing Management Association. A retired Lt. Col. in the USAF Ready Reserves, he holds the Meritorious Service Medal for his many years of military service. He is an honorary member of the Purchasing Management Association of Cleveland, Ohio, his hometown, where he now lives.

CONTENTS

FOREWORD

The U.S. Civil War propelled the growth of both the steel and railroad industries. By the 1880s, these were the two biggest industries the world had ever known. It was during this period that purchasing policies, practices, and procedures were first developed and codified. During the next 100 years, many firms adopted this code and the code stayed in place with only minor changes. Then in 1984, due to the Japanese holding our economic feet to the fire, the Just-in-Time (JIT) movement began in earnest in the United States. One of the results of the JIT movement was that purchasing and supply chain management got a lot of attention. Consequently, from 1984 to 1994 there were more changes in purchasing policies, practices, and procedures than during the previous 100 years. This process of change has continued to accelerate since then.

You may be thinking, "That was a nice history lesson, but what does that have to do with me?" Just this—in today's world of work, the only job security is your competency. Only those people who work hard at continuously developing their knowledge and skills will be relevant and sought after as we go forward in what is without a doubt a much more challenging world economy. That is why reading books like this one is so vital to your future. Knowledge is *power*!

If I were still a purchasing practitioner, I would read books like this one to determine whether I am using the best practices. And, if I wasn't, I would first list the changes that I wish to make and then prioritize them. Second, I would implement those changes that I could make on my own. Next, I would plan how I was going to persuade those whose help I needed to put the other items into practice. Finally, I would set about making those improvements. After that I would probably have to start over because we live in an ever-changing environment.

Proactive Purchasing in the Supply Chain is no silver bullet for solving *every* procurement problem, but it does contain an encyclopedic knowledge of most of the things that you face daily.

There is an old joke. Those who can't do, teach, and those who can't teach, teach teachers. I have known all three of these authors for more than 30 years. They were "doers" before they were professors, and they were authors of other successful books before they wrote this one.

David Burt was a Lieutenant Colonel in the Air Force and a Chief Procurement Officer. After his retirement, he became a professor of purchasing and supply management. In addition, he has been able to stay connected to the real world of procurement by acting as a consultant to many large and small companies.

Sheila Petcavage has had many years of purchasing and operations management experience in both distribution and manufacturing. She currently is an assistant professor of supply management. Sheila has also continued to stay connected with business and industry through many workshops she has conducted on procurement and supply management. She knows what works and what doesn't.

Like the other two authors, Dick Pinkerton worked in industry for a numbers of years. After that, he became a professor and a dean of a graduate school of administration. In addition to these positions, he was an Air Force officer and retired as a Lieutenant Colonel in Ready Reserves. And, like the other two, he has stayed connected to the real world of supply management through his consulting activities.

Why does this book deserve your consideration? Because it is a guide to what is new and best in procurement and supply management. It will help you by letting you know when you are using the best practice and when you are not—if not, it will let you know what you can do to make the improvements you need to make. This book can be a valuable resource for seasoned professional supply managers as well as a primer for the newly appointed buyer.

Ken Killen
Rocky River, Ohio

Dr. Killen is an ISM J. Shipman Gold Medalist and a retired professor of purchasing and management. He has also written a number of books and articles on purchasing and management.

ACKNOWLEDGMENTS

We would like to thank Mr. Steve Shawver, owner of Minuteman Press of Strongsville, in Strongsville, Ohio, for his tremendous support in the development of this book. We also wish to express our deep appreciation to Judy Bass, Senior Editor, Quality and Process Improvement, for McGraw-Hill, for her help and backing in getting *Proactive Purchasing in the Supply Chain* to press.

INTRODUCTION

This book is written for purchasing managers who want to know and use the latest best procurement practices to help their organizations maximize the effectiveness and efficiency of their supply chain. Purchasing has evolved from a transaction-based clerical activity to the present-day value-added function. The ability of purchasing to dramatically impact and increase return on investment and to reduce costs is demonstrated throughout this book. Certification programs, increased educational opportunities, and key leadership examples from individuals such as R. David Nelson of Honda, Tom Stallkamp of Chrysler, Gene Richter of IBM, Dean Ammer of Northeastern University, Victor H. Pooler, Jr., of Carrier, and other "champions" are some of the factors behind this evolution to strategic, proactive purchasing.

Buyers who view carefully selected suppliers as partners will develop long-term relationships based on trust that focus on total cost of ownership vs. price and early supplier involvement in new and improved product development. This kind of enlightenment also leads to the use of cross-functional teams and other best in class methods, tools, procedures, and analysis of price-cost-quality relationships described in the various chapters. We hope each chapter facilitates your ability to lower costs, improve quality, shorten the time to market, and become a key contributor to your organization's mission, strategy, and tactics. In addition, we cover essential "need to know" material including: legal issues, green purchasing, negotiation, ethics, production-inventory control, purchasing risk management, sourcing, how to fight price increases, demand-logistics management, and many other critical subjects.

The last chapter provides a detailed planning model on how to implement these world-class procurement methods. All chapters contain additional notes that provide real case history examples. We are very pleased to have received permission from the Institute for Supply Management to publish the winners of the Gene Richter Supply Management Excellence Awards to outstanding purchasing organizations. They are excellent examples of how they transformed their purchasing operations from reactive to proactive supply management.

CHAPTER 1

Purchasing: The Foundation of the Supply Chain

Purchasing is one of the basic functions common to all organizations. It is the process of acquiring goods, services, and equipment from another organization in a legal and ethical manner. Professional purchasing addresses five rights: purchase of the right item or service, in the right quality, in the right quantity, at the right price, at the right time. Purchasing provides the foundation of supply management, which tends to have a wider scope of activities. The focus shifts from price to the total cost of ownership. Supply management also puts more emphasis on helping a firm increase its profitable sales.

The term "materials management" was a popular expanded definition which grouped purchasing, production and inventory control, and incoming transportation under one director. The current popular term "supply chain management" reflects the expanded purchasing management functions.

Supply management (also known as procurement at many firms and governmental agencies) is a five-stage process that begins with the identification of the item or service required to meet the needs of the organization. During this stage, the need is translated into a statement describing the item or service required to satisfy the need. It is estimated that some 85 percent of the cost of an item or service is determined during this stage. In advanced organizations, supply management professionals, and frequently, pre-qualified suppliers are involved in this stage. The second stage of supply management involves identifying the supplier who will best satisfy the need. The third phase involves the process of establishing a fair and reasonable price for the item or service to be purchased. The fourth phase results in an enforceable agreement for the purchase that meets the needs of both parties. The fifth phase requires managing the relationship to ensure timely delivery of the required item or service, in the quality specified at the agreed time. During this final stage, the supply management organization may work with the supplier in an effort to improve the supplier's efficiency with the objective of improving quality or reducing costs, or both.

Several departments, including marketing, sales, and logistics, have begun to lay claim to the term "supply" or "supply chain management." Although this can cause confusion, it emphasizes supply management professionals' need to communicate with these groups so that supply chain management can be effective. As Johnson, Leenders, and Flynn write, "purchasing, supply management, and procurement are used interchangeably to refer to the integration of related functions to provide effective and efficient materials and services to the organization".[1] Throughout this text, the term "supply management" will be limited to the definition contained in this chapter with a focus on procurement issues.

Strategic sourcing represents increasing responsibility for supply management. Strategic sourcing formalizes three activities: (1) periodic analysis of an organization's spending (what is purchased and from whom); (2) analysis of the supply market (who offers what and what changes are taking place in the relevant component of the supply world); and (3) development of a sourcing strategy that supports the corporate strategy while minimizing risks and costs.

Supply Management and the Bottom Line

Supply management has an overwhelming impact on the firm's bottom line. It directly affects the two forces that drive the bottom line: sales and costs.

Historically, supply management has been considered important based on its impact on costs. At an increasing number of firms today, the procurement process is recognized as having a significant impact on sales and revenues. Supply management has an overwhelming impact on the firm's bottom line. It directly affects the two forces that drive the bottom line: sales and costs. Therefore, it must be a core competency of the firm, an expertise that is highly valued by the organization. For a typical manufacturing firm, purchasing and supply management are responsible for spending over half of every dollar the firm receives as income from sales. More dollars are spent for purchases of materials and services than for all other expense items combined, including wages, depreciation, taxes, and dividends. It is important for management to note that the cost of materials is approximately 2½ times the value of all labor and payroll costs and nearly 1½ times the cost of labor plus all other expenses of running the business.

For the typical services firm, supply management plays an equally important, although subtler, role. Millions of dollars are spent on marketing and advertising, communications and information technology. Those services can enhance or degrade any firm's efficiency and effectiveness. Management at both manufacturing and services firms ensures that supply management professionals are involved at all stages of the procurement of such services. Strategic supply chain management enables a company to maximize its bottom

Net Income

Increased Sales:
• Faster to Market
• Improved Quality
• Pricing Flexibility
• Innovation
• Enhanced Customer
Satisfaction
• The Supplier of Choice
• Customer Fulfillment Flexibility
• Shorter Cycle and Lead

Lower Total Cost:
• Better Product Design
• Acquisition Cost
• Processing Cost
• Quality Cost
• Downtime Cost
• Risk Cost
• Cycle Time Cost
• Conversion Cost
• Non-Value Added Cost
• Supply Chain Cost
• Post Ownership Cost

Figure 1.1   A graphic reproduction of supply management's impact on the bottom line.

line in an ethical manner. Figure 1.1 shows how supply management can drive sales up and costs down.

Increased Sales

Supply management has a significant impact on a firm's sales; principally in the following eight areas.

Faster to Market or Time-Based Competition   Thirty years of marketing research has demonstrated the importance of being early to market. In many cases, the first firm to introduce a successful new product or service will hold 40 to 60 percent of the market after competition enters the picture. This research also demonstrates that the profit margins enjoyed by the first firm to introduce a new product tend to be twice those of its competitor, as first reported in the Profit Impact of Market Strategy (PIMS) approach.[2] Firms that have embraced strategic supply chain management have reduced their new product development cycles by an average of 30 percent as a direct result of taking a cross-functional approach to product development (this is also known as concurrent engineering). Supply management and carefully selected suppliers are key members of these cross-functional teams. This topic is addressed in greater detail in Chapter 5, New Product Development.

Time-based competition also includes a firm's ability to meet unexpected surges in demand for its products. In many cases, a firm's ability to ramp up production is constrained by its suppliers' abilities to meet those surges in demand. The development and management of a competent, responsive supply base plays a critical role in the firm's ability to meet unexpected demand.

Improved Quality   We are all sensitive to the quality of the products and services we purchase. An automobile with a reputation for transmission problems will drive potential customers to its competitors. Conversely, a firm whose products or lines of products have a reputation for quality gains market share over its competitors and frequently is able to command premium prices. Some 75 percent of many manufacturers' quality problems can be traced back to defects in purchased materials. The percentage of quality problems that can be attributed to defective incoming materials for a service provider is usually less but still significant.

Thus, if a manufacturer or services provider is able to reduce defects in incoming materials, it can improve the quality of its products in the marketplace. Firms that embrace strategic supply chain management work with their suppliers to design quality into the suppliers' products and maintain quality during production. The result is virtually defect-free incoming materials, improved quality in the marketplace, more sales, and improved profit margins.

Pricing Flexibility   Research conducted by the University of San Diego indicates that a strategic approach to supply management will reduce the total cost of ownership[3] associated with purchasing and owning or leasing materials, equipment, and services an average of 25 percent. When the cost of producing an item or service is reduced, marketing receives the gift of pricing elasticity. Through the application of sound economic principles, marketing can estimate whether net income will increase more by (1) holding selling price and sales volume constant and increasing net profit per unit, (2) reducing the sales price and thus increasing sales volume, or (3) using a combination of increasing net profit while reducing sales price.

Innovation   The University of San Diego research study cited previously indicates that of 240 firms surveyed, approximately 35 percent of all successful new products were the result of technology gained from the supply base.[4] This leveraging of supplier technology is a major source of income for these firms. Collaborative and alliance relationships with the firm's supply base play a key role in ensuring and enhancing this technology flow. The development and management of supplier relationships is a key responsibility of supply management.

Enhanced Customer Satisfaction   Strategic supply chain management helps achieve shorter fulfillment lead times, consistent on-time delivery, high fill rates, complete orders, quicker responses to customers' requirements, and the ability to meet unique or special requests.

The Supplier of Choice   By providing the best value (a combination of quality, service, and price), a firm becomes the supplier of choice to another channel member or to the end customer.

Customer Fulfillment Flexibility   Strategic supply chain management provides the supply support that allows a firm to be responsive to customer desires for flexible lead-time and changes in product configurations.

Shorter Cycle and Lead Times   These benefits result from improved supplier relationships and involvement in supplier product and process improvements.

Lower Total Cost of Ownership

All members of the supply management system must be on the lookout for non-value added activities at any and all stages of the system.

The total cost of ownership is the summation of the costs of acquiring and owning or converting an item of material, piece of equipment, or service, and post-ownership costs, including the disposal of hazardous and other manufacturing waste and the cost of lost sales because of a reputation for poor product quality caused by defective materials or purchased services that are incorporated in the end product or service.

Better Product Design   We estimate 70 to 80 percent of the total cost of ownership is built into a requirement—whether for production materials, equipment, services, or maintenance repair and operation items—during the requirements development process. Early supply management and supplier involvement can reduce costs significantly during this critical stage.

Acquisition Cost   The acquisition cost or price paid for an item or service is normally a major component of the total cost of ownership. As will be seen in the following chapters, numerous actions can be taken to reduce the acquisition cost. A few of those activities are specification of the most cost-effective material or item of equipment, use of the appropriate specification, standardization, good sourcing and pricing practices, and professional contract and supplier relationship management.

Processing Cost   The cost of developing sourcing and pricing requirements and then ensuring they arrive on time in the quality specified can be reduced significantly by applying efficient supply management processes and techniques.

Quality Cost   Costs are incurred in ensuring that the buying firm receives the optimal level of quality. These costs may be reduced by applying progressive quality techniques such as the design of prototypes and statistical process control. Selecting suppliers capable of producing the desired level of quality and then certifying their design and manufacturing systems can improve incoming quality while reducing administrative quality costs.

Downtime Cost Downtime frequently is the largest component of the total cost of ownership for many items of production and operating equipment. One minute of downtime in a production line may cost $26,000.[5] At this rate, an hour of down time can cost $1,560,000. Thus, when purchasing equipment, the sourcing team must place as much—or more—emphasis on reliability[6] and maintainability[7] as on purchase price.

Risk Cost Many firms needlessly spend millions of dollars to minimize the risk of supply disruptions. These firms maintain needlessly large inventories or dual or even triple sources of supply to ensure continuity of supply. Carefully developed and managed relationships with appropriate suppliers can eliminate the need for most inventory or dual sources.

Cycle Time Cost While this is difficult to quantify, the shorter the cycle time for virtually all activities, the lower the cost. The shorter the cycle time to bring new products to market, develop a statement of work, or select a new source, the lower the total cost.

Conversion Cost Machine time, labor, process yield loss, scrap, and rework are examples of conversion costs. These costs are every bit as real as the purchase price of an item entering the production process. A pound of brass may cost twice as much as a pound of steel, but the higher acquisition price for the brass may be more than offset by savings in machine and labor costs during conversion of the brass to a component or end product because the brass may require less work to make it a usable product.

Non-Value Added Costs A careful analysis of all of the costs involved in bringing an item or service to market frequently reveals that 40 to 60 percent of the costs confer no added value to the finished good! Robert Handfield indicates that estimates of the amount of time spent on non-value added activities can be as high as 80 to 90 percent of the total time required to complete a cycle.[8] James P. Womack and Daniel T. Jones, in their book *Lean Thinking*, added indicate, that "it takes an average of 11 months for the can of cola in a domestic refrigerator to actually get there…During that 11 months, the time that the material is actually being converted as opposed to simply waiting is a mere three hours!"[9]

All members of the supply management system (e.g., design, manufacturing and quality engineering, manufacturing, and procurement) must be on the lookout for non-value added activities at any and all stages of the system.

Supply Chain Cost The development and management of supply chains and supply networks require a significant investment, primarily in the form of human resources. The proper selection, training, and educating of the individuals involved in these activities, together with the application of software systems, can reduce the necessary investments.[10]

Post-Ownership Cost   These costs frequently are overlooked but must be considered when addressing the total cost of ownership. They include the disposal of scrap and other waste, customer service, warranty costs, and the cost of lost sales resulting from customer dissatisfaction with the product.

Supply Management and Return on Investment (ROI)

…nearly a 50 percent increase in the firm's return on its investment, something most CEO's would die for!

Investors frequently evaluate top management's performance by calculating the return on the total capital invested in the business. Inventory, equipment, and other materials purchased constitute corporate assets attained through the investment of capital. The fact that supply management frequently is responsible for spending over half of most companies' total dollars highlights the profit-making possibilities of the purchasing and supply function. Every dollar saved in purchasing is equivalent to a new dollar of profit. Figure 1.2 illustrates this point by showing the relationships of basic elements that influence the return on investment (ROI). The figures in parentheses reflect a 5 percent reduction in the cost of materials for a manufacturing firm. Notice how, in this example, a 5 percent reduction in material cost increases ROI from 10 to 13 percent, a 30 percent increase!

As we have observed, supply management can have a significant impact on a firm's sales volume. The underlined numbers indicate the impact of a 5 percent increase in sales, holding all other variable ratios (including original material costs) constant. We see that ROI increases from 10 percent to 11.42 percent from increasing sales alone.

Now, let's look at the combined impact of a 5 percent increase in sales and a 5 percent reduction in the cost of all materials purchased for this volume of activity. The figures in parentheses and underlined in Figure 1.2 show the combined impact of these two forces. We can see that the combined impact of these two realistically obtainable achievements is to increase ROI to 14.52 percent. This is a nearly a 50 percent increase in the firm's ROI, something most CEO's would die for!

The Progression to Strategic Supply Chain Management

Supply Chains

Purchasing is the foundation of supply management, which in turn is the foundation of supply chain management. Through the process of acquiring goods, services, and equipment from other organizations, a chain of upstream suppliers is formed—*a supply chain* (see Figure 1.3).[11]

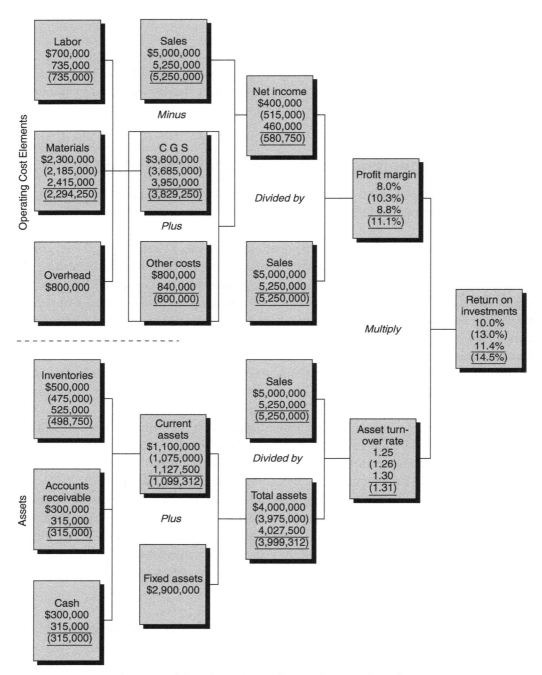

Figure 1.2 A graphic view of the relationships of basic elements that influence return on investment. The figures in parentheses reflect a 5% change in the number indicated. Changes in parentheses reflect a 5% decrease in material costs. Underlined figures show a 5% increase in sales. Numbers that are in parentheses and underlined reflect a 5% decrease in material costs *and* a 5% increase in sales. Note a nearly 50% increase in ROI resulting from these combined changes.

The Supply Chain

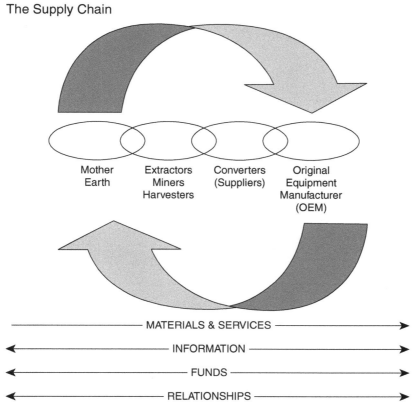

Figure 1.3 A firm's supply chain includes all internal functions plus external suppliers involved in the fulfillment of needs for materials, equipment, and services in an optimized fashion. The supply management system plays a key role in helping the firm satisfy its role in its supply chain.

The firm's supply system includes all internal functions (such as operations, engineering, production control and scheduling, inventory management, demand forecasting, and marketing) plus external suppliers involved in meeting the organization's needs for materials, equipment, and services. This supply system and the firm's supply chains play a key role in helping the firm fulfill its role in its value chains.

The *value chain* is a series of organizations that add value to goods and services flowing from Mother Earth to the end customer (see Figure 1.4). The value chain must be viewed as a whole, a single entity, rather than fragmented groups performing their own functions. Money enters the value chain only when the ultimate customer buys a product or service. Transactions within the value chain simply allocate the ultimate customer's money among the members of the chain.

The Value Chain

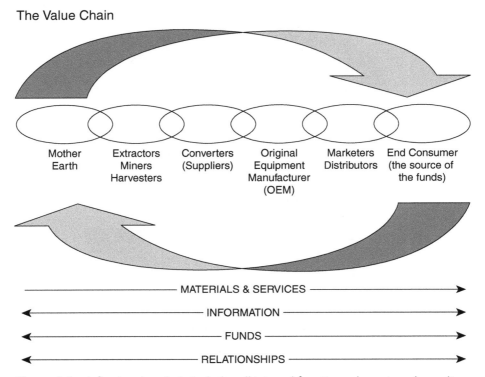

Figure 1.4 A firm's value chain includes all internal functions plus external suppliers involved in the identification and fulfillment of needs for materials, equipment, and services in an optimized fashion. Also included in the value chain are the firm's distribution channel and end consumers.

The supply chain is the *upstream portion* of the organization's value chain and is responsible for feeding the production or conversion process. Marketing and distribution are the principle components of the *downstream portion* of the value chain. Marketing takes appropriate action to identify customers' wants and needs and to facilitate sales to the end customer. Distribution manages the movement of finished goods from the original equipment manufacturers (OEMs) through the distribution channel to the end customer. Successful firms such as Toyota, Dell, Wal-Mart, and Procter & Gamble are aware that competition takes place between value chains. This awareness and resulting strategic and tactical activities result in leadership in their industries.

The Supply and Value Networks: The Next Phase of Supply Chain Management

Networks are flexible virtual systems that are linked by communication systems and alliances. Within the network, many things are happening simultaneously. Final consumers also provide input about wants and needs that are communicated throughout the network system. These systems optimize the flow of materials and services, information, and money. Supply and value networks focus on the ultimate customer. They are designed and managed so that one member does not benefit at the expense of another. World-class value networks are highly adaptive, focus on speed, are innovative, and are tightly integrated.

In comparison to integrated supply and value networks, the traditional approach to supply chain management is more linear in concept. This approach features independent decision making as a result of gaps between the entities that constitute the supply chain. Those gaps are caused by lack of communication and information sharing and can result in excess inventories, inflated lead-times, and increased costs throughout the value chain. Michigan State University (MSU) has long been regarded as one of the leaders in purchasing, supply management, and logistics research and education. Nick Little of MSU observed that the individual players in the chain are all seeking to deliver value for the end consumer. However, there are a number of elements in that value:

▲ Value creation—through the innovation, development, and launch of new products and services.
▲ Value delivery—through the order fulfillment process.
▲ Value maintenance—through processes to provide after sales service, support maintenance, and so forth.

"These three value processes need to span your company, your suppliers and your customers in order to successfully meet the needs of end consumers."[12] It is in seeking this value that simple linear supply chains evolve into more tightly knit supply and value networks.

Implementing Strategic Supply Chain Management

In order to gain the benefits of its supply chains, senior management must recognize the importance of supply chain management and support the required transformation to strategic status. One of the most visible ways of demonstrating its support of this transformation is to appoint a Chief Supply Officer at an organizational level equal to that of marketing, engineering, and operations. Then, senior management should realign the firm's internal resources with the objective of enabling the success of the firm's supply chains. The transformation must be carefully planned and executed. Getting top management's commitment and everyone's involvement are keys to success.

Successful firms must know where they are in relation to where they want to be. Benchmarking best-in-class practices and developing metrics or measurements enable firms to establish a baseline of where they are, develop an appropriate action plan and then track their progress toward strategic supply management. Appropriate action plans and metrics allow the firm to focus on its vision and continuously improve its contribution to the bottom line. Figure 1.5 provides a diagnostic that allows a firm to evaluate

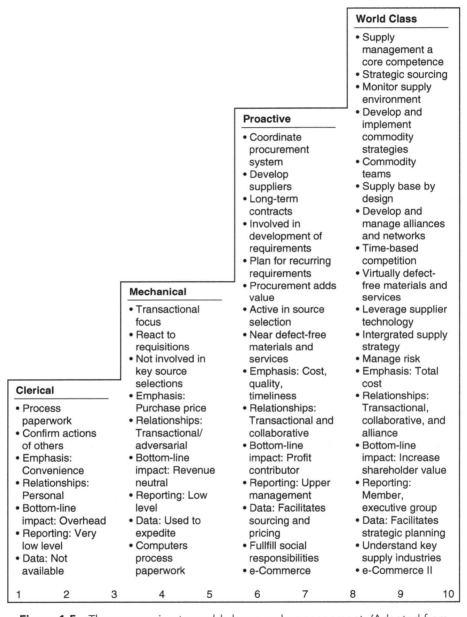

Figure 1.5 The progression to world class supply management. (Adapted from *The American Keiretsu* by David N. Burt and Michael F. Doyle, Homewood, IL: Business One-Irwin, 1993, 21.)

where it is on the progression to strategic supply chain management and can also serve as a road map to guide the implementation of strategic supply chain management. The four columns show the progression from a reactive clerical to a strategic focus. In each of these four stages, we see the focus of the supply management change as it develops.

Experience has shown that using the step chart in Figure 1.5 to evaluate a department's rating will initially result in a high rating. However, initial ratings will drop after attending purchasing seminars, credit classes, or simply reading solid books on proactive procurement and supply management. One reason is the tendency to know only "how we do it here," and lacking deep knowledge of best practices, we assume "how we do it here" is the best way to do it. The major reason for certification programs is to force practitioners to continually update their knowledge of the latest and best practices. This new knowledge influences where an organization feels it is on the step chart. It also highlights the reason the step chart ratings must be prepared by internal users, suppliers, and the purchasing department. Multiple ratings will help convince the CEO of the true score. See if your rating of your purchasing department changes when you rework it at the start of the final chapter, implementing proactive purchasing in the supply chain.

The Roles of a Supply Management Professional

Dr. Joseph Cavinato of the Institute for Supply Management provides the following thoughts on the roles of today's supply management professionals:

> A supply management professional has four key roles:
>
> One is a leadership role in seeking new opportunities in the (supply) marketplace and driving them for follow-through in the organization.
>
> A second role is being an identifier of outsource opportunities, finding the right outsource, and leading the charge to an efficient and effective relationship and oversight system for the organization.
>
> A third management role is also required on a higher level than before. This is the management of systems and relationships. Having the proper eyes, ears, and antennas in place with the proper interpretive mechanisms is an essential value-added need for the organization. It is an assertive contributing role with both outsiders and insiders.
>
> A creator role is called for in the form of identifying new opportunities and making them available to the organization. This means creating strategies, systems, and supply options of entire "packages" of value attributes that span many departments and groups. This also includes seeking and implementing top line revenue opportunities for the organization.[13]

Summary

Purchasing, one of the basic activities common to all organizations, is the process of acquiring goods, services, and equipment from another organization. Purchasing is the foundation of supply management, a process that has an overwhelming impact on the firm's bottom line. Supply management directly affects the two factors that control the firm's bottom line: sales and costs.

Supply chain management is the process of managing the flow of raw materials, from Mother Earth to the OEM. It is the upstream portion of the value chain. Marketing and distribution—the downstream side of the value chain—influence demand and sales and manage the movement of finished goods from the OEM through the distribution channel to the end customer. The value chain is a sequence of integrated activities that must be performed by various organizations to move goods from the sources of raw materials to ultimate consumers.

An organization's success is driven by its ability to compete effectively as a member of its supply and value chain communities, not as an isolated enterprise. The ability to interact quickly with customers, suppliers, and other partners is critical to the survival and success of a firm and its chains.

Chapter 2 addresses the relation between the organizational status of the supply management function and its ability to have an impact on the firm's success. Insight into this relationship is very useful to the professional who desires to be part of an organization in which his or her efforts will have a significant impact on the organization's success.

Appendix: An Overview of the Mechanics of Supply Management

The Typical Purchasing Cycle: Materials

A supply management department buys many different types of materials and services. The procedures used in completing a total transaction normally vary among the different type of purchases. However, the general cycle of activities in purchasing most operating materials, supplies, and services is fairly standardized. The following steps constitute the typical purchasing cycle:

- ▲ Recognize, define, and describe the need.
- ▲ Transmit the need.
- ▲ Investigate and select the supplier.
- ▲ Prepare and issue the purchase order, contract, or agreement.
- ▲ Follow up the order (including expediting and de-expediting).
- ▲ Receive and inspect the material.

▲ Audit the invoice.

▲ Close the order.

Figure 1.6 outlines these steps in operational form for a requirement for materials. (Chapters 8 and 9 address the process for the purchase of equipment and of services.) More important, Figure 1.6 details the minimum flow of communications required for a system to function smoothly and efficiently. These communications may be electronic messages or paper documents, depending on the type of system used. The precise form the electronic message or the documents take varies widely from one company to another. The important point to note, however, is that a properly controlled purchase requires extensive communication with numerous work groups. Procurement procedures constitute the framework within which this task is accomplished.

Recognition, Definition, Description, and Transmission of the Need

The need for a purchase typically originates in one of a firm's operating departments or in its inventory control section. The supply management department is notified of the requirement by one of two methods: a purchase requisition or a material requirements planning (MRP) schedule. The purchase description that is transmitted to the supplier forms the heart of the procurement and is detailed on the requisition form.

Standard Purchase Requisition

The purchase requisition is an internal document, in contrast with the purchase order, which is basically an external document. Most companies use a standard, serial numbered purchase requisition form for requests originating in the operating departments. The requisition communicates the user's needs. Essential information communicated includes a description of the service, material, quantity, and date required. Requisitions are often electronically transmitted through the approval system to the appropriate buyer. Firms that maintain their inventory records on a computer utilize a programmed inventory monitoring system that identifies the item whose inventory level has reached the reorder point. When the computer detects this condition, it automatically prints an inventory replenishment requisition that goes to purchasing for action—and the purchasing cycle is under way.

Material Requirements Planning Schedule

When a design engineer (or a design team) completes the design of a part or an assembly, he or she makes a list of all the materials (and quantity of each) required to manufacture

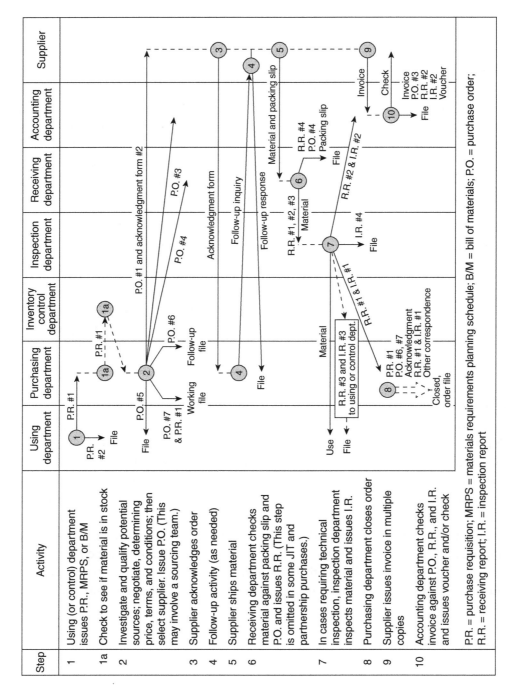

Figure 1.6 General procedure and document flow chart for a typical purchasing cycle.

the item. This list is called an *engineering bill of materials.* In firms using computerized production and inventory planning systems, such as an MRP (materials requirements planning) system, the engineering bill of materials is first reconfigured into a *structured multilevel* bill of materials. This structured bill of materials for each item being manufactured can then be used in determining specific material requirements for a given production schedule during a specific time period. The computer program utilizes the reconfigured "bills" along with the production schedules for all items as input—and calculates as output the precise time-phased requirements for each material that will be used in the manufacturing process. This schedule is then sent to purchasing for direct use in obtaining the required materials. It obviously eliminates the necessity of preparing numerous purchase requisitions—and it is ideally suited for use in a multiproduct intermittent manufacturing operation.

Supplier Selection and Preparation of the Purchase Order

As soon as a need has been established and precisely described, the procurement professional begins an investigation of the market to identify potential sources of supply. In the case of routine items for which supplier relationships have already been developed, little additional investigation may be required to select a good source.[14] The purchase of a new or a high-value item, on the other hand, may require a lengthy investigation of potential suppliers.

After qualifying a preliminary group of potential sources, the procurement professional can employ the techniques of competitive bidding or negotiation, or both. When competitive bidding is used, the procurement professional initiates the procedure by requesting quotations from a reasonable number of firms with whom the buying group is willing to do business. Although "request for quotation" forms vary widely among firms, typically they contain the same basic information that will subsequently be included on the purchase order. These requests may even be sent electronically.

Once a supplier has been selected, the supply management department issues a serial numbered purchase order. In most cases the purchase order becomes a legal contract document. For this reason, the procurement professional should take great care in preparing and wording the order. Quality specifications must be described precisely. If engineering drawings, statements of work, or other related documents are to be considered an integral part of the order, they should be incorporated clearly by reference. Quantity requirements, price, and delivery and shipping requirements must be specified accurately. The order should include all data required to ensure a satisfactory contract, and either party should word it in a manner that leaves little room for misinterpretation.

In addition to those provisions that are unique to each contract, most firms also include as a part of every contract a series of terms and conditions that are standard for all orders (typically called a "boilerplate"). These terms and conditions are designed to give legal protection to the buyer on such matters as contract acceptance, delivery performance, contract termination, shipment rejections, assignment and subcontracting of the order, patent rights and infringements, warranties, compliance with legal regulations, and invoicing and payment procedures. Each company develops its terms and conditions of purchase in accordance with its own unique needs. Consequently, much variation exists among firms. Chapter 17 describes some of the legal considerations that should be addressed when creating a contract.

After an order has been issued, changes in company requirements frequently require a change in the contract. In such cases, the supply professional issues a *change order,* following the same procedures as were followed for the original order. When accepted by the supplier, the change order either supplements or replaces the original order.

Acknowledgment and Follow-Up of the Order

In most cases, the original copy of the purchase order that is sent to the supplier constitutes a legal "offer" to buy. No purchase "contract" exists, however, until the seller "accepts" the buyer's offer. The seller's acceptance can take one of two forms: (1) performance of the contract or (2) formal notification that the offer is accepted.

The purpose of sending the supplier an acknowledgment form along with the purchase order is twofold. First, it is a form that can be completed conveniently and returned to the purchasing firm, acknowledging acceptance of the order. At the same time, the supplier can indicate whether it accepts the buying terms and is able to meet the desired delivery date. If a supplier ships the ordered item immediately from stock, it frequently disregards the acknowledgment form thereby accepting the buyer's terms and conditions for the contract.

If shipment is not made immediately, an acceptance should be sent to the supply management department. Although the acknowledgment form usually serves this purpose, some sellers prefer to use their own forms, which state their terms and conditions of sale. In either event, the procurement professional should check the acceptance closely to see that the supplier has not taken exception to any provisions of the order. If the seller's acceptance terms are different from those on the buyer's order, the law holds that they will automatically be incorporated in the contract unless they materially alter the intent of the offer, or unless the buyer files a written objection to their inclusion. In cases where the seller's and the buyer's terms are in direct conflict, the law omits such terms from the contract, leaving settlement of the differences to private negotiation or legal adjudication. In view of the posture adopted by the courts on this matter, it is amply clear that a buyer must review suppliers' order acceptances with great care.

The supply management department's responsibility for an order does not terminate with the making of a satisfactory contract. Supply management bears full responsibility for an order until the material is received and accepted.

Even though a supplier intends to meet a required delivery date, many problems can arise to prevent it from doing so. When there is a reasonable chance that the supplier may not stay on schedule, important orders with critical delivery dates should receive active follow-up attention. At the time such orders are placed, the procurement professional should determine specific dates on which follow-up checks are to be made.

Receipt and Inspection

The next step on the traditional purchasing cycle is receipt and inspection of the order. When a supplier ships material, it includes in the shipping container a packing slip which itemizes and describes the contents of the shipment. The receiving clerk uses this packing slip in conjunction with his or her copy of the purchase order to verify that the correct material has been received. The received materials are often entered electronically into the buyer's system by keyboard or scanning of a bar code.

Services

The procurement of services involves many of the same processes. Instead of a specification developed or adopted by engineering serving as the heart of the purchase order or contract, a statement of work serves this purpose. A team, including the user of the services and procurement professional, develops the statement of work. Frequently, one or more qualified potential suppliers are involved in this development effort. Receipt of services may require a technical inspection to verify services have been received in accordance with the contract. These inspections may be required at various points throughout the completion of the order or contract. Procurement of services can present challenges not experienced in materials buying. Chapter 9 addresses this challenging process.

The Invoice Audit and Completion of the Order

Occasionally, a supplier's billing department makes an error in preparing an invoice, or its shipping department makes an incorrect or incomplete shipment. To ensure that the purchaser makes proper payment for the materials actually received, sound accounting practice dictates that some type of review procedure precedes payment to the supplier.

A typical procedure involves a simultaneous review of the purchase order, the receiving report, and the invoice. By checking the receiving report against the purchase order, the purchaser determines whether the quantity and type of material ordered was in fact received. Then by comparing the invoice with the purchase order and receiving report, the firm verifies that the supplier's bill is priced correctly and that it covers the proper quantity of acceptable material. Finally, by verifying the arithmetic accuracy of the invoice, the correctness of the total invoice figure is determined.

Auditing invoices is a repetitive, time-consuming task that should be handled as efficiently as possible. It should also be conducted soon after receipt of the invoice to permit the accounting department to make prompt payment and obtain any applicable cash discounts. Prompt payment also supports the firm's efforts to establish and maintain good supplier relations. Because of the labor cost involved in auditing invoices, many companies do not verify the accuracy of low-dollar-value invoices.

Invoice auditing technically is an accounting function. When possible, it is prudent to separate the responsibility for authorizing payment for an order from the responsibility for placing the order. Theoretically, the purchasing department's job is completed when the material covered by a purchase order has been received in the plant and is ready for use. In practice, however, some firms assign the invoice auditing responsibility to accounting, while others assign it to purchasing.

In the purchase of complex or technical materials or services, operationally it makes sense to assign the auditing task to the professional who handled the order. This individual is familiar with the materials or services and their technical nomenclature, prices and contract provisions, and all ensuing negotiations. Invoices for such orders often are difficult to interpret and evaluate without a detailed knowledge of these things. Auditing invoices for the purchase of most standard materials, on the other hand, is a routine task that should be assigned to appropriate accounting personnel. A majority of most firms' orders fall in this category.

Figure 1.6 indicates that the supply management department closes its purchase order file before the invoice is audited. This is usually the case if accounting audits the invoice. When supply management audits the invoice, its records are closed after the audit. Closing the order simply entails a consolidation of all documents and correspondence relevant to the order; the completed order is then filed in the closed-order file. In most firms, a completed order consists of the purchase requisition, the open-order file copy of the purchase order, the acknowledgment, the receiving report, the inspection report, and any notes or correspondence pertaining to the order. The completed order file thus constitutes a historical record of all activities encompassing the total purchasing cycle.

The Small-Order Problem

Small orders are a perennial problem in *every* organization—and a serious problem in some. An examination of a typical company's purchase order files reveals that a sizable percentage (sometimes up to 80 percent) of its purchases involve an expenditure of less than $250. In total, however, these purchases constitute a small percentage (seldom more than 10 percent) of the firm's annual dollar expenditures.

For example, 75 percent of Conoco's purchase orders are for expenditures of less than $500, and 50 percent are for less then $100.[15] The Intel Corporation found that its purchasing department spent 66 percent of its time managing 1.7 percent of the firm's expenditures.[16] Kaiser Aluminum Chemical devised a system whereby blank checks were sent along with their orders. This allowed the supplier to fill the order and fill out the check for payment. This system reduced the number of invoices handled, and the amount of time and human capital required to process payments.[17]

Clearly, no manager wants to devote more buying and clerical effort to the expenditure of less than 10 percent of his or her funds than to the expenditure of the other 90 percent. Yet, this frequently is what happens. The very nature of business requires the purchase of many low-value items. Nevertheless, small orders are costly to buyer and seller alike. It costs a seller only a few cents more to process a $1,000 order than it does to process a $10 order. The following sections discuss various methods a purchasing manager can use to minimize the small-order problem.

Centralized Stores System

A stores system is the first approach typically used to reduce the volume of small-order purchasing activity. When experience shows that the same supply items are ordered in small quantities time after time, the logical solution is to order these items in larger quantities and place them in a centralized inventory for withdrawal as needed. An analysis of repetitively used production materials leads to the same action for the multitude of low-value items. If usage of an item is reasonably stable, an optimum order quantity can be computed using a basic economic order quantity approach. This will be discussed in detail in Chapter 18. There is, of course, a limit to the number of items and the financial investment a firm can place in inventory.

Blanket Order System

A store's system solves the small-order problem only for items that are used repetitively. A *blanket order* system helps solve the problem for the thousands of items a firm cannot carry in inventory, as well as some that it does carry.

Briefly, the general procedure used for this type of purchase is as follows. On the basis of an analysis of past purchases, the buyer determines which materials should be handled in this manner. After bidding or negotiating, the buyer selects a supplier for each item, or family of items, and issues a blanket order to each supplier. This order includes a description of each item, a unit price for each item when possible, and the other customary contract provisions. However, no specific order quantities are noted. The blanket order typically indicates only an estimated usage during the period of coverage (usually one to three years). It also states that requirements are to be delivered upon receipt of a release from the procurement professional or other authorized person. On receiving a requisition for one of the materials, the procurement professional merely sends a brief release form to the supplier. On the release form are noted the blanket order number, the item number, and the quantity to be delivered. Receiving reports are filed with the original order, and at the end of the month are checked against the supplier's monthly invoice. At the end of the period, the order may be renewed or placed with another firm, depending on the supplier's performance record.

Many companies develop their own unique modifications of the basic procedure. For example, instead of advising suppliers of order releases by means of a written form, some companies simply issue releases to local suppliers by telephone, fax, or electronically. By noting such releases on the order, the procurement professional still retains adequate control.

In the event that material is needed immediately (and the supplier is nearby), some firms allow the using department to pick up the material without notifying the supply management department. The employee obtaining the material simply endorses and enters the proper accounting charge on the sales receipt, a copy of which is sent to the supply management department. In many firms today, user administration of the blanket order is common place as it frees up time to work on more important tasks.

Benefits of Blanket Orders

▲ Fewer purchase orders/reduced clerical, purchasing, accounting, and receiving time.
▲ Less time spent on tactical work allows for more value added activities.
▲ Offers leverage on volume pricing.
▲ Reduces lead times and inventory levels.
▲ Develops longer-term relationships with suppliers.
▲ Allows supplier to plan more effectively, thereby reducing buyer's price.

To function effectively in the long run, however, any blanket order system must provide adequate internal control. Absence of the control element encourages petty fraud and poor supplier performance. The elements essential to effective control are:

1. A numbered purchase order, including proper internal accounting charge notations.
2. A record of authorized delivery releases.
3. Bona fide evidence of receipt of the material.

Despite the fact that blanket order systems offer both the buying and the supplying organizations a number of important benefits, organizations often fail to fully utilize this tool in dealing with the issues of small orders.

Systems Contracting

Frequently used as a basic purchasing strategy, as well as an approach for minimizing the small-order problem, systems contracting is an extension and more sophisticated development of the blanket order purchasing concept. Some firms call it "stockless" purchasing.

As its name implies, systems contracting involves the development of a corporate-wide agreement, often a one- to five-year requirements contract, with a supplier to purchase a large group or "family" of related materials. The items to be purchased are usually described in detail in a "catalog" that becomes part of the contract. Estimated usage usually is included, along with a fixed price for each item and an agreement by the supplier to carry a stock of each item adequate to meet the buyer's needs. Various types of supplies and commonly used operating items, typically purchased from distributors, are the materials most often covered by these types of agreements.[18]

In addition to the benefits of blanket order purchasing, a major objective of systems contracting is to minimize both the buyer's and the supplier's administrative costs associated with the purchases. The operating procedures of the two firms are integrated to the extent practical. For example, users in the buyer's various operating locations usually send their purchase requisitions directly to the supplier holding the contract for the item. The requisition thus serves as the purchase order. The supplier then simply maintains a list of such shipments on a "tally sheet," identifying each by the requisition number (or a supplier-assigned number), and periodically (monthly or semi-monthly) submits the tally sheet to the buyer for payment in lieu of an invoice.

These types of integrated procedures and shortcuts typically develop a closer relationship between the two firms and reduce paperwork and associated costs markedly. The buyer's inventories and carrying costs obviously decline as well.

Electronic Ordering Systems

The evolution of the Internet has created opportunities to purchase products and services more efficiently. Computers talking to computers replenish inventory of repetitively used items thereby expediting the purchasing process, reducing paperwork, and simplifying internal accounting and control.

A platform developed to facilitate electronic buying, the Trading Process Network developed by General Electric can help buyers collaborate with suppliers via e-mail, post information to suppliers via their TPN Office Website, and share engineering information and other information securely. Capable of interfacing with all aspects of the supply chain, GE touts that its TPN Business Services link the enterprise with the trading community by providing an electronic channel for distributing information around the world.[19]

Purchase Debit and Credit Cards

The use of corporate debit and credit cards by employees for MRO purchases and small-order buys has become commonplace. In addition to eliminating the need for most purchase orders, this buying technique reduces the purchasing cycle time, improves purchasing relations with operating departments, provides much faster payment to suppliers, and significantly reduces the workload in the accounts payable department.

The debit cards often offer the same protection to the buying firm without the interest that would be charged on credit transactions. Banks advertise the business debit card as being safer than cash, faster than writing a check, and easier to track. It can be used to pay recurring payments such as Internet service or insurance premiums. Both debit and credit cards offer detailed records of use to allow for control and protection against unauthorized use.

Internally, cards are issued to operating department personnel with preset spending limits, variable daily limits on purchases, or various tools to protect the account from fraud. Each card carries the appropriate departmental accounting charge number. The organization then receives a detailed monthly bank statement with the purchase and account number so that the account can be reconciled against expenses reported by the various users.

Many financial institutions today make it very inviting for companies to use debit and credit cards by offering perks such as purchase assurance and extended warranties, road side assistance, travel services and assistance, and in some cases, insurance coverage.

Supplier Stores/Consignment System

If a purchaser buys a large enough volume of certain materials from a single supplier, the supplier sometimes can afford to staff a small "store" at the purchaser's plant and operate it on a consignment basis. Some suppliers find that annual purchases of approximately $100,000 justify such a branch operation. Users then simply go to the store and sign for their purchases. At the end of the month, the company is billed for its purchases.

This system clearly is not a short-term arrangement. The purchaser, therefore, must take great care in selecting the supplier and in negotiating the terms of the agreement.

Supplier Delivery System

The supplier delivery system is somewhat similar to a supplier store system, but it is more feasible for firms with a smaller volume of purchases. Many suppliers who are not willing to set up a store at the buyer's plant are willing to stock numerous miscellaneous materials and make daily or semiweekly deliveries. Purchase requisitions for such materials are accumulated. The supplier's delivery person then picks them up on the specified day, and at the same time delivers the material ordered on the preceding batch of requisitions. This continuous shuttle service provides reasonably fast delivery and also reduces the purchaser's paperwork and inventory problems. Properly designed, the system can provide for adequate accounting control.

Concluding Remarks

The basic steps in the purchasing cycle are the same for any buy. The time and effort put into completing each step will vary depending on the importance of the need. With today's technology, many of these steps may be automated. The end result of this process may vary depending on the skill of the procurement professional or sophistication of the organization's buying process.

Endnotes

1. D. Fraser Johnson, Michiel R. Leenders, and Anna E. Flynn, *Purchasing and Supply Management*, 14th ed. (Burr Ridge, IL: McGraw-Hill Irwin, 2011), 4.
2. Robert D. Buzzell and Bradley T. Gaze, *The PIMS Principles: Linking Strategy to Performance* (New York: The Free Press, 1987), 183–184.
3. This research was reported at the 8th International Annual IPSERA Conference, London, UK, March 1998. Total cost of ownership is addressed in Chapter 15.
4. This research was reported at the 8th International Annual IPSERA Conference, London, UK, March 1998.
5. R. David Nelson, Patricia E. Moody, and Jonathan Stegner, *The Purchasing Machine* (New York: Free Press, 2001), 27.
6. Reliability is the degree of confidence or probability that an item will perform a specified number of times under prescribed conditions. See David N Burt, *Proactive Procurement* (Englewood Cliffs, N J; Prentice Hall, 1984), 24.
7. Maintainability addresses how easily an item can be repaired or restored to operational status.
8. Robert Handfield, *Reengineering for Time-Based Competition* Westport, CT: Quorum Press (1995).
9. From a book review by David Jessop, *European Journal of Purchasing and Supply Management* (December 1997): 241.
10. For an excellent collection of articles on cost reduction, see *Articles for C.P.M Exam Preparation* (Tempe, AZ: National Association of Purchasing Management, 2000), 225–249.
11. For a unique and interesting article addressing this topic, see Richard L. Pinkerton, "The Evolution of Purchasing to Supply Chain Management," in *The Purchasing and Supply Handbook*, eds. John A. Woods and The National Association of Purchasing Management (New York: McGraw-Hill, 2000), 3–16.

12. "Supply Chains and Supply Networks: How Do I Win?" This research was reported by Nick Little, MCIPS, Assistant Director Executive Development Programs, The Eli Broad Graduate School of Management MSU, at the 90th Annual International Supply Management Conference, Minneapolis, MN, May 2005, 3–4.

13. Joseph L. Cavinato, "An Analysis of the Expansion of the Purchasing Field into New Value-Added Roles in Organizations," Institute for Supply Management, Inc., August 15, 2001, 4.

14. In practical purchasing terminology, these types of purchases are termed "rebuys" or "modified rebuys."

15. Gordon Regan, "Conoco Procurement Card Program," presentation to the Executive Purchasing Roundtable, Phoenix, AZ, February 28, 1994.

16. Roger A. Whittier, "How Intel's Purchasing Now Uses Plastic to Generate Cost Savings," *Supplier Selection and Management Report,* (May 1994): 13.

17. Anna E. Flynn, "Evolution of a Profession and a Program," *Inside Supply Management,* 14(1)(January 2003): 29.

18. For a good discussion of systems contracting in MRO buying, see J.A. Lorincz, "Systems Contracts Put Control in MRO Buying," *Purchasing World,* (May 1986): 51–54 and Mary Lu Harding, "Designing Auto-Resupply Systems," presented at the 82nd Annual International Conference Proceedings 1997 (available on the ISM Website at www.ism.ws).

19. George L. Harris, "Revolutionary Ways to Transfer Purchasing Responsibilities to Users and External Organizations," presented at ISM's 82nd Annual International Conference Proceedings, 1997.

Suggested Reading

Andraski, Joseph C. "Out of the Silos," *Supply Chain Management Review* 11(3)(2007): 24–25.

Barnes, Dan. "Supply Chain Gains," *Banker* 157.973 (2007): 44–45.

Boyer, Kenneth, Markham Frohlich, and Thomas Hukt. *Extending the Supply Chain* New York: AMACOM, 2005.

Budzki, Robert A., Douglas A. Smock, Michael Katzorke, and Shelley Stewart, Jr. "Supply Management: How are You Really Doing?" *Supply Chain Management Review* 9(9) (2005): 10–15.

Budzki, Robert A., Douglas A. Smock, Michael Katzorke, and Shelley Stewart, Jr. *Straight to the Bottom Line: An Executive's Roadmap to World Class Supply Management,* Boca Raton, FL: Co-published by *Purchasing Magazine* and Supply Chain Management Review with J. Ross Publishing, 2005.

Cavinato, Joseph L., Anna E. Flynn, and Ralph C. Kauffman. *The Supply Management Handbook,* 7th ed., New York: McGraw-Hill, 2006.

Giannakis, Mihalis and Simon R. Croomm. "Toward the Development of a Supply Chain Management Paradigm: a Conceptual Framework." *The Journal of Supply Chain Management* Spring (2004): 27. Giannakis and Croom describe an interesting supply chain problem domain called the "3s model." This model describes: (1) the synthesis of the business and resources network, (2) the characteristics of synergy between the different actors in the network, and (3) the synchronization of all operational decisions involved in the production of goods and services.

Hammer, Michael. *The Agenda: What Every Business Must Do to Dominate the Decade*, New York: Crown Business, 2001.

Hanfield, Robert, and Ernest Nicholos, *Supply Chain Redesign: Transforming Supply Chains into Integrated Value Systems.* Upper Saddle River, NJ: Financial Times Prentice Hall, 2002.

Hugos, Michael. *Essentials of Supply Chain Management,* 2nd ed. Hoboken, NJ: John Wiley & Sons, 2006.

ISM Glossary of Key Supply Management Terms, 5th ed. Tempe, AZ: ISM, 2009. http://www.ism.ws/glossary/index.cfm?

Killen, Kenneth H., and John W. Kamauff. *Managing Purchasing: Making the Supply Team Work.* Burr Ridge, IL: McGraw-Hill/Irwin, 1995.

Killen, Kenneth, H., and Robert L. Janson. *Purchasing Managers' Guide to Model Letters, Memos and Forms.* Englewood Cliffs, NJ: Prentice Hall, 1991. Although most of these forms are now computerized, the information is still pertinent.

Lambert, Douglas M. "The Eight Essential Supply Chain Management Processes." *Supply Chain Management Review* Spring (2004): 18–26.

Little, Nick. "Supply Chains and Supply Networks: How Do I Win?" Proceedings of the 90th Annual International Supply Management Conference, Minneapolis, MN, May 2005.

Monczka, Robert M., and James P. Morgan *"Quantum Leap" Purchasing*, 131(10)(2002).

Nelson, Dave, Patricia Moody, and Jonathan, R. Stegner. *The Incredible Payback: Innovative Sourcing Solutions that Deliver Extraordinary Results,* New York: AMACOM, 2005.

Nelson, Dave, Patricia E. Moody, and Jonathan Stegner. *The Purchasing Machine.* New York: Simon & Schuster, 2001.

Pinkerton, Richard L. "The Evolution of Purchasing to Supply Chain Management." Forward to *World Class Supply Management: The Key to Supply Chain Management*, 7th. ed. by David N. Burt, Donald W. Dobler, and Stephen L. Starling, Burr Ridge, IL: McGraw-Hill Irwin, 2003.

Sollish, Fred, and John Semanik. *The Procurement and Supply Manager's Desk Reference.* New York: John Wiley & Sons, 2007.

Trent, Robert J. *Strategic Supply Management: Creating the Next Source of Competitive Advantage.* Fort Lauderdale, FL: J Ross Publishing, 2007.

CHAPTER 2

Organizational Issues

In any group activity, three principal factors largely determine the level of performance attained by the group as a whole.

- ▲ The capabilities of the individuals.
- ▲ The motivation of the individuals.
- ▲ The organizational structure within which the individuals function.

This chapter focuses on the last factor.

The first two factors are obvious to most business people, but to many people the impact of the third is less clear. In the case of supply management, the function's location in the management hierarchy of a firm is important, for this decision either facilitates or limits the influence supply management policies and actions can have on the firm's total performance. Within the department itself, the form of organization selected influences the types and levels of expertise developed and also, to a great extent, the effectiveness with which the talents of individuals are utilized.

A firm's organizational structure reflects management's basic attitudes toward the major activities involved in its operation. Where should the supply management function fit in a firm's organizational structure?

Placement of Supply Management Within the Organization

The location of the supply management department within an entity's organizational structure greatly influences that department's ability to function optimally and influence the decision-making process involved in effective procurement. The lower supply management is on the organizational chart, the less likely it is that it can influence corporate strategy significantly. In practice, the center for advanced purchasing studies (CAPS), in a study conducted in 2000 found "the supply organizational structure was forced to be congruent with the overall corporate structure," as opposed to selecting the most appropriate structure for large firms.[1]

The importance of supply management in any firm is determined largely by four factors:

1. *Availability of materials and services* Are the major materials and services used by the firm readily available in a competitive market, or are some key materials and services bought in volatile markets that are subject to periodic shortages and price instability? If the latter condition prevails, creative performance by analytical supply management professionals is required; this typically is a top-level group.

2. *Absolute dollar volume of purchases* If a company spends a large amount of money for materials and services, the magnitude of the expenditure means that effective supply management usually can produce significant profit. Small unit savings add up quickly when thousands of units are purchased. On the services side, a contract for IT (Information Technology) services can run into millions of dollars. Thus, a 10 percent savings contributes significantly to the bottom line.

3. *Percentage of product cost represented by materials and services* When a firm's materials and outsourced services costs account for 40 percent or more of its total operating budget, small reductions in material and service costs increase profit significantly. Well-executed supply management usually pays handsome dividends in such companies.

4. *Types of materials and services purchased* Perhaps even more important than the preceding considerations is the amount of control purchasing and supply personnel have over the availability, quality, and costs of purchased materials and services. Most large companies use a wide range of materials and services, many of whose price and service arrangements definitely can be influenced by creative purchasing performance. Some firms, in contrast, use a fairly small number of standard production and supply materials or services from which even a top-flight purchasing and supply department can produce little profit through the use of creative management, pricing, and supplier selection activities.

In her article "I'm Convinced: You've Got Value!", Mary Siegfried Dozbaba emphasizes the need for purchasing and supply management departments to demonstrate to the top executive their value and commitment to improving profitability. To attain commitment and respect from the top, purchasing and supply management must shift its focus from internal processes to the big-picture issues such as determining and defining the requirements, contract negotiation, supplier relations, and strategic, long-term goals of the organization. "You need to be willing to let responsibility flow out of the unit. Purchasing and supply does not need to approve every requisition...stop being the price police," she says. "You want to ....be called in to negotiate the multimillion dollar contracts, not just to buy pens."[2] Such activities will ensure that a purchasing and supply department is favorably positioned to support organizational needs in today's highly competitive marketplace.

Other Factors Impacting Organizational Structure

Classification of Responsibilities and Activities

The starting point in thinking about potential organization structures is a delineation and an analysis of the work to be done by the unit: the responsibilities and activities. The six classifications of work in a purchasing operation are as follows:

1. *Management* Management of the purchasing and supply function involves all the tasks associated with the management process, with an emphasis on the development of policies, procedures, controls, and the mechanics for coordinating purchasing operations with those of other departments. On an exception basis, it also involves the management of unique supplier and commodity problems.

2. *Buying/supply management* This includes a wide variety of activities, such as working with users to help develop requirements and specifications, reviewing requisitions, analyzing bids, negotiating, and selecting suppliers. Additional responsibilities involve continuing work with a supplier to improve the supplier's capability and performance in the areas of cost, quality, and service.

3. *Contract and relationship management* This responsibility ranges from monitoring purchase orders and working with accounts payable, to the application of project management skills, to a key procurement such as a construction project. Relationship management is appropriate for many major procurements in which the supplier's motivation, cooperation, and collaboration are essential factors in a successful relationship and a successful procurement.

4. *Strategic planning and research work* A well-developed purchasing and supply management operation has a large number of research projects and systems studies that require specialized knowledge and analytical ability. The more an organization has progressed toward a supply management focus, the more emphasis it places on these strategic activities. The core activities in this area include economic, industry, and supply market studies; the development of buying strategies for material or services buying; the development and implementation of supply base and partnering plans; product research and value analysis work; and operating and information systems analysis.

5. *Follow-up and expediting* Order follow-up activity involves various types of supplier liaison work, such as reviewing the status of orders and occasionally visiting suppliers.

6. *Clerical activities* Every department must enter orders and maintain working files, catalog and library material and records for commodities, suppliers, prices, and so on.

The precise manner in which purchasing work is subdivided and grouped depends on the size of the department, which depends on the size of the company.

Operational vs. Strategic Responsibilities

In small firms, responsibilities may be handled by a purchasing or supply manager and one or two assistants; everyone wears several hats. In large organizations, the department may consist of 100 to 300 purchasing and supply professionals.[3] In 1998, R. David Nelson, Vice President, Worldwide Supply Management at Deere and Company hired 94 supplier development *engineers* to accomplish supplier activities, resulting in $22 million in cost savings! A major activity or responsibility for any purchasing and supply management department is to provide an uninterrupted flow of materials and services. The tasks required to accomplish this goal often are referred to as tactical or *operational activities*. These activities are not a great source of cost savings yet are critical in avoiding tremendous losses resulting from potential disruptions in operations. Operational activities offer minimal value-added benefits but are often pressing and time consuming. The focus on such tactical activities results in less time being allocated to profit, generating *strategic activities*. Figure 2.1 demonstrates various operational and strategic characteristics and activities.

Supply managers have begun to see the need for two types of resources in their organizations: (1) a team of people who manage the operational and tactical activities of purchasing and materials management and (2) supply managers who are involved in the development of broader strategic aspects of the function. These organizations are in a position to separate operational and strategic responsibilities formally in their organizational structures. This ability to focus on the strategic sourcing process promises long-term increased profitable sales and cost savings; improving profitability and competitive advantage.

Organizational Authority

The placement of supply management decision-making authority affects the structure of the supply organization. *Centralized authority* exists when the decision-making process

Operational Responsibilities	Strategic Responsibilities
Placement of Purchase Orders	Supplier Development Responsibilities
Managing Contract and Blanket Order Releases	Coordinating the Procurement Systems
Expediting Inbound Orders	Developing Long-Term Contracts
Maintaining Continuity of Supply to Production Lines	Developing and Integrating Supply Strategy
Managing Supplier Relationships	Managing Risks in Supply Chain
Transactionally Focused Activities	Strategically Focused Activities

Figure 2.1 Operational and strategic characteristics and actities.

is the responsibility of a single person. This person is held accountable by top management for the proper performance of purchasing activities. *Decentralization of purchasing authority* occurs when personnel from other functional areas—production, engineering, operations, marketing, finance, and so on—make unilateral decisions on sources of supply or negotiate with suppliers directly for major purchases. As discussed in this book, this concept is concerned solely with the placement of purchasing *authority*. It has nothing to do with the location of buying personnel.

Generally speaking, in a single-site operation, to decentralize the purchasing function needlessly is to deny a firm some of its potential profit. Centralization of the purchasing function is essential for attainment of both optimum operating efficiency and maximum profit. Most companies today view the centralization of purchasing as a logical and desirable evolution of Frederick Taylor's basic concept of the specialization of labor. The extent to which the efficiencies of functional specialization are realized when a firm creates a supply management department, however, depends largely on the authority delegated to that department. When functioning properly, centralized purchasing produces the following benefits:

1. *Reduces potential duplication of effort.*
2. *Leverages volume purchases*   Volume discounts are possible when all company orders for the same and similar materials are consolidated. In addition, a firm is able to project a unified policy to its suppliers, gaining maximum competitive advantage from its total economic power.
3. *Consolidation*   Consolidation provides an opportunity to standardize and simplify parts. Additional benefits can be gained from value analysis and value engineering coordinated through a centralized function.
4. *Transportation savings*   Transportation savings can be realized through the consolidation of orders and delivery schedules.
5. *Allows for specialization*   Centralization develops purchasing specialists whose primary concern is purchasing. With training, purchasing specialists inevitably buy more efficiently than less-skilled individuals can.
6. *Reduction of suppliers' costs*   Suppliers are able to offer better prices and better service because their expenses are reduced. Their sales personnel make fewer calls, prepare fewer orders, make fewer shipments, prepare fewer invoices and do less recordkeeping.
7. *Improved inventory control*   More effective inventory control is possible because of companywide knowledge of stock levels, material usage, lead times, and prices.
8. *Lower administrative costs*   Fewer orders are processed for the same quantity of goods purchased reducing purchasing, receiving, inspection, accounts payable, and recordkeeping expenses.

9. *Centralized control*   Responsibility for the performance of the purchasing function is placed with a single department head, facilitating management control.
10. *Reduction in the costs of services*   Warren Norquist, while Vice President of Materials Management at Polaroid, centralized the purchase of marketing services such as advertising. As a direct result of this centralization, Polaroid had savings in excess of 25 percent.

Despite the general advantages of centralization, *complete* centralized purchasing is neither always possible nor always desirable. Four types of situations justify some decentralization. The first is found in companies that process *single natural raw materials*. Many of those firms separate the purchase of key raw material from the purchase of other materials. Firms in the textile, leather, food, beverage, and tobacco industries are good examples. In these industries, the raw materials are products of nature that are purchased in unstable markets in which prices fluctuate widely. Buying typically takes place at auctions or through commodity exchanges conducted in small local warehouses. In such markets, a practical knowledge of grades is equally as important as knowledge of prices. Buyers of these commodities usually guard their specialized know-how with secrecy, frequently handing it down from one generation to the next.

A second situation justifying some decentralization of purchasing authority exists in *technically oriented firms that are heavily involved in research*. In those firms, some exceptions to complete centralization are always desirable. Many one-time purchases in the research, design engineering, and related departments can be handled more effectively by professional personnel in these departments. Moreover, the dollar volume of such purchases is usually relatively small.

The third situation justifying a different type of decentralization is found in the *operation of multisite institutional and manufacturing organizations*. Decentralization in this circumstance can allow for faster response time for the requisitioner, a better understanding of the requirements unique to that plant, quicker support in product development projects, and ownership of the process and products.[4]

Fifth, the *purchase of nontechnical odds and ends* often calls for a partial decentralization of purchasing. Credit card and petty cash fund purchases of less than several hundred dollars are a good example. Decentralizing through the use of these approaches can be a money saver.

The danger of losing purchasing control does not stem from partial decentralization of the purchasing function per se. Some decentralization is necessary as a matter of common sense. The use of cross-functional teams in the decision-making process is increasing. Organizations are changing to compete in the global market. Companies are seeking to take advantage of both structures by forming a *hybrid structure* that includes both centralized and decentralized decision making. This results in the best of both structures, maintaining necessary control while meeting the unique needs of the other functions or divisions. It is important to state clearly in written purchasing policies, the purchasing manual, and supplier welcome booklets, who has the authority to buy what, that is, who are the agents authorized to sign purchase orders and purchase agreements?

e-Commerce

Technology has provided a solution to the debate about centralized vs. decentralized organizational structures. As computers began to appear on desktops in the late 1980s, purchasing departments were able to reduce administrative time and costs by placing purchase orders electronically. Electronic data interchange, or EDI as it was called, was the early stages of e-commerce. Electronic catalogs of approved supplies or services could be created through a centralized supply management department, and disseminated to decentralized locations for purchase. Electronic billing and remittance of invoices improved cash flow while reducing the cost of doing business. e-commerce allowed corporate policy and procedure to be distributed and monitored effectively while giving divisional supply management departments the authority to make purchases to meet their local requirements. e-commerce has been a catalyst enabling organizations to benefit from the hybrid supply management structure.

Organizational Structures

As was stated previously, the position of the supply management department on a firm's organizational chart can vary. However, as materials gained in importance, becoming a larger portion of the cost of goods sold, issues such as inflated inventories, poor quality, material stock-outs, long lead-times, and miscommunications provided the impetus for a movement to capture control of materials under one individual. This gave rise to the materials management approach to structuring the material portion of an organization.

The Materials Management Organization

The concept of organizing the functions that affect the acquisition, movement, and storage of materials under one manager evolved in the early 1960s. The term *materials management* was introduced to describe an integrated systems approach to the coordination of materials activities and the control of total material costs.[5] The purchasing, planning and scheduling, transportation, and warehousing functions were organized under the control and responsibility of one individual, the materials manager. The objective was to optimize performance of the materials system, as opposed to sub optimizing the performance of the individual functions that are parts of the materials system. The results were "great improvements in inventory levels, customer service, and communications, ultimately improving the bottom line."[6]

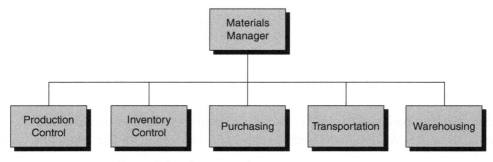

Figure 2.2 The materials management structure.

The materials management organization was popular during the 1960s and 1970s. This concept of integrating the functions involved in the management of materials paved the way for the more sophisticated concept of supply and supply chain management. Figure 2.2 illustrates the structure of a materials management department in an organization.

The Supply Chain Management Structure

In the late 1980s, Japan invaded the U.S. automobile market with cars assembled in Japan. This marked the beginning of globalization as we know it today. The early 1990s saw the appearance of personal computers on the desktops of organizations, putting data and information at the fingertips of managers. With the introduction of the Internet and World Wide Web, the information age began. Through these developments, the concept of managing materials for competitive advantage flourished. The materials management structure evolved beyond the idea of managing the corporation's inventory and those functions responsible for supply. The broader concept of *supply chain management* grew to encompass the planning and management of all activities involved in forecasting, sourcing and procurement, and all incoming logistics management activities. It broke down the functional silos, creating integrated processes throughout the total organization.[7] In describing the breadth of supply chain management, Marilyn Gettinger, C.P.M., President of New Directions Consulting Group, writes:

> Importantly, it also includes coordination and collaboration with channel partners such as suppliers, intermediaries, third-party service providers, and customers. In essence, supply chain management integrates supply and demand management within and across companies. It is a total system approach to facilitate coordination internally and with supply chain partners, often using enhanced communication and information technologies.[8]

Organizing with Cross-Functional Teams

Cross-functional teams have become a common approach to addressing many supply-management-related activities, including: new product development,[9] value analysis and value engineering, standardization and simplification, engineering change management, the development of statements of work describing services requirements, commodity teams, the acquisition of capital and operating equipment, make/buy and outsourcing analysis, source selection,[10] potential supplier field reviews, negotiation, post-award management and problem solving, supplier development, and the development of strategic alliances. Since cross-functional teams require a significant investment in human resources, their use commonly is limited to time-critical and high-monetary-value activities. The appendix to this chapter addresses cross-functional teams in greater detail.

Budgeting and Staffing

Obviously, all firms have a budget process that involves planning the operational activities and the resources to accomplish the tasks including people, facilities, software, equipment, office supplies, training, and travel expenses for the next operating period. This is why we recommend the chief procurement officer prepare an annual materials report that lists the annual spend by type of materials, the major contracts with key suppliers, the number of personnel and their training activities, on-time delivery percentage, quality record, cost savings, value analysis/value engineering projects, partnerships and alliances, variance analysis, cost-avoidance achievements, and any other activities to inform all concerned as to the total materials activities of the firm. There will be more detail regarding development of this report in the last chapter.

The most critical asset of any organization is its people. The days of purchasing personnel learning their jobs via on-the-job training (OJT) are long over. It is highly recommended that supply managers hire graduates with business degrees and preferably majors in purchasing or materials management from colleges with a focus in supply chain management such as Michigan State University, Bowling Green State University, Arizona State University, Florida State University, North Carolina State University, and the University of San Diego. Contact the Institute for Supply Management in Tempe, Arizona at www.ism.ws for a list of colleges and universities offering appropriate courses and degree programs in purchasing and supply management.

Typical positions include: MRO buyer, capital equipment buyer, buyer-planner for scheduling, inventory control analyst, tool crib keeper, raw materials buyer by type as the volume warrants, commodity manager, traffic specialist, sourcing expert, and other specialists as needed. The position of expeditor has been greatly reduced as we move to better sourcing. Many expeditors chasing delayed or lost shipments means the sourcing process is marginal.

We strongly recommend all purchasing personnel be members of ISM, APICS, and other appropriate societies. The dues should be paid by the organization. In addition, all personnel should be required to pursue certification programs appropriate to their positions such as CPSM and CPIM (APICS).

From our consulting experience, supply managers often fail to budget properly the necessary continuing education and travel funds for on-site visits to key suppliers. All of us need to keep up to date by attending continuing education classes, conferences, seminars, national conventions, and local professional meetings. (See the list of organizations in the appendix of this chapter).

Summary

Foremost in developing an organizational structure is finding a structure that will allow the organization's supply chain to function effectively and efficiently. ISM Professor of Supply Chain Management, Robert M. Monczka tells us in *Finding a Structure that Works*,

> Establishing appropriate organizational structure and governance processes for the supply management function is critical to effective supply chain management. Competitive and customer pressures, globalization, outsourcing and the need for innovation from external sources, combined with unrelenting pressure to achieve cost reduction, faster time to market and improved customer responsiveness, all increase the importance of this decision.[11]

Appendix A

Benefits Resulting from Cross-Functional Teams[12]

Synergy

The many activities identified previously have one thing in common: they all benefit from a variety of functional inputs. For example, during the new product development process, marketing has information on customers' wants and needs, their willingness to purchase at different prices, and present and potential competition. Design engineering has knowledge of current and future design processes and constraints. Manufacturing engineering has information on the firm's and its suppliers' manufacturing processes and their limitations. Supply management provides a window to the supply world and its capabilities and limitations, and the likely cost and availability of various materials and services under consideration. Customer service, quality, finance, IT, and carefully selected suppliers all have many additional contributions to make during the new product development process. When these professionals come together under the leadership of a capable team leader, the result is normally a synergy that, in turn, results in a far more profitable new product, far quicker than would have occurred with the traditional sequential approach to new product development.

Input from All Affected Functions

The cross-functional approach greatly increases the likelihood that all issues that should be considered are addressed. For example, customer support and service frequently were overlooked in the traditional sequential approach to new product development. Standardization efforts, which are conducted or controlled by a single functional area, such as manufacturing engineering, frequently overlook the procurement, manufacturing, and marketing implications of implementing new standards. The result is surplus purchased materials, manufacturing bottlenecks, or products that do not compete in the marketplace.

Time Compression

A hypothetical example may help in understanding the traditional sequential or functional approach to many of the activities described previously. Marketing at Alpha Corporation has identified a need for a new, complex transducer. Marketing describes this need to design engineering, which designs the transducer. On completion of the design, the resulting specifications and drawings are forwarded to manufacturing engineering, the function responsible for translating design engineering's specifications into production plans. Manufacturing engineering determines that certain tolerances cannot be met by the firm's production equipment. Design engineering is requested to revise the specifications. Design engineering contacts marketing to determine what impact the revised tolerances (ones which the firm's equipment can meet) will have on sales. If the impact is significant, manufacturing, and possibly plant engineering, may become involved. They may decide that new equipment is required. This process continues on through the quality assurance function. Quality assurance reviews the specifications and production plans to ensure that the required level of quality will be obtainable. The customer service function then reviews the specifications and manufacturing plans so that it may develop plans to support the transducers in the field. Obviously, each function along the sequential path leading to production of the desired transducer may question (or even challenge) the design specifications and manufacturing plans. This back-and-forth process ultimately leads to a product that is late to market and overpriced.

The traditional sequential approach as depicted in Figure 2.3 demonstrates how convoluted the process becomes when compared to the effective and efficient cross-functional team approach shown in Figure 2.4.

Overcoming Organizational Resistance

With the cross-functional approach, all functional areas are involved up front, which helps to reduce organizational resistance to decisions that will affect specific functional

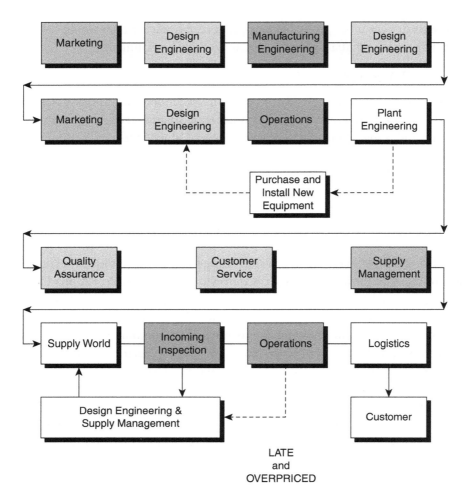

Figure 2.3 Sequential, functional approach to new product development.

areas. In contrast, decisions that impact multiple functional areas, which are made without representation of those areas, are likely to meet resistance. The representatives of each of the functional areas involved on the team constructively provide their input and are involved in the resulting agreement. In turn, each representative is responsible for ensuring acceptance by his or her functional area of the team's decisions. Experience indicates that once a team makes a decision, implementation of the resulting plan is much easier and faster than with the sequential approach.

Enhanced Problem Resolution

The cross-functional team approach is far more efficient and effective at solving problems than the traditional functional one. For example, if a supplier, in spite of its best

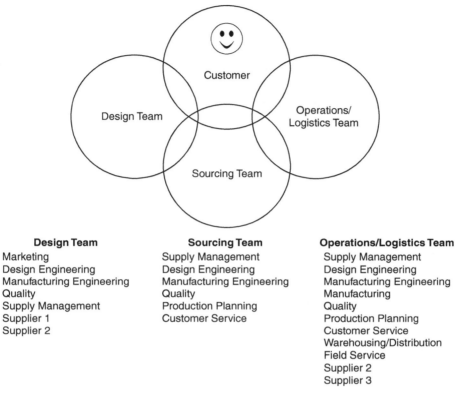

Figure 2.4 Cross-functional team approach to new product development.

Design Team	**Sourcing Team**	**Operations/Logistics Team**
Marketing	Supply Management	Supply Management
Design Engineering	Design Engineering	Design Engineering
Manufacturing Engineering	Manufacturing Engineering	Manufacturing Engineering
Quality	Quality	Manufacturing
Supply Management	Production Planning	Quality
Supplier 1	Customer Service	Production Planning
Supplier 2		Customer Service
		Warehousing/Distribution
		Field Service
		Supplier 2
		Supplier 3

efforts, is unable to meet the contract schedule or quality requirements, a cross-functional team representing supply management, manufacturing engineering, and quality may be formed to work with the supplier to resolve the problem. (The solution to a surprising number of such problems is found to be within the customer firm's control, not the supplier's!)

Negotiations

Negotiations for critical or large monetary value materials, services, supplies, or items of equipment are conducted much more effectively by a well-prepared and well-coordinated cross-functional team than by the finest supply professionals alone.

Improved Communication and Cooperation

The traditional functional approach to the activities listed in the opening paragraph of this appendix, normally results in efficiency within each department. However, this approach inhibits communication and cooperation among the departments involved in the activities.

Some 40 years ago, one of the authors was the Chief Procurement Officer (CPO) of a relatively small business unit. The organization's plant engineering department designed specifications for construction projects. These specifications served as the basis of invitations for bids (IFB) and the resulting contracts. Numerous questions and problems were encountered with potential suppliers during the bidding process and then with the successful bidder. The specifications were ambiguous and, in several cases, contained inconsistencies.

The CPO met with the plant engineer and offered to become involved in the development and review of the specifications. The plant engineer rejected the offer stating: "Development of specifications is my responsibility." Interestingly enough, the plant engineer subsequently expressed his desire to become involved in the sourcing process.

Challenges and Problems with the Cross-Functional Approach

There are several challenges and inherent problems with a cross-functional approach. They include additional investment in scarce resources, role conflict, overload for key team members, continuity and rewards.

Additional Investment in Scarce Resources

A single professional normally requires far fewer labor hours to accomplish a task than does a team. For example, a single supply management professional can accomplish the many actions involved in selecting a critical supplier in considerably fewer labor hours than can a cross-functional sourcing team. However, a team consisting of a design engineer, a manufacturing engineer, and a quality engineer together with a supply professional will do a far more thorough job of selecting the right source.

Role Conflict

Normally, cross-functional team assignments are additional duties for many or all of the individuals involved. In many cases, the team member's functional manager expects the individual to perform his or her normal functional responsibilities. Such responsibilities require about 40 hours per week, and the individual also is expected to satisfy his or her team responsibilities. A number of years ago, one of the authors directed thesis research at the Graduate Logistics Division of the Air Force Institute of Technology. The research focused on minimizing role conflict between functional and team assignments. The subjects were assigned to cross-functional teams developing the B-1 bomber. The researchers focused on a multi-matrix approach to project management as a means of avoiding or minimizing the inherent role conflict. While promising, the research results were inconclusive. Some 25 years later, management is still attempting to cope with the issue of role conflict resulting from part-time assignment of individuals to cross-functional teams. Despite these issues, the benefits of teams usually outweigh the resulting problems.

Overload for Key Team Members

Overload is an obvious result of the role conflict inherent in the additional duty assignment of key team members. Paradoxically, the most attractive team members are those individuals who are key contributors to their functional organizations. Management must be sensitive to the danger of such an overload in an effort to avoid burnout and the possible loss of such individuals.

Continuity

Obviously, once team members have been trained, developed, and learn to work together in a synergistic manner, continuity of membership becomes critical. Retirement, departure to become employed at another firm, promotion, and layoffs can all have a negative impact on the team's operation. Careful selection and assignment of team members can reduce, but not eliminate, such problems.

Rewards

By now, it should be apparent that individuals who are assigned to cross-functional teams as an additional duty should be rewarded appropriately. The greatest reward is the satisfaction associated with "making a difference" in the team's success. Senior professional managers ensure that functional managers (the chief procurement officer, director of R & D, director of manufacturing, etc.) recognize each individual's contributions both to their functional organizations and to their cross-functional team.

A few enlightened organizations have demonstrated success with team incentives. These incentives range from team dinners, to a team vacation in Hawaii, to the award of stock options. In some cases, the team will receive a bundle of rewards (such as stock options). The team members then allocate the options according to a consensus of the members on the relative contribution of each member. More information on this subject may be found in *Rewarding Teams: Lessons from the Trenches*.[13]

Prerequisities to Successful Cross-Functional Teams

While there are many such prerequisites, we will focus on three especially critical ones: executive sponsorship, effective team leaders, and qualified team members.

Executive Sponsorship

An absolute prerequisite for successful cross-functional teams is the support of an executive sponsor. "Top management team support and political factors may be even more critical to the success of cross-functional teams than the internal team processes."[14] The individual sponsor should either have all of the functional areas involved reporting to him or her, or have the informal ability to secure the cooperation and support of colleagues in obtaining the assignment of the appropriate human resources to the project. Additionally, the executive sponsor must track the cross-functional team's progress, run interference, and obtain additional resources as appropriate.[15]

Effective Team Leaders

Without skilled leadership, teams frequently become lost, flounder, get off course, lose sight of their goal, lose confidence, become mired in interpersonal conflicts, stop short of their goal, and never contribute their full potential. Surveys of highly effective teams have shown that their team members rated their leaders as highly skilled. Lower performing teams rated their leaders as being much less effective.

Ideally, a team leader has people skills, communication skills, technical knowledge, enthusiasm, and experience working with the people who will be on the team. The new role of the team leader is to build a team with vision, authority, accountability, information, skills, and commitment to assume more and better operational control of the team's work. The new leader expands the capabilities of the team members and the team itself. As a result, the team can perform some of the leader's traditional work roles, such as budgeting, scheduling, setting performance goals, and providing training. The team gradually assumes the day-to-day operations, thereby allowing the leader to manage resources, ideas, technologies, and the work processes. The most challenging aspect of this new role is that the leader must give up a part of his or her former, more authoritarian, role. Such a shift of roles allows the leader more time to take on strategic roles. The result of the shift is that the team is able to contribute more and with greater speed.

Team leaders must assume a number of roles. They must understand people so they can influence them. They should encourage and maintain open communication, help the team develop and follow team norms. The team leader needs to step back from his or her management role of directing employees, and assume a more collaborative role as a facilitator. The leader should help guide the team and allow the team to identify problems, develop solutions, and then implement the solutions. The team members must be free to express themselves as long as such action is not destructive in nature. The team leader should support the expression of conflicting points of view. "Team synergy begs for a conflict of ideas. Conflict can bring into being the creative tension where paradigm-shifting ideas are born."[16] While the team leader may need to retain some of the final decision making in the early stages, it should be the leader's goal to develop the team so that it can assume responsibility for the decision making entirely.

The team leader helps the team focus on the task and removes obstacles that stand in the way of the team's performance. Helping the team to focus will ensure that the team progresses through productive stages of team development, and will reduce the tendency for it to revert back to one of the less productive stages. Effective team leaders help minimize turf issues and keep the team focused on the good of the organization. Additionally, the leader needs to make sure that the team members have all the resources that they may need so that they do not become distracted. The team leader should remove any obstacles to the team's success.

Additionally, team leaders help establish a vision, create change, and unleash talent. The leader helps the team establish a mission statement and define its goals. Leaders create change within both the organization and the team. They force people to think outside the box and help develop creative solutions to problems. Leaders need to have good people skills in order to identify and draw out hidden talents of the team members.

Former American League relief pitcher David Baldwin, writing in the *Harvard Business Review*, addresses the issue of "blame." He focuses on how managers in Major League Baseball employ blame. Baldwin contends that blame plays an important role in shaping an organization's culture. He proposes five important rules of blame, which we believe apply to most or all team leaders:

1. Know when to blame—and when not to.
2. Blame in private and praise in public.
3. Realize that the absence of blame can be far worse than its presence.
4. Manage misguided blame.
5. Be aware that confidence is the first casualty of blame.[17]

Lippincott's book, *Meetings: Do's, Don'ts, and Donuts* (Lighthouse Point Press, 1994), offers the following suggestions to team leaders:

▲ Decide whether a meeting is the best way to accomplish this. Consider circulating routine information via e-mail. If a meeting is required, distribute an agenda at least two days in advance.

▲ State in one or two sentences what you would like your meeting to accomplish.

▲ Set ground rules to maintain focus, respect, and order during the meeting.

▲ Take responsibility for the outcome of the meeting. For example, help keep the meeting on track and help resolve conflicts.

▲ If your meeting isn't working, try other tools, such as brainstorming techniques or computer software that helps you create the agenda.[18]

Qualified Team Members

Experience indicates that the most critical variable influencing one's ability to be a "high" contributor is a willingness and desire to contribute. Baxter Health Care of Paramatta, New South Wales, Australia, ensures that team members are "willing" participants by announcing forthcoming team projects to all employees. Individuals are encouraged to volunteer for the additional assignments as team members representing their functional areas. In many cases, competition for a team position is intense. Thus, the team leader is in the enviable position of being able to select members from a pool of volunteers. Obviously, team membership has the potential for satisfaction, and intrinsic and tangible rewards at Baxter.

The Wisconsin Department of Revenue has identified the following communication skills as being significant to employee success. (If an individual is deficient in one or more areas, he or she can attend training offered by the department, management development programs, pursue self-study, or obtain a mentor.)

Listening   The ability to understand, organize, and analyze what we hear.

▲ Actively attend to and convey understanding of the comments and questions of others.
▲ Identify and test the inferences and assumptions we make.
▲ Overcome barriers to effective listening (semantic, psychological, physical).
▲ Summarize and reorganize a message for recall.
▲ Keep the speaker's intent, content, and process separate.
▲ Withhold judgment that can bias responses to the message.

Giving clear information

▲ Assess a situation, determine objectives, and give information that will best meet the objective.
▲ Construct and deliver clear, concise, complete, well-organized, and convincing messages.
▲ Keep on target—avoid digressions and irrelevancies, and meet the aim of the communication.
▲ Determine how to use persuasion effectively.
▲ Maintain a climate of mutual benefit, trust, rapport, and a win-win outcome.

Getting unbiased information   Minimize the filtering and editing that takes place when information is transmitted from person to person.

▲ Use direct, non-direct, and reflective questions.
▲ Identify forces that may bias the information.
▲ Confirm understanding and obtain agreement and closure.

Foster Open Communication

▲ Create an atmosphere in which timely, high-quality information flows smoothly between self and others.

▲ Encourage open expression of ideas and opinions.[19]

Team Development and Training

Each team will develop its own personality, but the key objective of all teams must be willingness to subordinate personal and functional interests to the team's goals. Having a competent leader and well-qualified team members are two critical first steps. The third step is team development and training. Team development and training call for investments that should pay a high return. For example, Southern California Edison (SCE) has used team development and training to create one of the best supply management systems in the utility industry. Under the leadership of Emiko Banfield, 24 cross-functional supply management teams have been established to manage supply issues. Each SCE team receives three days of team development and training as a foundation for its activities. As a result, SCE has taken over $250 million out of a spend of approximately $1 billion. Working through cross-functional teams, Banfield discovered that internal barriers could be reduced, setting the stage for successful collaboration if teams were trained properly.[20]

The Supply Chain Management Institute at the University of San Diego is pioneering an alternative approach to team development. Four-person cross-functional teams from client firms undergo interactive training on selected supply management topics. A one-hour workshop is conducted after each one-hour training module. During the workshop, each team conducts a gap analysis, comparing one of its processes with the world-class processes presented in the previous training module. The team then develops a preliminary action plan to close the gap. The plan identifies key actions, a time line, and an estimate of the bottom line impact of the team's proposed plan. Preliminary findings indicate that cohesive teams evolve with this approach to training and organizational transformation. Although the findings are preliminary, this approach to team development appears to be very cost effective.

Adequate Time

Unrealistic deadlines are major problems that block the success of many cross-functional teams. As Burt and Pinkerton wrote, "Too much pressure for results too soon will almost always force a team to premature and less effective decisions."[21] The tendency for management to act now rather than allowing time for good analysis is an old habit in the United States. Many of America's global competitors have the patience to allow time to nurture participative management. The results of nurturing participative management are well known. Just ask American automotive manufacturers about the cost of quick reactions without fostering participation.[22]

Interfirm Teams

When buying and supplying organizations recognize the interdependence and the benefits to both parties of a collaborative or alliance relationship, development of an interfirm team should be considered. In effect, a super-ordinate cross-functional team will result. Dan Mohr, Director of Supplier Relations for GTE, observes: "Relationship teams are the building blocks upon which the relationship prospers. Team meetings provide a forum to jointly discuss new ways to reduce process costs, improve service to our customers, and enhance time to market, which ultimately expands market share for both organizations."[23] As with cross-functional teams within each firm, assignment of the "right" individuals and team training are essential prerequisites for success. One significant difference is that two executive sponsors will be required, one at each firm. The interfirm's first task, after receiving appropriate training, is the development of a customized effective and efficient communication system.

As we will see in Chapter 4, many progressive organizations are working with selected collaborative suppliers to develop and manage supply alliances. One of the keys to success with such efforts is the development and use of interfirm teams. Experience indicates that the basis of such interfirm teams must be the existence of cross-functional teams at both the buying and the supplying organizations. The development and use of interfirm teams is more challenging than are in-house cross-functional teams, but the benefits are even greater!

Supply Management's Roles on Cross-Functional Teams

Timothy M. Laseter, Vice President, Booz-Allen & Hamilton Inc. in New York, identifies four principle roles for supply management professionals who are members of cross-functional teams:

1. Provide the process expertise of supply management in areas such as supply base research, supplier cost modeling, or (more typically) negotiation;
2. Provide content knowledge of a specific supply market or commodity area that the supply management individual directs;
3. Serve as the liaison with the supply management organization to ensure project needs obtain priorities among other staff in the corporate organization;
4. Represent the supply management point of view in considering tradeoffs, setting priorities, and making decisions affecting policy.[24]

In 210 BCE, Arbiter Petronius of the Greek Navy wrote, "We trained hard…but it seemed that every time we were beginning to form up into teams we would be reorganized, and I was to learn later in life that we tend to meet any new situation by reorganizing…"[25] Fortunately, great progress has been made in the design and use of teams, especially in the areas of new product development, project management, source selection, and negotiation.

Appendix B: The Professional Organizations of Supply Chain Management

Institute for Supply Management (ISM): www.ism.ws, 2055 E. Centennial Circle, Tempe, AZ 85284, (800) 888-6276. Founded in 1915, ISM is thought to be the oldest and largest purchasing society in the world and is a member of the International Federation of Purchasing and Supply Management (IFPSM). Originally known as the National Association of Purchasing Agents (NAPA), then the National Association of Purchasing Management (NAPM), and finally its current title to reflect the expansion of its service to over 40,000 supply professionals in 75 countries. ISM's primary focus is on purchasing and offers a wide range of training and educational materials, seminars, workshops, and an annual international supply management conference. Their certification program has evolved from the Certified Purchasing Manager (CPM) to the Certified Professional in Supply Management (CPSM).

ISM publishes an annual supply management school (college) directory and salary survey, both obvious sources for recruiting. Their annual educational resource catalog is extremely useful. There are local chapters or "affiliates" as ISM calls them.

ISM publications include the esteemed *Journal of Supply Chain Management* and the monthly magazine, *Inside Supply Management* which includes the highly respected ISM Report on Business that is widely quoted in various business publications, and in government documents. The ISM yearly salary survey usually appears in the May issue of *Inside Supply Management*.

The Center for Advance Purchasing Studies (CAPS) is a joint venture between ISM and Arizona State University in Tempe, AZ. CAPS conducts best practice projects, industry benchmarking studies, trends, and other relevant research sponsored by more than 300 large, multi-national firms.

American Production and Inventory Control Society (APICS): This Association for Operations Management's focus is obvious from their name. www.apics.org. The U.S. corporate office is located at 8430 West Bryn Mawr Ave., Suite 100, Chicago, IL 60631, ph. 800-444-2742. Their European office is in Brussels, Belgium, fax +3227431550. APICS offers a complete bookstore of educational materials, seminars, workshops, and a Certified Supply Chain Professional (CSCP) program with both general and specific areas such as transportation and logistics. There are over 34,000 members and many local chapters.

American Society for Quality (ASQ): www.asq.org, 600 North Plankinton Ave., Milwaukee, WI 53203, 800-248-1946. Former name American Society for Quality Control

(ASQC). The name provides the focus. ASQ has extensive books and publications in the field of quality control, standards, seminars, conferences, and certification programs in several areas such as: biomedical auditor, calibration technician, lean manufacturing, six sigma master black belt, pharmaceutical professional, etc. ASQ has over 100,000 members.

Council of Supply Chain Management Professionals (CSCMP): www.cscmp.org, 333 East Buttefield Rd., Suite 140, Lombard, IL 60148, ph. 630-574-0985. This is the former Council of Logistics (COL), and their focus is transportation and logistics; but, they seem to be expanding their mission. They offer educational events, research and resources, round tables, publications—bookstore and an annual global conference. CSMP has about 8,000 members.

Supply Chain Council, Inc. (SCC): www.supply-chain.org, 12320 Barker Cypress Rd., Suite 600, PMB 321, Cypress, TX 77429, ph. 202-962-0440. This is the nonprofit organization of about 750 corporate members that developed several IT models such as Supply Chain Operations Reference (SCOR) that define, measure, and improve the supply chain. SCC runs the SCOR Professional (SCOR-P) certification program and also offers a bookstore, research, and other tools. See: Peter Bostorff and Robert Rosenbaum, *Supply Chain Excellence: A Handbook for Dramatic Improvement Using the SCOR Model,* 2nd ed., NY, NY. The American Management Association (AMACOM), 2007. This text also includes the design chain operations model (DCOR) and the customer chain operations reference (CCOR) model. There are six special industry groups within the council including: lean six sigma, aerospace, defense, automotive, electronics, and chemicals.

National Contract Management Association (NCMA): www.ncma.org, 2170 Beaumeade Circle, Suite 125, Asburn, VA 20147, ph: 571-382-0082.

Founded in 1959, NCAMA is an association of government contract procurement officers and defense contractors. They offer extensive training tools, publish a journal, offer seminars, conferences, and several certification programs, including: Certified Federal Contracts Manager (CFCM), Certified Commercial Contracts Manager (CCCM), and the Certified Professional Contracts Manager (CPCM). The obvious focus of NCMA is on contract negotiation.

National Institute of Governmental Purchasing, Inc. (NIGP): www.nigp.org, 151 Spring St., Herndon, VA 20170-5223, ph: 703-736-8900. NIGP is an organization of city, county, state, and federal buyers-purchasing personnel. NIGP offers the usual assortment of conferences, code of ethics, consulting and professional services, publications, training seminars, an online contract administration course, and two certification courses: CPPB & CPPO administered by the Universal Public Procurement Certification Council (UPPCC).

International Computer Negotiations, Inc. (ICN): www.caucusnet.com, Drawer 2970, Winter Park, FL 32790-2970, ph: 407-740-0700. This organization focuses on IT procurement, software issues, vendor management, and negotiations training. ICN offers an extensive list of seminars and workshops they call "The Association of Caucus Technology Acquisition Professionals."

Grainger Center for Supply Chain Management: www.graingercenter.com, The University of Wisconsin School of Business, 3450 Grainger Hall, 975 University Ave., Madison, WI 53706-1323, ph: 608-262-1941. Founded in 1991, with a very generous gift from the Grainger Foundation of Lake Forest, IL (W.W. Grainger, Inc., the leading industrial distributor of maintenance, repair, and operating (MRO) products in the U.S., Canada, plus expansions in Japan, Mexico, India, China, and Panama). The Grainger Institute offers graduate supply chain courses-degrees, research projects, SAP university alliance, certificate programs in sustainability and entrepreneurship, special seminars, industry partnerships, executive speaker series, site visits, and a placement service. Grainger is an excellent recruiting source for MBA's with previous job experience and course work in state of the art supply chain subjects including procurement. Contact Dr. John M. McKeller, Senior Lecturer, Marketing and Procurement/Supply Management.

Project Management Institute (PMI): 14 Campus Blvd., Newtown Square, PA 19073-3299, ph: 855-746-4849. This is a huge (320,000 member) organization open to anyone with an interest in or actually working as a project director. PMI has offices all over the world, publishes a project guide book (PMBOK™ Guide–4th ed.) and has an online certification program, the PMP.

Supply Chain Management Institute (SCMI): www.sandiego.edu/business/centers/supply_chain_management, School of Business Administration, University of San Diego, 5998 Alcala Park, San Diego, CA 92110-2492, ph: 619-260-2791. Founded by our co-author, David N. Burt, this private Roman Catholic Institution is located on a lovely bluff overlooking beautiful San Diego Bay. SCMI offers a master of science in supply chain management (MS-SCM) that is designed for high performing managers and executives and has a unique two phases. Phase I is a graduate certificate in supply chain management and for those who want the MS degree, phase II which includes an optional "real world" graduate team project sponsored by a leading U.S. corporation. See the Raytheon case study in Appendix C, Chapter 22. Phase II also includes a learning portfolio. There are also BA and MBA degree programs in supply chain management and a MS-Global Leadership. In addition to on-campus resident degree programs, the distance learning program is available for the Certificate Program. All courses are web-based and delivered via web CT. In addition to the certificate and degree programs, SCMI conducts applied research and collaborates with industry via three annual forums: supply management, operations, and logistics. Another fine recruiting source.

Health Care

Association for Healthcare Resource & Materials Management (AHRMM): www.ahrmm.org, 155 N. Wacker Dr., Chicago, IL 60606, ph: 312-422-3840. An affiliate of the American Hospital Association, AHA, AHRMM was founded in 1951 and had many names including Hospital Purchasing Agents, and the American Society for Hospital Materials Management (ASHMM). About 4,000 members participate in various procurement educational programs, conferences, and the Certified Materials & Resource Professional (CCMRP) program. There are sub groups within the American Hospital Association, see www.aha.org.

Buyers Co-operative: HPS Infonet, www.hpsnet.com. Organized in 1949, HPS leverages the buying power of over 2,600 organizations in the typical group consolidation buying association with its partnership with MedAssets. There is an annual conference and trade show.

Hospital Buyer: On hold since January 2009, information from back email newsletters is still available. See www.hospitalbuyer.com.

National Association of Educational Procurement (NAEP): 5523 Research Park Drive, Suite 340, Baltimore, MD 21228, ph: 443-543-5540, fax: 443-543-5550. Former title, National Association of Educational Buyers. www.naepnet.org. This 85 year old organization has the full range of educational programs and buyer's co-operatives.

Endnotes

1. Michiel R. Leenders, P. Fraser Johnson, Anna E. Flynn, and Harold E. Fearon, *Purchasing and Supply Management: With 50 Supply Chain Cases*, 13th ed. (Burr Ridge, IL: McGraw-Hill Irwin, 2006), 39.
2. Mary Siegfried Dozbaba, "I'm Convinced: You've Got Value," *Purchasing Today* 10(5) (May 1999): 44.
3. Large multiplant firms may employ up to 1000 or more buyers and engineers.
4. Robert M. Monczka, Robert B. Handfield, Larry C. Giunipero, and James L. Patterson, *Purchasing & Supply Chain Management*, 4th ed. (Mason, OH: South-western Cengage Learning, 2009) 164–170.
5. For the classic reference, see Dean S. Ammer, *Materials Management and Purchasing*, 4th ed. (Homewood, IL: Richard D. Irwin) 1980.
6. Marilyn Gettinger, "Strategic Thinking: Movin' On Up From Supply Management to the Supply Chain." 91st Annual International Supply Management Conference, May 2006.
7. Ibid.
8. Ibid.
9. Laura M. Birou, Stanley E. Fawcett, and Gregory M. Magnam in their study, "The Product Life Cycle: A Tool for Functional Strategic Alignment," found that "...companies are striving to break down functional barriers that inhibit effective product and process design." *International Journal of Purchasing and Materials Management* (Spring 1998).

10. "Chrysler used a team approach and chose suppliers before the parts were even designed, which meant virtually eliminating traditional supplier bidding." James Bennet, "Detroit Struggles to Learn Another Lesson From Japan," *The New York Times*, June 19, 1994, F5.

11. Robert M. Monczka, "Finding a Structure That Works," *Inside Supply Management*, 17(12) (December 2006): 10–11.

12. Many of the ideas contained in this section were first introduced by one of the authors' mentors, Professor Norman Maier of the University of Michigan, during the mid-1960s.

13. Glen Parket, Jerry McAdams, and David Zielinski, *Rewarding Teams: Lessons from the Trenches* (Hoboken, NJ: Jossey-Bass, 2000).

14. Michael A. Hitt, "Corporate Entrepreneurship and Cross-Functional Fertilization: Activation, Process and Disintegration of a New Product Design Team," *Entrepreneurship: Theory and Practice* (Spring 1999): 145.

15. For additional insight, see David N. Burt and Richard Pinkerton, *Strategic Proactive Procurement* AMACOM, 1996, 33, and James W. Dean, Jr. and Gerald I. Susman, "Organizing For Manufacturing Design," *Harvard Business Review* (January-February 1989): 28–36.

16. Tom Schulte, quoted in "Conflict Resolution, A Required Skill for Engineering Team Managers," IOMA's report on managing design engineering, January 2000, 2.

17. David G. Baldwin, "How to Win the Blame Game," *Harvard Business Review* (Jul–Aug. 2001): 57.

18. Cited in "Relationship-Building Skills" by David J. O'Shea, *NAPM InfoEdge* (September 1998): 9. Used with permission of the National Association of Purchasing Management.

19. O'Shea, op. cit., 7.

20. Emiko Banfield, *Harnessing Value in the Supply Chain* (New York: John Wiley & Sons, 1999): 29–30.

21. David N. Burt and Richard L. Pinkerton, *Strategic Proactive Procurement*, AMACOM, 1996, 195.

22. Diane Brown, "Supplier Management Teams," *NAPM Insights* (August 1994): 33.

23. Mary Crews, "Relationship Management Yields Results," *Purchasing Today* (June 2000): 8–9.

24. Timothy M. Laseter, "Overcoming Conflicting Priorities," *Purchasing Today* 9(12) (December 1998): 37.

25. Cited in David M. Moore and Peter B. Antill, "Integrated Project Teams. The Way Forward for UK Defence Procurement," *European Journal of Purchasing & Supply Management* (Sept. 2001): 57.

Suggested Reading

Dischinger, John, David J. Closs, Eileen Mcculloch, Cheri Speier, William Grenoble, and Donna Marchall. "The Emerging Supply Chain Management Profession," *Supply Chain Management Review,* January/February (2006): 62–68.

Ellram, Lisa. "Supply Chain Management: The Industrial Organization Perspective," *International Journal of Physical Distribution & Logistics Management,* 21(1)(1991): 13–22.

Gattorna, John. *Living Supply Chains: How to Mobilize the Enterprise Around Delivering What Your Customers Want* (Harlow: Pearson Education Ltd., 2006).

Giannakis, Mihilas, and Simon R. Croom. "Toward the Development of a Supply Chain Management Paradigm: A Conceptual Framework," *The Journal of Supply Chain Management,* Spring (2004): 27–37.

Guertin, Ron and Vince Scacchitti. "Managing Difficult Teams," Presented at the 90th Annual International Supply Management Conference, May 2005.

Handfield, Robert B., and E.L. Nichols. *Introduction to Supply Chain Management,* Englewood Cliffs, NJ: Prentice Hall, 1999.

Harris, George L. "Missing Links in Strategic Sourcing," *Inside Supply Management,* August (2005): 27–31.

Zenger, John. *Leading Teams: Mastering the New Role.* Homewood, IL: Irwin, 1994.

Katzenbach, Jon R., and Douglas K. Smith, *The Wisdom of Teams–Creating the High-Performance Organization,* New York: McKinsey & Company, 1993.

Killen, Kenneth H., and John W. Kamauff. *Managing Purchasing; Making the Supply Team Work,* Tempe, AZ: Irwin Professional Publishing, 1995.

Lambert, D.M., and M. Cooper. "Issues in Supply Chain Management," *Industrial Marketing Management,* 29 (2000): 65–83.

Larson, Carl E., and Frank M.J. LaFasto. *TeamWork–What Must Go Right/What Can Go Wrong,* Newbury Park, CA: Sage Publications, 1989.

Larsson, P.D., and A. Holldorsson. "What is SCM? And, Where Is It?" *Journal of Supply Chain Management,* 38(4) (2002): 36–44.

Lummus, R.R., and R.J. Vokurka. "Defining Supply Chain Management: A Historical Perspective and Practical Guidelines," *Industrial Management & Data Systems* 99(1) (1999): 11–17.

Hitt, Michael A. "Corporate Entrepreneurship and Cross-Functional Fertilization: Activation, Process, and Disintegration of a New Product Design Team," *Entrepreneurship: Theory and Practice,* 23(3)(1999): 145–168.

Monczka, Robert M., and Robert J. Trent. "Cross-Functional Teams Reduce New Product Development," *NAPM Insights®* 5(2)(1994): 64.

Monczka, Robert M., and Robert J. Trent. "Effective Cross-Functional Sourcing Teams: Critical Success Factors", *International Journal of Purchasing and Materials Management* 30(4)(1994): 3.

Parker, Glenn M. "Success Strategies for Cross-Functional Teams: Management's Role," Presented at the 80th Annual International Conference Proceedings, Anaheim, CA, 1995.

Trent, Robert J. "Understanding and Evaluating Cross-Functional Sourcing Team Leadership," *International Journal of Purchasing and Materials Management,* Fall (1996).

Trent, Robert J. "Individual and Collective Team Effort: A Vital Part of Sourcing Team Success," *International Journal of Purchasing and Materials Management,* 34(4) (1998): 46.

Trent, Robert J., and Robert M. Monczka. "Purchasing and Supply Management: Trends and Changes Throughout the 1990's," *Journal of Supply Chain Management,* 34(4) (1998): 2–11.

CHAPTER 3

Supply Management: An Organization Spanning Activity

"Purchasing is changing like it never has before in its history, (and) so are all of the other areas of the organization."[1] As businesses cope to compete in the ever-changing landscape of this global market, internal business relationships become critical.

Supply Management's Relations with Other Departments

A supply management department is the hub of a large part of a company's business activity. By its very nature, supply management has continuing relationships with all other departments in the firm as well as with the firm's suppliers. Supply management operations cut across all departmental lines. Figure 3.1 provides a graphic illustration of supply management's many interfaces within the organization.

Supply Management and Engineering

Design engineers have traditionally played key roles throughout the supply chain management process ranging from their roles as new product development team leaders to members of off-spec (incoming materials that deviate from the relevant specifications) review teams. However, with the advent of outsourcing engineering tasks and projects, engineers are actively seeking ways to demonstrate value-add to the organization.[3]

Supply management, engineering, and operations have many problems in common. Design engineering greatly influences the amount of time supply management has to handle a procurement assignment. Engineering has the initial responsibility for preparing the technical specifications for a company's products, services, and the materials that go into them. To exercise this responsibility effectively, engineering should have the assistance

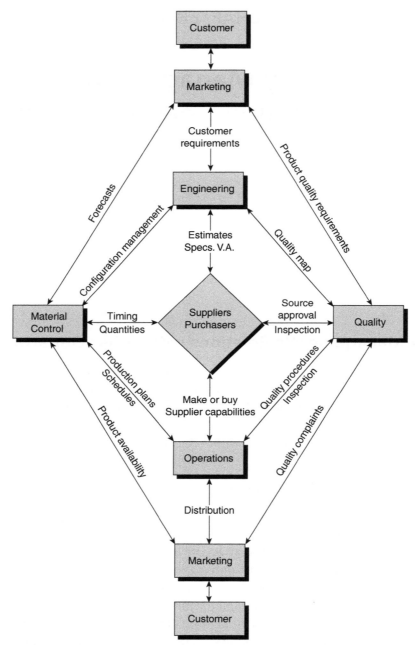

Figure 3.1 The many internal interfaces of the supply management function. (David N. Burt and Richard L. Pinkerton, *A Purchasing Manager's Guide to Strategic Proactive Procurement*, New York: American Management Association, 1996, 5.)

of supply management and operations. A number of firms have initiated early supply management and early supplier involvement programs to ensure that supply management and suppliers contribute to the development of new products and services as they work together with the engineers in the development of the product. Some firms refer to this process as concurrent engineering vs. sequential design work where engineers design a product and then look to supply management and outside suppliers for input.

The product costs associated with quality, material, fabrication, and production are inextricably related to the design specifications. Similarly, specifications can be written in a manner that reduces or enlarges the number of firms willing to supply specific items. If costs are to be controlled and profit maximized, the materials specified by engineering must be economical both to purchase and for fabrication. Ideally, materials should be available from more than one efficient, low-cost producer. Obviously, the quality of the materials must satisfy operations and the ultimate customer.

Supply management and engineering occasionally differ in their concepts of materials problems. The differing views are understandable. Engineers naturally tend to design conservatively; hence, their specifications may provide amply for quality, safety, and performance. By training, the engineer may be inclined to seek the "ideal" design, material, or equipment without complete regard for cost, availability, or functional need. A supply professional is more concerned with commercial issues such as cost and availability while meeting the functional need of the customer. Several situation-specific questions by supply management to engineering usually can help integrate cost and availability considerations into the design process. For example, supply management may ask, "Is it possible to reduce the designer's performance goals and safety margins and to work closer to actual performance requirements? Is an expensive design with a high safety factor necessary if a less costly design with a lower but acceptable safety factor will do the job? Why use costly chrome plate if brushed aluminum is adequate?" Clearly, conflicting functional interests cannot always be resolved easily. The answers to such problems are seldom clear-cut. Mutual understanding and a willingness to give and take are required from both sides if mutually satisfactory solutions are to be reached. The key consideration *must* be the best interests of the firm, not any single functional area.

Supply Management and Manufacturing and Operations

The supply management-manufacturing relationship begins during new product development and intensifies when manufacturing transmits its manufacturing schedule or materials requisitions to materials control, which translates these documents into a procurement schedule. Purchase timing is often a cardinal difficulty in making that translation. When the user does not give supply management sufficient time to purchase wisely, many needless expenses inevitably creep into the final costs of a company's products. When supply management has inadequate time to qualify suppliers properly, develop competition, or negotiate properly, premium prices are likely to be paid for materials. Costly special production runs, premium transportation costs, and quality problems are three common results of inadequate purchasing lead time.[4]

A production shutdown is the most serious problem stemming from insufficient procurement lead time. In most process types of operations (chemicals, cement, paint, flour, etc.), either equipment runs at nearly full capacity or it does not run at all. Consequently, material shortages in these industries can be catastrophic, resulting in complete production stoppage. Losses resulting from material shortages in nonprocess industries are not always so disastrous or apparent. A production shutdown in a metal fabricating shop, for example, can be piecemeal. The indirect costs of such shortages, consequently, are often hidden in production costs. One or two machines from a large battery of 50 can be shut down as a routine occurrence. Conventional accounting records fail to reveal the financial impact of this kind of slow profit-draining inefficiency. Late delivery of equipment, services, or supplies also can affect the efficiency of a services organization. Imagine a dentist without the appropriate drill bit or an information technology (IT) supplier whose time to market is affected by late delivery of necessary software.

Coordination between supply management and manufacturing pays off in many ways. For example, a more expensive alternative material that will save the company money can be selected on occasion. This may sound like a paradox. Pay more and save more—how can this happen? Savings in manufacturing and assembling costs often can exceed the increased purchase costs. In the normal manufacturing operations of casting, forging, machining, grinding, stamping, and so on, some materials are much more economical to work with than others. For example, government suppliers have saved thousands of dollars by using bronze instead of steel extrusions in aircraft elevator and rudder counterweights. Bronze costs more than steel, but savings in machining time more than offset the increase in material cost. In this case not only is the direct cost reduced, but also skilled machinists and expensive machine tools are freed to do other high-priority work. See Table 3.1 for a comparison of these costs.[5]

Going beyond these day-to-day operational interfaces, supply management and manufacturing must coordinate effectively to achieve some of a firm's key strategic goals. For example, manufacturing management strives to achieve faster "time to market" performance, and reduce the time required for product change-overs and tool and line setup work. Supply management must be able to assist in these efforts by obtaining faster responses from suppliers, working with suppliers to improve their capabilities, and so on. In these types of activities, it is imperative that manufacturing and supply management work together closely.

Table 3.1   Effect of Different Materials on Productivity and Cost

		Costs Totals (steel)		Costs Totals (bronze)
Sales		$100		$100
Costs of goods sold	Man-Hours	Costs	Man-Hours	Costs
Raw material cost		5		10
Direct labor (machining)	2	30	1	15
Variable overhead		6		3
Fixed overhead		50		50
Total cost		$91		$78
Operating income		$9		$22

Note 1: Productivity Improvement: [(.5 − 1.0)/.5] = 100% improvement
Note 2: Profit Improvement: [(9 − 22)/9] = 144% improvement

IT is simplifying the relationship between supply management and manufacturing. Computers and sophisticated software allow the firm's MRP (materials resource planning) system to communicate seamlessly with the counterpart systems at the firm's suppliers, and ERPs, (enterprise resource programs) add to the necessity for data accuracy and interconnectivity. With this approach, supply management is not involved in the day to day tactical activity of placing orders.

As pressure continues for reduction in manufacturing costs, more and more products and process will be outsourced and the supply management-manufacturing relationship will take on greater importance. Organizations will try to control the quality and quantity of outsourced materials without investing capital in equipment or having responsibility for labor. According to Joseph Cavinato, "All of this plays well into supply managers' roles, because higher and higher level relationships and arrangements are needed for these all to work efficiently and effectively. Instead of buying raw materials, they are now arranging for complete products to be produced and in many cases distributed."[6]

Supply Management and Quality

Quality professionals should be involved in supply management from the development of new products, to involvement in sourcing, and on through supplier development with the objective of minimizing quality problems throughout the supply chain. Quality's role and responsibilities change significantly when the manufacturing function is outsourced. Quality is involved in qualifying the potential supplier. Then it becomes responsible for monitoring the supplier's quality system and providing technical assistance if quality problems occur.

Supply Management and Marketing

Supply management *should* be marketing's best friend. As described in Chapter 1, supply management has a major impact on the firm's sales. The quality of the firm's products is often dependent on the quality of its suppliers. Marketing's success in generating sales is in part attributable to the firm's ability to introduce new products in a timely manner— new products based on technology obtained from the firm's supply base and pricing flexibility resulting from reductions in the cost of goods sold. Many companies recognize the direct relationship between marketing excellence and profitability. In their enthusiasm to increase sales, however, companies may overlook the leaks in profit that can occur when sales activity is not properly meshed with supply and production activities. The sales/supply/production cycle has its genesis in a sales forecast. The forecast is the basis for the production schedule, which in turn is the basis for the materials schedule. The sales forecast also influences a firm's capital equipment budget as well as its advertising campaigns and other sales activities.

Prompt communication of changes in the sales forecast to manufacturing and supply management permits these departments to modify their schedules as painlessly and economically as possible. Likewise, changes in the production schedules should be communicated immediately to sales representatives. This action permits marketing to alter its distribution schedule in a manner that will not alienate customers. Supply management must immediately transmit to marketing, as well as to other management groups, information concerning increases in material prices. The information permits marketing to evaluate the effect of rises in price estimates given for future sales quotations on current selling prices and on plans for future product lines.

Supply management and marketing must wisely blend their interests in the delicate area of reciprocity (buying from customers). If satisfactory legal reciprocal transactions are to be developed, they must be pursued with an understanding of the true costs of reciprocity. Buying from friends can be good business, but not when it is done at the expense of product quality or higher prices for purchased materials or services. In its desire for increased sales, a company can lose sight of the fact that increased sales do not always result in increased profit. Increased sales may result in decreased profit if they simultaneously require an increase in purchase prices.

A supply management department can be of major help to its marketing or sales department by serving as its practical sales laboratory. A firm's supply management department is the target for many manufacturers' sales operations. Supply management's files are replete with sales literature, policies, and promotional approaches of a broad range of manufacturers and distributors. Supply management professionals are aware of the personal selling methods sales representatives have used most effectively on them. They are equally aware of sales practices that fail or irritate them. Therefore, a company's supply professionals can be an excellent source of information for developing and refining the company's sales policies and procedures.

Many marketing departments spend considerable amounts of money on advertising and promotion. In many cases, the focus of these expenditures is the impact of the advertising, not its cost. Several years ago Warren Norquist, former Vice President of World Wide Purchasing and Materials Management of Polaroid, demonstrated that the application of sound supply management practices to the purchase of advertising resulted in an average savings of 24%.[7]

Supply Management and Finance

The finance department is charged with two principle responsibilities: obtaining funds and overseeing their use. Poor financial planning and execution is *the* major cause of business failure. Supply chain management is responsible for as much as 80% of many firms' financial resources. Thus, the CFO (Chief Financial Officer) and his or her key subordinates have a vested interest in a cost-efficient supply chain management system. Although supply managers often view finance as a department with a "low-price mentality," finance has the means of identifying costs in order to help purchasing gain a better understanding of the total cost of ownership. Marilyn Gettinger, President of New Directions Consulting, tells us supply should collaborate with finance because "each has the same focus: investing resources wisely with the greatest return on investment".[8]

Regardless of the price advantage available, the right time to buy from the standpoint of business conditions is not always the right time to buy from the standpoint of the company's treasury. If the supply management department makes commitments to take advantage of unusually low prices without consulting the finance department, the company could find itself paying for these purchases with funds needed for other purposes. On the other hand, if the finance department does not strive diligently to make funds available for such favorable buying opportunities, the company may have to pay higher prices later for the same material.

Finance's willingness to pay suppliers in a timely manner affects both supply management's ability to obtain low prices and to forge and maintain collaborative relationships. During the 1970s, for example, Timex had a policy of paying its suppliers the day the supplier's invoice and the receiving report arrived in the accounts payable office. As a result, Timex became a preferred customer of many of its suppliers. During two material shortage periods of the 1970s, Timex's preferred customer status allowed it to avoid the shortage problems experienced by most firms. Because of enlightened supply management practices, Timex never missed a beat! We see that a cooperative relationship between purchasing and finance can impact the development of good supplier relations.

As discussed in Chapter 1, supply management has a major impact on 10 major components of a firm's costs. An efficient and effective supply management function significantly reduces the funds required to operate the firm. The timing of purchasing expenditures can be of significant importance to a finance department that is working diligently to protect the firm's financial ratios and solvency. Supply management and finance should coordinate on expenditures that may significantly affect the firm's cash position. Finance should be represented on cross-functional teams that are purchasing major equipment or construction services due to the magnitude of the expenditures involved.

Investors properly are concerned with the firm's return on investment (ROI). As shown in Figure 3.2, supply management has a major impact on this key indicator. Notice how a 5% reduction in a hypothetical firm's expenditures increases its ROI by 30%!

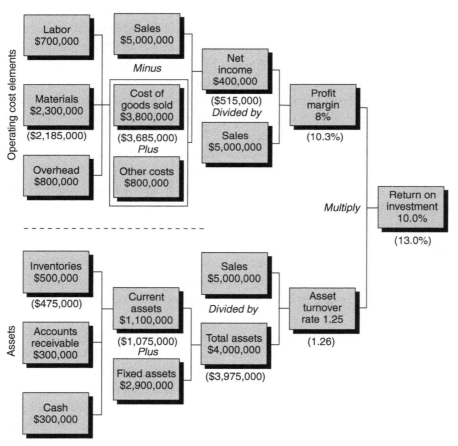

Figure 3.2 The relationships of basic elements that influence return on investment. the figures in parentheses reflect a 5% reduction in the cost of materials.

Supply Management and Information Technology (IT)

Supply management and IT have an increasing number of interdependencies. In many cases, IT is outsourced. The director of IT and a supply management professional must work closely to develop the appropriate statement of work and the sourcing and pricing processes and to manage the resulting contract and relationship.

Many firms are purchasing business-to-business (B2B) e-commerce buy-side software systems from firms such as Ariba and Oracle. Such software systems have a major impact on the firm's procurement processes. End users are empowered to purchase directly from the firm's and its approved supplier's electronic catalogs. The transition from a paper-based system of requisitions, manual approval, manual citation of budgetary authorization, requests for proposals, purchase orders, receiving reports, and payment checks to electronic purchasing must be planned and implemented carefully.

Electronic communication for production materials—whether over the Internet or through electronic data interchange (EDI)—requires coordination and cooperation between supply management, IT, and manufacturing or material control.

Another example of the interdependency between purchasing and IT is the development of a database that provides timely and accurate input to supply management for strategic planning and tactical activities. Relatively few firms have developed such an information system.

Obviously another interface exists between supply management and IT: the procurement of software, software services, and equipment. It is our observation that when the two departments operate collaboratively, a synergy results. On the other hand, when IT operates in a Lone Ranger mode and does its own procurement, much waste frequently results.

Supply Management and Logistics

At one time, logistics and purchasing were relatively tactical. Both have progressed to recognition as critical functions. Today, logistics spends approximately 10% of a manufacturer's income. Purchasing (supply management) spends some 60%.

Logistics is concerned with the movement of goods. In many cases, logistics is responsible for both incoming goods and the distribution of goods to the next member of the supply chain and frequently to the end customer itself. In virtually all cases, logistics professionals design and manage the firm's distribution system, consisting of warehouses, distribution points, and freight carriers.

The relationship between supply management and logistics tends to vary from firm to firm. In some cases, supply management plays a dominant role in sourcing and pricing logistics services. In other cases, the logistics department performs these services with little or no supply management involvement. The critical issue should not be one of jurisdiction; it should be one of professionalism and excellence. It should not matter whether supply management or logistics plays the key (or dominant) role. What does matter is that professional supply management practices are employed. Today, professional practices include hiring third-party firms to store, handle, and transport assets. "This option allows what would be fixed system costs to be incurred and paid for by the client organization on a per unit basis."[9]

Many organizations are experiencing tremendous results from collaboratively managing the logistics portion of their supply chains. In the service industry in particular, strategies are being utilized to improve the process of getting information and product through the supply chain. "Deerfield, Illinois-based Baxter Healthcare is in the process of leveraging its global logistics costs."[10] Their collaborative approach to logistics is working to improve forecasting, and a collaborative approach to the demand planning processes is occuring.

Supply Management and Accounts Payable

The proliferation of corporate purchasing cards has had a major impact on both supply management and accounts payable. The use of such purchasing cards has had several beneficial effects: (1) it empowers end users of standard and low value requirements to purchase directly from distributors, (2) it reduces tactical, non-value adding purchases by the supply management department, and (3) it significantly reduces accounts payable activities.

Supply management and accounts payable frequently have conflicting interests and drivers in the area of timely payment to suppliers. Accounts payable commonly reports to the CFO. Finance is responsible for obtaining funds and their productive use. Finance professionals frequently take considerable pride in seeing their "idle" funds invested at returns of 6 to 12%. This superficially "logical" thinking causes many finance professionals to keep the money entrusted to their safe-keeping as long as possible. One means of achieving this apparently laudable objective is to delay paying suppliers as long as possible. For example, many suppliers to hospitals must wait for six months to be reimbursed for materials, equipment, or services provided. (We assume that the accounts payable/finance people must feel good that they have earned investment income on the backs of helpless suppliers!)

Ignoring the ethical implications of such unilateral action for the moment, such action is in conflict with supply management objectives. The sophisticated supplier who has experienced such delays in payment simply increases its selling price to such customers. Nobody wins. Of greater importance, such non-responsive payment often conflicts with supply management's efforts to become a preferred customer or to develop collaborative and even alliance relationships. Obviously, such conflicts can and must be overcome through open discussions between supply management and finance professionals.

Supply Management and Lawyers

Legal professionals frequently are actively involved in contract negotiations and contract formation. In other cases, their role is one of review and approval of contracts developed by supply management professionals. Value adding attorneys who are involved in supply management issues normally must embrace a collaborative approach to dealing with the firms' suppliers. It is the senior author's observation that a legal education is *not* a good background for members of most negotiating teams. This is equally true of the contract formation process. Unfortunately, most attorneys are trained to look for risk and worst case scenarios. Collaborative, pie-enhancing negotiations frequently are disrupted by attorneys who are obsessed with risk avoidance!

This is not to say that lawyers cannot add value: They most certainly can! But care must be exercised in inviting the involvement only of lawyers who are concerned with the best interests of both parties and with enabling pie-enhancing negotiations.

Supply Management in Non-Manufacturing Organizations

Supply management has as much—and sometimes more—impact on the success of non-manufacturing organizations as it does on manufacturing firms. The timely availability of reliable equipment, supplies, and services at the right total cost of ownership affects the ability of such organizations to provide timely quality services at a profit.

In a manufacturing setting, design and manufacturing engineering professionals normally lead the new product development effort and the development of the appropriate specifications describing what is to be purchased. In a non-manufacturing setting, supply management often must lead or facilitate the requirements process due to the absence of engineering or other qualified requirements professionals. Thus, a supply professional may need to assume the responsibility for leading a cross-functional team that is identifying or describing a non-manufacturing need that will be satisfied through a procurement.

While the total costs of purchases compared with net income or budget authorizations may be proportionately less at a non-manufacturing firm than for a manufacturer, such expenditures still are very significant. Supply management's impact on sales can be every bit as significant as in manufacturing firms. Quality implications, time-to-market, pricing elasticity (based on reductions in the cost of goods sold), technology inflow, and continuity of supply combine to have a major impact on a non-manufacturing firm's sales. Thus, we see that supply management also has a major impact on a non-manufacturing firm's bottom line.

Supply Management in Government

Supply management (frequently called "procurement" in government circles) has a major impact on the efficient and effective use of our tax dollars at all levels of government. Virtually all of the problems present in manufacturing organizations are present in government procurement. Not surprisingly, many of the advances in the art and science of supply management originated in the federal government.

Supply Management and the External Environment

Business Relationships

All phases of supply management involve relations with external suppliers: early supplier involvement in the development of requirements, strategic sourcing, pricing (including cost analysis and negotiations), and post-award activities. These interfaces are explained in detail throughout the book.

Monitoring the Supply Environment

Supply managers are responsible for protecting their firms from unexpected threats or shocks from their supply world in the form of price increases or supply disruptions. These threats include material and labor shortages, which affect one or more industries that supply the firm. Shortages will affect both the price and availability of purchased materials, supplies, and services. The firm should take action to minimize the impact of such shortages by monitoring changes in the supply environment such as the following:

▲ Changes in legislation that may affect the workplace. Such changes can impact both price and availability. An example is a new Environmental Protection Agency regulation on toxic wastes that affects one or more suppliers.

▲ Wars and other conflicts, which may disrupt the availability of materials and services the firm or its suppliers require. Firms that proactively monitor the environment take defensive action in anticipation of the resulting material and labor shortages and price increases.

▲ A consolidation among suppliers. The extreme case is consolidation to the point of monopoly. Such changes may require a change in the firm's supply strategy.

▲ Wages, projected wages, and possible labor relations issues.

Supply managers should have early information that will allow them to take advantage of favorable market conditions. Opportunities result both from additional capacity coming online and from reductions in demand for required materials, equipment, or services.[11]

The responsibilities of protecting their firms from unexpected threats or shocks motivate supply professionals to develop supply monitoring systems. One of the challenges confronting today's supply professional in monitoring is the abundance of data.

Today purchasers are literally inundated with bits of data concerning their suppliers, the markets in which those suppliers participate, and the functioning of the economy as a whole. Turning such data into meaningful, useful information—supply market knowledge—is one part of the supply professional's tasks, but so is leveraging supply information into knowledge that increases the competitive advantage of the firm. Before attempting to understand the supply market, supply professionals must first possess a clear understanding of what is meaningful to their own organizations.[12]

Monitoring supply markets is a fascinating and challenging activity. In the late 1980s, Warren Norquist, former Vice President of World Wide Purchasing at Polaroid, assigned three researchers the responsibility of monitoring Polaroid's supply environment and then advising Polaroid buyers of potential threats and opportunities in their supply world. Frank Haluch, writing in the August 2000 issue of *Purchasing Today*, outlined a six-step environment monitoring strategy:

▲ Determine the cost, supply, and technology drivers of the materials and services that a supply manager is watching.
▲ Identify the major suppliers and customers of the materials and services.
▲ Determine the sources of information for those drivers.
▲ Build a model that predicts the material (or service) behavior.
▲ Monitor the model to determine its accuracy.
▲ Continuously make improvements as new relationships are understood and additional data become available.[13]

Supply environment monitoring coupled with timely reaction to the threats and opportunities that are identified is a key strategic activity, which has significant impact on the firm's success and survival.

Completing the Supply Chain Linkage: Supplier Integration with the Customer

As the authors have stressed in previous chapters, supply management has great impact on sales as well as costs. Suppliers should be major sources of innovation that result in profitable new products or services. Early supplier involvement can reduce the time required to bring a new product to market, resulting in greater market share. Supplier quality affects the quality of the firm's products, its image in the marketplace, and its sales volumes. Suppliers can help "lower costs, improve delivery, lower inventory, and problem-solve capabilities during the (customer order) fulfillment stage of production."[14] Suppliers and end-user connections emerge to meet a wide variety of needs.

Summary

The purchasing process spans both internal and external organizational boundaries. As the process evolves to supply management, both the complexity and the strategic importance of the process increase dramatically. As the panorama of global business environments continues to unfold, the relationships supply managers have both inside and outside of the organization will continue to evolve.

Endnotes

1. Joseph L. Cavinato, "Business Change: It Isn't Just in Purchasing and Supply," *Purchasing Today*[R] 12(12)(2001): 38.
2. David N. Burt and Richard L. Pinkerton, *A Purchasing Manager's Guide to Strategic Proactive Procurement*, (New York: American Management Association, 1996) 5.
3. Joseph L. Cavinato, "Business Change: It Isn't Just in Purchasing and Supply," *Purchasing Today*[R] 12(12)(2001): 38.
4. Supply management has the responsibility to keep users informed concerning supply lead times for all categories of production materials.
5. David N. Burt and Richard L. Pinkerton, *A Purchasing Manager's Guide to Strategic Proactive Procurement* (New York: AMACOM, 1996) 9.
6. Joseph L. Cavinato, "Business Change: It Isn't Just in Purchasing and Supply," *Purchasing Today*[R] 12(12)(2001): 38.
7. Personal discussions with Mr. Norquist, 1989.
8. Marilyn Gettinger, "Supply Management and Finance—Building Bridges," *Inside Supply Management* 17(1)(2006): 6.
9. Personal discussions with Mr. Norquist, 1989.
10. John Yuva, "Collaborative Logistics: Building a United Network," *Inside Supply Management* 13(5)(2002): 42.
11. The Institute for Supply Management™ (formerly the National Association of Purchasing Management) publishes a comprehensive report on business in *Purchasing Today* each month. The report shows macro trends in both manufacturing and non-manufacturing sectors and featured reports on select industries.

12. Richard R. Young, Ed. "Knowledge of Supply Markets," in *The Purchasing Handbook*, 6th ed. Joseph L. Cavinato and Ralph G. Kauffman, Editors-in-Chief, New York: McGraw-Hill, (2000). For more on this report, see Ralph G. Kauffman, "Indicator Qualities of the NAPM Report on Business," *The Journal of Supply Chain Management* Spring (1999).

13. Frank Haluch, "Taking the Market's Pulse," *Purchasing Today*, August 2000, 6.

14. Robert Monczka, Robert Trent, and Robert Hanfield, *Purchasing & Supply Chain Management*, 3rd ed. South-Western, Mason, Ohio, 2005.

Suggested Reading

Ayers, James B. *Handbook of Supply Chain Management*, 2nd ed. Boca Raton, FL: Auerbach Publications, 2006.

Humphreys, P. "An Inter-Organizational Information System For Supply Chain Management," *International Journal of Production Economics* 70(3)(2001): 245–255.

Lambert, Douglas, M. *Supply Chain Management: Processes, Partnerships, Performance*, 2nd ed. Sarasota, FL: Supply Chain Management Institute, 2006.

Moore, Kevin R. "Trust and Relationship Commitment in Logistics Alliances: A Buyer Perspective," *International Journal of Purchasing & Materials Management* 34(1) (1998): 24–37.

Mentzer, John T., Matthew B. Myers, and Theodore P. Stank. *Handbook of Global Supply Chain Management*, Thousand Oaks, CA: Sage Publications, 2007.

Mentzer, John T. *Supply Chain Management*, Thousand Oaks, CA: Sage Publications, 2000.

Mentzer, John T., William DeWitt, James S. Keebler, Min Soonhoong, Nancy W. Nix, Carlo D. Smith, Zach G. Zacharia. "Defining Supply Chain Management," *Journal of Business Logistics* 22(2)(2001): 1–25.

Stank, Theodore P., Scott B. Keller, and Patricia J. Daugherty. "Supply Chain Collaboration and Logistical Service Performance," *Journal of Business Logistics* 22(1)(2001): 29–48.

CHAPTER 4

A Portfolio of Relationships

A Transformation in Relationships

As supply management progresses from reactive and mechanical purchasing to proactive procurement and on to strategic supply chain management, a similar transformation is evolving in relationships between buyers and suppliers. During the dark days of reactive purchasing, relations between salespersons representing suppliers and their counterparts in purchasing were reasonably cordial, but frequently adversarial. A gain for one resulted in a loss for the other, which is often called a win-lose outcome. The interaction between supplier and purchaser was often characterized by highly manipulative tactics by both parties designed to maneuver the other side into a position where one's gain would be the other's loss. As purchasing became more professional, buyers and suppliers began to see benefits of more collaborative relationships, where the outcome could result in a win-win relationship for both parties.

During the late 1980s and early 1990s, suppliers saw many advantages in developing "partnerships" with customer firms. Sales managers charged their salespeople with becoming "partners" with their key customers. However, due to lack of clarity from management, many of the salespeople did not understand the implications of the term "partner." Additionally, the legal implications of the term "partner" raised further implications and concerns. For example, is a buyer "partner" free to solicit prices from competing suppliers? Is a supplier "partner" required to give all of its "partners" the same price and services?

While the term "partnership" is still relatively common, we avoid use of the term preferring the terms "collaborative relationship" and "strategic alliance." With this new perspective, purchasing's role has changed "from a provider of the right components at the right time and lowest costs, to a manager of the supply base responsible for the generation of competitive advantages for the company."[1] Scholars such as J. Dyer and H. Singh believe that a firm's potential competitive advantage is more than its assets, technology,

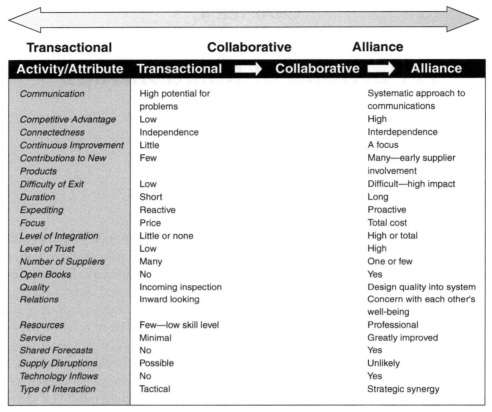

Activity/Attribute	Transactional ➡	Collaborative ➡	Alliance
Communication	High potential for problems		Systematic approach to communications
Competitive Advantage	Low		High
Connectedness	Independence		Interdependence
Continuous Improvement	Little		A focus
Contributions to New Products	Few		Many—early supplier involvement
Difficulty of Exit	Low		Difficult—high impact
Duration	Short		Long
Expediting	Reactive		Proactive
Focus	Price		Total cost
Level of Integration	Little or none		High or total
Level of Trust	Low		High
Number of Suppliers	Many		One or few
Open Books	No		Yes
Quality	Incoming inspection		Design quality into system
Relations	Inward looking		Concern with each other's well-being
Resources	Few—low skill level		Professional
Service	Minimal		Greatly improved
Shared Forecasts	No		Yes
Supply Disruptions	Possible		Unlikely
Technology Inflows	No		Yes
Type of Interaction	Tactical		Strategic synergy

Figure 4.1   Three types of buyer supplier relationships.

and resources. The emerging relational view recognizes that a firm's competitive advantage can evolve from the relationships between buyers and suppliers.[2] Figure 4.1 portrays the three levels of relationships between buyers and sellers as they develop from transactional arm's length to the close working relationship of an alliance. Note the shift in attitude to activity/attributes as the barriers are reduced.

Transactional Relationships

We call the most common and most basic type of relationship "transactional." Such a relationship is neither good nor bad. Transactional simply describes an arms-length relationship wherein neither party is especially concerned with the well being of the other and moves from one contract to another rather than develop any real long-term relationship. Virtually all buying firms will have transactional relationships. Most will have collaborative ones and some will have strategic alliances.

Transactional relationships have several characteristics:

- An absence of concern by both parties about the other party's well being. With transactional relationships, there is little or no concern about the other party's well being. What one party wins, the other loses.
- One of a series of independent deals. Each transaction is entered into on its own merits. There is little or no basis for collaboration and learning.
- Costs, data, and forecasts are not shared. Arms-length transactions, not openness, are characteristics of transactional relationships.
- Price is the major focus of the relationship. Getting the best price is the focus of the transaction. Ideally total cost analysis, as described later in this textbook, precedes any procurement transaction.
- Quality of the relationship. Since there is little or no concern for the other's well being, neither buyer nor supplier will rush to the other's assistance in bad times or when problems arise.
- A minimum of purchasing time and energy is required to establish prices. Market forces normally establish prices in transactional relationships. Thus, little purchasing time and energy are required to establish prices.
- Transactional purchases lend themselves to e-procurement and, in some cases, reverse auctions.

The advantages of transactional relationships include:

- Relatively less purchasing time and effort are required to establish price. As we have noted, the market forces of supply and demand establish the price with transactional procurements. Therefore, little purchasing time and effort are required to establish price.
- Lower skill levels of procurement personnel are required. Much less judgment and managerial expertise are required with the vast majority of transactional procurements.

The disadvantages of transactional relationships include:

- The potential for communication difficulties is much greater with transactional relationships than with collaborative or alliance ones.
- Considerable investment in expediting and monitoring incoming quality is required to ensure timely delivery of the right quality.
- Transactional relationships are inflexible when flexibility may be required. Changing technology and changing market conditions can require flexibility in buyer/supplier relationships.
- Transactional procurements tend to result in more delivery problems than do collaborative and strategic alliance ones. Friends look out for friends, not opportunistic buyers or suppliers.

▲ Quality with transactional relations will be only as good as required. Far more incentive and opportunity exist to improve quality in a collaborative or alliance relationship.

▲ Transactional suppliers tend to provide the minimum service required.

▲ Buyers tend to experience less effective performance by their transactional suppliers than do those employing collaborative or strategic relationships. Transactional suppliers have much less to lose from a customer who is dissatisfied than do collaborative and strategic relationship suppliers.

▲ Transactional customers are subject to more supply disruptions than are collaborative or alliance ones. Buyers who maintain continuing, collaborative relations with their suppliers are much less subject to supply shortages than are opportunistic ones.

▲ Since the supplier recognizes the transactional and price nature of the relationship, it is not motivated to invest time and energy in the development of the potential buyer's products.

▲ Buyers seldom know the total cost of ownership of the items and services they are purchasing through transactional relationships.

▲ Frequent changes in suppliers result in hidden switching costs.[3]

Collaborative and Alliance Relationships

Ginni Rometty, a Senior Vice President with IBM Global Business Services, recently reported results of a Global CEO Study on innovation conducted by IBM. One key finding showed "76% of CEO's think external collaboration with business partners and customers is key to innovation."[4] For the procurement of non-commodity[5] items and services, innovative, collaborative, and alliance relationships tend to result in lower total costs than do transactional relationships for several reasons. Such items may require process improvements and the adoption of technical innovations. This is difficult to achieve in transactional relationships. The risks and uncertainties present with transactional relationships reduce the likelihood of investments in R & D and training as well as the procurement of new, more efficient equipment focused on the customer firm's needs. Thus, major opportunities for cost reduction within supplying organizations may be lost with transactional relationships.

Cost reductions resulting from value engineering and value analysis (VE/VA) are much more likely with collaborative and alliance relationships.[6] Suppliers are more likely to take the initiative to reduce costs through VE/VA when they are involved in long-term relationships than with short-term transactional ones.

Longer-term performance agreements allow suppliers an opportunity to reduce their costs. The extended learning curve effect[7] with both production and services allows collaborative and alliance suppliers to reduce their costs and share these savings with customers.

Collaborative and alliance relationships replace the market forces employed by transactional procurement with controlled competition, benchmarking, and advanced supply management pricing practices. The results are lower total costs, higher quality, reduced time to market, and reduced risk of supply disruptions.

Researchers Stanley and Pearson found that the three most important factors in a successful buyer-supplier relationship are: (1) two-way communication, (2) the supplier's responsiveness to supply management's needs, and (3) clear product specifications.[8] Further research has shown that better communication and shared information regarding products purchased and new products in development stages result in better quality, response time, improved cost savings, and greater efficiencies.[9] Over the last 5 to 10 years, we have seen a move to improve communication, increase information sharing, and a willingness to reduce the barriers that have marred relationships in the past.[10]

A good example of collaboration is Caterpillar Engineers teaming up with its catalytic converter manufacturer Tenneco. At Caterpillar's Mossville, IL engine center, the team created a mock production line to work out the kinks in the manufacturing process. Using the joint design, assembling the component at a Tennaco plant in Nebraska cut Caterpillar's costs on the part by 20%.[11]

Collaborative Relationships

The key difference between collaborative relationships and transactional ones is an awareness of the interdependence and necessity of cooperation. "The focus on relationship management will require that all elements of relationship management, including trust building, communications, joint efforts, and planning and fostering interdependency, will be increasingly studied and managed to achieve competitive advantage."[12] Recognition of interdependency and need for cooperation provides many benefits. Both parties are aware that money enters their supply chain (or supply network) only if the chain's end products are cost competitive. Recognizing the need for interdependence and cooperation, the customer's firm enjoys the benefits of early supplier involvement (ESI). The results are improvements in cost, quality, time to market, and the leveraging of supplier technology.

Continuous improvement is far easier to implement and manage when both sides recognize their interdependence and cooperate. The end objective with continuous improvement is a reduction in total costs as they strive for greater value together than each could create individually.

The likelihood of disruptions in supply is greatly reduced. Collaborative suppliers look out for their friends, not their opportunistic customers. Collaborative relationships help cushion bad times. Both customers and suppliers who value each other, based on long-term relationships and respect, are more likely to come to each other's aid during times of adversity. Blockbuster, the video rental company, has become a leader in the market by forming collaborative relationships with its suppliers. In the 1990s, Blockbuster revolutionized the way distributor/retailers purchased and managed the revenue of movie sales and rentals. Traditionally, a high price was paid to the supplier with sales for the video store being generated by rentals. The cost to purchase the new tapes was so high that few tapes were purchased and video stores were often out of stock on some of the latest releases. Blockbuster decided to collaborate with its suppliers by agreeing to share the sales revenues from rentals equally with the suppliers. The suppliers responded with a small fixed price for new videos. The availability of new releases dramatically increased as did revenues for Blockbusters and its suppliers. Within less than a decade, the entire industry operated in this collaborative method. It is estimated that the video rental industry's total profits have increased by 7%. Other industries have followed Blockbuster's lead.[13]

Lower total costs are the common result of collaborative and alliance relationships. The level of certainty and continuity of demand in collaborative and alliance relationships increases the likelihood of investments in R&D, training, and the procurement of new, more efficient equipment focused on the customer firm's needs. Cost reductions resulting from value engineering and value analysis are enhanced, and the extended learning curve effect with both production and services activities allows collaborative and alliance suppliers to reduce their costs and share these savings with their customers.[14]

The major disadvantage of collaborative and alliance relationships is the amount of human skill, time, and energy required to develop and manage the relationships. However, firms such as Honda and Deere & Co. have demonstrated that the required investments provide very attractive returns.

Supply Alliances

The fundamental difference between collaborative relationships and supply alliances is the presence of institutional trust[15] in alliances. The failure to develop and *manage* institutional trust is the principle reason that so many supply alliances fail.

Supply alliances reap incredible benefits because physical asset specialization and human specialization. Dyer defines *physical asset specialization* as

> relationship–specific capital investments (e.g., in customized machinery, tools, information systems, delivery processes, and so forth) that allow for faster throughput and greater product customization. Physical asset specialization allows for product differentiation and may improve overall quality by increasing product integrity. *Human specialization* refers to relationship-specific know-how accumulated by individuals through long-standing relationships. In other words, individuals across companies have substantial experience working together and

have accumulated specialized information and language that allows them to communicate and coordinate effectively with each other. They are less likely to have communication breakdowns that result in errors; this, in turn, results in higher quality, faster development times, and lower costs.[16]

The primary benefits of supply alliances include:

▲ Lower total costs.   Synergies can be created in alliances that cannot happen in transactional or even collaborative relationships. The synergies result in reductions of direct and indirect costs associated with labor, machinery, materials, and overhead.

▲ Reduced time to market.   Reducing the time to design, develop, and distribute products and services is a key driver that leads to improved market share and better profit margins.

▲ Improved quality.   The use of both the design of experiments[17] and supplier certification[18] are the norm with supply alliances. These two activities design and manufacture quality rather than inspect for errors. The result is improved quality at lower total cost.

▲ Improved technology flow from suppliers.   Openness and institutional trust enhance the inflow of technology from alliance partners that leads to many successful new products. In 1999, Dell and IBM formed an alliance worth some $16 billion over seven years. In effect, Dell was harnessing IBM's vast research, development, and production abilities. Dell was purchasing storage devices, custom logic chips, static random-access memory, and other components. The two companies were cross-licensing patents; that is, they were sharing relevant technologies.[19]

▲ Improved continuity of supply.   Alliance customers are the group least likely to experience supply disruptions.

Alliances share several attributes:

▲ The focus of most supplier alliances is achieving the simultaneous objectives of continuous improvements while squeezing out cost. Most alliances put major emphasis on the inflow of innovation from the supplier partner. Many have implemented programs of co-creation.

▲ A high level of recognized interdependence and commitment is present.

▲ An atmosphere of cooperation exists. Potential conflicts are addressed and resolved openly. When problems occur, the focus is a search for the root cause, not the assignment of blame.

▲ The alliance is controlled through a complex web of formal and informal interpersonal connections, information systems, and internal infrastructures that enhance learning.

▲ Openness exists in all areas of the relationship including cost, long-term objectives, technology, and the supply chain itself.

- ▲ The alliance is a living system that progressively evolves with the objective of creating new benefits for both parties.
- ▲ The alliance partners share a vision of the future in how they will work together.
- ▲ Ethics are more important than expediency.
- ▲ The relationship is adaptable in the face of changing economics, competition, technology, and environmental issues.
- ▲ The design of experiments and supplier certification are the norm with supply alliances. These two activities design and manufacture quality. The result is improved quality at a lower total cost.
- ▲ Negotiations and re-negotiations occur in a co-creative manner wherein the parties focus on enlarging the pie and then dividing it.
- ▲ Executive-level commitment and alliance champions protect and nourish the alliance.
- ▲ Supplier and buyers agree to meet current competition in terms of cost and technology developments and avoid conditions in long-term contracts that could prove economically destructive.

If supply alliances are so attractive, why aren't they the way to conduct all business? Alliances are a very resource-intense approach to supply management and tend to be reserved for the most critical relationships such as the Quaker Oats–Graham alliance described in the opening case and in the appendix of this chapter.

During the 1990s, Chrysler became a fascinating example of the power of alliance relationships. In 1989, Chrysler had some 2500 suppliers in its production supply base. Chrysler was rated as the least desirable customer of the big three U.S. auto assemblers. With a surprising amount of assistance from Honda of America Manufacturing, Chrysler transitioned itself from a transaction-based buyer to a collaborative one. Results? Time to market was reduced by 30%. Profit margins per vehicle increased an average of 750%. Chrysler employed many of the principles introduced in this text: cross-functional teams, early supplier involvement, target costing, value analysis (with incentives in the form of additional profits or additional sales volumes for suppliers), improved communications through techniques such as co-location of supplier engineers at Chrysler design centers, and a supplier advisory board.[20]

Unfortunately, when Daimler acquired Chrysler in the late 1990s, such a level of cooperation and coordination was not in its play book. Chrysler's approach to supply management was so far ahead of that of Daimler's Mercedes Benz (M-B) division that some of us thought that M-B would have the Chrysler procurement operation become responsible for all of the new corporation's procurement! With hindsight on this disastrous acquisition, we now wonder how profitable Daimler-Chrysler would have been had the firm embraced Chrysler's visionary approach to supplier relations.

Which Relationship Is Appropriate?

How does a supply management executive determine whether a relationship should be transactional, collaborative, or a strategic alliance? Several key questions should be asked to determine the "strategic" elements of a relationship:

1. Are there many relatively undifferentiated suppliers providing what amounts to interchangeable commodities? If so, a collaborative alliance or relationship would not be appropriate. Try a transactional relationship instead.

2. Does the potential supplier possess economic power that it is willing to employ over its customers? A transactional or very carefully developed and managed collaborative relationship is usually appropriate and may be the only method the supplier will accept.

3. If there is recognition by both parties of the potential benefits of an alliance, but adequate qualified human resources are not available at one or both firms, a collaborative relationship is usually more appropriate.

4. A collaborative relationship frequently is an appropriate first step on the road to a strategic alliance.

5. Is one supplier head and shoulders above the rest in terms of the value it provides, including price, innovation, ability to adapt to changing situations, capacity to work with your team, task joint risks, etc? If so, an alliance may be in order, assuming that the supplier is willing to enter into an interdependent, trusting relationship.

6. Are some suppliers "strategic" to your business? In other words, do they have a major impact on your competitive advantage in the marketplace? Are you highly reliant on them to provide a unique product, technology, or service? If so, an alliance may be vital.

7. Would your company benefit greatly if the supplier were more "integrally connected" with your company, perhaps with their engineers working side by side with yours or co-locating their manufacturing facilities adjacent to or within yours? If yes, consider an alliance. (The appendix details a classic example of this.)

8. Do your customers require high degrees of flexibility and speed of responsiveness, causing you to demand the same performance from your suppliers? This is a classic alliance driver.

Trust is another key factor differentiating the three classes of relationships. The simplest definition of trust is "being confident that the other party will do what it says it will do." Some level of trust must be present in all three of our types of relationships. But the level of trust increases with collaborative relationships and becomes an essential characteristic with strategic alliances. Colleague Robert Porter Lynch has become a leading authority on trust. His 2011 co-authored article "Trust to Lead—the Structure of Trust" appeared in the May 2011 edition of the *European Business Review* and is recommended to all readers interested in this vital subject.

Few of these relationships are pure: a transactional relationship may have one or more collaborative characteristics while a collaborative relationship may have one or more transactional as well as some alliance characteristics.

The Supplier's Perspective

The competition for world-class suppliers is well underway. As a result, the most attractive supplier may decide that a collaborative or alliance relationship with a potential customer firm is not in its best interest. In effect, the supplier may possess economic power that it desires to exercise in an effort to maximize its net income; the classic case is the monopolist. Or, the preferred supplier may be unavailable because the buying firm's key competitor may have already established an exclusive relationship. World-class suppliers are careful in their selection of customer firms. The supplier will be very concerned with the potential customer firm's finances, especially as they affect its ability to pay. The customer firm's finances also provide insight into prospects for a long-term relationship and continuing demand for the supplier's products or services. Quality suppliers want customers who have good growth prospects.

The potential customer's demand pattern for the supplier's product is of great interest, especially in the areas of stability and fit. World-class suppliers want to ensure that potential customers' quality requirements are within their capabilities. The supplying firm is also very concerned with the buying firm's approach to problems. Does it discipline suppliers who encounter problems or does it help solve problems together in a mutually beneficial way, deriving critical learning along the way, thus making the problem a "learning foundation" from which to improve? Suppliers are attracted to customers who have a reputation for working collaboratively with suppliers who experience a problem to identify and correct the root cause of the problem.

Suppliers want "good" customers. Several issues affect a customer firm's rating as a "good" customer including:

▲ Does the customer have a reputation for timely payment? Cash flow is a major concern of all suppliers.

▲ Is the customer secretive? Suppliers prefer customers who are open and approachable.

▲ Are the buyers honest? Are they ethical, truthful, and reliable?

▲ Are the customer's procurement personnel responsive? Suppliers prefer customers that are available (not one whose supplier hours are 11:30–1:00, Monday–Friday).

▲ Are the customers known as professionals? World-class suppliers conduct themselves professionally, and expect to be treated professionally.

For suppliers and buyers who comprehend the value of shifting from tactical, transactional-based relationships to strategic, value-based alliance relationships, it will be essential to engage the multi-dimensional assessment of the elements of total cost of ownership (see Chapter 11) to determine exactly where cost can be reduced, value enhanced, and substantial competitive advantage created.

Questions to Be Addressed Before Proceeding

While strategic supplier alliances receive a great deal of media coverage and discussion within the supply management community, are they for everyone? Will the benefits of an alliance outweigh the effort, risk, and resources required? For those supply management professionals and organizations that are investigating the possibility of strategic supplier alliances, it can be helpful to ponder the following questions:

- Is there a danger that the supplier may act in an opportunistic manner over time?
- Do electronic systems at the purchasing and supplier organizations allow for optimum communication and sharing of information?
- Is the potential strategic alliance supplier well equipped, in terms of knowledge, expertise, and resources to stay current in the industry?
- Are both the purchasing and the supplier organizations willing to keep attention focused on the joint customer in order to establish supply chain objectives and goals?
- Are there other suppliers in the marketplace who are more accessible through e-procurement, who are worth investigating before committing to a strategic alliance?
- Has the supply manager been thoroughly trained in managing an alliance relationship?
- Is the purchasing organization proud to be aligned and associated with the supplier organization as they present a joint marketing front for the links further downstream in the supply chain?
- Is the purchasing organization comfortable with the level of risk associated with reducing the supply base?[21]
- Are both supplier and buyer aligned in what their ultimate customer considers to be valuable?
- If there is substantial risk for the supplier to develop new technologies, sub-systems, products, processes, or service support, is the buying firm willing to share or reduce the risks or pay for them? Are there "meet competition" provisions?
- Are both supplier and buyer aligned in their respective visions to be able to make long-term commitments to each other?
- If an alliance is in order, are there sufficient operational points of interaction where the supplier can engage with the buying firm, such as joint development programs, just-in-time inventory, electronic communication, or co-location of service personnel?

Developing and Managing Collaborative and Alliance Relationships

Situations Wherein Alliances May Not Be Appropriate

Quite obviously, alliances are not always appropriate. Professor Ralph Kauffman has identified 14 such situations and has developed them into five major categories.[22]

1. *Stability* of the prices, market, and buyer's demand
   ▼ *Price Volatility:* Commodities traded on open markets that have significant price volatility. The problem for a partnership/alliance is how to share risks and benefits that may result from price volatility. Some arrangements can be made to mitigate this problem including price adjustment mechanisms based on costs or indexes and, for some commodities, hedging in futures markets.
   ▼ *Demand Volatility:* Materials or services that have significant volatility in individual buyer demand. If the buying firm's needs are not predictable, the supplier must deal with the likelihood of overstock or stockout, or erratic productions schedules. To do this may generate additional costs for the supplier that must be built into the price that the buyer pays.
   ▼ *High Switching Likelihood with High Switching Costs:* Situations with high switching costs that also have a high likelihood of switching being desirable. Purchases that involve changing technology, critical quality, or other characteristics and where there are no strong suppliers may indicate a high likelihood of needing to switch in spite of high switching costs. In such cases, maximum flexibility is desirable.

2. *Capability* of potential suppliers
   ▼ *No Partnership/Alliance-Capable Supplier for the Item:* Items for which there is no full-service, world-class supplier capable of a partnership/alliance relationship. The lack of a capable supplier would dictate some other form of supply relationship. No partnership or alliance would be preferable to one with an inept supplier.
   ▼ *No Partnership/Alliance-Capable Supplier in the Geographic Area:* Areas of the world where there is no full-service, world-class supplier capable of a partnership/ alliance relationship. Depending on the material or service required, there may be regions where a partnership or alliance is not possible due to lack of a competent supplier in that region.
   ▼ *Rapid Technological Change:* Situations of rapid industry-wide technological change where the buyer would be disadvantaged if locked into one supplier. The buyer must have assurance that technology (either that which is being purchased or that which the supplier uses in its production, or both) is maintained at the

state of the art required by the buyer's industry. Not all suppliers have the capability to remain technologically competitive.

▼ *Mismatch of Rates of Technological Change:* If the buying firm's industry is changing and developing more rapidly than that of the supplier's industry, it may be difficult to arrive at a partnership or alliance that is fully beneficial to one or both parties.

3. *Competition* in the supply market

▼ *Non-Competitive Market:* Non-competitive markets where the supplier partner may be in a position to take advantage of the buying firm. Generally, a partnership or strategic alliance will reinforce the supplier's power relative to the buying firm.

▼ *Supplier Dependency Creation:* Situations where extreme dependency on a particular supplier would be created by a partnership or strategic alliance. If the buying company is relatively small compared to the selling company and the buyer's business is not vital to the seller, the buyer may be at risk of future supply. For example, a buyer becomes totally dependent on a seller through a partnership or alliance for a material vital to the supplier and for which there are few, if any, alternative suppliers. If the supplier determines at a future time that the business is not compatible with its business objectives, it may terminate the agreement and cause supply difficulties for the buyer.

▼ *Neglected Areas:* Situations where purchases have been mismanaged or not managed for years, for example many types of indirect purchases. Andrew Cox et al indicate that for these types of purchases, in order to obtain the lowest total cost, a relationship that leverages the free market should be used. They advocate partnerships for such situations only if there are few supply alternatives available and if there are high supplier switching costs.

▼ *Suppliers Seeking to Reduce Competition:* Situations where suppliers appear to be using partnerships/alliances as a marketing ploy to eliminate competition and reduce industry capacity. These may save cost in the short run, but if the supplier's strategy truly is to reduce capacity and competition, costs may increase in the long run. Being locked into such a supplier would not be desirable.

4. *Benefits* to the buying firm from the relationship

▼ *No Leverage from Partnership:* Situations where there is nothing to leverage with a partnership/alliance. Typically, a partnership or alliance will leverage some aspect of the exchange involved. Leveraged items include volume, total cost, process or procedural cost, inventory, or innovation. If there are no leverage possibilities, there may not be a justification for the work involved in establishing and maintaining a partnership or alliance.

▼ *No Hard Savings from Partnership:* Situations where hard savings are not present as a result of a partnership/alliance. Soft savings such as non-quantifiable quality improvements and partial-person staff reductions are nice, but unless they result in some other cost avoidance, they never show up on the bottom line. To justify the work involved in establishing and maintaining partnerships and alliance, there must be some hard savings.

5. *Internal buy-in* to partnership

▼ *No Internal Customer Buy-In:* Situations where the internal customers of the buying organization do not have joint ownership with supply management of the partnership/alliance arrangement. Most purchasing organizations use a cross-functional team approach to develop, implement, and maintain partnership and alliance agreements. If that is not done, or if the internal customer members of the team do not agree with all the terms of the agreement, the partnership or alliance will likely fail.

The Role of Power

Recent books and articles on supply management have largely ignored the impact and role of power. Thus, unfortunately, relatively little is known about the role of power in collaborative or alliance relationships. At the Third Annual North American Research Symposium on Purchasing and Supply Chain Management hosted by the Richard Ivey School of Business, University of Western Ontario in May 2000, two professors from the Netherlands observed that: "Power has an ideological tinge and produces negative associations." Pfeffer recognized that power is a topic that makes people uncomfortable. Andrew Cox argues that, "Power is at the heart of all business to business relationships."[23] Using power is often seen as unethical. However, power, influence, and dependence do exist. Ignoring them will not make them less important for understanding buyer-seller relationships. Power obviously is not always being used, although it can still influence decisions and strategies, simply because it is recognized by both trading partners. There is always a threat of (mis)use of power to which parties respond in advance. Studying networks, argued that power is the central concept in networks analysis because its mere existence can condition others.[24] Thus, it is highly likely that the use and role of power in supply chain management will be the subject of research over the coming years.

Power plays a key role in two important subclasses of buyer-supplier relationships: the captive buyer and the captive supplier. In a captive buyer relationship, "the buyer is held hostage by a supplier free to switch to another customer."[25] The captive supplier makes investments in order to secure a portion of the buyer's business, with no assurance of sufficient business to recoup the investment. Cooperation and some level of recognized interdependence suggest that these two classes of relationships are subsets of the broad heading of collaborative relationships.[26]

A Portfolio Approach

No single approach to relationship management is inherently superior. "Successful supply chain management requires the effective and efficient management of a portfolio of relationships..."[27] The portfolio approach has its roots in 1950s financial investment theory. One of the most useful approaches to selection of the appropriate relationships is advanced by Bensaou, who proposes three environmental factors to consider: (1) the product exchanged and its technology, (2) the competitive conditions in the upstream market, and (3) the capabilities of the suppliers available.[28] One of the most interesting phenomenon of modern-day business is that of multidivisional firms which are simultaneously buying some items and selling others to each other. Even more interesting is the fact that an array of buyer-supplier relationships may be present and appropriate.

New Skills and Attitudes Required

Developing and managing collaborative and alliance relationships requires skilled professionals who recognize the benefits of collaboration. These individuals must be able to identify and obtain necessary data and use the data to exploit and enhance relationships. As Smith-Doerflein and Tracey point out, these professionals must "learn to work in and adjacent to chaos, unpredictability, and uncertainty. They must be agile, flexible, and highly adaptive."[29]

Smith-Doerflein and Tracey support the use of role playing, sensitivity training, diversity training, and live cases to develop the professionals who work in this area. These researchers go beyond individual development to advocate the creation of a learning organization, as introduced by Peter Senge. The learning organization is one "where people continually expand their capacity to create results they truly desire, where new and expansive patterns of thinking are nurtured, where collective aspiration is set free, and where people are continually learning how to learn together. A learning organization excels at advanced, systematic, and intentional collective learning and is effective, productive, adaptive, and very good at setting and achieving goals."[30]

e-Commerce and the "Right" Type of Relationship

Frequently we are asked: "How does B2B eCommerce affect our selection of the 'right' type of relationship?" Selection of the "right" type of relationship and of the right supplier must be a function of the requirement, not of the Internet! B2B eCommerce is an enabler, a tool that is most effective when adopted by firms that embrace strategic supply chain management. It is unfortunate that some software sales personnel and some members of customer firms mistakenly assume that B2B eCommerce will eliminate the need for professional supply management. B2B eCommerce must be seen for what it is: a powerful enabler! In the last few years, there has been an evolution from paper-driven transactions to electronic transactions with a transition to integrated supply management processes. Now, Internet offerings will prompt even more transaction reduction initiatives and a greater transformation from paper to electronic tools. Reverse auctions are one such tool.[31] Held online, reverse auctions are a popular way to enhance competition. Companies use these auctions to encourage suppliers globally to bid on orders in the hopes of reducing the price and cycle times for processing orders.[32] However, reverse auctions are not effective for all products and the question of how to apply this tool without damaging supplier relationships arises. Tools are not strategies. The supply professional of the future will be, first and foremost, a strategic thinker and a creator of competitive advantage.

Unfolding before us is a massive array of new opportunities that will forever redefine the role of supply in the corporation, facilitated by a new set of tools ranging from the worlds of computer software, electronic commerce, and the Internet. However, these tools are simultaneously alluring and deceiving.

The allure comes from the multitude of new possibilities created by e-commerce and all of its siblings, including e-procurement, e-bidding, reverse auctions, e-payment, and e-business. But these tools can also be deceptively attractive to the uncritical eye and used inappropriately with little thought given to the solution and the net effect. Therefore, three traps must be steadfastly avoided.

First, we must avoid the trap of "guilding the pig," by which we take an archaic, cumbersome procurement process and "webbize" it—the process remains ugly and inefficient, despite its newfound digital disguise.

Second is the temptation to seek the holy grail of the "magic pill," the one solution that can be used to solve every procurement situation, thus circumventing design of a more thoughtful strategy and solution tailored to fit specific goals and needs.

Third is the trap of "supplier equality." Equality may be the law in civil rights, but it makes for poor procurement. Some suppliers are simply commodity vendors and should be treated in a tactical, transactional manner, pushed on price, and managed through e-bidding auctions, while other suppliers are far more strategic to a corporation's business and should be treated as close alliance partners, integrated into a company's business processes, and challenged to create powerful value streams, such as product and process innovations and fast time to market. Alliance partners are a company's assets in creating competitive advantage for the ultimate customer. The world of procurement is rapidly splitting into supplier-based and alliance-based relationships, and procurement professionals must know how to differentiate the two.

As discussed in Chapter 2, while B2B e-commerce may eliminate the day-to-day sourcing and pricing aspects of procurement when a transactional relationship is appropriate, cross-functional development and sourcing teams still have many vital responsibilities with such procurements.

The Internet greatly facilitates communication and, when properly employed, will enhance the development and maintenance of collaborative and alliance relationships. However, the Internet does not terminate the need for personal relationships based on face-to-face contact. Michael Dell states that, "The real potential lies in its (the Internet's) ability to transform relationships within the traditional supplier-vendor-customer chain and to create value that can be shared across organizational boundaries. The companies that position themselves to build 'information partnerships' with suppliers and customers and make the Internet an integral part of their strategy—not just an add-on—have the potential to fundamentally change the face of global competition."[33]

The new world of electronic commerce holds the promise of a bold, new, and enticing future by focusing on speed, connectivity, and innovation. But foremost, we must understand and differentiate our supply base to understand what suppliers bring to the corporate table. The evolutionary integration of electronic commerce requires a new set of strategies and tools that will help to make supply professionals treasured assets in the corporation. The evolution will continue to require a new type of thinking, a new set of rewards and measures, and a new set of skills. Let us not forget that it will be a challenging time, designed for those who are willing to learn, adapt, change, and take a leadership role. We must seize the Internet opportunities being offered and take risks in the electronic commerce arenas. Understanding that it is acceptable to make a mistake, it is often from trial and error that we learn. As leaders, we must continue to challenge "it won't work here" and "this too shall pass" attitudes that have prevailed in the past.

Continuous learning and capability building will be the norm, not the exception. These two activities will need to be more cross-functionally oriented and work together more to select and implement the optimum buying process tools. Supply management thinking, now more than ever, needs to encompass the whole supply process, the end-to-end "need it, design it, buy it, receive it, transform it, deliver it, service it" process.

Greater focus is needed to have supply professionals become change leaders and problem resolvers rather than followers and problem identifiers. Supply personnel, who over the years have been tagged as non-value added paper pushers having a "green eyeshade" mentality, will need to overcome this stigma and change their focus to become value added creators and analysts to support the supply strategy.

What's more, supply professionals will need to shift to a new level of understanding: they have a strategic role in the corporation. They each must keep a business focus in the forefront, understanding that long-term competitive advantage, speed to market, and supplier management for cash conservation and profit maximization are far more critical to long-term success than the historic role of squeezing suppliers for lower prices and other concessions.

Relationships of the Future

The terms used today—supply chain, tiers, and channels—all imply a static, rigid set of links that do not change position relative to one another. In these models, information, materials, and money flow sequentially, one link at a time, and only between adjacent links. This environment creates friction, inertia, and inflexibility. Every exchange increases the risk of error and multiplies cycle times. Incompatibility within and between supply chain member IT applications prevents fluid information flow. Compounding these problems is the fact that most companies are part of multiple supply chains. The resulting uncertainty creates mistrust among partners and leads to withholding of critical information. High inventories, uncoordinated schedules, and dissatisfied customers are the result.

As the corporate world develops more Internet competence, there may be an uncomfortable convergence of professionals from supply management, accounts payable, information systems, and e-commerce. For the uninitiated, these professionals will interact with a set of traditions oftentimes filled with conflicting points of view. Instead, professionals from each functional area will need to reinvent the way they have traditionally interacted. Supply Management will need to enter the information age in a powerful, proactive, and innovative manner. Accounts Payable will need to design and link its paying processes to the buying processes in a creative way. Information Systems will need to be more functionally adept and user friendly. Tools such as ERP (Enterprise Resource Planning) Systems and e-commerce will need to intensity their focus on totally integrated supply process solutions instead of promoting the latest electronic solution searching for a problem or a need.

If the 1990s reengineered the internal business processes of corporations, this and the next decade will be the era of reengineering entire value chains and value networks—from the initiation of the purchase request through design, manufacturing, logistics, and service to the final customer. The winners in the future marketplace will be those networked companies who can combine their internal advantages into powerful value networks that are faster, more efficient, more agile and innovative, and ultimately more profitable than other competing supply networks. The wise supply executive will recognize this shift early, knowing that supply chain/network reengineering may well determine the fate of the corporation.

Summary

The modern-day supply professional is blessed with an array of relationships from which to choose. Selection of the appropriate type of relationship for a specific purchase is a matter of sound business judgment. The type of relationship selected will affect total cost, quality, timeliness, the level of cooperation, the inflow of innovation, and other variables of concern to the purchasing firm.

The next section of the book addresses the requirements process, the most critical phase of procurement. This is the phase during which 70 to 80% of a product's or service's cost is built in. In Chapter 5, we describe the process of developing a new product and describe the role that supply managers must play.

Appendix

Institutional Trust

Research, including personnel interviews, and much of the current literature demonstrate the importance of trust between alliance partners.[35] Dyer points out that "Trust is critical for partner success because without it suppliers and customers will spend considerable resources negotiating, monitoring, and enforcing inflexible contracts."[36] As previously noted, when institutional trust is present, the parties have access to each other's strategic plans in the areas of the interface between the companies and their respective cross-functional teams. Relevant cost information and forecasts are shared. Risks and rewards are addressed openly. Informal agreements are as good as written ones. Institutional trust is measured and managed.

Traditional organizations that memorialize their supply agreements with detailed legal contracts often overlook the value of trust. In an ever-changing world that is moving faster and faster, trust creates a major competitive advantage in that fast decision-making can only be done in an environment of trust. Further, in an ever-changing, highly uncertain business world, trust based on integrity, commitment, and common values is the only thing that remains stable, while everything else—strategy, technology, people—move. An ever-changing world requires frequent renegotiations between alliance partners. If there is no trust, the re-negotiations are likely to be degenerative, antagonistic, and often result in a win-lose relationship, at which point the alliance dies.

In addition, innovation between buyer and supplier is typically a major strategic driver, and one of the most powerful elements of competitiveness. Co-created innovation is stymied

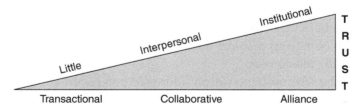

Figure 4.2 Spectrum of supply relationships and institutional trust.[34]

when distrust is rampant. People communicate better in a trusting environment, and cross-functional coordination is more facile. What's more, distrust causes companies to engage in additional non-value-added work, such as time wasted in legal contracts, burdensome paperwork, redundancy, and so forth. Thus, trust translates directly to more profit.

Trust in alliances is not at all "blind trust," but rather a "prudent trust" that is carefully designed, planned, and mutually agreed upon. Initially, as alliances are first formed, this trust is typically established interpersonally, usually between the alliance champions and senior executives that create the alliance entity.

This interpersonal trust, based on the vision and values of the founders, is critical at the inception of the alliance. When problems occur (and inevitably they will), the trust will be the foundation for the problem resolution. Similarly, as strategic or technological conditions change, as will occur in fast moving environments, trust will be a critical ingredient in re-negotiating and strategically repositioning the alliance.

However, for the alliance to survive in the long term, and for supplier networks to evolve, interpersonal trust is not enough—institutional trust, a higher order of trust, must prevail. The following list is the result of six years of research on the issue of institutional trust by faculty and students at the University of San Diego.

▲ Institutional trust is developed over time—it is part of a process.
▲ Internal trust is developed before external trust.
▲ Institutional trust is based on individual and institutional integrity.
▲ Institutional trust is greater than individual trust.
▲ Trust and the relationship are viewed as worthwhile investments.
▲ The partners have access to each other's strategic plans in the area of the interface.
▲ Relevant cost information and forecasts are shared.
▲ When key individuals leave, fingerprints are left behind that hold the relationships together.
▲ Trust is visible—something for others to see, feel, and emulate.
▲ Informal agreements are as good as written contracts.
▲ Both parties are sensitive to changes that might affect the cultural bridge between the two firms.
▲ The relationship is adaptable in the face of changing needs by either party.
▲ Personnel at both firms recognize the interdependent nature of the relationship.
▲ Personnel at both firms consider the sharing of information to be a means of developing trust in the relationship.
▲ Conflict in the relationship is openly addressed and resolved.
▲ Both parties consider the rights, desires, and opinions of their partners during internal discussions.
▲ The firms have mutual goals in the area of the interface.

▲ Firms build a bank account of trust based on many deposits.

▲ Trust has different meanings in different cultures.

Antecedents of Trust

In 2001, Fawcett and Magnan reported on a comprehensive study addressing supply chain management. The study was based on both a literature review and 52 in-depth interviews with representatives of leading companies at each stage of the supply chain.

The authors observed that

trust has numerous antecedents including open and honest information sharing, commitment, clear expectations, and follow through. The passage of time, high levels of actual performance, and the fulfillment of promises also precede trust. Finally, real trust exists only when both sides agree that it does. Relationships that one party describes as trust-based are often viewed as less friendly and less mutually advantageous by the other side.[37]

Actions that are taken to develop and manage trust include:

▲ Both CEOs make a personal investment.

▲ Appropriate senior managers from both firms commit to nurturing the relationship.

▲ Either party may bring up ethical issues without fear of retribution.

▲ An ombudsman is assigned at both firms.

▲ An inter-firm team consisting of representatives of relevant functional areas is appointed to develop and manage the relationship.

▲ Discussions between the two firm's personnel are conducted in an atmosphere of respect.

▲ The inter-firm team receives guidance and training in the implementation of practices that facilitate the development of strategic alliances before addressing specific project details.

▲ Listening, understanding, time, and energy are invested to develop and maintain trust.

▲ Senior leaders at both firms act as a champion to teams.

▲ Members of the inter-firm team share in the development of the communication system between the firms.

▲ Members of the inter-firm team take specific actions to develop and measure trust in the relationship.

▲ Risks and rewards are addressed openly.

▲ Negotiation is used as a trust-building opportunity.

▲ Within the parameters of the relationship, appropriate members of both firms work together on technology plans for the future.

▲ Technical personnel from both firms visit their partners to learn and observe how the other's products are developed, manufactured, and utilized.

▲ Contractual relations are designed to enhance trust (and lawyers who engage in the contractual negotiations must be acutely tuned to the necessity of building trust into their legal interactions).

▲ Contractual relations focus on continuous improvement.

▲ Team and relationship skills are developed early.

▲ Both company leaders create a formal relationship agreement—an agreement that memorializes a commitment to work well together.

▲ A contracting philosophy and a legal infrastructure are designed to enhance buyer-supplier relationships rather than the legal mechanism (t's and c's) to protect each firm.

▲ Institutional trust is measured and managed.

Supply management is one of the keys to Toyota's incredible success. Toyota has perfected the art and the science of supplier relationships, with trust playing a key role. Jeffrey Dyer points out that

> …Toyota's ability to develop trust with suppliers is not based primarily on personal relationships between Toyota and its suppliers. Nor is it based primarily on the stock ownership it holds in its keiretsu suppliers. Rather, this trust is linked to the perceived fairness and predictability of Toyota's routines for managing external relationships.[38]

Supply Management in Action

A Supplier Alliance at Quaker Oats[39,40]

This section describes how Quaker Oats Company of Chicago, Illinois, formed an alliance with Graham Packaging of York, Pennsylvania. Graham is a leading global manufacturer of custom blow-molded plastic containers. Plastic bottles are the largest single quantity and cost item purchased by Quaker Oats. Topics include how to select an alliance partner, how to negotiate a massive joint-building construction and equipment project, and what kind of contract facilitates a "win-win" relationship based on trust and sharing of cost information. This alliance actually became a joint venture with Quaker building an addition for the plastic bottle plant and Graham investing in the equipment and people required to operate the plant at the Quaker Gatorade facilities in Atlanta, Georgia.

Introduction

Founded in 1901, Quaker Oats had sales of $4.6 billion in 1998, produced by 11,860 employees. One of Quaker's successful brands, Gatorade, had $1.7 billion in worldwide sales and is far and away the number one brand in the sports beverage category with 82% share of market in the United States. Other well-known brands include Cap'n Crunch cereal, Life cereal, Chewy Granola Bars, Quaker Fruit and Oatmeal breakfast bars, Rice-A-Roni, Near East, and Aunt Jemima syrups and pancake mixes.

Plastic Bottles

Quaker is one of the world's largest purchasers of plastic bottles for sports drinks, with Gatorade as one of the great global brands. Total U.S. case volume is roughly 100 million cases purchased mostly from April through September. The incumbent bottle supplier had been supplying Quaker since 1986. Demand had grown from 100 million to over 1,400 million bottles by 1998. Because the initial emphasis had been on bottle supply and performance, the supplier had negotiated a price based on new investment and defined quantities. Over the 12-year period new negotiations had taken place, but the supplier had managed to increase or retain its high start-up margins.

The incumbent supplier did offer various means of cost reduction. These were deemed insufficient.

Quaker after Snapple and Supply Chain Management

By 1997, new purchasing management was engaged to find supply chain cost efficiencies. The bottle was a logical first place to look (i.e., the low-hanging fruit). A quick analysis indicated that the material cost was less than 40% of the bottle price to Quaker, a ratio that indicated there was tremendous margin for price reductions. After paying what the company thought was a reasonable price for many years, it was now immediately apparent to Quaker that the bottle price was too high.

A cross-functional team from accounting, engineering, technical, and the purchasing staff was formed to model the bottle's manufacturing cost. After numerous trips to plants and discussions with consultants, an independent cost model confirmed Quaker's initial belief that bottle prices could be reduced substantially. The current supplier was not willing to reduce the price even though Quaker demonstrated that profit was excessive. Competition for Quaker's business would be the strategy.

The Alliance Options

Several options were examined with the objectives of reducing cost and improving quality and service:

1. Merchant supplies the total product.
2. Self-manufacture with inputs from key raw material suppliers.
3. In-house plant operated by a supplier. This option would require an alliance due to long-term commitment needed, ongoing cost containment, and contractual issues regarding having another company operate on Quaker's premises.

Merchant supply was rejected due to:

1. Absence of lower cost alternative to merchant supply (freight cost hurdle).
2. No known way to gain effective cost understanding/cultural improvement with arm's-length relationship (lack of both parties' commitment).

Self-manufacture was rejected due to:

1. Not a core competency.
2. Supplier's cost of capital was generally lower than Quaker's—best to use a supplier's money.

In-house plant was chosen because:

1. Best cost—no freight, direct feeding of filling line eliminated palletizing, and fresher materials.
2. Best opportunity to institutionalize continuous improvement. Alliance relationship comes from the open book need to drive improvement. Quaker and Graham (the selected supplier) agreed to act as one company on each other's behalf.

Negotiating an Alliance

The Quaker Negotiation Team included:

> The Director of Packaging Purchasing
> Senior Manager of Purchasing
> Supervisor of Finance
> Industrial Engineer
> Manager, Supply Chain Planning

The Graham Packaging Company team included:

> Senior Vice President and General Manager, Food & Beverage Business Unit
> Director of Sales, Beverage and Business Unit
> Director of Finance, Beverage Business Unit

Quaker visited eight bottle companies, two consulting companies, one practitioner of in-plant molding, two machinery suppliers, and two resin manufacturers. This effort took approximately one year and involved many, many sessions. In addition, Quaker refined the performance "should cost model" with the cross-functional team. Finally, trust-building sessions were held with two potential suppliers (the finalists). These sessions included senior management dinners, use of an outside consultant as a facilitator on partnering, and frequent visits to each other.

In the survey to understand the market, it was determined that Graham Packaging of York, Pennsylvania, had the capability to meet Quaker's needs because of its focus on Quaker's type of bottle and its proven record with customers. A key to begin developing an alliance was to build trust by disclosing bottle costs.

Graham and Quaker held a series of meetings to define what each company wanted from a potential business relationship. It was determined that both were aligned on views of how to jointly create value. Quaker discussed Graham's target ROI needs and Graham acknowledged and understood Quaker's needs for low-cost bottles. A key Quaker issue was how to find a supplier who would manage the risk of unused bottle-making capacity after the substantial Gatorade seasonal peak and do it to the benefit of both parties. Graham agreed early in the discussions to price the bottle on a highly utilized machine basis in spite of Quaker's inability to commit to fill the equipment 100% of the year. If needed, Graham would find other customers to keep the plant running and to maintain the full economic benefit for both companies.

Although the basis for the agreement ended up to be partnering, it did not begin that way. As both parties discussed how to work to manage cost and find efficiencies, the only apparent way was to create an open-book accounting relationship based on trust.

At the core of supply chain management, the reason why trust fails to develop between buyer and supplier, and why maximum value in the relationship is not created, is because parties hide information from each other. The largest source of hidden information is cost. Opening the books is the best way to eliminate distrust and truly find a better relationship that would warrant any company's investment in developing a relationship.

The Contract

Without divulging confidential information, the arrangement is not long-term if Graham does not perform adequately. The incentive for the relationship to be constantly renewed has been built in through periodic "re-awarding" of the business based on performance. While those close to the relationship know this is not needed, the board of directors at Quaker Oats needed these safeguards.

Construction on the in-plant facility at the Gatorade plant in Atlanta, Georgia, has been completed. Quaker spent $10 million in plant expansion; Graham spent $28 million on equipment, and will own and operate the plant and be responsible to meet certain quality and efficiency standards. If Graham fails to meet these standards, it will bear the cost of non-performance.

The type of contract is:

1. Evergreen—from one fixed period to another.
2. Completely open book—Quaker pays all expenses and a fixed return on invested capital (which was mutually costed).

3. Cancelable for failure to perform.

4. Volume sensitive—Quaker will compensate for volume shortfalls. However, Gatorade has not failed to grow at least 9% per year over the last 18 years. Never bet against your own business.

How is the Alliance Working? What About Future Alliances?

The bottles produced in the in-house plant are the lowest cost bottles in the Quaker system. The alliance has produced much better (more accurate) forecasting that helps lower costs by schedule stabilization. While this is not the only cost reduction benefit in the endeavor, it demonstrates the need to capture and analyze total cost of ownership. The few start-up quality issues were resolved very quickly and, as one would expect, the delivery and other service aspects are "perfect." Joint quality improvement initiatives enabled by the alliance are generating impressive results on the Quaker Oats filling lines as improved quality translated to increased productivity, reduced scrap, and lower costs. Joint cost savings initiatives and the "open book" approach are also already delivering bottle cost savings to Quaker that exceed project objectives by over $1 million annually.

As to other alliance possibilities, Quaker would like more in other packaging materials. Unfortunately, the packaging industry is very "old school" and adversarial. Consolidations result in frequent management changes, which negate building long-term relationships. In the field of packaging, costs are driven be machinery utilization so huge volumes are required for economies of scale. This is especially true for flexible film and folding cartons. Perhaps Quaker can partner with other non-competitors to consolidate and achieve the volume…but that's another story.

Epilogue

In a follow-up conversation between Dr. Pinkerton and Mr. Reider in March 2007, it was learned that this supplier alliance contract has been extended and the concept applied to other sites at various bottling plants. Mr. Reider states that one of the key aspects of the contract was the "meet competition clause." This assures that if Quaker Oats can buy bottles cheaper, Graham, or any current supplier will have to meet the price or risk the execution of a cancellation provision.

Endnotes

1. Christopher Jahns, Roger Moser, Evi Hartmann, and Martin Lockstrom, "Rent-Based Supplier Management Behavior," The Eighteenth Annual North American Research and Teaching Symposium on Purchasing and Supply Chain Management, Educational Resources Committee of ISM, CAPS Research, Institute for Supply Management, and the Purchasing Management Association of Canada, Tempe, AZ, March 29-31, 2007.

2. Christopher Jahns, Roger Moser, Evi Hartmann, and Martin Lockstrom, "Rent-Based Supplier Management Behavior," The Eighteenth Annual North American Research and Teaching Symposium on Purchasing and Supply Chain Management, Educational Resources Committee of ISM, CAPS Research, Institute for Supply Management, and the Purchasing Management Association of Canada, Tempe, AZ, March 29-31, 2007.

3. The cost of switching to a new supplier may include higher initial costs because the present supplier has enjoyed the benefits of learning, the costs of working with the new supplier's production and quality people, and costs within the buying firm, including receiving, accounts payable, etc.

4. Ginni Rometty, *Leadership Excellence,* 24(2)(2007): 3–4.

5. Commodity items/services are those that are highly interchangeable with another supplier's offering. School writing tablets and non-critical fasteners (nuts, bolts, rivets, etc.) are examples of commodities.

6. Value engineering and value analysis are addressed in Chapter 5.

7. For more on the effects of the learning curve, see Chapter 12.

8. Linda L. Stanley and John N. Pearson, "Buyer-Supplier Strategies and Their Impact on Purchasing Performance: A Study of the Electronics Industry," Conference 2000, Richard Ivey School of Business, University of Western Ontario, London, May 24-27, 2000.

9. Antony Paulraj and Injazz J. Chen, "Strategic Buyer-Supplier Relationships, Information Technology and External Logistics Integration," *The Journal of Supply Chain Management,* 43(2)(2007): 2–14.

10. Gregory M. Magnan and Stanley E. Fawcett, "Supply Chain Collaboration: Analyzing of Five Years of Change," The Eighteenth Annual North American Research and Teaching Symposium on Purchasing and Supply Chain Management, Educational Resources Committee of ISM, CAPS Research, Institute for Supply Management, and the Purchasing Management Association of Canada, Tempe, AZ, March 29-31, 2007.

11. Shruti Singh, "Caterpillar Looks for a Few Close Friends: It Hopes Teaming up with Suppliers will Boost Earnings," *Bloomberg Business Week* October 21, 2010. www.businessWeek.com/print//magazine/content 10-44b420102225909.html

12. P.L. Carter, J.R. Carter, R.M. Monczka, T.H. Slaight, and A.J. Swan, "The future of Purchasing and Supply: A Ten-Year Forecast," *The Journal of Supply Chain Management,* Winter (2000): 14–26.

13. Stephan M. Wagner and Exkhard Lindemann, "Pie Sharing in Buyer-Supplier Relationships: What Determines Each Party's Share?" The Eighteenth Annual North American Research and Teaching Symposium on Purchasing and Supply Chain Management, Educational Resources Committee of ISM, CAPS Research, Institute for Supply Management, and the Purchasing Management Association of Canada, Tempe, AZ, March 29-31, 2007.

14. See Chapter 12 for insight into the learning curve.

15. The term "trust" has such a wide variety of meanings and interpretations that we have chosen to coin the term "institutional trust." Institutional trust is the key element that differentiates supply alliances from collaborative relationships. With institutional trust, the parties have access to each other's strategic plans in the area of the interface. Relevant cost information and forecasts are shared. Risks and rewards are addressed openly. Informal agreements are as good as written ones. Institutional trust is measured and managed. The issue of institutional trust is addressed in grater detail in the appendix to this chapter.

16. Jeffrey H. Dyer, *Collaborative Advantage* (New York: Oxford University Press, 2000): 42–43.

17. The design of experiments is a quantitative approach of analyzing deviations from desired outcomes during the design process. Identifying deviations during this stage—before progressing to production—has a powerful positive effect on an item's quality. See *World Class Quality* by Keki R. Bhote, American Management Association, AMACOM, 1991 for an excellent description of Design of Experiments (DOE).

18. Supplier certification requires the customer to carefully review a supplier's manufacturing operation to ensure that it has the necessary processes in place to meet the customer's quality requirements. Items produced by a supplier whose system has been certified for the specific item can flow directly into the customer's production without being inspected.

19. "Dell to Buy $16 Billion in IBM parts," *Los Angeles Times*, p C-1, March 5, 1999.

20. For more detailed insight into Chrysler's transition, please read "How Chrysler Created an American Keiretsu" by Jeffrey H. Dyer, *Harvard Business Review*, July/Aug (1996): 42–53.

21. Roberta J. Duffy with input from Dr. Joseph L. Cavinato, "Align to be Strategic," *Purchasing Today,* September (2000): 40.

22. Ralph Kauffman, "Supplier Partnerships and Alliances, Is That All There Is? When to Consider Other Relationship Strategies," *difOrientering* 38(2001):5. Published by the Danish Purchasing and Logistics Forum, Copenhagen.

23. Andrew Cox, "Understanding Buyer and Supplier Power: A Framework for Procurement and Supply Competence," *The Journal of Supply Chain Management* Spring (2001).

24. J. Pfeffer, *Power in Organizations* (Marston Mills, MA: Pitman Publishing Inc., 1981) and H.B. Thorelli, "Networks: Between Markets and Hierarchies," *Strategic Management Journal,* 7(1986): 37–51.

25. M. Bensaou, "Portfolios of Buyer-Supplier Relationships," *Sloan Management Review,* 40(4)(1999): 35–44.

26. The Spring 2001 issue of *The Journal of Supply Chain Management* addresses power in buyer-supplier relationships. The devoted student of supply management is encouraged

to read the thought-provoking articles authored by Andrew Cox, Joe Sanderson, Chris Lousdale, and Glyn Watson.

27. M. Bensaou, op. cit., 37.

28. M. Bensaou, op. cit., 42.

29. Kimberly A. Smith-Doerflein and Michael Tracey, "Training as a Component of Supply Chain Management," The Third Annual North American Research Symposium on Purchasing and Supply Chain Management, Richard Ivey School of Business, University of Western Ontario, May 24–27, 2000.

30. Calvert, G. Moble, and Marshall, S.L. "Grasping the Learning Organization," *Training and Development*, 48(5): 38–43.

31. Reverse auctions are contrary to traditional auctions in that the seller is competing for the business and the end result is to drive the price down rather than up. This tool is used mostly in industrial B2B transactions. Sheila D. Petcavage, "Reverse Auctions—Should You Be a Player," *ISM Materials Management News* 3(1)(2007).

32. Srinivas Talluri, and Gary L. Ragatz, "Multi-Attribute Reverse Auctions in B2B Exchanges: A Framework for Design and Implementation," *The Journal of Supply Chain Management*, 40(1)2004: 52.

33. *Industry Week*, November 16, 1998.

34. Institutional trust: The term "trust" has such a wide variety of meanings and interpretations that we have chosen to coin the term "institutional trust." Institutional trust is the key element that differentiates supply alliances from collaborative relationships. With institutional trust, the parties have access to each other's strategic plans in the area of the interface. Relevant cost information and forecasts are shared. Risks and rewards are addressed openly. Informal agreements are as good as written ones. Institutional trust is measured and managed.

35. Nirmalya Kumar, "The Power of Trust in Manufacturing–Retailers Relationships," *Harvard Business Review*, November-December (1996): 92–106 and R.M. Monczka, K.J. Peterson, R.B. Hanfield, and Gary L. Ragatz, "Success Factors in Strategic Supplier Alliances: The Buying Company Perspective," *Decision Science*, 29(3)(1998): 553–575.

36. Jeffrey H. Dyer, *Collaborative Advantage* (New York: Oxford University Press, 2000): 38.

37. Stanley E. Fawcett, and Gregory M. Magnan, "Achieving World-Class Supply Chain Alignment: Barriers, Bridges, and Benefits," Center for Advanced Purchasing Studies (CAPS) Focus Study, 2001, Tempe, AZ.

38. Jeffrey H. Dyer, *Collaborative Advantage* (New York: Oxford University Press, 2000): 15–16.

39. Richard L. Pinkerton, Ph.D., C.P.M., Chair and Professor Emeritus of Marketing and Logistics, The Sid Craig School of Business, California State University, Fresno, and Richard G. Reider, Director, Packaging Purchasing, Quaker Oats Company. Published

in the Proceedings of the 85th Annual International Purchasing Conference, April 30 to May 3, 2000.

40. This case study has been updated based on information gathered by Richard L. Pinkerton in a March 2007 interview with Mr. Richard Reider, Director, Packaging Purchasing, PEPSICO.

Suggested Reading

Carr, Amelia S. and John N. Pearson. "Strategically Managed Buyer-Supplier Relationships and Performance Outcomes," *Journal of Operations Management* 17(1999): 497–519.

Cox, Andrew, Daniel Chicksand, Paul Ireland, and Tony Davies, "Sourcing-Indirect Spend: A Survey of Current Internal and External Strategies for Non-Revenue Generating Goods and Service," *The Journal of Supply Chain Management,* (2005):39–51.

Dyer, J., and H. Singh. "The Relational View: Cooperative Strategy and Sources of Interorganizational Competitive," *Academy of Management Review* 23(4)(1998): 660–679.

Ellram, Lisa M., "The Supplier Alliance Continuum," *Purchasing Today,* February (11/2) (2000): 8–10.

Emmett, Stuart, and Barry Crocker. *The Relationship-Driven Supply Chain: Creating a Culture of Collaboration throughout the Chain,* Hampshire, United Kingdom: Gower Publishing Limited, 2006.

Kwon, Ik-Whan G., and Taewon Suh. "Factors Affecting the Level of Trust and Commitment in Supply Chain Relationships," *The Journal of Supply Chain Management* 40(2)(2004): 4–13.

McHugh, Marie, Paul Humphreys, and Ronan McIvor. "Buyer-Supplier Relationships and Organizational Health," *The Journal of Supply Chain Management,* 39(2)(2003): 15–25.

Pinkerton, Richard L., "Business-to-Business Procurement—The Current Revolution Within Supply Chain Management," *World Markets Series Business Briefing: European Purchasing & Supply Chain Strategies,* London: World Markets Research Centre, 2001. www.wmrc.com

CHAPTER 5

New Product Development

Overview

This chapter addresses four key issues: (1) early supply management and supplier involvement; (2) the process of designing and developing new products, with emphasis on supply management's role in the process; (3) several approaches to increasing supply management's role in the new product development process; and (4) a description of supply management professionals who interface successfully with engineers during the new product development process.

World-class firms excel at a crucial triad of activities: new product development, the design of the required production process, and development of the optimal supply chain. This chapter addresses the first member of this triad.

"Rapid changes in technology, the emergence of global industrial and consumer markets, increasing market fragmentation and product differentiation, and the increasing options for developing and producing products have increased the pressure on all firms to more effectively and efficiently develop new products."[1] In many progressive firms, the design of new products is conducted by a team representing a number of functional areas. Marketing, product planning, design engineering, reliability engineering, supply management, manufacturing engineering, quality, finance, field support, and, frequently, carefully selected suppliers and customers are involved, as appropriate. If effectively done, new product development (NPD) can be a source of competitive advantage for a firm and a competitive strategy for the internal and external partnerships of the supply chain.[2]

Anecdotal evidence indicates that the development of new products by such cross-functional teams and the use of concurrent engineering[3] have the potential of significantly improving three key objectives: time to market, quality, and total cost.[4] The turnaround of many troubled manufacturers during the past decade was the result of replacing departmental walls with teamwork among those who should be part of the design process. Supply management professionals and carefully selected suppliers are moving to earlier involvement in the new product development process because of the important contributions they can make in the areas of quality, cost, and timely market availability. This early involvement commonly is referred to as *early supply management involvement* and *early supplier involvement* (ESI). German authorities Arnold and Essig conclude that "by involving supply management and suppliers in the simultaneous engineering process (as members of cross functional teams) at an early stage, R&D gets the chance to increase efficiency... In fact, early supply management involvement helps to shorten engineering time and increase engineering quality."[5]

The lack of effective, cooperative teamwork among the functions just noted frequently has been accompanied by quality problems, cost overruns, forgone all-in-cost savings,[6] major scheduling problems, and new products that are late to enter the marketplace. Further, early recognition of problems is difficult or impossible in the absence of cooperative teamwork. Extensive redesign, rework, and retrofit operations are common when a company is operating in the traditional functional mode. Ultimately, the absence of teamwork results in products that are a continuing burden to the firm's long-term competitiveness.

Cost overruns and forgone cost savings frequently result when the designers (or the design team) fail to consider the supply base's design, manufacturing, quality, and cost capabilities. For example, during the early 1980s, design engineers at General Electric's Jet Engine Division frequently designed materials to be purchased from outside suppliers under the mistaken belief that the outside suppliers had the same manufacturing and process capabilities as GE. In fact, this was not the case; the outside suppliers frequently did not have the same equipment, processes, and quality capabilities. The results were cost growth and schedule slippages as the suppliers, using a trial-and-error process, attempted to meet GE's specifications. Frequently, it became apparent that those specifications could not be met and that a costly and time-consuming process of reengineering would be required.[7]

A similar example of costs resulting from the failure to consider supply implications during design involves IBM. In 1993, IBM's PC units' sales were just over $8 billion, with earnings of about $200 million (2.5%). By contrast, Compaq's profits were $462 million on sales of $7.2 billion (6.4%). According to *Business Week*, "At least one reason . . . seems clear, IBM still does not use common parts across its product families." Another contributor to lower profits was IBM's failure to shift away from pricey Japanese components as the value of the yen rose.[8] It was noted that IBM recognized that its supply management system had been the source of significant cost overruns and forgone dollar savings. In 1994, that recognition resulted in the appointment of a new Chief Procurement Officer: Mr. Gene Richter. Under Richter, three-time Purchasing Man of the Year, IBM procurement has undergone an incredible transformation and is now approaching "World Class" status.

Scheduling problems frequently result from late delivery of required parts. For example, earnings at Apple Computer fell nearly 30% in the third quarter of 1999, largely as a result of supply shortages. Apple received only 45% of the G4 chips its suppliers originally had promised. This, in turn, led to significant reductions in the sales of Apple's Power Mac G4 computers.[9] When supply considerations are not addressed during new product development, unique nonstandard components may be specified. Those components frequently require longer lead times than do standard items. The use of nonstandard items often leads to the inability of the manufacturer to react quickly to changes in market demand, frequently resulting in lost sales. To reduce reaction time to changes in demand, firms are replacing unique components with standard "commodity" ones. In addition to being more readily available, commodity components tend to be far less expensive than the unique items they replace.

The global marketplace and global competition, coupled with advanced communication systems, computers, and sophisticated software, have generated an environment where "time to market" and first to market have significant competitive advantages. Clearly, the need to reduce development time has forced companies to look for new methods to compete. The use of supply professionals and suppliers earlier in the product development cycle is a key way to reduce time to market. The advantages of an integrated approach to new product development no longer can be ignored.[10]

In the early 1990s, the Chrysler Viper went from concept to production in 36 months, in contrast to an industry norm of 60 months. Chrysler did not achieve that goal by itself; it got a lot of support from its suppliers. "They were as much a part of the Viper team as anyone . . . suppliers are an integral part of the team," said Dave Swietlik, the man in charge of procurement for the Viper program. "Their processes drive design."[11]

When a cross-functional team has the responsibility for the development of new products, a concurrent approach to the myriad of tasks involved is taken. This avoids the traditional (and time-consuming) passage of a project from concept development, to design, to manufacturing engineering, to supply management, to manufacturing, to marketing, to field support. That sequential approach requires even more time and personnel resources when changes have to be made in the product's design. The cross-functional team uses a concurrent approach in which the team members work together and collaborate throughout the process.

The Design Process[12]

Design is the progression of an abstract notion or idea to something that has a function and a fixed form. The desired levels of quality and reliability must be "engineered in" during the design phase of the new product. "Suppliers must have access to product design as early as humanly possible in the design process to assure optimal use of any special skills or processes they can contribute."[13] The design stage is also the optimum point at which the vast majority of the cost of making an item can be reduced or controlled. If costs are not minimized during the design stage, excessive cost may be built in permanently, resulting in expensive, possibly noncompetitive, products that fail to realize their profit potential.

The new product development process is a series of interdependent and frequently overlapping activities that transform an idea into a prototype and on to a marketable product. The process is much more fluid and flexible than is portrayed in the forthcoming flow diagrams in this chapter. As the original idea progresses through the development process, it is refined and constantly evaluated for technical and commercial feasibility. Trade-offs between the various objectives (price, cost, performance, market availability, quality, and reliability) are made throughout the process. These days, one hears a great deal about designing for manufacturability; however, invariably, the focus is on the firm's internal manufacturing process. However, when those responsible for design ignore the manufacturing process and technological capabilities of outside suppliers, problems with quality, time-to-market, configuration, control, and cost are inevitable. If optimal design performance is to be achieved, suppliers must be active from the beginning, when they can have a major impact on performance, time, cost, and quality. Selected suppliers should participate in feasibility studies, value engineering, and prototype, failure, and stress analysis, among other product development tasks.

There is a growing trend among manufacturers to develop an "envelope" of performance specifications for suppliers. For example, instead of determining the materials, manufacturing processes, and engineering drawings for a seat for one of its motorcycles, in the 1980s Kawasaki began specifying the environmental conditions and the maximum weight the seat had to withstand together with a drawing showing how the seat was to attach to the motorcycle frame. The suppliers' engineering and CAD/CAM[14] tools, not the buying firm's, were then dedicated to designing selected components. This approach allows engineers at the buying firm to focus on the development of more sophisticated core technologies and proprietary systems. The customer firm's engineers do not prepare engineering drawings for nonstrategic components. However, they review and approve the supplier's designs. That not only redirects critical engineering resources to higher-value activities but also places responsibility for manufacturability and quality with the supplier.

To involve suppliers effectively and early, manufacturing companies invite carefully selected suppliers' engineers into their own engineering departments. In a 1995 *Harvard Business Review* article, management guru Peter Drucker described William Durant as the inventor of the keiretsu, a set of companies with interlocking business relationships. Durant designed and built General Motors during the early 1900s. "Durant deliberately brought the parts and accessories makers into the design process of a new automobile model right from the start. Doing so allowed him to manage the total cost of the finished car as one cost stream."[15] Manufacturers should allow key suppliers to review the design of the entire subassembly before committing to it. Not only does this tease out new ideas, it also helps the supply partner understand the customer's real needs—and likely future needs.

Involving suppliers in the new product development process is more challenging than one might imagine. Handfield and Ragatz observe that, "Successful supplier integration initiatives result in a major change to the new product development process. Further, the new process must be formally adapted by multiple functions within the organization to be successful. One of the most important activities in the new development process is understanding the focal supplier's capabilities and design expertise, conducting a technology risk assessment, weighing the risks against the probability of success."[16]

The changing competitive environment requires that much more planning, coordination, and review take place during the design and development process than previously was the case. Complexity of product lines must be addressed. Lower levels of complexity result in higher schedule stability, a prerequisite to just-in-time manufacturing. Feasibility studies, computer simulations, prototype analysis, failure analyses, stress analyses, and value engineering all must be conducted in an effort to develop producible, defect-free products quickly at the lowest possible total cost.

The new product development process has undergone a tremendous change during the last several years. The process is described in Figures 5.1, 5.2, and 5.3 and is discussed next.[17]

The Investigation or Concept Formation Phase

There are several types of new product design. The first is the one used for a totally new product. This is the least common approach because completely new products are the exception. Most new product design is actually an adaptation or an expanded feature set for a previous design. Advancing technology, process improvements, and market expansion drive the majority of new product design activity. The process described is equally applicable to a totally new product or a "new and improved one."

Defining the New Product

The design and development process begins with the investigation phase. First, the product is defined. This function is normally performed with considerable marketing involvement. Intel carries marketing to its logical extreme: it emphasizes design ethnography, which focuses on understanding the customer and the culture in which a product is to be used.[18] The design and development process has been formally titled "customer focused product and process development" at some firms, or "quality function deployment" at others.[19] Marketing authority Regis McKenna is quoted as saying:

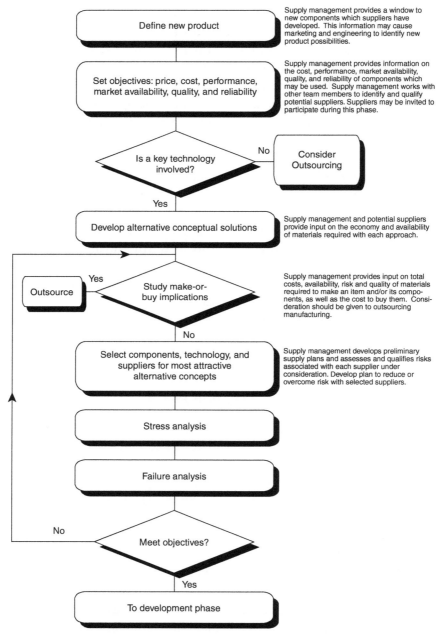

Figure 5.1 Team and supply management activities during new product development *(NPD) investigation phase.* Adapted from David N. Burt and Richard L. Pinkerton, *A Purchasing Manager's Guide to Strategic Proactive Procurement.* New York: AMACOM, 1996, 27.

Companies need to incorporate the customer into product design. That means getting more and more members of an organization in contact with the customer-manufacturing and design people, as well as sales and marketing staff. You can, for example, have customers sitting in on your internal committee meetings.[20]

Designers can make up for some of the shortcomings of consumer input because they usually understand more about future technological possibilities and look at a longer timeline. They are also in a better position to know what competitors may offer. For example, consumers may desire a "user-friendly" personal computer that is easy to get started with, but the designer realizes the computer should also meet longer-term needs. Therefore, designers should have the freedom to create innovative product designs that not only meet current user requirements but are also up to the demands of future consumer expectations. This give-and-take requires a delicate balance between designers and consumers because research has shown a high correlation between inadequate feedback from users and the failure of new products containing technical innovations.[21]

One of a supply management professional's key responsibilities is to acquire, assimilate, digest, and share information concerning new or forthcoming developments in the supply markets for which he or she is responsible. Interviews with present and potential suppliers, visits to suppliers (with emphasis on their research and development and production activities), attendance at trade shows, weekly reviews of relevant literature, and discussions with colleagues at local supply chain management organizations and American Production and Inventory Control Society meetings help the professional remain current. Through such activities, the supply management professional will become aware of new products and new technologies that may be of interest. This information may help product managers in marketing and senior design personnel responsible for identifying and developing new products. While being careful to screen out inappropriate information, the buyer should share potentially attractive information with marketing and engineering.

Statement of Objectives

Next, a statement of needs, desires, and objectives is developed. Needs are based on marketing's perception or knowledge of what customers want (or the customer's direct input if the customer is a member of the design team), balanced against the company's objectives and resources. Needs that are potentially compatible with the firm's objectives (profit potential, sales volume, and so on) and resources (personnel, machines, and management) are considered for development. Product objectives, including performance, price, quality, and market availability, are established and become the criteria that guide subsequent design, planning, and decision making. A well-informed procurement professional is the key source of information on the cost, performance, market availability, quality, and reliability of supplier-furnished components that may be used in the new product.

Establishing a realistic target cost at this stage of the new product development process is mandatory at world-class firms. (Target Cost = Targeted Market Selling Price – Targeted Profit.)

Purchasing authority Lisa Ellram writes, "By establishing the target cost up front, purchasing, the supply base, designers, and marketing can all work toward a common goal in the value engineering, design for purchasing, and early supplier involvement processes."[22] The planned product lifecycle typically includes not only the original product but also several future products that will incorporate improvements in design, function, features, and so on. These new products are driven by advances in technology, design, or materials; competitive offerings; and customer expectations. These desired advances frequently are known at the time of the original product design, but they are not included in the design because the technology does not exist or requires additional development to be production-ready. This product feature design "wish list" is very important to the design engineer because he or she most closely understands the design trade-offs and compromises that were included in the original design. This "wish list" of technology requirements is extremely important. Unfortunately, most firms do not document these technical interests that eventually drive a subsequent iteration through their product development process. Not only should these data be documented, but they must become an important focus for a supply partner's R&D efforts. Quick development will drive new product offerings that add additional sales volume, frequently at premium prices, for both the manufacturer and supplier.

Key Technology

The development team should determine whether a key technology is involved. If it is not, the team may decide to have an outside supplier develop both the technology and the product.

Development of Alternatives

Alternative ways of satisfying these needs, desires, and objectives should be developed and then evaluated against the criteria established in the preceding step.

There is an unfortunate tendency to proceed with the first approach that appears to meet a need even though less obvious alternatives may yield more profitable solutions. Alternative approaches should be evaluated on the basis of suitability, producibility, component availability, economy, and customer acceptability.

▲ *Suitability* refers to technical considerations such as strength, size, power consumption, capability, maintainability, and adaptability. Engineering has primary responsibility for these issues.

▲ *Produceability* is the ease with which a firm can manufacture an item. In the past, designs needed to be changed to accommodate the firm's or its suppliers' ability to produce the item economically. Problems arose when the needed changes were implemented. Early manufacturing engineering involvement in the design is needed to ensure the produceability of items made internally, and early supplier involvement helps ensure the producability of items furnished by suppliers.

▲ *Component availability* is the time at which components are available, while component economy describes the cost of the item or service. Component availability and economy are the responsibility of purchasing.

▲ *Customer acceptability* is defined as the marketability of an item to potential customers.

The selection of components, technologies, and suppliers for the most attractive conceptual solutions is a complex process. At progressive firms such as GE, Hewlett-Packard, and Deere and Company, this selection process is a team effort, with design engineering providing the majority of the staffing and the team leadership.

Often, an engineer has a need that must be filled—a power transmission gear ratio, a structural component, a capacitance, a memory requirement. This need usually can be met in more than one way, and yet many times the engineer may not be aware of the options available. In such a case, a supply management professional may be able to offer suggestions. A gear, for example, might be machined of bronze or steel, die cast in aluminum or zinc, molded from plastic, or formed by powder metallurgy. All these options may meet engineering's constraints while offering a wide range of cost, availability, and reliability choices. Supply management and potential suppliers can provide information on the economy and availability of the materials and subassemblies to be purchased under each approach.

The Internet is playing an increasingly critical role in compressing development time. In 2000, Spin City began providing its customers a service that allowed customers to search electronically for current data sheets, free symbols, pricing, and availability for more than 1 million components. Electronics parts suppliers were able to communicate product and service information directly to the engineer's desktop in real time, reducing costs and development time.[23] Practiced by leading firms today, ESI is a key contributor to the product development process. With ESI, suppliers are carefully prequalified to ensure that they have both the desired technology and the right management and manufacturing capability. Before inviting an outside supplier to participate in the development of a new product, the cross functional development team will ask the following and related questions:

▲ Will the supplier be able to meet our cost, quality, and product performance requirements?

▲ Does the supplier have the required engineering capability?

▲ Will the supplier be able to meet our development and production needs?

▲ Does the supplier have the necessary physical process and quality capabilities required?

▲ Does the supplier have both the resources and the reputation of being able to overcome problems and obstacles as they arise?

▲ Is the supplier financially viable?

▲ Are the supplier's short- and long-term business objectives compatible with ours?

▲ If a long-term relationship appears desirable, are the technology plans of the two firms compatible?

▲ If a long-term relationship appears desirable, is it likely that we can build a trusting relationship?

When a component or subsystem is to be developed by an outside supplier under an ESI program, normally two or three potential suppliers will be requested to design and develop the required item. Potential suppliers are given performance, cost, weight, and reliability objectives and are provided information on how and where the item will fit (interface) in the larger system. These potential suppliers must develop quality plans during the design of the item to ensure that the item can be produced in the quality specified. Selection of the "winning" supplier is a team effort, with supply management, design engineering, reliability engineering, product planning, quality, manufacturing, finance, and field support participating. Performance, quality, reliability, and cost are all considered during the selection process. When a carefully crafted strategic alliance for the item or the commodity class (e.g., fasteners, resistors, safety glass) exists, the alliance supplier alone will be invited to design and develop the required item.

Early supply management and supplier involvement can reduce the well-known start-up problems that occur when the design and the supplier's process capability are poorly matched. Ideally, supplier suggestions will be solicited and the matching of design and manufacturing process will take place during the investigative phase of the design process. The support suppliers provide in the early stages of design is a critical factor in squeezing the material costs out of a product, improving quality, and preventing costly delays.

Make-or-Buy and Outsourcing Analysis

The make-or-buy and outsourcing issues should be addressed for all new items that can be either purchased or produced in-house. Every job release and every purchase request implies a decision to make or to buy. Supply management plays a key role in the make-or-buy process by providing information on the cost, quality, and availability of items. (For more on this critical issue, see Chapter 10.)

Select Components, Technologies, and Supplies

Several options may meet engineering's constraints while offering a wide range of cost, availability, and reliability choices. Supply management professionals and selected suppliers provide information on the availability of the materials and subassemblies to be purchased under each approach. The Internet allows engineers to check for component capability and product attributes in real time. The early involvement of quality engineers allows advanced quality planning to commence in a timely manner. Quality standards are developed to ensure that components and products that are being designed can be produced at the quality specified.

The selection of required standard components is facilitated by the availability of a current internal catalog of standard items and sources that have been prequalified.[24] The use of such a catalog simplifies the design engineer's job while supporting the efforts of engineering or materials management to standardize the items used. The use of standard materials, production processes, and methods shortens the design time and lowers the cost of designing and producing an item. In addition, standardization reduces quality problems with incoming materials, inventories, administrative expenses, inspection, and handling expenses, while achieving lower unit costs.

The selection of technologies is a complex issue due to inherent cost/benefit trade-offs and functional orientations. Engineers are eager to incorporate the latest technology. The marketplace often richly rewards those who are first to market with innovative products; therefore, there is a strong case for incorporating new technology or processes before they are perfected. The cost of such a decision can be high. Not only does such an approach result in a proliferation of components to be purchased and stocked, but it frequently results in the use of items whose production processes have not yet stabilized; quality problems, production disruptions, and delays frequently result, increasing project risk. Engineering, quality, supply management, and manufacturing personnel must ensure that both the costs and benefits of such advanced developments are properly considered. The design team should design new products to the requirements of the customer, not necessarily to the state of the art.

Stress Testing and Failure Analysis

Once candidate component and subsystem items have been identified, they are subjected to stress testing and failure analysis. Failures are caused by failure mechanisms, which are built into the item and then activated by stresses. Studying the basic stresses and the failure mechanisms they activate is fundamental to the design of effective reliability tests. The correct design approach is to find and eliminate the fundamental causes of failure. This means that the most successful stress tests are ones that result in failures. Successful tests are also tailored to look for particular failure mechanisms efficiently by selectively accelerating the tests.

Every failure has a cause and is a symptom of a failure mechanism waiting to be discovered. The tools of failure analysis are both statistical and physical; used together, they are a potent means for detecting the often unique fingerprint of the underlying source of the failure.

The Development Phase

Rapid advances in computer technology and software have made large-scale, complex computer simulations possible. Manufacturers typically conduct extensive computer simulations to identify interferences, fit issues, functionality, algorithmic logic accuracy, and so forth, before the development of prototypes. As the technology continues to advance, computer modeling and simulation may replace prototype development.

Despite these technical advances, breadboard or hardware prototypes commonly are developed so that the design team may conduct tests on the integrated system to eliminate performance and quality problems. The selected approach is reviewed in detail for feasibility and likely risk. Efforts are made to reduce risk to acceptable levels by developing and testing prototypes.

Prototypes

As shown in Figure 5.2, the first complete prototypes of the new product are designed, built, and tested. Documentation such as materials lists, drawings, and test procedures is created. It is not unusual to repeat this phase more than once, perhaps building the first prototype in the laboratory to test the design and the second prototype in manufacturing as a test of the documentation. The design should not exit this phase until a prototype has met all the design goals set for it, although it may not be possible to demonstrate the reliability goal because of the small number of prototypes available for testing.

Design Reviews

The design review is the point at which the new design can be measured, compared with previously established objectives, and improved. Supply management participates in design reviews and provides information on the effect of specifications and the availability of items that are standard production for, or are inventoried by, suppliers. The supply professional must ensure that the specification or other purchase description is complete, is unambiguous, and provides necessary information on how items furnished under it are to be checked or tested. He or she should be satisfied that the purchase description is written in terms relevant to and understandable by potential suppliers.

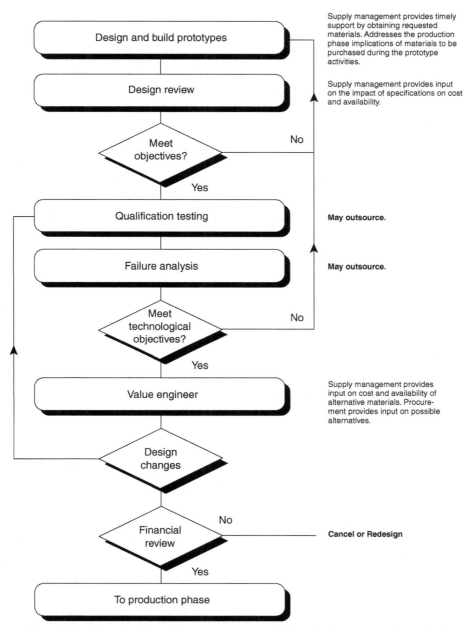

Figure 5.2 Team and supply management activities during new product development *(NPD) development phase*. Adapted from David N. Burt and Richard L. Pinkerton. *A Purchasing Manager's Guide to Strategic Proactive Procurement*. New York: AMACOM, 1996, 28.

Boeing provides an example of how computers have an impact on new product development. It can now design a commercial aircraft entirely by computer. It solves nearly all of its design issues through computer animation, avoiding the need to build physical prototypes. This approach reduces the cost of making design changes during production. The practice cut the time required to design the 777 by 50%.[25]

Qualification Testing

Qualification tests are conducted on the prototype equipment. There are two different types: (1) margin tests and (2) life tests. *Margin tests* are concerned with assuring that the threshold of failure—the combination of conditions at which the product just begins to malfunction—is outside the range of specified conditions for the product's use.

Life tests are intended to find patterns of failure that occur too infrequently to be detected by engineering tests on one or two prototypes. These tests differ from margin tests primarily in the number of units tested and the duration of the test.

Failure Analysis

The stress testing and failure analysis techniques described in the investigation phase are applied to the prototype.

Does the Prototype Meet Objectives?

The design team determines whether the prototype meets the objectives established in the investigation phase. If the prototype fails this analysis, the project reenters the design process and a new or upgraded prototype is developed.

Value Engineering

During World War II, many critical materials and components were difficult to obtain, and most manufacturers were required to incorporate numerous substitutions in their design and production activities. Harry Erlicher, then Vice President of Purchasing for the General Electric Company, observed that many of the substitutions required during that period resulted not only in reduced costs but also in product improvements. Consequently, Mr. Erlicher assigned to Larry Miles the task of developing a systematic approach to the investigation of the function/cost aspect of existing material specifications. Larry Miles not only met the challenge successfully but subsequently pioneered the scientific procurement concept General Electric called "value analysis (VA)."[26]

In 1954, the U.S. Navy's Bureau of Ships adopted a modified version of General Electric's VA concept in an attempt to reduce the cost of ships and related equipment. In applying the concept, the Navy directed its efforts primarily at cost avoidance during the initial engineering design stage and called the program "value engineering" (VE), even though it embodied the same concepts and techniques as GE's VA program. In an operational sense, however, the two terms typically are used synonymously in industry today—only the timing differs. Hence, throughout this book when the term "value analysis" is used, it carries the same conceptual meaning as the term "value engineering," except for the practical matter of timing.

Value Engineering vis-à-vis Value Analysis

As practiced in U.S. firms for many years, VA techniques were most widely used in programs designed to engineer unnecessary costs out of existing products. Finally, the more progressive firms began to follow the Navy's lead by establishing what they too called "value engineering" programs—programs that applied the VA concept during the early stages of the new product design process. Clearly, this is the first point at which it should be applied. This is where the greatest benefits are produced for both the firm and its customers.

What is the mix of VA and VE applications in American industry today? No one really knows. However, the number of both programs has grown markedly in the last decade, with VE programs setting the pace.

The VE concept finds its most unique use in two kinds of companies: those that produce a limited number of units of a very expensive product and those that mass-produce products that require expensive tooling. In these types of companies, VA of an item already in production is often impractical because it is too late to incorporate changes in the product economically. In manufacturing certain electronic instruments used in defense systems, for example, the production run is often so short that it precludes the effective use of VA after production has been initiated. In fact, the Federal Acquisition Regulations now stipulate that most major defense procurement contracts be subjected to VE studies before initial production.[27]

A somewhat different situation that produces similar operating results is used in firms that mass-produce automobiles. For example, in manufacturing the body panel for a car, once the design is fixed and the dies are purchased, it is normally too costly to change them even though VA studies subsequently may disclose design inefficiencies.

VE utilizes all the techniques of VA. In practice, it involves very close liaison work between the supply, production, and design engineering departments. This liaison is most frequently accomplished through the use of product design teams or supply and production coordinators who spend considerable time in the engineering department studying and analyzing engineering drawings as they are produced. Once coordinators locate problem areas, VA techniques are employed to alleviate them.[28]

VE is a systematic study of every element of cost in a material, item of equipment, service, or construction project to ensure that the element fulfills a necessary function at the lowest possible total cost. Ideally, the VE thought process is instilled in all members of the new product design team through appropriate training. (World-class firms provide 40–50 hours of VE training per year to those who would benefit!) The team members (including selected suppliers) apply VE as the product development project evolves. In some instances, value engineers are assigned to the development teams to ensure that these powerful tools are applied.

The inclusion of the VE step in Figure 5.2 is a safeguard: if VE thinking has been incorporated throughout the development process, a separate VE review may not be necessary. However, experience indicates that a VE review at the indicated point will result in significant savings and improved quality and performance. Two tools aid those involved in the VE process:

▲ Design analysis
▲ The VE checklist

Design Analysis

Design analysis entails a methodical step-by-step study of all phases of the design of a particular item in relation to the function it performs. The philosophy underlying this approach is not concerned with appraisal of any specific part per se. Rather, the appraisal focuses on the function that the part or the larger assembly containing the part performs. This approach is designed to lead the analyst away from a traditional perspective, which views a part as having certain accepted characteristics and configurations. Instead, it encourages the analyst to adopt a broader point of view and to consider whether the part performs the required function both as effectively and as efficiently as possible. Both quality and cost are objects of the analysis.

A technique many firms use in analyzing component parts of a subassembly is to dismantle, or "explode," the unit and then mount each part adjacent to its mating part on a pegboard or a table. The idea is to demonstrate visually the functional relationships of the various parts. Each component can thus be studied as it relates to the performance of the complete unit rather than as an isolated element. Analysis of each component in this fashion is done to answer four specific questions:

1. Can any part be *eliminated* without impairing the operation of the complete unit?
2. Can the design of the part be *simplified* to reduce its basic cost?
3. Can the design of the part be changed to permit the use of simplified or less costly *production methods*?
4. Can less expensive but equally satisfactory *materials* be used in the part?

Design simplifications frequently are more apparent than is possible under the original design conditions when viewed from the standpoint of the composite operation. (This in no way reflects unfavorably on the work done initially by the design engineer.) The discovery of such potential improvements is simply the product of an analysis with a substantially broader orientation than that of the original designer. An organized VE study usually utilizes a number of individuals with different types of backgrounds, experience, and skill impossible to combine in the person of a single designer. The resulting design changes often permit the substitution of standardized production operations for more expensive operations that require special setup work. In some cases, considering the volume of parts to be produced, an entirely different material or production process turns out to be more efficient than the one originally specified. Figure 5.3 shows the logic underlying a VE study.

The Value Engineering Checklist

Most companies develop some type of checklist to systematize the VE process. Literally hundreds of questions and key ideas appear on those lists. Some of the checklists are highly specialized for particular types of products. Illustrative of the more general questions is the following checklist.

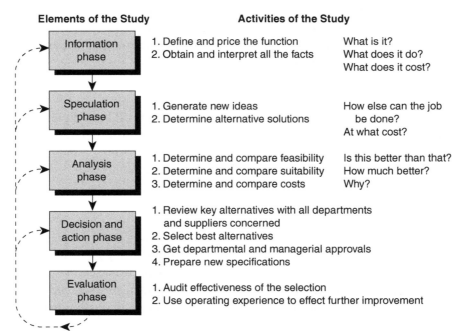

Figure 5.3 A generalized procedural model of the VE process.[29,30]

First, determine the function of the item, then determine:

1. Can the item be eliminated?
2. If the item is not standard, can a standard item be used?
3. If it is a standard item, does it completely fit the application, or is it a misfit?
4. Does the item have greater capacity than required?
5. Can the weight be reduced?
6. Is there a similar item in inventory that could be substituted?
7. Are closer tolerances specified than are necessary?
8. Is unnecessary machining performed on the item?
9. Are unnecessarily fine finishes specified?
10. Is "commercial quality" specified? (Commercial quality is usually more economical.)
11. Can you make the item less expensively in your plant? If you are making it now, can you buy it for less?
12. Is the item properly classified for shipping purposes to obtain the lowest transportation rates?
13. Can cost of packaging be reduced?
14. Are suppliers contributing suggestions to reduce cost?[31]

In using this or similar checklists, those involved evaluate the component under investigation with respect to each item on the checklist. When a question is found to which the answer is not entirely satisfactory, this becomes a starting point for more detailed investigation. The checklist focuses the analyst's attention on those factors that past experience has proved to be potentially fruitful cost-reduction areas.[32]

Viability

Before proceeding to production, a careful business analysis must be completed. In effect, the development team asks: "Will the product provide our firm's required return on its investment?"

The Production Phase

Manufacturing and Production Plans

In the production phase, as shown in Figure 5.4, the manufacturing plan and the procurement plan (frequently in the form of a bill of materials) are finalized. As a result of its early involvement in the design and specification development process, supply management also should have been able to develop contingency plans that will satisfy the firm's needs if the first source doesn't work out. The appropriate plans are now formalized and implemented.

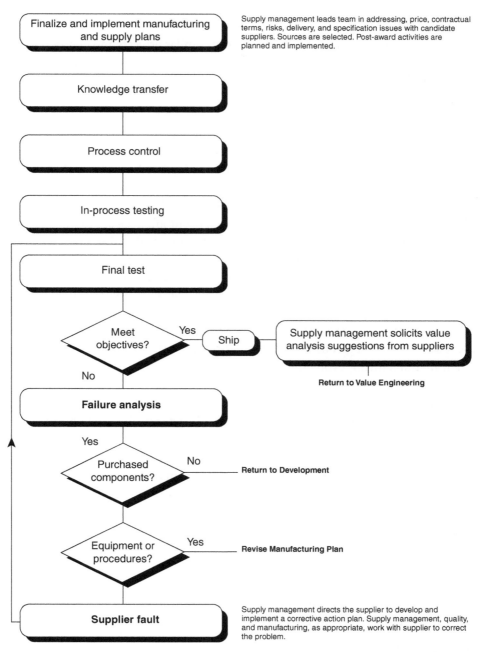

Figure 5.4 Team and supply management activities during new product development (NPD) production phase. Adapted from David N. Burt and Richard L. Pinkerton, *A Purchasing Manager's Guide to Strategic Proactive Procurement.* New York: AMACOM, 1996, 30.

Knowledge Transfer

Manufacturing engineering applies experience from similar projects and new developments from other manufacturers to the firm's production process. Manufacturing engineers also work with suppliers to share new and improved production techniques.

Process Control

Contrary to popular opinion, the design is not finished when the transfer from development to production takes place—quite the contrary. Unfortunately, changes at this stage of product development are very costly and tend not to be evaluated with the same thoroughness as the original alternatives. Finding a quick fix typically is the order of the day, preferably a fix that does not require extensive retooling or scraping. Still, there are some legitimate reasons why changes in the design occur after release to production. For example, there may be phenomena that occur so infrequently that they are not discovered until a large number of products are manufactured. Another reason for changes at this point is the pressure to develop new products in a shorter time. This time compression frequently results in concurrent engineering. This means that a new manufacturing process is developed simultaneously with a new product using that process, as opposed to the more traditional sequential approach. This is a risky approach, but one that is gaining popularity because it saves time and results in earlier new product release.

When manufacturing problems arise, whether in the buying firm or the supplier's manufacturing operations, there is a tendency to look for a quick fix. One type of solution is to adjust the manufacturing process to minimize the problems, rather than to change the design. Perhaps this approach is taken because the process documentation is internal and not shipped to the customer along with the product. More likely, such changes in manufacturing processes are made because the process is under the jurisdiction of production and consequently the change does not require design engineering's approval.

Such an approach can create problems. The situation can deteriorate to the point where there is a customized process for each product—nothing is standard and the process is out of control most of the time. When design rules and process parameters are both being varied at the same time, the situation quickly becomes too complex to understand or control, and quality suffers.

The correct solution is to optimize the process, get it under control, and keep it that way. Then the designs can be modified so that they fit the standard process, producing stable and predictable yields day after day.

In-Process and Final Testing

There are two objectives for in-process testing: (1) to adjust or calibrate the performance in some way and (2) to eliminate defects before much value is added to the product. Final product testing ensures that the item meets its performance objectives.

Every failure has a cause and is a symptom of a failure mechanism waiting to be discovered. For example, if failure analysis identifies a purchased component as the source of the failure, further analysis is required to determine whether: (1) faulty equipment or procedures are to blame or (2) the problem resides with the supplier. Failure analysis also may identify a latent defect in the product's design, requiring redesign.

Engineering Change Management

Any changes in components or the product itself may have profound effects on its cost, performance, appearance, and acceptability in the marketplace. Changes, especially at the component or subassembly level, can have a major impact on manufacturing. Unless changes in the configuration of an item or its components are controlled, manufacturers may find themselves in trouble. They may possess inventories of unusable raw materials or subassemblies. They may possess materials that require needlessly expensive rework to be adapted to a new configuration. They may produce an end item that will not meet the customer's needs. Uncontrolled changes generally mean that quality and reliability requirements have been compromised without appropriate retesting.

Engineering change management, a discipline that controls engineering changes, has been developed to avoid such problems.[33] How often engineering change management is required is a matter of managerial judgment, but for most modern technical items, engineering change management is a necessity. In some cases, it will be imposed on the manufacturer by its customer. When engineering change management is used, changes are controlled and recorded. Marketing and all activities involved in the purchase, control, and use of purchased materials are told of any proposed changes to the item's characteristics. These organizations then comment on the effect of the proposed change. Such control and coordination is especially important when production scheduling and the release of purchase orders are controlled by a material requirements planning system.

There are many ways to organize an engineering change management group. Ideally, an engineering change management board is established with engineering, manufacturing, marketing, production planning, inventory management, and supply management represented. When there is a materials management organization in the firm, a senior representative of production planning and inventory control is a logical candidate to chair this board. It is crucial that supply management and the function responsible for materials control be involved in the review of proposed engineering changes for three reasons: (1) to provide input on the purchased material's implications of a proposed change, (2) to discuss the timing of proposed changes in order to minimize costs associated with unusable incoming materials, and (3) to be aware of forthcoming changes so that appropriate action can be taken with affected suppliers.

Adherence to this or a similar design process is key to the firm's success in the development of new products. Product quality, cost, and availability all must receive proper attention. Engineering, manufacturing, marketing, quality assurance, and supply management all have vital roles to play in the design process.

How to Expand Supply Management's Contributions

This chapter has described the design and development process. In too many firms, the design engineer attempts to address not only the technical and functional issues of design and development, but also manufacturing considerations, marketing implications, and the commercial considerations of economy and availability. Many of these individuals enjoy interacting with suppliers on both technical and commercial issues, and most believe they are serving their employer's best interests even when making sourcing and specifications decisions that turn out to be sub-optimal in the long run.

It is important to note that the effectiveness of early supplier involvement appears to be a function of the industry involved. Researchers McGinnis and Vallopra found that "supplier involvement is not a panacea for every new product development effort." They also found that the potential for supply management's

> …contribution to new product development is substantial. These potentials can be realized if the supply management staff has the abilities to successfully participate in (and lead) multi-functional teams; has the skills needed to identify, screen, and select suppliers to include in new product development; and the competence to manage, control, and coordinate supplier involvement in a multi-functional team environment.[34]

Supply management professionals have their work cut out for them. They must develop and maintain cooperative relations with engineering that protect the profitability of the firm. Early supply management involvement is an essential ingredient of the program to maximize a firm's profitability.

Supply management professionals must understand the orientation and dedication of the typical design engineer. Obviously, an ability to speak the engineer's language (i.e., *engineeringese*) is very helpful. In a study conducted by one of the authors, it was found that supply management personnel who think in the same manner as engineers have a much higher success rate when dealing with engineers than do other individuals. Such thought processes can be identified through established testing procedures.

Whenever feasible, supply professionals should provide advice on the commercial implications of designs under consideration in a positive and constructive manner. They must learn to co-opt their engineering counterparts by providing value and service. Supply management then is seen as a partner who takes care of business problems, thereby allowing engineers to concentrate on technical issues. Several successful approaches to obtaining the desired level of supply management input during the design process are described now.

Design or Project Teams

When the importance of a project or program warrants, a dedicated project team is the ideal means of ensuring early supply management involvement. These teams are often referred to as cross-functional design teams.

Materials Engineers

Individuals with an engineering background are good candidates for supply management positions whose responsibilities require involvement with design engineering. Some supply management organizations divide buying responsibilities into two specialties: (1) materials engineering and (2) the supply management activities of sourcing, pricing, and negotiating. The materials engineer is responsible for coordinating with design engineering, for prequalifying potential sources (usually with the assistance of quality assurance), and for participating in value management.

Co-Location

This approach calls for the placement of members of the supply management staff in locations where design engineering and development work is done. These individuals are available to collaborate with design engineers and others by obtaining required information from prospective suppliers and advising designers on the procurement implications of different materials and suppliers under consideration. When Harley-Davidson opened its product design center in 1997, it co-located design engineering, supply management, manufacturing, marketing, and key supplier personnel. Cross-functional teams are the order of the day. The result? Faster to market, reduced total cost, and improved quality.[35]

Supply Management Professionals Who Interface Successfully with Engineers

The supply professional is the key to successful early supply management involvement in the new product development process. Management directives, policies, and procedures supporting early supply management involvement all help. But it is only when design engineers realize that the early involvement of a supply professional is a productive asset, and not a nuisance or an infringement on their territory, that early supply management involvement makes its full contribution.

A supply management professional who recognizes the importance of being involved early in the process must acquire the necessary skills and knowledge to be seen and accepted as a contributor. Courses in the development and interpretation of engineering drawings, as well as in a wide variety of technologies, can be taken via correspondence, night school, or a few degree-granting programs.[36] Sales personnel love to talk and will gladly help a willing listener gain technical insight into their products. Visits to suppliers' operations provide further insight and understanding.

The design and development of new products is one of a manufacturing firm's most crucial activities. Profitability and even survival are affected. Supply management and the firm's suppliers have major contributions to make during this process. An increasing number of successful firms involve supply management and suppliers up front because of contributions they can make in the areas of quality, cost, and time to market.

Appendix

Simon Croom of the Warwick Business School, Coventry, U.K., and the University of San Diego establishes the need for two types of competencies during product development: operational and relational. He defines operational competencies as those related to the design, manufacture, and delivery of a product. Relational competencies are involved in communication, interaction, problem resolution, and relationship development. Croom demonstrates the importance of relational competencies in achieving successful collaborative product development.[37]

European researchers Finn Wynstra, Bjorn Axelsson, and Arjan van Weele argue that authors in the fields of purchasing, procurement, and supply management have failed to identify many of supply management's activities in product development. These highly respected researchers identify 21 supply management activities related to product development. The interested reader is encouraged to review their provocative articles contained in the summer 1999 and autumn 2000 issues of the *European Journal of Purchasing and Supply Management.*

Handfield, Ragatz, Petersen, and Monczka address the complex issue of evaluating the capabilities of suppliers that are being considered for early involvement in new product development. The interested reader is encouraged to review this insightful article.[38]

Charles Fine, author of *Clockspeed,* advocates that firms design their supply chains strategically and concurrently with their products and production process.

> When firms do not explicitly acknowledge and manage supply chain design and engineering as a concurrent activity to product and process design and engineering, they often encounter problems late in product development, or with manufacturing launch, logistical support, quality control, and production costs. In addition, they run the risk of loosing control of their business destiny.[39]

We agree with Dr. Fine.

Endnotes

1. Michael McGinnis and R. Vallopra, "Purchasing and Supplier Involvement: Issues and Insights Regarding New Product Success," *The Journal of Supply Chain Management*, Summer (1999): 4–15.
2. John E. Ettlie and Paul A. Pavlou, "Technology-Based New Product Development Partnerships," *Decision Sciences*, 37(2)(2006): 117–147.
3. Concurrent engineering is a process in which functional specialists execute their parts of the design as a team concurrently instead of in separate departments serially.
4. Charles O'Neal, "Concurrent Engineering with Early Supplier Involvement: A Cross-Functional Challenge," *International Journal of Purchasing and Materials Management*, Spring (1993): 3–9.
5. Personal interview with Dr. Uli Arnold and Dr. Michael Essig, Stuttgard, Germany, July 1999.
6. "All-in-cost" is a summation of purchase price, incoming transportation, inspection and testing, storage, production, lost productivity, rework, process yield loss, scrap, warranty, service and field failure, and customer returns and lost sales associated with the purchased item. The term "all-in-cost" is similar to "total cost of ownership" and simply "cost." All recognize that the purchase price of an item is merely one component of the total cost of buying, owning, and using a purchased item.
7. Personal interview with Gene Walz, Materials Manager, General Electric, Jet Engine Division, June 1983.
8. Ira Sager, "IBM: There's Many a Slip. . . ," *Business Week*, June 27, 1994, 26–27.
9. "Apple Net May Fall Up to 30% in Quarter," *Wall Street Journal*, September 21, 2000.
10. Chong Leng Tan and Michael Tracey, "Collaborative New Product Development Environments: Implications for Supply Chain Management," *The Journal of Supply Chain Management*, 43(3)(2007): 2–15.
11. Ernest Raia, "The Chrysler Viper: A Crash Course in Design," *Purchasing*, February 20, 1992, 48.

12. Portions of this section are based on *Proactive Procurement: The Key to Increased Profits, Productivity, and Quality* by David N. Burt, Englewood, NJ: Prentice Hall, (1984) and *Proactive Procurement* by David N. Burt and Richard L. Pinkerton (New York: AMACOM, 1996).

13. John A. Carlisle and Robert C. Parker, *Beyond Negotiation: Redeeming Customer-Supplier Relationships,* (Chichester, UK: John Wiley & Sons, 1989): 127.

14. Computer-assisted design/computer-assisted manufacturing.

15. Peter F. Drucker, "The Information Executives Truly Need," *Harvard Business Review* Jan-Feb (1995): 56.

16. Robert B. Hanfield and Gary L. Ragatz, "Involving Suppliers in New Product Development," *California Management Review* Fall (1999): 60.

17. Some world-class firms prepare a report at the end of a new product development project to document the major lessons learned—ones that can be applied to future projects.

18. Peter Tarasewich and Saresli Nair, "Designing for Quality," *Industrial Management,* Jul-Aug (1999): 18.

19. The interested reader is referred to "The House of Quality" by John R. Hauser and Don Clausing, *Harvard Business Review,* May-June (1988): 63–67.

20. Interview with Regis McKenna by Anne R. Field, "First Strike," *Success,* October 1989, 48.

21. Peter Tarasewich and Suresh K. Nair, "Designing for Quality," *Industrial Management,* Jul/Aug (1999): 18.

22. Lisa Ellram, "Cost Reduction: Match the Tool to the Purchase," *Purchasing Today,* October (1998). For more insight into this process, see "Purchasing and Supply Management's Participation in the Target Cost Process," by Lisa M. Ellram, *The Journal of Supply Chain Management,* Spring (2000): 39–51.

23. Personal interview: Pat Guerra, CEO, Spin City, October 10, 2000.

24. This catalog is developed and maintained by the joint efforts of design engineering, reliability engineering, supply management, and manufacturing engineering. It reflects the technical and commercial implications of the items included. The catalog typically classifies components as low, medium, or high risk, in an effort to dissuade design engineers from using high-risk components in new products. The internal catalog is in contrast to a supplier's catalog, which, while simplifying the engineer's efforts to describe an item, places the firm in an unintentional sole-source posture.

25. "The Economy," *Fortune Magazine,* October 2, 2000.

26. The classic book on value engineering and analysis is *Techniques of Value Analysis & Value Engineering,* 2nd ed., by Lawrence D. Miles, McGraw-Hill, 1972. This book is still the best on the subject.

27. For an interesting discussion of how the Department of Defense utilizes value engineering, see "DOD Honors ASD Value Engineering Program," *Skywriter*, August 1991, 7.

28. For a complete discussion of this topic, see D. W. Dobler, "How to Get Engineers and P.A.'s Together," *Purchasing World*, November (1980): 48–51; D. N. Burt, *Proactive Procurement*, (Englewood Cliffs, NJ: Prentice-Hall), 1984, chap. 2; and David N. Burt and Richard L. Pinkerton, *A Purchasing Manager's Guide to Proactive Procurement*, (AMACOM, 1996), chap. 11.

29. Most items perform more than one function—usually a basic function plus several supporting functions. Experience has shown that often the basic function constitutes 20 to 25% of the cost of the item and supporting functions account for the rest of the cost. Consequently, it is important to clearly identify these two types of functions. Use of the FAST (function analysis system technique) diagram approach provides an easy way to organize functions and subfunctions in their logical relationships. Details are available in Carlos Fallon, *VA*, Wiley Inter-Science Publishers, 1991; and Gary Long, *VA/VE Workshop Workbook*, Society of American Value Engineers, September 24, 1993, Phase One and Phase Two.

30. The development of alternative materials and processes is the most challenging, but perhaps the most stimulating phase of VE. Creativity and brainstorming should be encouraged and supported. Professor Alvin Williams and his colleagues suggest a number of other techniques that readers may find helpful. For details see Alvin J. Williams, Steve Lacey, and William C. Smith, "Purchasing's Role in Value Analysis: Lessons from Creative Problem Solving," *The International Journal of Purchasing and Materials Management*, Spring (1992): 37–41.

31. Basic Steps in Value Analysis, a pamphlet prepared under the chairmanship of Martin S. Erb by the Value-Analysis-Standardization Committee, Reading Association, NAPM, Tempe, AZ, 4–18.

32. For an interesting list of suggestions, see Dave A. Lugo, "Boost Your Creativity with Divergent Thinking and Checklists," *NAPM Insights*, May (1994): 12.

33. Engineering change management controls the changes to a product's design—specifically, its form, fit, and function.

34. Michael A. McGinnis, "New Product Development with and without Supplier Involvement: Factors Affecting Success in Manufacturing and Nonmanufacturing Organizations," Purchasing 2000 Conference, Richard Ivey School of Business, 455–461.

35. Personal interviews with Leroy Zimdars, former Director of Product Purchasing, Harley-Davidson, 1997 and 1998.

36. A small but growing number of universities now offer an integrated procurement and engineering management program.

37. Simon R. Croom, "The Dyadic Capabilities Concept: Examining the Process of Key Supplier Involvement in Collaborative Product Development," *European Journal of Purchasing* & *Management*, 7(2001): 29–37.

38. Robert B. Handfield, Gary L. Ragatz, Kenneth J. Petersen, and Robert M. Monczka, "Involving Suppliers in New Product Development," *California Management Review*, 42(1)(1999): 59–82.

39. Charles H. Fine, *Clockspeed* (Cambridge, MA: Da Capo Press, 1998) 133.

Suggested Reading

Tan, Chong Leng, and Michael Tracey. "Collaborative New Product Development Environments: Implications for Supply Chain Management." *The Journal of Supply Chain Management* 43(3)(2007): 2–15.

Crawford, Merle, and Anthony Di Benedetto. *New Products Management*, 7th ed. New York: McGraw-Hill, 2003.

Parker, Delvon B., George A. Zsidisin, and Gary L. Ragatz. "Timing and Extent of Supplier Integration in New Product Development: A Contingency Approach." *Journal of Supply Chain Management* Jan. (2008): 71–83.

Ettlie, John E., and Paul A. Pavlou. "Technology-Based New Product Development Partnerships." *Decision Sciences* 37(2)(2006): 17–147.

Kahn, Kenneth B., Ed. *PDMA Handbook of New Product Development*, 2nd ed. Hoboken, NJ: John Wiley & Sons, 2005.

Karol, Robin, and Beebe Nelson. *New Product Development for Dummies*. Hoboken, NJ: Wiley, 2007.

Lakemond, Nicolette, Christian Berggren, and Arjan vanWeele. "Coordinating Supplier Involvement in Product Development Projects: A Differentiated Coordination Typology." *R & D Management* 36(1)(2006): 55–66.

McGinnis, M.A., and R.M. Vallopra. "Purchasing and Supplier Involvement in Process Improvement: A Source of Competitive Advantage." *The Journal of Supply Chain Management* 35(3)(1999): 42–50.

Nellore, Rajesh. "The Impact of Supplier Visions on Product Development." *The Journal of Supply Chain Management* 37(1)(2001): 27–36.

Petersen, K.J., R.B. Handfield, and G.L. Ragatz. "Supplier Integration into New Product Development: Coordinating Product, Process and Supply Chain Design." *Journal of Operations Management* 23(3-4)(2005): 371–388.

Rainey, David L. *Product Innovation: Leading Change Through Integrated Product Development.* Cambridge, U.K.: Cambridge University Press, 2005.

Rogers, D.S., D.M. Lambert, and A.M. Knemeyer. "The Product Development and Commercialization Process." *International Journal of Logistics Management* 15(1) (2004): 43–56.

Kono, Toyohiro, and Leonard Lynn. *Strategic New Product Development for the Global Economy.* Basingstoke, Palgrave Macmillan, 2007.

Schilling, M.A., and C.W. Hill. "Managing the New Product Development Process: Strategic Imperatives." *Academy of Management Executive* 12(3)(1998): 67–81.

CHAPTER 6

Purchasing Descriptions and Specifications

Specifications and Standardization

Participation by both critical suppliers and supply management in the development of clear specifications and comprehensive standardization is required for an organization to evolve to strategic supply chain management.

Proactive development of specifications and standardization can help an organization reduce the total costs of a product or service developed either in-house or externally. The importance of including supply management in the design process was established in Chapter 5.

In a manufacturing firm, when specifications for a tangible product are fixed, the final design of the product is also fixed. The final design of the product often dictates fixing other costs, such as packaging and required service for the product. Therefore, when the final design is fixed, the product's competitive position and its profit potential are also fixed. As was stated in Chapter 5, it is estimated that 75 to 85% of avoidable total costs are controllable at the design stage. Consequently, early involvement of supplier professionals is essential in a firm's effort to reduce total cost.[1]

Specifications and standardization are two related topics in the field of supply management. Specifications form what is called the purchase description. Standardized parts, components, and services may be included in the purchase description, but standardization goes beyond mere inclusion in a description. Standardization is treated in many companies and supply chains as a philosophy for creating competitive advantage. As will be discussed in this chapter, the development of specifications and standardization requires strategic action as well as tactical vigilance. This chapter discusses specifications and then standardization.

Purposes of Specifications

The purchase specification forms the heart of a procurement. Whether a purchase order or contract will be performed to the satisfaction of the buying organization frequently is determined at the time the specification is selected or written. Purchase specifications serve a number of purposes, including the following:

▲ Communicate to professionals in the supply management department what to buy.
▲ Communicate to prospective suppliers what is required.
▲ Establish the tangible goods to be provided.
▲ Establish the intangible services to be provided, such as warranty, maintenance, and support.
▲ Establish the standards against which inspections, tests, and quality checks are made.
▲ Balance the specification goals of individual departments, relevant suppliers, desired product or service performance, and cost.

Recognition that procurements should be made with an understanding of total cost of ownership (as discussed further in Chapter 13) requires supply managers to consider specifications that go beyond the tangible good or primary service needed. For example, laptop specifications should include the desired warranty and support levels.

Collaborative Development

Development of specifications should be done with cross-functional teams whenever economically justified. Through collaborative interactions of various departmental representatives and relevant suppliers, the specifications output can balance goals that often conflict with one another. Performance goals, such as quality and delivery, should be balanced against cost. Individual department goals should be balanced. Supplier goals should be considered. The balancing concept is illustrated in Figure 6.1. The balancing process is best done in an atmosphere of collaboration and mutual desire to develop specifications outcomes in which "win-win" opportunities are maximized.

As suggested by Figure 6.1, multiple goals are balanced simultaneously. For example, in the design of a DVD player, high quality and timely delivery goals may conflict with cost-containment goals. The objective in collaboratively developing the specifications would be to achieve the quality, delivery, and cost goals simultaneously. Perhaps a supplier suggests that a standard part that the buying firm was unaware of could be used where the original specification used a non-standard part. The standard part would decrease production time, improve quality, and cost less than the non-standard part. Unfortunately, many companies do not pursue balanced specifications through collaborative efforts.

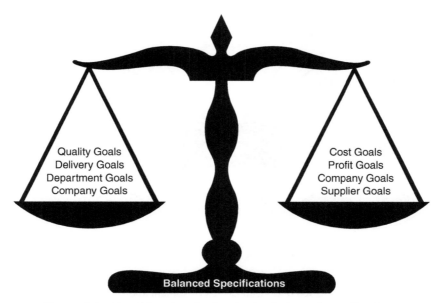

Figure 6.1 Considerations in developing balanced specifications.

Categories of Specifications

Purchase specifications can be classified into two broad categories of simple and complex, also referred to as low detail and high detail. The classification of simple or complex reflects the specification development itself, not the complexity of the product or service or the fulfillment of the specification. Both simple and complex specifications require a balancing of departmental differences, along with quality, delivery, and cost. However, in most cases, simple specifications require less balancing than do complex specifications.

In contrast to simple specifications, complex or detailed specifications are used when a simple specification is not possible or preferable. A complex specification requires more resources and more time to develop. Complex specifications are discussed after simple specifications in this chapter. All categories of specifications are presented in Figure 6.2 because combinations of categories of specifications are possible.

Simple Specifications

Simple specifications require fewer resources and less time to develop than do complex specifications. In many cases, simple specifications are completed in one sentence and have little need for collaboration between functional areas or supply chain members. For example,

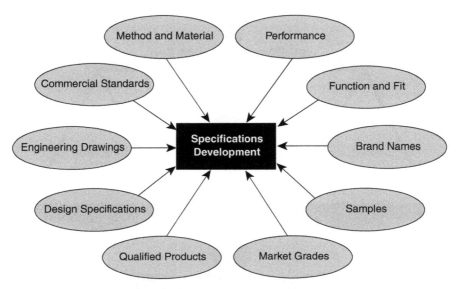

Figure 6.2 Categories of specifications.

the specification of an accounting department for supply management to purchase "12 Dell Latitude, D 630 laptops with their Intel Core 2 duo processor and warranty" is a complete, yet simple specification. Nothing other than the brand name, model and package type, and warranty is needed. The astute reader will recognize that supply management could contribute to the specification by working with accounting to meet its needs with a lower-cost manufacturer or prequalified supplier. The six categories of simple specifications are desired performance, function and fit, brand or trade names, samples, market grades, and qualified products.

Performance Specifications

A performance specification, in theory, is the perfect method of describing a requirement. Instead of describing an item in terms of its design characteristics, performance specifications describe in words—and quantitatively where possible—what the item is required to do. This type of description is used extensively in buying highly technical military and space products. For example, the product wanted could be a missile capable of being launched from a submarine with a designated speed, range, and accuracy. Potential suppliers are told only the performance that is required. Although performance is specified in precise detail, suppliers are not told how the product should be manufactured or what material should be used in its manufacture.

Performance specifications are not limited to complex items such as spacecraft. Electronics, aircraft, and automobile companies, for example, frequently use this method to buy common materials such as electrical wire, batteries, and radios. A performance specification for wire may require it to withstand a given temperature, have a designated resistance to abrasion, and have a particular conductivity capability. No mention is made in the specifications of what materials are to be used or how the wire is to be manufactured or insulated to give it the required characteristics. Manufacturers are free to make those choices as they see fit.

Industry uses performance specifications extensively in buying expensive, complicated machines and machine tools. Today, more production machines are replaced because of technological obsolescence than because of wear. Therefore, in buying such a machine, a firm should make every effort to obtain the ultimate in technological advancement. Often this can be done best by using performance specifications. To reduce and control the expense associated with this approach to describing requirements, descriptions should be written as explicitly as possible. Also, the product being purchased should be sectionalized into the greatest practical number of distinct components, with potential sellers required to provide a quote for each component. This practice helps solve the difficult problem of comparing sellers' prices by allowing comparison of individual components.

There are two primary advantages to describing quality by performance specifications: (1) ease of preparing the specifications and (2) assurance of obtaining the precise performance desired. For complex products, it is by far the easiest type of specification to write. It assures performance, and if the supplier is competent, it assures the inclusion of all applicable new developments. The clarity of a performance specification also brings clarity into any legal or liability issue that may ensue if the supplier does not meet the specification as agreed. A potential disadvantage occurs when the performance specification is out of date given current technology. For example, a late-model computer hard drive will often cost more than a current technology hard drive that stores more, has faster access rates, and a lower price.

Proper supplier selection is essential when performance specifications are used. In fact, the ability to select capable and honest suppliers is a prerequisite to the proper use of performance specifications. Because the supplier assumes the entire responsibility for designing and making the product, quality is entirely in its hands. If the supplier is not capable, it cannot apply the most advanced technical and manufacturing knowledge. If it is not honest, materials and workmanship may be inferior. When using performance specifications, supply managers must solicit competition among two or more capable sellers. Capable suppliers ensure quality; competition ensures reasonable prices.

Function and Fit Specifications

These purchase descriptions are a variation of performance specifications and are used in *early supplier involvement (ESI)* programs. With this approach, the design team describes the function to be performed and the way the item is to fit into the larger system (e.g., automobile, computer), together with several design objectives (cost, weight, and reliability).

Robert May supports the argument that ESI best meets the needs of companies by giving suppliers performance specifications. According to May, "The optimal use of supplier's special skills and processes is experienced when suppliers are provided with a set of performance specifications."[2]

As ESI becomes more common, this approach to describing requirements undoubtedly will increase in popularity. With careful prequalification of suppliers, there are no significant disadvantages with this approach.

Brand or Trade Names

When manufacturers develop and market a new product, they must decide whether to brand it. Branding or differentiating a product is generally done to develop a recognized reputation and thus gain repeat sales, protect the product against substitutes, maintain price stability, and simplify sales promotion.[3] The primary reason most manufacturers brand their products is to obtain repeat sales. Consumers develop a preference for brands. Therefore, branded products generally can be sold at higher prices than unbranded products of similar quality. A brand represents the manufacturer's pledge that the quality of the product will be consistent from one purchase to the next. A supply manager can be certain that a reputable manufacturer will strive to keep that pledge.

Brand name products are among the simplest to describe on a purchase order. Thus, they save time and reduce supply management expense. Inspection expense is also low for branded products. The only inspection required is sight verification of the brand labels. The brand is the quality ordered. The higher prices usually paid for name brands thus are offset to some extent by reduced description preparation and inspection costs.

A supplier's success in maintaining a consistent quality level is greatest in situations in which production and quality control are under its own supervision. If a supplier buys an item from several manufacturers, the quality variation in all probability will be larger than would be the case if the supplier made the item or bought it from a single source. For this reason, it is important for supply managers to know who is responsible for the production and quality control of all branded products that the company buys. In situations where tight quality control is essential, multiple sources of production should be avoided if possible.

It is often said that when a supply manager purchases by brand name, he or she eliminates competition by limiting the purchase to a single source of supply. If a supply manager had to limit purchases to a single brand from a single source, that would represent a major disadvantage of purchasing by brand name. In fact, however, there are very few situations in which only one brand is acceptable for a particular purpose. A profitable market for any item in a competitive enterprise economy attracts other manufacturers to make that item. Competition, therefore, is available by brands just as it is by other types of quality descriptions. In addition, the same branded product may be available from different wholesalers or jobbers who are willing to compete on price and service to get a buying company's order.

Making a specification of "brand A or equal" on the bid forms usually ensures competition among brands. What does "or equal" mean? This question generates many arguments. Realistically, it means materials that are of equal quality and are capable of performing the function intended. Equal quality means similar quality of materials and similar quality of workmanship. Comparing the quality of materials is relatively easy, but comparing the quality of workmanship is particularly difficult. Here nebulous considerations such as precision of production, fit with and matching of adjacent parts, types of finish, and shades of color must be resolved. The key to the "or equal" consideration is "Can the 'equal' perform the function for which the specified brand is desired?" If it cannot, it certainly is not equal; if it can, it is equal.

One practical way of resolving the "or equal" problem is to let the using department decide which products are equal before prices are solicited. Only companies whose products are accepted by using departments as equal are requested to submit prices. This technique helps avoid wounded feelings among potential suppliers. It also permits requisitioning departments to make more objective decisions.

In some situations, purchasing by brand name can be made more effective by including additional references or limitations in the purchasing description. For example, if the supply manager suspects that other materials can perform the desired function, reference in the description should give prospective suppliers an opportunity to offer such other materials for consideration. When limitations concerning physical, functional, and other characteristics of the materials to be purchased are essential to the buying company's needs, they should be set forth clearly in the brand name description. For example, in many purchases of equipment, interchangeability of repair parts is essential. When this is the case, the limitation just described should be spelled out in the brand name description. The invitation for bids or requests for proposal should include the right to examine and test the proffered item should an "or equal" product be offered.

For small quantities, brand buying is excellent.[4] The primary disadvantage of purchasing by brands frequently is higher price. Many categories of branded items sell at notoriously high prices. Antiseptics and cleaning compounds are common examples of such items. For these products, another type of purchase description is preferable. When purchased with detailed or performance specifications, savings often exceed 50%. In recent years, buying drugs by generic name rather than by brand has resulted in spectacular savings for many hospitals; savings up to 70% are not uncommon.

Samples

Samples have been called the lazy person's method of describing requirements. When samples are used, the supply manager does not have to look for an equal brand, pick a standard specification, or describe the performance wanted. Samples are neither the cheapest nor the most satisfactory method of purchase. Usually the money spent on inspection costs substantially exceeds the money saved in description costs. It usually is difficult to determine by inspection that the product delivered is in fact the same as the sample. Quality of materials and quality of workmanship are generally exceedingly difficult to determine from routine inspection. Therefore, in many cases, acceptance or rejection becomes a matter of subjective judgment.

Samples generally should be used only if other methods of description are not feasible. Color and texture, printing, and grading are three broad areas in which other methods of description are not feasible. A precise shade of green, for example, is difficult to describe without a sample. Proposed lithographic work is best judged by the supplier's proofs. Establishing grades for commodities such as wheat, corn, and cotton through the use of samples has proved to be the best method of describing these products.

Market Grades

Grading is a method of determining the quality of commodities. A grade is determined by comparing a specific commodity with previously agreed on standards. Grading is generally limited to natural products such as lumber, wheat, hides, cotton, tobacco, food products, and so on. The value of grades as a description of quality depends on the accuracy with which the grades can be established and the ease with which they can be recognized during inspection. There are, for example, thirteen grades of cotton, each of which must be determined from an examination of individual samples. Trade associations, commodity exchanges, and government agencies all expend great effort in establishing and policing usable grades.

In buying graded commodities, industrial supply managers often use personal inspection as a part of their buying technique. Just as individuals select by inspecting the shoes, dresses, and shirts they buy, industrial supply managers select by inspecting some of the commodities they buy in primary markets. There can be a significant difference between the upper and lower grade limits of many commodities. The difference is so great in some cases that materials near the lower limit of the grade may be unacceptable. Hence, inspection is critically important in buying many materials by market grade. Brewers and millers, for example, usually inspect all the grains they buy. Inspection is necessary if they are to obtain raw materials of the quality needed to produce a finished product of consistent quality.

Beef is an excellent illustration of the wide quality spread that can exist within a grade. Normally, 700-pound steers dressed and graded as "U.S. Prime" have a spread of roughly 40 pounds in fat content between the beef at the top of the grade and beef at the bottom of the grade. Such a wide spread may be a minor consideration to the purchaser of a 1-pound steak. However, to the industrial food service manager buying millions of pounds of beef, the difference can be thousands or hundreds of thousands of dollars.

Qualified Products

In some situations, it is necessary to determine in advance of a purchase whether a product can meet specifications. These situations normally arise when: (1) it takes too long[5] to conduct the normal post-purchase inspections and tests that are required to ensure quality compliance, (2) inspection to ensure compliance with the quality aspects of the specifications requires special testing equipment that is not commonly or immediately available, and (3) the purchase involves materials concerned with safety equipment, life survival equipment, research equipment, or materials described by performance specifications.

When advance qualification is indicated, suppliers are prequalified by a thorough review and test of the entire process by which they ensure compliance with their specifications. After qualification, the products of the approved suppliers are placed on what is called a *qualified products list (QPL)*. Trade name, model number, part number, place of manufacture, and similar identifying data describe approved products on the QPL.

Complex Specifications

Complex or detailed specifications are descriptions that tell the seller exactly what the buyer wants to purchase. A simple specification for buying ketchup might be "12-ounce plastic bottle of Heinz tomato ketchup." In contrast, ketchup specifications become complex if the actual recipe is given with ingredients and production procedures. A complex specification often goes beyond the design of a product to include specifications regarding methodology, packaging, transport, delivery schedules, warranty, and service.

There are four principle types of complex specifications: commercial standards, design specifications (generally accompanied by engineering drawings), engineering drawings, and material and method-of-manufacture specifications.

Commercial Standards

Recurring needs for the same materials have led industry and government to develop commercial standards for those materials. A commercial standard is nothing more than a complete description of the item that has been standardized. The description includes the quality of materials and workmanship that should be used in manufacturing the item, along with dimensions, chemical composition, and so on. It also includes a method for testing both materials and workmanship. Commercial standards are a cornerstone of the mass production system; therefore, they are important to efficient supply management and to the standard of living in the United States.

All nuts, bolts, pipes, and electrical items that are made to standard specifications can be expected to fit all standard applications regardless of who manufactured the item. Materials ordered by standardized specifications leave no doubt on the part of either the buyer or the seller as to what is required. Standard specifications have been prepared for many goods in commercial trade. National trade associations, standards associations, national engineering societies, the federal government, and national testing societies all contribute to the development of standard specifications and standard methods of testing. Commercial standards are applicable to raw materials, fabricated materials, individual parts and components, and subassemblies.

Purchasing by commercial standards is somewhat similar to purchasing by brand name. In both methods, the description of what is wanted can be set forth accurately and easily. Commercial standards are more complex because they require greater detail in the description. With the exception of proprietary products, most widely used items are standard in nature; hence, they are highly competitive and readily available at reasonable prices. There are many users of standard products; therefore, manufacturers that make them can safely schedule long, low-cost production runs for inventory. They do not need specific sales commitments before production. They know that materials will be ordered under these standard specifications when they are needed.

Inspection is only moderately expensive for materials purchased by commercial standards. Commercial standard products require periodic checking in addition to sight identification to assure firms that they are getting the quality specified.

Commercial standard items should be used whenever possible. They contribute greatly to the simplification of design, supply management procedures, and inventory management, as well as to cost reduction. Copies of standard specifications can be obtained from a number of government, trade association, and testing association sources. In fact, the easiest way to get a particular specification is to ask a manufacturer to provide a copy of the standard specification of the material or product that it recommends for the supply manager's intended need.

Design Specifications

Not all items and materials used in industry are covered by standard specifications or brands. For many items, therefore, a large number of buying firms prepare their own specifications. By doing this, those firms broaden their field of competition. All manufacturers capable of making the item described in the firm's specifications are potential suppliers.

By preparing its own specifications, a company often can avoid the premium prices of brand name items and the sole source problems of patented, copyrighted, and proprietary products. When preparing its own specifications, a company should attempt to make them as close as possible to industry standards. If any special dimensions, tolerances, or features are required, every effort should be made to attain those "specials" by designing them as additions or alterations to standard parts; this will save time and money.

Describing requirements with chemical or electronic specifications, or with physical specifications and accompanying engineering drawings entails some risk. For example, if a buying company provides the exact chemical specifications of the paint it wants, it assumes complete responsibility for the paint's performance. If the paint fades in the first month, it is the buyer's responsibility. If a buying company specifies for a metal fabricator the exact dimensions wanted in a part, the buyer assumes all responsibility for the part's fitting and functioning. If it happens that to fit and function properly, a part must be 26.045 in. long rather than 26.015 in. as specified in the purchase order, the responsibility for failure rests solely with the buying firm.

The very nature of the materials purchased under this method of description tends to require special inspection. The cost of inspection to assure compliance with company-prepared specifications can be high.

Engineering Drawings

Engineering drawings and prints occasionally are used alone, but more typically are used in conjunction with other physical purchase descriptions. Engineering drawings may be part of the design specifications described previously. Where precise shapes, dimensions, and spatial relationships are required, drawings are the most accurate method of describing what is wanted. Despite their potential for accuracy, exceptional care must be exercised in using them. Ambiguity, which sometimes is present in this method of description, can produce costly repercussions. All dimensions, therefore, must be completely covered, and the descriptive instruction should be explicit.

Engineering drawings are used extensively in describing quality for construction projects, foundry and machine shop work, and a myriad of special mechanical parts and components. There are four principle advantages to using drawings for description: (1) they are accurate and precise, (2) they are the most practical way of describing mechanical items that require extremely close tolerances, (3) they permit wide competition (what is wanted can easily be communicated to a wide range of potential suppliers), and (4) they clearly establish the standards for inspection.

Material and Method-of-Manufacture

When this method is used, prospective suppliers are instructed precisely about the specific materials to be used and how they are to be processed. The buying firm assumes full responsibility for product performance. Further, the buying firm assumes that its own organization has the latest knowledge concerning materials, techniques, and manufacturing methods for the item being purchased. In this case, purchasers see no reason to pay another company for that knowledge.

Material and method-of-manufacture specifications are used extensively by the armed services and the Department of Energy. A modified version of those specifications is used by industry. Large purchasers of paint, for example, frequently ask the manufacturers of a standard paint to add or delete certain chemicals when producing paint for them. Purchasers of large quantities of steel make the same type of request when purchasing special steels. Chemical and drug buyers, for reasons of health and safety, sometimes approach full use of the material and method-of-manufacture technique in describing quality. Also, these specifications are used most appropriately in situations in which technically sophisticated large companies deal with small suppliers that have limited research and development staffs. Normally, however, this technique is used infrequently in industry because it puts such great responsibility on the buying firm. It can deny a company the latest advancements in both technical development and manufacturing processes. Specifications of this type are expensive to prepare; inspection generally is very expensive.

There are two important features of this method of description. First, the widest competition is possible, and thus good pricing is assured. Second, since the product is nonstandard, the provisions against discrimination of the Robinson-Patman Act pose no barrier to obtaining outstanding pricing and service.

Combination of Methods

Many products cannot be described adequately with the use of a single method. In such cases, a combination of two or more methods should be used. For example, in describing the quality desired for a space vehicle, performance specifications could be used to describe numerous overall characteristics of the vehicle, such as its ability to withstand certain temperatures, perform certain predetermined maneuvers in space at precise time sequences, and stay in space for a specific period of time. Physical specifications could be used to describe the vehicle's configuration as well as the television cameras and other instruments it will carry. Commercial standards or brand names might be used to describe selected pieces of electrical or mechanical hardware used in the vehicle's support systems. A chemical specification could be used to describe the vehicle's paint. Finally, a sample could be used to show the color of that paint.

Few products are as complex as space vehicles; nevertheless, an increasing number of industrial products require two or more methods of quality descriptions. For instance, something as commonplace as office drapes could require chemical specifications to describe the cloth and fireproofing desired, physical specifications to describe the dimensions desired, and samples to describe the colors and texture desired.

Development of Specifications

Developing specifications can be a difficult task because it involves many variables, including the problem of conflicting human sensitivities and orientations. Many departments are capable of contributing to specifications development; however, they frequently are thwarted from fully doing so because of conflicting views. Before the optimum in design can be achieved, these major conflicting views must be reconciled.

Organizational Approaches

Several approaches to developing balanced specifications are used individually or jointly by most companies. The approaches, in order of collaborative orientation from lowest to highest, include informal, supply management coordinator, early supply management involvement (formerly EPI), early supplier involvement (ESI), consensus development, and cross-functional team (CFT).

Informal Approach

The informal approach emphasizes the concept of a supply manager's responsibility to "challenge" materials requests. At the same time, top management urges designers to request advice from supply managers and work with them on all items that may involve commercial considerations. The emphasis at all times is on person-to-person communication and cooperation between individual supply managers and designers. When this approach is used, a company-oriented, cost-conscious attitude is developed at the grass roots level throughout the organization.

There are two potential problems with the informal approach. The first and most obvious is that the lack of formalization through corporate policy or organizational structure may render the supply manager powerless and make the approach completely ineffective. The second problem is that the supply manager may create animosity when it is inappropriate to challenge a specification.

Supply Management Coordinator Approach

One or more positions are created in the supply department for individuals, frequently called *materials engineers*,[6] to serve in a liaison capacity with the design department.

Typically, the materials engineer spends most of his or her time in the engineering department reviewing design work as it comes off the drawing boards. The materials engineer searches for potential supply management problems in an attempt to mitigate them before the specifications have been completed.

The supply management coordinator approach is highly structured as well as expensive. It also is very effective. Therefore, it should be used whenever coordination problems stemming from the technical nature of a firm's product or the magnitude of its cost justify the investment.

Early Supply Management Involvement

As we discussed in Chapter 5, progressive firms increasingly are creating design policies to involve supply management in the early stages of new product development. Early supply management involvement was popularized in industry through the now-dated term EPI, which stands for early purchasing involvement.

Too often design engineers and production engineers resolve between themselves all four of the major departmental considerations of specifications preparation without consulting supply management. This is regrettable because professional engineers seldom have the commercial experience and market information required to resolve the supply management considerations of specifications. In their attempts to do so, they frequently develop stringent specifications that do not provide sufficient latitude to allow effective competition.

Early Supplier Involvement

ESI is used widely in industry. To implement ESI properly, a buying company should start by establishing the policy of involving supply management in the design process. After that policy is enacted, ESI can be actively engaged. ESI coupled with early supply management involvement can improve product quality and reliability, while compressing development time and reducing total material cost.[7]

Consensus Development Approach

Consensus development calls for specifications to be agreed on by the department managers. This collaborative approach falls short of developing a formal team. Although department managers disagree occasionally, compromise and consensus usually can be worked out when the various aspects of the problem are understood and the organizational mechanism for reaching consensus has been established. When specification conflicts arise and consensus cannot be reached, final authority for the decision should rest with the department that has responsibility for the product's performance.

Cross-Functional Team Approach

The CFT approach recognizes that a good specification is a compromise of basic objectives. A specifications CFT is established, with representatives (as appropriate) from design engineering, production engineering, supply management, marketing, operations (including production control), quality, and standards. As described in Chapter 5, members of the design team are involved as appropriate throughout the development of the product and its specifications. A common variation to this approach is for the development of the specification to be delegated to an appropriate technical expert with the resulting specification being reviewed and approved by the CFT.

Supply Management Research

Once a need has been identified and functionally described and when the size of the contemplated purchase warrants, supply management research and analysis should be conducted to investigate the availability of commercial products able to meet the need. This research and analysis also should provide information to aid in selecting a strategy appropriate to the situation. Supply management research and analysis involves obtaining the following information, as appropriate:

▲ The availability of standard products suitable to meet the need (with or without modification).

▲ The terms, conditions, and prices under which such products are sold.

▲ Any applicable trade provisions, restrictions, or controlling laws.

▲ The performance characteristics and quality of available products, including quality control and test procedures followed by the manufacturers.

▲ Any costs or problems associated with integration of the item with those currently used.

▲ Industry production practices, such as continuous, periodic, or batch production.

▲ The distribution and support capabilities of potential suppliers.

Writing Specifications

After the design of a product is determined, the next step is to translate the individual part and materials specifications into written form. Optimal performance in all departments is contingent on good specifications. To meet the needs of all departments, a specification must satisfy many requirements:

▲ Design and marketing requirements for functional characteristics, chemical properties, dimensions, appearance, etc.

▲ Manufacturing requirements for workability of materials and manufacturability.

▲ Inspection requirements to test materials for compliance with the specifications.

▲ Store requirements to receive, store, and issue the material economically.

▲ Supply management's requirement to procure material without difficulty and with adequate competition from reliable sources of supply.

▲ Production control's and supply management's requirement to substitute materials when such action becomes necessary.

▲ The total firm's requirements for suitable quality at the lowest overall cost.

▲ The total firm's requirement to use commercial and industrial standard material whenever possible, and to establish company standards in all other cases where non-standard material is used repetitively.

Common Problems

Several common problems frequently arise in companies after specifications have been developed. Three of these problems are discussed next. If present, they should be addressed in the specifications development process, before the specifications have been completed.

Lack of Clarity

Specifications should be written in clear and unambiguous terms. Clarity in written expression is not always easy to achieve. One company lost $65,000 on a closed-circuit television installation because its written specifications misled the supplier into believing that a more expensive installation than the buyer really wanted had been specified.

Limiting Competition

In addition to achieving clarity, care must be exercised to ensure that specifications are not written around a specific product in a way that limits competition. Several years ago, a fire chief wrote into the specifications for a new fire truck the requirement that the supplier of the truck manufacture the truck's twelve-cylinder engine. This completely restricted competition because only one supplier of fire engines manufactured twelve-cylinder engines in its own plant. If the fire chief specified what was wanted in terms of performance characteristics, such as speed and acceleration, competition would have been plentiful. This example typifies one of industry's most common forms of specifications abuse; slanting specifications to one supplier's product, thus reducing or precluding competition. In this case, fortunately, the situation had a happy ending. The management department challenged the specifications, and the fire chief agreed to rewrite them in a form that permitted maximum competition. A significant savings resulted from that change.

Unreasonable Tolerances

Specifying an unreasonable tolerance is another common specification difficulty. Unnecessary precision pyramids cost! It costs more to make materials to close tolerances, it costs more to inspect them, and more rejects typically result. The best method of avoiding such unnecessary costs is to adhere to the most economical method of manufacture while using standard specifications wherever possible. For example, in procuring 1000 drive pulleys for use in vacuum motors, the first decision would be to determine whether a casting process could manufacture the pulleys satisfactorily. Although this method dictates the use of looser tolerances, in large volumes the unit cost is considerably lower than that of a second alternative: machining the pulley from bar stock. The second decision would be to select an industrial standard for the part regardless of the method of manufacture used. This decision leads directly to a consideration of standardization.

Unreasonable Tolerances Example

Ben Rogers, a Ph.D. student in Production Operations Management, was excited to start gathering research data at a manufacturer of heavy construction equipment in Illinois. The goal of the research was to develop a model to forecast product costs on the basis of specifications given in designs. Yesterday, he met the managers in operations, design, and quality to decide what data to accumulate. Today, he hoped to start the data accumulation process with a senior design engineer, Keith Sampson.

As Ben walked up to the engineering department, he heard a loud and angry argument ensuing between Keith and Gary Hamm, a production manager. In an agitated tone Gary said, "Keith, we are on the floor reworking another non-standard bore from our supplier that once again did not meet the tolerances your group specified. For four years these tolerances have given us headaches. The tighter tolerances on the bores sure haven't reduced the complaints from customers about the road grader. Heck, there weren't any complaints that I can recall that were ever related to the old tolerances for this part anyway. I still don't understand why you never consulted our department on the change to begin with. One thing is for sure, the tighter tolerance has increased the frustration on the factory floor, angered our supplier who we keep charging for the reworks, and screwed up my schedule so I continuously miss due dates. I am fed up, it's time you changed this tolerance back to the old standard!"

Keith replied in a terse voice, "Gary, you know that to compete we need to continuously improve quality and that means tightening tolerances. There is no way I am going to change the design again! If your supplier is incapable of producing to our specifications, then dump the supplier. You are the operations manager, so start managing your operations and deal with it!"

Gary bumped Ben as he stormed out of Keith's office. Ben quietly walked on by, deciding that this was not the time to ask Keith for data. As Ben walked away, he wondered if he had just had a foretaste of the problems he would discover in the next six months.

This true story[8] illustrates what can happen when internal functions do not work together to develop specifications. The over-specification problem is a common one that has been discussed for the last 30 years, but still exists today in most companies. Regardless of the method used to describe specifications, only the minimum quality needed for the product to perform the function intended should be specified. Over-specifying and including restrictive features in purchase descriptions cause delays and increase costs.

The importance of developing balanced specifications and standards though interfunctional and relevant chain member participation is paramount for companies that need to improve their competitive position. The balancing act is accomplished by meeting the needs of the functional areas while balancing performance measures, such as quality and delivery, against cost.

Standardization

A uniform identification that is agreed on is called a *standard*. In business practice, the concept of standardization is applied in either industrial or managerial standardization. *Industrial standardization* can be defined as the process of establishing agreement on uniform identifications for definite characteristics of quality, design, performance, quantity, service, and so on. *Managerial standardization* deals with things such as operating practices, procedures, and systems.

History of Standardization

Eli Whitney contributed to the development of standardization in 1801, when he accepted a contract to furnish 10,000 muskets to the U.S. government. When it appeared that Whitney had fallen behind on his contract, he was summoned to Washington by Thomas Jefferson to explain the delay. Whitney took with him a box containing the parts of ten muskets. On a table before his congressional interrogators, he separated these parts into piles of stocks, barrels, triggers, firing hammers, and so on. He asked a representative to pick a part from each pile. Whitney then assembled these parts into a finished musket, repeating the process until all ten muskets had been assembled. After his demonstration, it was easy for Whitney to explain the apparent delay. Rather than furnishing a proportional number of guns each month, as an artisan gunsmith would have done after individually making the parts for each gun and then assembling each gun in turn, Whitney had been working to design machine tools and dies with which he could mass produce parts that were interchangeable with each other. He had standardized the parts. When his machine tools were completed, he was able to produce the muskets in a period during which an artisan gunsmith could have produced only a few muskets.

Whitney discovered that when parts were standardized, the skills of artisans could be transferred to machines that could be operated by less-skilled labor. This in turn reduced the need for highly-skilled labor, which at the time was in extremely short supply. His discovery revolutionized technology and led to the development of industrial standards.[9] Even more important, this practice introduced mass production and sizable industrial growth to the United States.

The burning to the ground of Baltimore's business district in 1904 clearly illustrates the need for standards in urban living. Like the battle that was lost for the lack of a horse-shoe nail, Baltimore was lost for the lack of standard fire hose couplings. Washington, New York, and Philadelphia all responded to Baltimore's cry for help. When their pumping equipment arrived in Baltimore, however, the rescuers just stood by helplessly because there was no way to connect the different-sized hose couplings to Baltimore's fire hydrants.

Eli Whitney introduced mass production in the United States; Henry Ford made it universal. Ford, however, misinterpreted one important relationship of standardization to mass production: He visualized mass production to mean a standard product produced on an assembly line. Ford thought he was speaking correctly when he said, "The customer can have any color car he wants as long as it is black." Actually, he missed the full implication of mass production. Mass production is the production of many diverse products, assembled from standardized parts that have been mass-produced.

Standardization has become a way of creating competitive advantage through mass customization. Perhaps no company today exemplifies this more than Dell, the largest assembler and seller of personal computers in the world. Dell works with many suppliers to design and produce parts, components, and modules that can be used with multiple models. For example, several laptop models use the same DVD player module. A supply chain standard design was developed to allow interchangeability of several suppliers' modules and ease of installation and support.

Types and Sources of Standards

In industry, there are three basic types of materials standards: (1) international standards, (2) industry or national standards, and (3) company standards. If a designer or user cannot adapt a national or international standard for his or her purpose, the second choice is to use a company standard. If the required part is truly a non-repetitive "special," then use of a standard is impossible.

Where can one get standard specifications? Specifications for items that have been standardized can be obtained from the organizations that have developed them, such as the following:

▲ International Organization for Standardization
▲ National Bureau of Standards
▲ American National Standards Institute
▲ American Society for Testing and Materials
▲ American Society for Quality
▲ Society of Automotive Engineers
▲ Society of Mechanical Engineers
▲ American Institute of Electrical Engineers
▲ Federal Bureau of Specifications
▲ National Lumber Manufacturers' Association

The European-based International Organization for Standardization (ISO) has several hundred specialized committees that develop a wide variety of standards that are promulgated by ISO and usually are accepted worldwide. Many of these standards are adaptations of standards from the American National Standards Institute (ANSI), the German Institute for Standards, the British Standards Institute, and other national standards organizations around the world.

A catalog of U.S. standards, international recommendations, and other related information is published annually and distributed without charge by the ANSI.[10] The Institute is a federation of more than 100 nationally recognized organizations, trade associations, and technical societies, or groups of such organizations. Its members can gain ANSI assistance in developing any standard desired. Recommendations for establishing a standard can be made at any time. If, after appropriate research and debate, ANSI approves the recommended standard, it will be adopted as a U.S. standard.

Both the civilian and military departments of the U.S. government participate in standardization work that greatly assists industry. For example, the National Bureau of Standards (NBS), among other things, was established to serve any firm, corporation, or individual in the United States engaged in manufacturing or other pursuits regarding the use of standards.

The need for international standards is fundamental; by eliminating technical trade barriers, international standards facilitate increased international trade and prosperity. The ISO 9000 series of quality standards, now used voluntarily worldwide, is a good illustration. The economic stakes associated with the development of international standards are so high in terms of increased international trade and prosperity that progress, albeit slow, is inevitable. Because private organizations, national and regional governments, and other international organizations are all involved in the adoption process, political infighting is inescapable.

Metric system measurements are among the important international standards. In December 1975, Congress passed the Metric Conversion Act, which provided for only voluntary action. The voluntary conversion appears to be working, but at a very slow pace. In many industries, the United States has already gone metric. The shutters of thousands of 8-, 16-, and 35-mm cameras daily click across America. Work is done daily in hundreds of repair shops on thousands of automobiles manufactured to metric standards. U.S. pharmaceutical companies went metric over 20 years ago, and the electronics industry has used the metric system since 1954.

More recently, in 1996, the International Organization for Standardization adopted ISO 14000 to establish environmental performance standards. Conflicting environmental regulations across national borders have long been a problem for international supply management. Like the ISO 9000 series, the ISO 14000 series focus on processes, not outcomes and both involve audit by a third party.[11]

Benefits of Standardization

Standardization benefits an organization in a variety of ways: it enables mass production, customization, simplification, and delayed differentiation, and improves supplier coordination and quality, and because of many other benefits, it lowers inventories.

▲ *Enables mass production* As the example of Eli Whitney showed, mass production becomes possible through the creation of interchangeable parts. Standardized parts and components enable management to stabilize production processes and focus on continuous improvement, thereby reducing costs.

▲ *Enables customization* Standardized parts and modules enable manufacturers to make a wide variety of finished products from a relatively small number of parts. With standardization, a wide variety of finished products may be assembled when ordered, thereby reducing inventory carrying costs and increasing flexibility to meet specific consumer demands. Dell exemplifies this in its ability to customize computers for customers on the same day the order is placed. Dell accomplishes customization largely based on standard components and modules.

▲ *Improves supplier coordination* Standardized parts and components provide a very clear specification for the supplier. The dimensions, characteristics, and performance of a standard part or component improve the ability to communicate between the buying and selling companies.

▲ *Improves quality* Standard parts and components are repetitively manufactured to the same design, enabling investment by the producing company in better machinery, training, and materials. The result is a significantly lower defect rate.

▲ *Enables simplification* Once standard parts are identified, simplification can be used to identify redundant standard parts that can be eliminated. Simplification is discussed in the next section.

▲ *Enables delaying differentiation* When customization of the product is accomplished as close to customer demand as possible, the differentiation of the product or service is delayed. For example, suppose a customer purchases a computer online with a customized configuration of standard parts and modules. The manufacturer has two choices to meet the demand. The manufacturer can pre-assemble hundreds or even thousands of computer configurations that customers may want so they are ready to ship when the demand occurs. The second choice is delayed differentiation where the manufacturer stocks standard components and modules that can be quickly assembled into customized configurations. Delayed differentiation results in carrying much lower inventory levels.

▲ *Lowers inventories* The lowering of inventories is primarily the result of carrying fewer parts because the number of distinct parts being carried has been reduced. Several other reasons for lower inventories due to standardization exist. Better quality resulting from greater use of standard parts and components reduces safety stock. Delayed differentiation reduces the need to carry as many finished goods in stock, thereby reducing overall inventory levels. Standard parts and modules usually have more certain and shorter supplier order lead-times. Reduced uncertainty in production lead-time reduces the need for additional inventories required for unreliable lead times. Shorter lead times directly translate into smaller order quantities.

The use of standards permits a firm to purchase fewer items, in larger quantities, and at lower prices. Thus, fewer items are processed and stocked. This reduces supply management, receiving, inspection, and payment costs. Stocking fewer items makes controlling inventories easier and less costly. The use of standardized approved items drastically reduces the number of defects in incoming materials. Consequently, the purchase of standardized materials reduces total costs in four ways: lower prices, lower processing costs, lower inventory carrying costs, and fewer quality problems. The benefits of standardization are presented in Figure 6.3.

Simplification

Simplification, a corollary of standardization, is another term for which recognized authorities have varying definitions. Most frequently, simplification means reducing the number of standard items a firm uses in its product design and carries in its inventory. For example, one company formerly used 27 different kinds of standard lubricating greases in the maintenance of its machinery. Analysis showed that in some cases the same grease

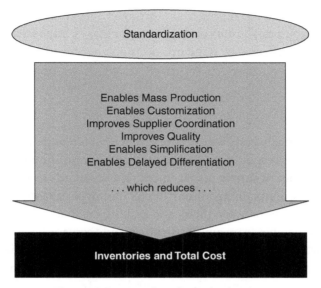

Figure 6.3   Benefits of standardization.

could be used for several different applications and that only six kinds of grease were needed. Hence, through simplification the number of standard greases used was reduced from 27 to 6. Similar analysis showed that the number of standard bearings and fasteners used in production could be reduced by about 50%. Reductions of this scope are commonplace.

Simplification savings result primarily from reduced inventory investment, more competitive prices, greater quantity discounts (because of larger-volume purchases and the use of blanket orders), and reduced clerical and handling costs (because fewer different items have to be handled and controlled).

Some authorities consider simplification an integral part of standardization, rather than a corollary of it. They visualize the simplification process as taking place primarily at the design level, rather than at the stocking level. They think in terms of simplifying (or reducing) the number of related items that are approved as standards in the first place.

Developing a Standardization Program

The benefits of standardization cannot be fully realized when they are developed internally only by design engineers. The next level up is to involve the CFT with internal members from supply management, marketing, quality, and other relevant functional areas. However, to gain all of the benefits of standardization presented earlier, standardization should be addressed across the supply chain through CFTs containing chain members. A standardization program can be approached in various ways, but because so many departments and suppliers are affected by standards decisions, a team effort is the most appropriate approach.

Standards Team

A standards team normally consists of representatives from engineering, supply, operations, marketing, and transportation. Relevant suppliers also should be included on the team under the guidance of supply management. The standards team typically is charged with the responsibility of obtaining input from all user departments and relevant suppliers, reconciling differences between them, and making the final standards decisions.

Theoretically, a member from any department could serve as head of the team. Supply management is particularly well qualified to head the team in companies where materials that comply with national standards or maintenance, repair, and operating (MRO) items form a large portion of the company's total purchases. In companies that manufacture highly differentiated technical products assembled from parts made to company standards, engineering is well qualified to head the team.

Importance of Supply Management

Regardless of the organization employed, the supply management department occupies a focal point in the process. Only in supply management are duplicate requests for identical (or nearly identical) materials, overlapping requests, and "special buy" requests from all departments visible. Hence, no program for standardization can be optimally successful unless supply management is assigned a major role in the program.

Materials Catalog

Once the decision to implement a standardization program has been made, the most common approach is to work toward developing a comprehensive materials catalog. A current, easily accessible materials catalog or database of approved standard items is the logical output of a standardization program. The catalog greatly aids the firm's design efforts. Many benefits of developing a materials catalog exist; the most obvious are:

▲ *Improved quality* The documentation of materials is the first step toward accumulating data to determine which materials have quality problems. The availability of such a catalog virtually eliminates the possibility that designers will incorporate materials that previously caused problems.

▲ *Reduction in design time*   Access to a materials catalog provides designers with a resource that will shorten the materials selection process in the design stage, reducing the total design time.

▲ *Reduction of non-standard parts*   The exercise of developing a materials catalog facilitates the use of standard parts.

▲ *Reduction of standard parts*   Simplification is easier because the standard parts documentation is centrally maintained. See the discussion on simplification given earlier in this chapter.

▲ *Reduction of inventory*   Through standardization and simplification, two activities enabled by the development of a materials catalog, inventories are reduced. The reduction is primarily due to the decrease in the variety of parts carried and the improved quality.

▲ *Benefits of centralization*   Development of a materials catalog for a company with several physically separated facilities provides the opportunity to take advantage of centralization benefits, including pricing leverage.

Electronic Materials Catalog

An emerging trend in companies is to move the materials catalog from a hardcopy form to an electronic file. A simple electronic materials database can be created in virtually any software package capable of organizing and maintaining data. The complexity of the software is directly related to the complexity of the data. Included in the term "data" are graphical images such as photos, drawings, and designs. The designs may be quite complex. For example, the rail industry (railroads) chose to include vector drawings in its electronic catalogs. Vector drawings are created using a computer-aided design software package in which arcs and lines are generated from mathematical formulas. The rail industry decided that converting a vector drawing to an image resulted in a loss of underlying intelligence.[12]

A simple database software package such as Microsoft Access is sufficient and allows for relatively sophisticated data maintenance, centralization, queries, and graphics. On the more complex end of the materials database continuum are enterprise resource planning (ERP) systems, which can centrally locate materials catalog information for access, updating, and utilization by all departments. Most ERP systems have already converted over to Web-based interfaces, allowing maximum access to centralized materials information in even the most remote regions of the world.

Benefits of Electronic Catalogs

Materials catalogs that are electronically maintained are superior to hardcopy versions because they can be centralized in one location, updated easily, electronically disseminated, queried using search techniques, and used to provide links to activate ordering and obtain additional information. The increasing dissemination of information using technologies such as secured intranets and extranets, data warehouses, and database-driven Web sites enables greater growth of electronic materials catalogs. As companies move much of their intellectual property to cyberspace, the use of electronic materials catalogs is on the rise.

Summary

Both specifications and standardization play important roles in the search for the right quality and the right value. They also assist in resolving the design conflicts between engineering, manufacturing, marketing, and supply management. As presented in this chapter, specifications are the heart of the resulting procurement. The strategic dimension of specification development cannot be ignored in today's globally competitive environment. Long-term planning through organizational change and philosophical transformation must occur in most companies so that balanced specifications contribute to the viability of the firm's supply chain.

Many firms still do not fully appreciate the concepts embraced in standardization and its corollary, simplification. Nevertheless, the philosophies underlying these concepts play an important role in creating competitive advantage. It seems highly probable that these same philosophies will continue to be important in the future. Aided by improved information technology and increased automation coupled with computer-aided design and computer-aided manufacturing systems, standardization is part of the answer to meet this decade's desire for customized products and services at low cost.[13] By further standardizing component parts, processes, and operations, companies can refine and streamline their systems. That refinement should permit the production of low cost, high quality, differentiated products that will be competitive in the global marketplace.

Endnotes

1. David N. Burt and Michael F. Doyle. *The American Keiretsu* (Homewood, IL: Business One-Irwin, 1993), 158.
2. Robert E. May, C.P.M., Sr. Consultant, Harris Consulting, Inc., "Top Ten Approaches to Cost Reduction" (presentation at the 1998 NAPM International Purchasing Conference, May, 1998). See napm.org for full transcript of the proceedings paper.
3. Manufacturers also produce merchandise for wholesalers and retailers who market it under their own private brand labels. In such arrangements, the manufacturer is relieved of marketing and promotional responsibilities.
4. Brand buying is mandatory in some situations. Common examples are when a supplier's production process is secret, when its workmanship exceeds all competitors, or when testing competitive items is too costly.

5. The federal government and some large industrial firms have defined "too long" as a period exceeding 30 days.

6. The reader should not infer from this discussion any intent to denigrate the work of the design engineer. None is intended. Often, for reasons of policy, tradition, or expediency, the design engineer is required to make decisions alone that could be made more effectively in collaboration with others. Nevertheless, billions of dollars are lost annually through the adoption of unnecessarily stringent specifications at the design stage.

7. Burt and Doyle, op. cit., 116.

8. The actual names of the individuals and company involved in the story are not given within the case. The story itself is true with respect to the actual events that occurred.

9. Robert, Schwarzwalder. "Searching the Industrial Standards, *Database Magazine* 20(2) (1997): 69.

10. ANSI, 1430 Broadway, New York, NY 10018

11. Montabon F. et al., ISO 14000: Assessing its Perceived Impact on Corporate Performance," *Journal of Supply Chain Management*, National Association of Purchasing Management, Spring (2001).

12. Dennis Smid and Jo-Anne, Kane. "Another Link in the Chain—Electronics Parts Catalogs," NAPM 81st Annual International Conference Proceedings, 1996. For proceedings paper, see www.ism.ws.

13. Recall the discussion in Chapter 1 describing the direct relationship between reductions in material costs and increases in profit margins and return on investment.

Suggested Reading

Brunsson, Nils, Benqt Jacobsson, and Associates. *A World of Standards*. New York: Oxford University Press, 2000.

Hoyle, David. *ISO 9000 Quality Systems Handbook*. Burlington, MA: Elsevier Butterworth-Heinenmann Publications, 2006.

Kerzner, Harold. *Project Management Best Practices: Achieving Global Excellence*. Hoboken, NJ: John Wiley & Sons, 2006.

Pacelli, Lonnie. *The Project Management Advisor: 18 Major Project Screw-Ups and How to Cut them Off at the Pass*. Upper Saddle River, NJ: Prentice Hall/Pearson, 2004.

Schwarzwalder, Robert. "Searching the Industrial Standard," *Database Magazine,* 20(2) (1997).

Spivak, Steven M., and F. Cecil Brenner. *Standardization Essentials: Principles and Practice*. New York: Marcel Dekker, 2001.

CHAPTER 7

Managing for Quality

Evolution of Quality Management

Firms have taken many different pathways to world-class quality management, but the end destination has been very much the same. A frequently overheard statement is "same wine, different bottles." The wine in this case is quality, and the different bottles are the different approaches and pathways. The fact that new bottles continue to sell is proof that quality is still and will always be critically important.

In historically understanding the evolution of the quality management movement over the last 50 years, it is useful to reflect on the multiple pathways that companies have taken to achieve competitive capability on a quality basis. The beginning of almost all pathways can be traced to a relatively small number of individuals who essentially "preached" about quality when very few companies recognized quality as a source of competitive advantage. Instead, many companies, such as General Motors, did not initially heed the call of the preachers and waited until quality was required for survival. Some companies, such as American Motors Company, never heeded the call and simply went out of business. In companies that decided to survive, the chief executive officer administered one or more management approaches that adhered to the philosophical understanding of the "religion" he or she had accepted. The management approach then determined which methodologies and tools the companies used to implement, improve, and maintain quality. The remainder of this chapter follows the pattern of influence by presenting philosophers, approaches, methodologies, and tools.

Philosophies of the Gurus

Quality management has been blessed with many visionaries who have helped develop the field to its current importance within companies. These visionaries are often referred to as gurus or preachers. The reason for this reverence is rooted in the fact that quality management investment in the past was often decided upon within companies on the basis of belief or "gut feelings," not hard costs. (The issue of costs of quality is addressed later in the chapter.)

The influences of gurus such as Deming, Crosby, Juran, Taguchi, and Imai are still evident in corporate cultures. A supply manager can identify which guru influenced any specific company the most by simply listening to managers and observing their processes. From a supply management perspective, understanding the influence of one or more gurus on a supplier aids in evaluating, selecting, understanding, and helping that supplier. Although their teachings were directed at managers who were trying to improve their own internal quality, most of the concepts are general enough to apply across supply chains.

W. Edwards Deming

W. Edwards Deming was perhaps the most influential quality guru of the last century. Ironically, Western countries ignored his message until the 1980s. Before that time he spent most of his career in Japan as a consultant, helping the Japanese rebuild their industries after World War II. Deming is best known for his 14 points summarizing the philosophy of quality management that he developed over time in Japan.[1] The 14 points are general enough that they are still applicable, and many of the concepts are directly transferable to modern supply management.

Statistical Methods

In his seminars on the 14 points, Deming always stressed the use of statistical methods in order to identify when a process is becoming unstable or unpredictable so that the problem can be identified and the process can be prevented from actually producing defects. This important topic is discussed later in this chapter in the section on statistical process control (SPC). Showing that he was far ahead of his time in supply management (it was called purchasing back then), he advocated the use of SPC at suppliers' facilities as well as at the buying firm's facilities.

Causes of Defects

Deming contended that the vast majority of defective or poor products produced in processes are directly traceable to poor-quality input materials, parts, and components. These, according to Deming, are often the result of poorly coordinated design specifications and poorly managed processes after the design is in production. From a supply management perspective, this means that better coordination with suppliers in specifications development and usage will improve quality and decrease costs. In addition, developing suppliers to use statistical techniques such as SPC can help reduce costs and maintain high quality after the design stage. Deming also felt that buying materials primarily based on price contributed to the selection of suppliers with quality problems.[2]

Philip Crosby

Philip Crosby is another American quality guru who rose to international fame as a management consultant. Before he became a consultant, he worked his way up from line inspector to corporate vice president and quality director of ITT. Crosby authored several seminal books; among the best known are *Quality is Free* and *Quality without Tears*.[3]

Zero Defects

Crosby is best known, and misunderstood, for championing the zero defects standard and popularizing many slogans, such as *"Do it right the first time."* The focus of the zero defects standard is on defining quality from the customer's perspective as conformance to requirements and then improving processes through prevention activities to meet the requirements. Crosby pushed the idea of measuring the costs of quality to support efforts toward zero defects, which culminated in his zero defects management approach (presented later in this chapter).

Crosby intended the zero defects approach to be a management performance standard, not a motivational program. Despite his intentions, it was treated as a motivational program in many companies; those companies eventually gave up on the program because they could never reach perfection. Deming believed the zero defects standard created anxiety, fear, frustration, and mistrust of management when it was not accompanied by the means to actually achieve the standard.

Masaaki Imai

Like Deming and Crosby, Masaki Imai became one of the world's leading management consultants. Imai introduced the world to continuous improvement through his book *Kaizen: The Key to Japan's Competitive Success*.[4] In a working environment, kaizen means continuous process improvement involving everybody.

Kaizen and Supply Management

Kaizen calls for everyone in an organization to work for constant and gradual improvement in every process. Because processes span supply chains, we believe that Kaizen should be extended to calling for everyone in the chain to work for constant and gradual improvement in every process. When a new standard is achieved, management should make certain it is maintained and conditions are present that ensure the attainment of even higher standards. Kaizen improvement is by Imai's definition, a long-term and long-lasting improvement resulting from team efforts focused on processes. Since it draws from existing employees, it usually requires less investment than do other management approaches, but great internal effort to maintain.[5]

Genechi Taguchi

Genechi Taguchi served as the director of the Japanese Academy of Quality and is a four-time recipient of the Deming Prize. Taguchi notes that as the level of conformance moves out toward the upper and lower limits, there is a quadratic increase in costs. Taguchi referred to this as the "quadratic loss function."

Taguchi advocates identifying target values for design parameters and producing robust designs by using statistical experimentation. The approach focuses on consistency in hitting the target values rather than being within a band of tolerance. Taguchi's "Loss to Society" model is presented in greater detail in the Tools and Methodologies section of this chapter.

Joseph Juran

Joseph Juran is perhaps best known for his *Quality Control Handbook*. First published in 1951, the book is revised periodically to remain relevant. Like Deming and Crosby, Juran is an American consultant with international fame. Also, like Deming, Juran worked in post World War II Japan, conducting seminars for top-level and middle-level executives.

The main principles of Juran's message are to focus on planning, organizational issues, creating beneficial change (breakthrough), preventing averse change (control), and management's responsibility.[6] He recommends a formula for results that has four important stages, as follows:

1. *Establish specific goals to be reached*—identify what needs to be done, the specific projects that need to be tackled.
2. *Establish plans for reaching goals*—provide a structured process.
3. *Assign clear responsibility*—for meeting the goals.
4. *Base the rewards on results achieved*—feed results information back and utilize the lessons learned and experience gained.

Planning for quality is seen by Juran as an indispensable part of what he calls the quality trilogy—quality planning, quality control, and quality improvement. He believes that objectives should be set yearly for increased performance and decreased costs. To develop the habit of always striving for those yearly goals, a company needs quality planning and a quality structure. Juran believes that the development of the goals, plans, and structure are the responsibilities of top management.

Juran countered Crosby's approach by stating that simplistic slogans and exhortations do not constitute a structure. According to Juran, "There are no short cuts to quality." The emphasis should be put on the results and the experience gained from those results, not on the quality campaign itself. Juran insists that "the recipe for action should consist of 90% substance and 10% exhortation, not the reverse!".[7]

Juran's approach to quality received widespread acceptance because of a clear setting of responsibility and detailed focus on planning. Juran's approach is especially popular with managers who feel the teachings of Deming and Crosby are vague and difficult to evaluate.

Management Approaches

Six Sigma, Total Quality Management (TQM), continuous improvement, zero defects, Quality Management System (QMS), and just-in-time (JIT) are management systems that continue to make large contributions to the improvement and maintenance of quality internally in companies and across supply chains. As we pointed out earlier, the system a company or chain chooses is greatly dependent on the philosophical views of upper management as influenced either directly or indirectly by the gurus. It should be noted that the type of production or service a company provides also influences which system is chosen. In addition, companies may implement several approaches sequentially or simultaneously. In the automobile industry, for example, continuous improvement in a JIT framework coupled with TQM is fairly common. In companies that bought into the zero defects movement, many now use the similar "set an ambitious goal" approach of Six Sigma. All of the systems may be used in manufacturing or service environments, but service operations are less likely to have a comprehensive quality management system because of the lack of physical outputs.

Management usually believes it needs a new approach to "rally the troops" much in the same way that psychologists discovered that painting the walls of a factory temporarily increases productivity. New quality management systems will always appear in corporations. In most cases, the latest approach will incorporate lessons learned from past systems, and so new initiatives should not be ignored. For example, Six Sigma incorporates the best tools and methodologies of the other approaches, such as cause-and-effect analysis and statistical process control, while advancing several newer tools and methodologies, such as balanced scorecards and project charters. The most recent incarnation of a quality system is the Quality Management System (QMS) as defined by the International Organization for Standardization.

Role of Supply Management

So how does a supply professional fit into a comprehensive system such as TQM or Six Sigma? In most cases, supply managers play a critical role in making the entire program work effectively. Most quality experts agree that poor quality of incoming materials causes approximately 75% of the problems and related costs associated with final product quality.[8] It is clear that quality consultant Joseph Juran is correct when he writes: "The assurance (for good quality) must come from placing the responsibility on the supplier to make the product right and supply proof that it is right."[9] Consequently, supply management becomes the "point" player or "the playmaker" in a firm's quality program.

Total Quality Management

TQM received recognition and adaptation in the 1980s because of the increasing need for firms to compete on a quality basis. By the late 1980s and early 1990s, publications were writing about the reasons why TQM had succeeded or failed in many organizations. The reviews were mixed, but generally were in favor of a comprehensive quality management system. Most quality experts agree that the reasons for failure of TQM are usually poor management execution of the system.

By the late 1990s, TQM essentially had matured and management needed new "bottles for its wine." Six Sigma appears to be the latest "new bottle" or management system. Although Motorola initiated Six Sigma in the 1980s, the management approach did not gain widespread recognition until the late 1990s. Six Sigma is discussed later in this section. Although the terminology of TQM will eventually pass, the management approach will never die because quality management in everything a company does is a requirement for competing today and in the future. In fact, this entire chapter is essentially about TQM if one goes by the vague definitions that once were in vogue. For example, the International Organization for Standardization offered the following definition of TQM:[10]

> A management approach to an organization centered on quality, based on the participation of all its members and aiming at long-term success through customer satisfaction, and benefits to the members of the organization and to society.

A generalized TQM model is presented in Figure 7.1.[11]

Continuous Improvement

The concept of continuous improvement was introduced in an earlier section on Imai, the champion of kaizen. Kaizen is synonymous with continuous improvement. Although

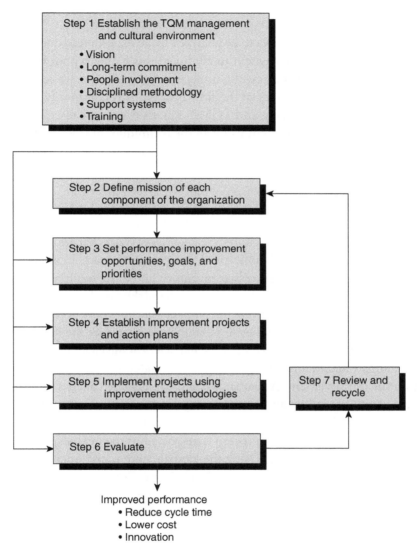

Figure 7.1 A generalized total quality management model.

Imai brought continuous improvement to the world through his books, Taiichi Ohno, the pioneer of the kanban system at Toyota, developed continuous improvement into a viable, tangible management approach. In this section, continuous improvement is presented within the JIT framework. Continuous improvement is a concept independent from JIT. It can be implemented without JIT, but it has received its most publicized adopters in companies that run JIT systems.

Kaizen Events

A kaizen event is a selected work area to be analyzed by an experienced and trained manufacturing team for process improvement to eliminate waste due to poor travel flow from one work station to another, defect rework, unnecessary duplication, wasted motion, overproduction, and other inefficiencies of the flow process. It is commonly associated with the concept of Lean manufacturing and TQM.[12] Kaizen events are operational real-time tests to seek continuous improvement using all the techniques in this and other chapters. Root cause analysis of problems is always the ultimate goal in order to identify the origin of problems, waste, and defects.

Just-in-time

Two major tenets form the basis for JIT as practiced at Toyota: respect for people and elimination of waste. Ohno is credited with identifying categories of waste that are avoidable: waste of overproduction, waste of unneeded motion and transportation, waste in needless processing and machine time, waste in holding excessive inventory, and waste resulting from defects.[13] A supply manager with a solid grasp of these wastes can quickly see them on a factory floor of a supplier. Often managers working within the supplier's facilities have become so complacent about their systems and their inherent wastes that they no longer see them. A knowledgeable supply manager needs to be able to see what the managers of the facilities cannot.

Ohno demonstrated that most waste can be eliminated if a kanban system is coupled with continuous improvement. A kanban system can also be called a "pull" system. In a pull system, units needed by an upstream stage of production are transported to that stage only when needed and in the exact amount needed. This is in contrast to traditional "push" production systems where units are transported to the next stage as soon as they have completed the previous stage. Using the kanban system, the operator of each stage is responsible for the collection of the units from the previous stage; hence, the operator "pulls" the material through the system. A kanban (signpost) is attached to the box of parts as they go to the assembly line, and the same kanban is returned when the parts are all used, to serve as an order for more and as a record of work done.

Supply managers must understand that JIT can be extended to sub-contractors and external suppliers. Extending JIT across a supply chain requires supply professionals and suppliers to work collaboratively so that only enough units to meet demand (that is, small lots) are produced and moved from one stage to another. The supply chain members should work together to assure that the need for speculative, JIT production is eliminated.

Elimination of Waste

In eliminating waste within a continuous improvement methodology, a company may choose to use the following steps. First, form a small group team that will go through kaizen training using the tools and methods discussed later in this chapter. The people in the group are usually factory floor workers, hence the tenet of respect for people. Second, the team will gather data on existing systems to discover where waste is and what the wasteful activity is costing the company. Many managers do not believe in assigning cost to waste because they believe it does not matter—waste is waste—no matter how small or large, and it should be eliminated. Juran and Crosby would argue that we need to know the cost of the waste to prioritize projects to eliminate the waste. Third, once a project for eliminating a waste is identified, then the P-D-C-A (Plan, Do, Check, Act) cycle is often used. After the P-D-C-A substeps are completed, the group returns to step two again and selects a new project.

Six Sigma

In the book *The Six Sigma Way*, Six Sigma is defined as a broad and comprehensive system for building and sustaining business performance, success, and leadership.[14] The key focus of Six Sigma is on processes, but with measurement of both processes and products. With Six Sigma, companies strive to achieve the statistical six sigma goal of near-perfection as measured at 3.4 defects per million opportunities (DPMO). The corresponding yield from the six sigma process is 99.9997%, as shown in Figure 7.2. The lofty goal, very reminiscent of the Zero Defects management philosophy, is used by Six Sigma advocates as a driver of organizational change.

Six Sigma was developed at Motorola in the late 1980s, as a way to accelerate the company's rate of improvement, provide a clear and objective goal for improvement, and create a stronger focus on the customer. The success of the six sigma measure as a mantra for change led Motorola to institutionalize the goal into a comprehensive management system. While the objective of statistical "six sigma" is the overarching goal, Six Sigma as a management system shifts the attention to the rate of improvement in processes and

Sigma	Corresponding Yield	DPMO
1	30.9%	690,000
2	69.2%	308,000
3	93.3%	66,800
4	99.4%	6,210
5	99.98%	320
6	99.9997%	3.4

Figure 7.2 Sigma quality levels and corresponding yields.

products. According to supporters of the Six Sigma management approach, there are some clear measurement benefits.

1. *Six Sigma starts with the customer.* Measures demand a clear definition of customer's requirements.
2. *Six Sigma provides a consistent metric.* Once you've defined the requirement clearly, you can define a "defect" and measure almost any type of business activity or process. For example, some measures from a supply management perspective could be late deliveries, incomplete shipments, or parts shortages.
3. *Six Sigma links the effort to an ambitious goal.* Having an entire organization focused on a performance objective of 3.4 DPMO can create significant momentum for improvement.[15]

DMAIC Cycle

Similar to the P-D-C-A cycle given in the previous section, Six Sigma offers a cyclical improvement model called DMAIC that focuses on improving processes. DMAIC stands for Define, Measure, Analyze, Improve, and Control.[16] In contrast to P-D-C-A for continuous improvement, the Six Sigma model has elements of both continuous and discontinuous (reengineering) improvement. Six Sigma's focus on processes is evident from the combination of process improvement and design (including redesign), "incorporating them as essential, complementary strategies for sustained success." The process design approach is reminiscent of reengineering, where the objective in Six Sigma is not to fix but rather to replace a process (or a piece of a process) with a new one. Six Sigma also calls for addressing product and service design through what the approach calls "Six Sigma Design." Six Sigma Design principles are intended to link the creation of new goods and services to customer needs and validate the linkage by data and testing.[17]

Six Sigma Blackbelts

A differentiator of the Six Sigma management system from other quality management approaches is the designation of levels of understanding of Six Sigma into belts: green, black, and master. The excitement that the belts cause is very real and, to put it bluntly, shrewd marketing. Taking the term from martial arts, a color of belt is a level attained through intensive training and experience. Although no solid definition of the levels exists, a master blackbelt is essentially an individual who has attained the highest level of understanding of Six Sigma and is capable of training others and working as an internal consultant in a corporation. Regular blackbelts become the primary drivers of Six Sigma improvements within companies by essentially managing specific projects. A greenbelt is a training level that is encouraged by Six Sigma advocates for all workers.

At GE, blackbelts receive three weeks of training, with follow-up exams and continued learning through conferences and other forums. A greenbelt at GE is the lowest commitment, which is training for a minimum of two weeks in Six Sigma. At GE, every management employee is required to at least become a greenbelt in Six Sigma.[18]

Six Sigma Themes

In the book *The Six Sigma Way*, the authors present six themes (or principles) of Six Sigma, summarized as follows:

▲ *Theme One: Genuine Focus on the Customer*   In Six Sigma, customer focus becomes the top priority. For example, the measures of Six Sigma performance begin with the customer. Six Sigma improvements are defined by their impact on customer satisfaction and value.

▲ *Theme Two: Data- and Fact-Driven Management*   Six Sigma discipline begins by clarifying *what* measures are key to gauging business performance; then it applies data and analysis to build an understanding of key variables and optimize results. Six Sigma helps managers answer two essential questions to support fact-driven decisions and solutions. What data/information do I really need? How do we use that data or information to maximum benefit?

▲ *Theme Three: Process Focus, Management, and Improvement*   Whether designing products and services, measuring performance, improving efficiency and customer satisfaction, or even running the business, Six Sigma positions the process as the key vehicle of success.

▲ *Theme Four: Proactive Management*   Proactive management means making habits out of what are, too often, neglected business practices: defining ambitious goals and reviewing them frequently; setting clear priorities; focusing on problem prevention vs. firefighting; questioning why we do things instead of blindly defending them as "how we do things here."

▲ *Theme Five: Boundaryless Collaboration*   The opportunities available through improved collaboration within companies and with their suppliers and customers are huge. Billions of dollars are left on the table (or on the floor) every day because of disconnects and outright competition between groups that should be working for a common cause: providing value to customers.

▲ *Theme Six: Drive for Perfection; Tolerance for Failure*   No company will get anywhere close to Six Sigma without launching new ideas and approaches, which always involves some risk. If people who see a possible path to better service, lower costs, new capabilities, etc. (i.e., ways to be closer to perfect) are too afraid of the consequences of mistakes, they will never try.

Successes with Six Sigma are being published in a variety of industries. One of the most outspoken advocates of Six Sigma is former GE Chairman John F. Welch. According to

Welch, "Six Sigma has forever changed GE. Everyone—from the Six Sigma zealots emerging from their Black Belt tours, to the engineers, the auditors, and the scientists, to the senior leadership that will take this company into the new millennium—is a true believer in Six Sigma, the way this company now works."[19] At GE, the best Six Sigma projects begin "…not inside the business but outside it, focused on answering the question—how can we make the customer more competitive? What is critical to the customer's success? . . . One thing we have discovered with certainty is that anything we do that makes the customer more successful inevitably results in a financial return for us."[20]

Quality Management System

The International Organization of Standardization (ISO) released its ISO 9000:2000 standards with radical revisions that justify the inclusion in this section of what ISO calls a Quality Management System (QMS). The principles are intended for use as a framework to guide organizations toward improved performance. According to ISO, the principles are derived from the collective experience and knowledge of international experts. They form the foundation of a QMS approach to managing quality. The principles are as follows:[21]

- ▲ **Principle 1: Customer focus.** Organizations depend on their customers and therefore should understand current and future customer needs, should meet customer requirements, and strive to exceed customer expectations.
- ▲ **Principle 2: Leadership.** Leaders establish unity of purpose and direction of the organization. They should create and maintain the internal environment in which people can become fully involved in achieving the organization's objectives.
- ▲ **Principle 3: Involvement of people.** People at all levels are the essence of an organization and their full involvement enables their abilities to be used for the organization's benefit.
- ▲ **Principle 4: Process approach.** A desired result is achieved more efficiently when activities and related resources are managed as a process.
- ▲ **Principle 5: System approach to management.** Identifying, understanding, and managing interrelated processes as a system contributes to the organization's effectiveness and efficiency in achieving its objectives.
- ▲ **Principle 6: Continual improvement.** Continual improvement of the organization's overall performance should be a permanent objective of the organization.
- ▲ **Principle 7: Factual approach to decision making.** Effective decisions are based on the analysis of data and information. Applying the principle of factual approach to decision making typically leads to ensuring that data and information are sufficiently accurate, reliable, and accessible.

▲ **Principle 8: Mutually beneficial supplier relationships.** An organization and its suppliers are interdependent and a mutually beneficial relationship enhances the ability of both to create value.

Tools and Methodologies

The quality movement, whether embraced through TQM, continuous improvement, Six Sigma, or some other management approach, is accomplished in the "trenches" on the floors of factories, warehouses, offices, and wherever business is accomplished. To accomplish quality, a plethora of tools and methodologies have been developed that are used whenever needed by middle management and the general workforce. Improvement-focused tools and methodologies all contributed and continue to contribute to quality improvement.

Terminology and Usage

The term "tool" emerged from the use of the methodologies in small group improvement efforts, such as quality circles, wherein the group would focus on solving a problem by using a mix of methodologies. Some of the methodologies would be used in the same problem-solving session, kind of like a screwdriver being used for many phases of a construction project. Some of the tools are better used to discover where focus should be placed. For example, a cost of quality analysis helps clarify what areas of investment may require attention. If downtime resulting from breakdowns exists, perhaps increasing investment in maintenance or training is warranted. We can view these "tools" as available in our "quality toolbox" for use when appropriate to help us achieve quality. Not all tools will be used on every project. Some may never be used.

Implications for Supply Management

Why should a supply professional need to be aware of the tools and methodologies of quality management? Some companies talk a great deal about quality at the upper-management level, but real improvements in quality occur in the trenches where the tools and methodologies are used. Physically seeing that these activities are occurring is reassurance that the company is investing in its quality system and the people running the system. When a supply manager does "the quality walk" through a supplier's facility, he or she often will discover meeting rooms or factory walls pasted with diagrams, charts, and lists that are outputs of using these tools and methodologies.

Common Tools

The most commonly used tools of quality are Pareto charts, cause-and-effect diagrams, process flow charts, run plots, frequency histograms, correlations plots, and control charts. These are often called the "seven tools of quality."

1. *Pareto Charts* are used to distinguish between critical and trivial problems. For example, a company could use Pareto charts with checklists to identify the number of occurrences or quality costs for each variable that causes their products to be scrapped. The variables themselves could be identified using cause-and-effect diagrams.

2. *Cause- and Effect-Diagrams* show possible causes of a problem. Also called a fishbone diagram, the cause-and-effect diagram is an aid to brainstorming and hypothesis generation. For example, the causes of defects, causing a product to be scrapped, could be identified using this method. Always try to find the underlying causes of causes. Ask, "Why is this a cause?" several times, until the true underlying cause is discovered.

3. *Process Flow Charts* are useful for showing linkages among parts of a process. This is a good tool for identifying bottlenecks and non-value added activities in processes. Simulation is a natural extension of flow and run charting to increase knowledge about the flows of a process.

4. *Run Plots* graph samples of parts/materials over time of some variable that is thought to be important. A run plot can be examined to see whether the process is subject to change or behaves consistently over time. Noting when changes occur may suggest hypotheses about their cause.

5. *Frequency Histograms* show the distribution of some variable that is thought to be important. Frequency histograms are useful for hypothesis testing. For example, how do the distributions of process times for manufacturing processes compare with ones you would expect?

6. *Scatter Diagrams* show correlations between two variables, typically a problem and a potential cause. To examine interactions between more than two variables, multiple regression should be used. For example, a negative correlation between increased quality and increased total costs could be investigated using a simple scatter diagram.

7. *Control Charts* are similar to run plots, but they are used for operational control. Control charts show the upper and lower allowable limits for a process variable. When the process exceeds those limits or shows a recognizable pattern, action should be taken to adjust the process. Once set up, control charts are effective tools for day-to-day monitoring and management of a process. They are discussed in greater detail later in this chapter.

Example Using the Tools

Consider a supplier who has recently installed a new production process and assigned several operators to a team to manage improvement projects. The team decides to map the process stages using a flow chart to identify where data should be gathered. The decision is to gather data at each of the inspection points. After data is gathered on where failures are occurring and estimating the costs of the failures (costs of quality), a Pareto chart may be used to identify which problem is generating the greatest failure costs. A cause-and-effect diagram could then be used to help identify the root causes of the highest cost failure. Through the cause-and-effect analysis and brainstorming, the team determines that higher levels of maintenance of the tools used in the production process are needed. To prevent the problem from occurring in the future, the team decides to implement statistical process controls to measure the process outputs and identify when the process is in need of additional maintenance before new defects are produced. Data is accumulated on the new solution and a decision is made at a later date as to whether the project has been a success. If it has been a success, the team moves on to the next highest priority project. If the project has not been a success, then the team restarts the improvement process. Regardless of the success or failure of the project, the team maintains detailed documentation of the effort, so that the gains can be maintained, lessons can be learned, and the information can be disseminated to other production processes within the company.

The successful use of the seven tools has partly been the result of the simplicity and ease of use they facilitate, while revealing much about a process. They are designed so that little training is required. Still, they do require commitment on the part of management to provide time and resources for small group members to utilize them properly.

Design of Experiments (DOE)

A brilliant quality control engineer at Motorola, Keke Bhote, used the DOE techniques developed by Dohan Shainin, another brilliant quality control engineer who worked for United Aircraft (now United Technologies). DOE tools discover key variables in process/product design and then find solutions to substantially or totally reduce the variations they cause while also reducing tolerances on the lesser variables to reduce costs.[22]

Costs of Quality

Since the introduction of the traditional quality cost model in the 1950s, managers have been urged to base quality-related decisions on the hypothesized tradeoff between the costs of prevention and appraisal and the costs of internal and external failure.[23,24] From a supply management standpoint, understanding the costs associated with quality decision making enables better understanding of both the total cost of ownership and how suppliers make decisions in their production facilities. (See Chapter 13 on the total cost of ownership.)

According to the traditional costs of quality (COQ) model, a company producing poor quality products can reduce nonconformance or failure costs by investing in prevention and appraisal activities. The model resolves a hypothesized tradeoff by specifying a nonzero optimal level of defects at the point where the marginal cost of increased prevention and appraisal activities equals the marginal benefit from failure cost reductions. An important note here is needed with respect to the optimum point. Within many progressive quality systems, the optimal levels of defects are very close to zero.[25] A Six-Sigma process, for example, should have 3.4. The cost categories in the COQ model are as follows:[26]

▲ *Prevention costs* are costs of all activities specifically designed to prevent poor quality in products or services. Examples are the costs of maintaining equipment and supplies, new product reviews, quality planning, supplier capacity surveys, process capability evaluations, quality improvement team meetings, quality improvement projects, quality education, and training.

▲ *Appraisal costs* are costs associated with measuring, evaluating, and auditing products or services to assure conformance to quality standards and performance requirements. These costs include the costs of incoming and source inspections, tests of purchased material, in process and final inspection, product, process, or service audits, calibration of measuring and test equipment, and the cost of associated materials and supplies.

▲ *Failure costs* are costs resulting from products or services not conforming to requirements or customer or user needs. Two categories of failure costs exist: internal and external. Internal failure costs are costs occurring prior to delivery or shipment of the product, or the furnishing of a service, to the customer. Examples are the costs of scrap, disposing of scrap, rework, redoing inspection, redoing testing, material review, and down grading. External failure costs are costs occurring after delivery or shipment of the product, and during or after furnishing a service, to the customer. Examples are the costs of processing customer complaints, customer returns, warranty claims, and product recalls. Opportunity costs of lost customers due to poor quality are sometimes included as an external failure cost, but are extremely difficult to estimate.

Loss to Society

In continuing the COQ discussion, Taguchi taught that a traditional "conformance to specifications" definition of costs underestimates failure costs. The conventional conformance to specification definition assumes that no loss occurs as long as output lies within upper and lower specification limits. The view has also been referred to as the "goalpost philosophy."

Tolerance Stacking

One problem with the goalpost philosophy is tolerance stacking. Tolerance stacking occurs when two or more parts are to be fit together; the size of their tolerances often determines how well they will match. Should one part fall at a lower limit of its specification and a matching part at its upper limit, a good fit is unlikely.

Taguchi developed a way to confront the stacked tolerances conformance problem. Taguchi begins with the idea of "the loss function," a measure of losses from the time a product is shipped. The theoretical notion is that "losses to society" occur whenever output deviates from its target value.[27] In more tangible terms, these losses include warranty costs, nonrepeating customers, and other problems resulting from performance failure. Taguchi then compares such losses to two alternative approaches to quality. The first approach is simple conformance to specifications. The second approach is a measure of the degree to which parts or products diverge from the ideal target or center. Taguchi demonstrates that tolerance stacking is worse when the dimensions of parts are more distant from the ideal target of a specification than when they cluster around it, even if some parts fall outside the tolerance band entirely. Knowing that the problem of stacked variances exists and understanding Taguchi's loss function causes managers to focus on reducing the variances of parts and processes.

Process Capability Analysis

No operations activity can produce identical results time after time. This is true even for machine-based production processes. Every process possesses some natural variability due to such things as machine part clearances, bearing wear, lubrication, variations in operator technique, and so on. In the language of statisticians, these are "chance" or "common" causes that produce *random variations* in the output. Over time, this natural variability in the output of a process will produce a distribution of outputs around the mean quality level. In many cases, this distribution approximates the normal bell-shaped curve. The difference between the two extremes of the curve, the high and low values, is defined as the *natural tolerance range* of the process. As long as the process is properly adjusted and is not affected by any outside nonrandom forces—as long as the process is "in control"—the distribution it produces is predictable, as shown in Figure 7.3.

If a buying firm's desired range of quality for a given purchased part is compatible with the natural tolerance range of a potential supplier's production process, the supplier should have little difficulty in providing the buying firm with parts that meet specifications. In contrast, when the buying firm's required quality range is narrower than the natural capability range of the process, the supplier is bound to produce some unacceptable parts.

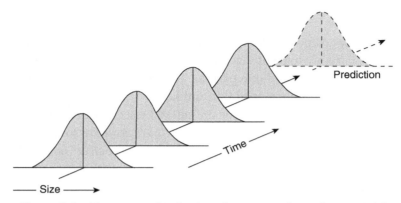

Figure 7.3 The output distribution of a process that is "in control."

Implications for Supply Management

Alert supply managers recognize the direct economic relationship between their specified quality requirements and the producer's ability to perform consistently at the specified level. In the case of a nonstandard item, before selecting a supplier, a capable supply manager must determine (1) whether the potential supplier in fact knows what the natural capability range is for its production process, (2) if so, whether the buying firm's *desired* range of quality is compatible with the supplier's natural capability range, and (3) if so, how the supplier plans to monitor the process to ensure that it stays in control so that it will produce satisfactory output consistently. The following examples should clarify this concept.

In Figure 7.4*a*, assume that a supply manager wants to purchase 100,000 metal shafts that are 1 in. in diameter with a tolerance of ± 0.005 in. Assume further that the supplier

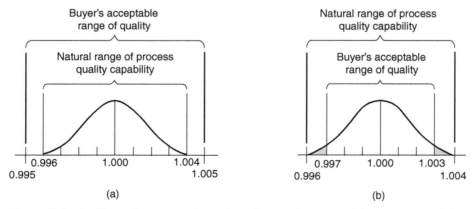

Figure 7.4 An illustrative comparison of quality requirements with process capability (process frequency distribution curves for shafts are shown).

has studied its process, stabilized it, and knows its natural capability for this type of job to be 1 in. ± 0.004. Examining Figure 7.4*a*, the supply manager sees that as long as the supplier's process operates normally and remains centered on 1.0 in., every piece produced will fall within his or her acceptable range of quality. In this example, then, if the supplier is able to keep the process in control, this situation appears to represent a sound purchase for the buying firm from a quality point of view. In this case, the supplier's process is said to be "capable."

Assume now that the supply manager wants to purchase 100,000 shafts that are 1 in. in diameter with a tolerance of ± 0.003 in. Examination of Figure 7.4*b* reveals that, dealing with the same supplier under normal operating conditions, the production process cannot entirely satisfy the buying firm's requirement. Some shafts will be produced with diameters less than 0.997 in. and some with diameters larger than 1.003 in. It is important for the supply manager to understand this type of situation. Unusable shafts, whether reworked or scrapped, create cost that eventually must be recovered by the supplier. In the long run, it will be the buying firm who pays. Assuming that the buying firm's requirements cannot be compromised, he or she is faced with two alternatives:

1. Negotiate with the supplier to determine whether the natural range of process capability can be narrowed economically to a point nearer the buying firm's requirements.
2. Seek another supplier whose process can meet the requirements more economically.

Process Capability Index

Another way to express a process's capability relative to a buying firm's specific design requirement is by means of a *process capability index (Cp)*. This index is defined as

$$Cp = \frac{\text{buying firm's absolute design tolerance}}{\text{natural capability range of the process}^{[28]}}$$

A *Cp* value of 1 indicates that the capability range of the process matches *exactly* the quality range required by the buying firm. From a practical point of view, this is a very marginal fit for the buying firm. A value of more than 1 reveals excess process quality capability, while a value of less than 1 indicates insufficient process capability. Consider the examples in Figure 7.4.

$$Cp(a) = \frac{1.005 - 0.995}{1.004 - 0.996} = \frac{0.01}{0.008} = 1.25$$

$$Cp(b) = \frac{1.003 - 0.997}{1.004 - 0.996} = \frac{0.006}{0.008} = 0.75$$

In case (a), $Cp = 1.25$, indicating that the quality capability of the process, if it stays centered and stable, exceeds the buying firm's requirement. In case (b), $Cp = 0.75$, showing clearly that the process is not capable of satisfying the buying firm's quality requirement. In case (a), the process is "capable," whereas in case (b), the process is "incapable."

The preceding discussion makes the important assumptions that the manufacturer's process is stable and the average can be adjusted to line up with the center point (the target value) of the buying firm's specification (as is the case in Figure 7.4). In some operating situations this is not always possible. Suppose, in the situation discussed in Figure 7.4a, that after a great deal of experimentation the best process average the supplier was able to achieve was 0.999 in. This condition is depicted in Figure 7.5. One can see by inspection of the sketch that there is adequate excess quality capability at the upper end of the scale (1.005 – 1.003), but at the lower end there is no excess (0.995 – 0.995). Any movement of the process distribution to the left will produce out-of-specification shafts.

Thus, we see that one additional factor is important in making a process capability analysis—the location of the process average relative to the buying firm's target specification value. That means that the previously calculated capability index (Cp) must be adjusted for the off-center location of the process average. The adjusted Cp is called the *process capability/location index* (Cpk). It can be calculated as follows:

$$Cpk = Cp(1 - k)$$

where $k = \dfrac{\text{buying firm's target value} - \text{process mean}}{\text{buying firm's absolute design tolerance}/2}$

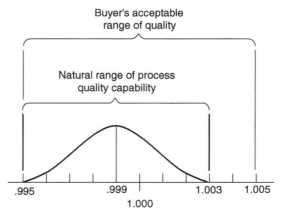

Figure 7.5 An illustrative comparison of quality requirements with process capability when the process mean is not centered.

For the situation in Figure 7.5, the location index is:

$$k = \frac{1.000 - 0.999}{(1.005 - 0.995)/2} = \frac{0.001}{0.005} = 0.2$$

$$Cpk = 1.25(1 - 0.2) = 1.25 \times 0.8 = 1.0$$

The *Cpk* value is interpreted just like the *Cp* value; less than 1 indicates an incapable process, and a value greater than 1 indicates a capable process. In this case, the fit is marginal because of the absence of any leeway on the lower end of the distribution.

Statistical Process Control

The preceding discussion about the natural process capability range was based on the premise that the supplier could keep the process operating in a stable manner—that is, "in control."

At this point in the discussion, it is necessary to point out that in addition to the random variations that occur naturally in any process, during the course of operation some *nonrandom* variations caused by external factors also occur. At times a machine goes out of adjustment, cutting tools become dull, the hardness or workability of the material varies, human errors become excessive, and so on. When these things happen, they usually take the process "out of control." What really happens is that *the quality capability of the process changes because of these unplanned events.* The process output distribution changes. It may spread out, increasing the range; it may shift up or down, altering the mean and the extreme values; or a combination of these things may occur. The end result is that the characteristics of the distribution no longer can be predicted—statistically, *the process is out of control.*

The statistical process control technique has the ability to detect these process shifts due to outside forces, or assignable causes, as they occur. When the changes are detected, the process is stopped and an investigation is initiated to find the cause of the problem. An operator who is familiar with the process and the equipment typically can locate the problem fairly quickly. This technique thus enables the operator to detect the problem, make necessary corrections, and continue operation, with the production of few, if any, defective products.

After a supply manager locates a potential supplier whose process capability matches his or her needs as discussed in the previous section, the final step for a supply manager is to persuade the supplier to use an SPC system that will keep the process in control and ensure the consistency of the process's output quality.

The following paragraphs discuss the basics of SPC theory and application.

Control Charts

Several different types of control charts are used for different kinds of applications. The most common, however, are the $\overline{X}$ and R control charts for variables. Typically these are applied in situations where the quality variable to be controlled is a dimension, a weight, or another measurable characteristic. Operationally, the two charts are used together. The $\overline{X}$ chart monitors the absolute value, or the location, of the process average; the R chart monitors the spread (range) of the output distribution, that is, the piece-to-piece variability.

Figure 7.6 illustrates an $\overline{X}$ and an R chart in simplified form. Before the charts are constructed for operating use, *the process must be studied carefully to determine that it is, in fact, stable*—that it is in statistical control. This is usually done by quality assurance and maintenance specialists and often is a time-consuming experimental task. Once the process has been stabilized—that is, its operation is influenced only by natural, random variations—its natural capability range can be determined and process control limits can be calculated.

At this point, the operating procedures are established. Usually the operator will inspect a small sample of output units every 15 minutes or so. Sample size typically runs between 3 and 5 units. The idea is to sample sequentially produced units in a manner that tends to minimize the quality variation within a given sample subgroup, and to maximize variation between the periodic sample subgroups.

The measurements are recorded as they are taken in the upper portion of the table shown below the R chart. (For illustrative purposes, hypothetical shaft dimension data from the example in Figure 7.6 are included in the first three subgroup columns.) The measurements for the *subgroup sample* are then summed and the average, $\overline{X}$, is computed. The range for the subgroup is also determined and entered in the table. After 20 to 25 subgroups have been inspected, enough data are available to provide good estimates of the process average and spread. At this point, an average of the subgroup $\overline{X}$ values is computed as $\overline{\overline{X}}$ (the average of the averages), and an average of the subgroup R values is computed as $\overline{R}$. The $\overline{\overline{X}}$ value represents the mean value of the average shaft diameter sizes determined in each subgroup; *this is the value used as the process average on the $\overline{X}$ chart.* The $\overline{R}$ value represents the mean value of the range of the shaft diameter sizes found in each subgroup; *this is the value used as the average range on the R chart.*

In most applications, $\overline{X}$ *control chart limits are* set at $\overline{\overline{X}} \pm 3$ standard deviations of the $\overline{X}$ values (3 sigma limits). The reader should note that the frequency distributions used in constructing control charts are distributions of averages, not distributions of the individually measured values produced by the process. This fact ensures the existence of a normal distribution of control chart values when the process is in control.[29] Thus, the 6 sigma range of the chart encompasses 99.7% of the $\overline{X}$ values that will result from the process operation as long as only natural random process variations are occurring.

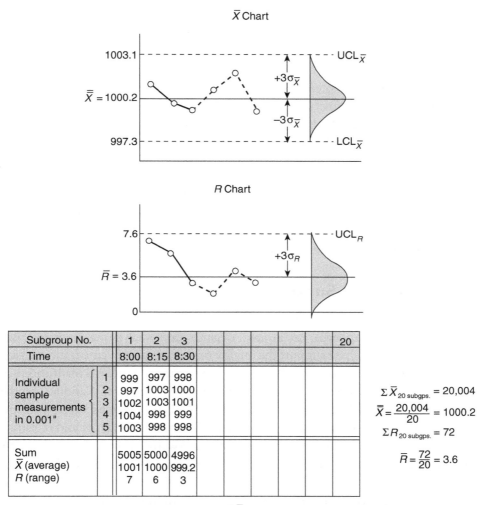

Figure 7.6 An illustration of $\overline{X}$ and R charts in simplified form.

The 3 sigma control chart limits can be determined by calculating the standard deviation of the $\overline{X}$ values, multiplying by 3, and subtracting and adding to $\overline{\overline{X}}$. In practice, however, this calculation has been simplified by the construction of a table of constant factors (for various sample sizes) that can be applied to $\overline{R}$ (and $\overline{\overline{X}}$) to obtain the control limit values directly. This can be done quickly and easily.

The same rationale and procedure apply to the construction of R chart control limits. In this case, however, it should be noted that only the upper control limit is of practical significance in operation.

After the control charts are constructed, *as the operation continues data are plotted and the charts are interpreted.* As shown in Figure 7.6, the operator plots both the $\overline{X}$ value and the R value on their respective charts after his or her inspection and computations for each subgroup sample.

In the majority of cases, interpretation of the charts is a relatively simple matter. Any points falling outside the control limits usually indicate the existence of nonrandom variation. (If the process were actually in control, this would happen only 3 times in every 1000 subgroup inspections.) The appearance of other nonrandom patterns may also be observed. For example, an unusually large number of points in sequence on the same side of the average line may indicate a shift in the process average or an expansion of the spread. As a rule of thumb, a run of seven points usually indicates an out-of-control situation. Clear-cut trends of points in one direction may indicate a tool wear or adjustment problem. As noted previously, when the operator suspects an out-of-control situation, he or she stops the process and investigates the potential problem.

Quality Movement Support

Support for the quality movement over the last several decades has come from three sources: organizations, standards, and awards. Organizations such as the American Society for Quality (ASQ) have helped support the quality management movement through seminars, local meetings, conferences, certifications, and research. Standards institutions such as the International Organization for Standardization (IOS), American National Standards Institute (ANSI), and the British Standards Institute (BSI) have helped support the quality management movement through development of common standards and registration bodies for certifying the quality of company systems. Awards such as the Deming Prize, Malcolm Baldrige National Quality Awards, and the European Quality Awards all support the quality management movement through documentation and recognition of companies achieving world-class status. These awards serve to motivate others and disseminate the best practices of the companies to society.

Organizations

Worldwide there are many organizations that support quality initiatives and education. In Europe, the European Organization for Quality and the European Foundation for Quality Management (EFQM) are two excellent sources. The main quality-focused organization in the United States is the American Society of Quality (ASQ), formerly called ASQC, The American Society For Quality Control.

Most supply professionals receive quality training, education, and networking through a variety of sources. In the United States, most supply management professionals that receive quality training or education have done so through the Institute for Supply Management,[30] which integrates almost everything discussed in this chapter into its teachings. The Educational Society for Resource Management, more commonly referred to as APICS,[31] is another source.

Standards

Standards in quality management referred to in this section are process-based standards or norms that do not deal with technical specifications of the products and materials, but with how they are made. Globally, there are quite a few standards-developing institutes and organizations. The ISO 9000 system is currently the most dominant standards system used for quality control.[32]

International Organization for Standardization

The ISO headquartered in Geneva, Switzerland, is an international body composed of members representing standards organizations.[33] The objective of the organization is to promote the development of standards, testing, and certification in order to encourage the international trade of goods and services.[34] The global emphasis on economic competitiveness, the unification of the European market, and a broad array of different quality standards among countries led to concentrated ISO work in the quality area over the last 20 years. The main product of this work was the issuance of the ISO 9000 series of *quality system* standards in 1987, which are periodically revised to maintain their relevancy. The ISO 14000 series addresses aspects of environmental management. The intent is to provide a framework for a holistic strategic approach to an organization's environmental policy, plans, and actions.[35]

A firm that adopts one or more of the standards is required to document what its quality management procedures are for each element in the standards. The firm is then required to be able to prove to an ISO auditor that the procedures are in fact followed in practice. Recall the underlying eight principles of the ISO 9000:2000s series presented earlier in this chapter in the section on the Quality Management System. ISO 9001:2000 requires that you plan and manage the processes necessary for the continual improvement of your quality management system. Continual improvement as defined by the ISO as "a process of increasing the effectiveness of your organization to fulfill your quality policy and your quality objectives." ISO 9004:2000 even goes beyond ISO 9001:2000 to improving the efficiency of your operation. ISO 9001:2000's focus on continual improvement of the quality management system is a major departure from the past reputation of the ISO 9000 series where a commonly heard joke was "As long as you document it and make it as your documentation says you make it—you can make cement life preservers and still get ISO 9000 certified!" The new standards do not allow companies to pass registration without showing tangible efforts to improve products. As with the ISO 9000 certifications, there is a series of ISO 14000 standards aimed at managing the business environment.

Additional ISO Standards

The ISO has created thousands of standards during its existence. Among some of the topic-specific standards that may be applicable to supply management are: ISO 10006 for project management, ISO 10007 for configuration management, ISO 10012 for measurement systems, ISO 10013 for quality documentation, ISO/TR 10014 for managing the economics of quality, ISO 10015 for training, ISO/TS 16949 for automotive suppliers, and ISO 19011 for auditing. A large number of sources are available to learn more about ISO's many standards. The best starting point is the ISO Web site at www.iso.org.

Impact on Supply Managers

Since their inception in 1987, the ISO 9000 standards have had implications for supply professionals—but the revisions to ISO 9000:2000 create even greater focus on supply management.

▲ The new standards promote the adoption of a process approach compared to the procedural approach described in the 1994 versions. This is important to supply management because processes are what span supply chain boundaries and are, as a result, a responsibility for supply professionals.

▲ Changes have also occurred in terminology. The most important changes concern the use of the term "organization" instead of "supplier," and the use of the term "supplier" instead of "subcontractor." According to ISO, these changes respond to the need of being more consistent and friendly with the normal use and meaning of the words.

▲ The number of requirements for documented procedures has been reduced in ISO 9001:2000, and the emphasis is placed on the organization demonstrating effective operation. This reduces the barriers to suppliers in achieving ISO 9000 registration and brings greater relevancy to the outcomes.

▲ The standards have become a competitive weapon in global business. Consequently, if a firm wants to do business in the global marketplace, the chances are good that it will have to be ISO 9000:2000 registered.

▲ While the standards do not require that a registered firm's suppliers also be registered, the revisions make it clear that registration of supply chain members is preferable.

▲ The implications of the "continual improvement" emphasis in the standards includes suppliers, which places greater importance on supply management's role in meeting the ISO 9000:2000 series challenge whether or not suppliers are registered.

Awards

Malcolm Baldrige Award

In 1987, President Ronald Reagan signed into law the Malcolm Baldrige National Quality Act. The act called for establishment of a national quality award that would provide a comprehensive framework of guidelines an organization could use to evaluate its quality program and its quality improvement efforts. Additionally, the award was designed to provide recognition for U.S. organizations that had demonstrated excellence in the attainment and management of quality. Each year firms compete for this recognition in three categories where up to two firms per year can be recognized in each category:

1. Large manufacturing companies or subsidiaries
2. Large service companies
3. Small manufacturing or service companies

On balance, the program has been extremely popular, and certainly has elevated the awareness and interest of American business people in the importance of quality. Most managers recognize that quality improvement is a developmental process that takes time, and that there are seldom any quick fixes. But, clearly, the Baldrige Award format provides a helpful road map. Part of the significance of the Baldrige has been in how companies have used it to audit their company to see where they stand and target the weak areas uncovered.[36] Additionally, previous award winners are required to share their success stories with other interested firms.

European Quality Award

Following the trend of establishing geopolitically driven awards, the EFQM in partnership with the European Commission and the European Organization for Quality announced the creation of the European Quality Award in 1991.[37] The award consists of two parts: The European Quality Prize and the European Quality Award. The prize is given to companies that demonstrate excellence in quality management practice by meeting the award criteria. The award goes to the most successful applicant. The award is very similar to the Malcolm Baldrige criteria, with the exception of results criteria in the award that address people satisfaction and impact on society.

Deming Prize

In appreciation for Dr. W. Edwards Deming's contributions to Japanese industry, in 1951 the Japanese technical community established this prestigious award that bears Dr. Deming's name. The Deming Prize is designed to recognize both individual and organizational achievements in the field of quality, with a unique emphasis on the use of statistical techniques. Any firm that meets the requirements is eligible to receive the award, including individuals and firms outside of Japan. Past recipient companies include Nissan, Toyota, Hitachi, Nippon Steel, and Florida Power and Light.[38]

Supply Management Issues

Supply managers trying to make their efforts more proactive came to realize in the late 1990s that working with suppliers to improve their quality management systems yielded greater benefits than trying to control an existing system fraught with waste. Collaborative quality management will enable supply chains to gain competitive advantage over other chains. Supply management is central to such collaboration.

Generally speaking, four factors determine the long-run quality level of a firm's purchased materials:

1. Creation of complete and appropriate specifications for quality requirements.
2. Selection of suppliers having the technical and production capabilities to do the desired quality/cost job.
3. Development of a realistic understanding with suppliers of quality requirements, and creation of the motivation to perform accordingly
4. Monitoring of suppliers' quality/cost performance—and exercise of appropriate control

Supply management is directly responsible for factors 2 and 3, and it should play a strong cooperative role in factors 1 and 4.

Requirements Development

Quality is often perceived in as many ways as those pondering its meaning. Quality can be defined in three ways:

1. In absolute terms
2. Relative to a perceived need
3. As conformance with stated requirements

In *absolute terms*, quality is a function of excellence, intrinsic value, or grade, as determined over time by society generally or by designated bodies in specialized fields. Hence, most people consider gold to be a high-quality precious metal. In a more utilitarian sense, "prime" beef is generally considered to be among the highest-quality meats in the meat market. While few absolutes endure the test of the ages, for a given period of time in a given culture, most people hold absolute views about the quality of many things.

In business and industrial activities, generally quality is first defined in terms of *relationship to a need* or a function. In these cases, the important thing is not the absolute quality of an item, but the suitability of the item in satisfying the particular need at hand. Thus, design engineers, users, and supply managers attempt to develop a material specification in which the quality characteristics of the specified material match closely with the quality characteristics needed to satisfactorily fulfill the functional requirements of the job. Consequently, in the *development* of product or material specifications, quality is defined relative to the need.

Once the specifications have been finalized, the specific requirements have been set for those who subsequently work with the specifications. For these people, especially supply management personnel and suppliers, quality is defined very simply—*conformance with the stated requirements*. A supply manager's responsibility, and a supplier's job, is to deliver material whose quality conforms satisfactorily to specification requirements.

The last two definitions are the ones with which supply managers regularly contend. As supply management works with users and suppliers in the development of material specifications, it is concerned with quality relative to a functional need. In dealing with suppliers' output quality levels, supply management is concerned with quality in the sense of its conformance to requirements.

A sound material specification represents a blend of four different considerations: (1) design requirements, (2) production factors, (3) commercial supply management considerations, and (4) marketing factors. When dealing with commercial considerations, supply management personnel should make the following investigations with respect to quality:

▲ Study the quality requirements.
▲ Ensure that quality requirements are completely and unambiguously stated in the specifications.
▲ Investigate their reasonableness, relative to cost.
▲ Ensure that specifications are written in a manner that permits competition among potential suppliers.
▲ Determine whether existing suppliers can build the desired quality into the material.
▲ Ensure the feasibility of the inspections and tests required to assure quality.

For some materials and components, such investigations are relatively simple. For others, they are extremely complex, involving highly technical considerations. Some firms, for example, include "reliability or quality engineers" on the supply management staff to assist with analysis of the more complex problems. When technical quality problems arise, a reliability engineer and supply manager jointly review the specifications to determine the appropriateness of the quality requirements. Working in a coordinating capacity, they make their recommendations to the design engineer, directing to his or her attention the potential quality problems arising from commercial considerations.

As discussed earlier, many firms utilize a cross-functional product development team in the overall design process. This approach is ideal for integrating the views of supply management, as well as the other appropriate functions, in the specification development process. In some cases, it is desirable to involve appropriate designers or application engineers from the supplier's organization in the specifications development process before the specs are finalized. Early cooperative involvement of these individuals frequently provides technical and manufacturing input from the supplier's perspective that is useful in reducing costs or in avoiding subsequent processing and quality problems in the supplier's operation. This type of cooperative activity is becoming more common in progressive buying organizations today. Clearly, supply management is responsible for planning and coordinating this type of supplier involvement.

Supplier Quality Analysis

Most firms can minimize their material quality problems simply by selecting competent and cooperative suppliers in the first place. The following paragraphs discuss briefly some of the methods used to achieve this objective.

Product Testing

One practical approach used in determining potential suppliers' quality capabilities is to test their products before purchasing them. The quality of most purchased materials can be determined by *engineering tests* or by *use tests*. Usually such tests can be conducted by the buying firm; when this is not practical, commercial testing agencies can be used.

The object of product testing is twofold: (1) to determine that a potential supplier's quality level is commensurate with the buying firm's quality needs, and (2) where feasible, to compare quality levels of several different suppliers. This permits the development of a list of qualified suppliers that the supply manager can compare on the basis of the quality/cost relationship.

Supply managers frequently utilize their own operating departments for the performance of use tests. It is not uncommon, for example, for a manager to test several brands of tires on his or her firm's vehicles during the course of regular operations. Although still feasible, use testing of most *production* parts and components is more difficult because tests often cannot be conducted until the buying firm's finished product is placed in service. In many cases, though, the buying firm's sales force is able to obtain feedback data on operating performance from customers. In such cases, various kinds of use tests can be conducted to compare the performance of different suppliers' components. Regardless of the method used, supply management usually needs the cooperation of other departments in setting up and conducting the tests. Consequently, the supply manager typically functions as an organizer and an administrator in coordinating the efforts of others.

A word of caution is appropriate, however, for the practitioner unskilled in experimental testing. Test results for products being compared must be obtained through well-designed experiments that permit *valid* comparisons. For example, mileage data on two sets of tires, each taken from a different truck, are not comparable if the two trucks were operated under significantly different conditions. The variables in the testing situation must be controlled to the extent that the results are truly comparable. A comparison of apples with oranges is of little value in the decision-making process.

Unfortunately, in conducting use tests without the benefit of controlled laboratory conditions, it may be difficult to generate truly comparable test data. Consequently, experience and judgment play an important role in determining the extent to which test results are comparable and, hence, useful. At times, interpretation of test data may also require a basic knowledge of statistical inference. While many supply managers have this background, it is essential to recognize when the help of an experienced statistician is required.

In firms that have testing laboratories for their engineering and research work, the supply management department has an additional resource to support its quality management activity. Such laboratories can take much of the experimental and interpretive burden off supply management's shoulders. In addition, they are usually equipped to conduct more precise and sophisticated tests with greater speed and ease than use testing allows. An example will illustrate the point. One of IBM's computer-manufacturing plants has an electronic testing laboratory used primarily for research and development work. The laboratory, however, also serves the supply management department. Before new electronic components are purchased for assembly into IBM products, a sample of each new component is subjected to rigorous tests in this laboratory to determine whether its performance characteristics meet the company's quality specifications. Firms without testing laboratories frequently use one of the many commercial testing laboratories located throughout the country. A directory of these laboratories, indicating their locations and their types of services, is published annually by the U.S. Department of Commerce.

First, article inspection involves requiring the first part to be shipped and inspected by the buying plant. If the part passes, then full production is authorized. The problem is determining if the trial part is made by the same methods and procedures as the actual production run. The question becomes, "is the trial part "hand made" by special craftsmen and not on production machines by ordinary workers?" If this is the case, production parts to follow may not meet the expectations of the buying firm.

Proposal Analysis

A second point at which supply management can assess a potential supplier's quality capabilities is in the proposal analysis. Firms indicate in their proposals, either directly or indirectly, how they intend to comply with the quality requirements of the purchase. The supply manager must be especially alert in detecting areas of misinterpretation or possible areas of overemphasis by the prospective supplier that could result in excessive costs. In purchases where quality requirements are critical, the trend today is to *require* potential suppliers to state *explicitly* how they plan to achieve the specified quality level with consistency.

For those potential suppliers whose written proposals survive the supply manager's analysis, the next step in evaluating quality capabilities is an on-site capability survey. Because of the time and expense involved, most companies conduct this survey only for their more important purchases and for government contracts requiring it. In inspecting a prospective supplier's facilities and records, and in talking with management and operating personnel, the buying firm's investigating team attempts to answer questions such as these:

- What is the firm's basic policy with respect to product quality and quality control?
- What is the general attitude of operators and supervisors toward quality? Does the firm utilize a specific quality program that focuses attention on the attitudes and responsibilities assumed by each individual for high quality work?[39]
- Does the prospective supplier use statistical methods to reduce process variation?
- What is the prospective supplier's engineering/production experience and ability with respect to this specific type of work?
- Is the production equipment capable of consistently producing the quality of work required?
- Exactly how is the firm organized to control quality, and to what extent does quality receive management support?
- What specific quality measurement techniques and test equipment does the prospective supplier employ? Is statistical process control utilized effectively?
- Is quality training continuous? Is the quality environment a way of life?

These questions make it clear that a supply manager alone usually cannot conduct an effective survey; rather, the endeavor must be a *team effort.* Specialists from design and process engineering, quality assurance, and sometimes maintenance are needed to give professional interpretation to the facts the survey uncovers. The supply manager's major responsibility in such an endeavor is to organize and coordinate the efforts of these specialists and to make a composite evaluation of the potential supplier, considering the findings of all team members.

Inspection Dependence

Even with the continuing improvement of quality systems to prevent defects, most firms maintain a traditional inspection department. In addition, some selected operations in both the buying firm's and the supplying firm's plants may still require either 100% or sampling inspection. This is particularly true in some cases following certain types of assembly or final assembly operations. In any case, to some extent inspection activities are still part of a supply manager's world—and for this reason, the topic is discussed briefly in the next few paragraphs.

Receiving and Inspection Procedure

If a shipment is coming from a certified or a JIT supplier, it may go directly to the point of use, bypassing the traditional receiving and inspection operations. In a majority of cases, however, when a purchase order is issued, the receiving and inspection departments both receive electronic or paper copies of the order, which specifies the inspection the material is to receive. When a shipment arrives, receiving personnel check the material against the supplier's packing slip and against the purchase order to ensure that the firm has actually received the material ordered. This is the basis for subsequent invoice/payment approvals. The receiving clerk then visually inspects the material (looking for shipping damage and so on) to determine its general condition. Finally, a receiving report is prepared on which the results of the investigation are noted. In some cases, no further inspection is required. In the case of more complex materials, a copy of the receiving report is forwarded to the inspection department, advising it that material on a given order has been placed in the "pending inspection" area and is ready for technical inspection.

The inspection department performs the specified technical inspection on a sample or on the entire lot, as appropriate, and prepares an inspection report indicating the results of the inspection. If the material fails to meet specifications, a more detailed report is usually completed describing the reasons for rejection. Rejected material, in some cases, is clearly useless to the buying firm, and the supply management department immediately arranges with the supplier for its disposition. In other situations, the most desirable course of action is less clear-cut. The buying firm often has three alternative courses of action:

1. Return the material to the supplier.
2. Keep some of the more acceptable material and return the rest.
3. Keep all the material and rework it to the point where it is acceptable. (From a strategic point of view, this is not a good alternative. It says to the supplier that it is permissible to ship off-spec material, and it can be interpreted as an invitation to do it again.)

Cases involving rework may be sent to a materials review board for study and decision. A typical board is composed primarily of personnel from production, production control, quality, and supply management. After the board reaches its decision, the appropriate papers are sent to the supply manager, who concludes final cost negotiations with the supplier. Rework costs should be charged back to the supplier.

Technical Inspection

Before a contract or a purchase order is issued, quality control personnel, in conjunction with engineering and supply management, should decide what type of inspection the incoming material will require to ensure that it meets specification requirements. If the supplier is certified, perhaps no inspection will be required. If the supplier is using SPC in the item's production, it may be sufficient simply to review the supplier's control charts for selected processing operations to determine if further inspection is required. In other cases, sampling inspection may be desirable[40], and in still other situations, 100% inspection may be required.[41] The most common sampling methods are single, double, and sequential. In all cases, quality personnel should prepare a technical inspection plan. Key information from the plan should then be communicated to the supplier in the purchase order or a related contractual document. In this way, appropriate people in both organizations are fully aware of the procedures to be followed and can work together toward that end.

Defect Detection System

To control the quality of production materials entering a manufacturing or assembly operation, companies have historically utilized defect detection systems using inspection. That is, after a batch of items has been produced at one step in the process, the items are inspected to identify the ones that do not meet the design specification. Those that do not are reworked or scrapped, and the good items pass on to the next processing operation. Frequently, although not always, a similar inspection activity is also conducted after this operation, and so on through the entire manufacturing process until a finished product is produced.

Three basic problems are inherent in the defect detection type of operation. First, there tends to be some duplication of inspection activity, both within the supplier's manufacturing operation and between the buying firm's and the supplier's operations. Second, a very large number of items are inspected. Third, *and most important*, defective items are found only after they are finished (or semi-finished)—after the mistakes have been made and after substantial processing costs have been incurred.

Defect detection systems are inefficient and expensive. Because they require time and resources to execute, there is a tendency to cut back on inspection and to reduce the standards, resulting in the acceptance of some off-spec items. Cutbacks usually occur in the interest of cost control and maintenance of production schedules. The bottom line is that in many firms using this system, quality levels tend to be inconsistent and perhaps lower than originally planned.

In Deming's 14 Points discussed earlier in this chapter, one of the points is to cease dependence on inspection. To alleviate the problems inherent in most defect detection systems, Deming proposed the use of a *defect prevention system*. The idea is to identify operating (process) problems that may produce defective items, preferably before any defective products are produced. This approach monitors the output of a process as it occurs and identifies unacceptable process changes soon after they occur. When an unsatisfactory situation is identified, the process is stopped and the operating cause (tool wear, machine adjustment, operator error, and so on) is determined. Appropriate corrective action on the operating system is then taken to prevent the production of more defectives. The specific technique Deming suggested to use to detect such process changes was *SPC*. A defect prevention system utilizing SPC is shown in simplified graphic form in Figure 7.7.[42]

Another approach to preventing defects and reducing the need for inspection in a repetitive manufacturing environment is to implement *jidohka*, championed by Ohno while he was with the Toyota. Jidohka uses an automatic mechanism to stop the entire production system whenever a defective part is found along the process line. Appropriate adjustments then take place so that major problems are prevented from arising in the future. This concept, apart from adhering to the principle of prevention rather than cure, can substantially

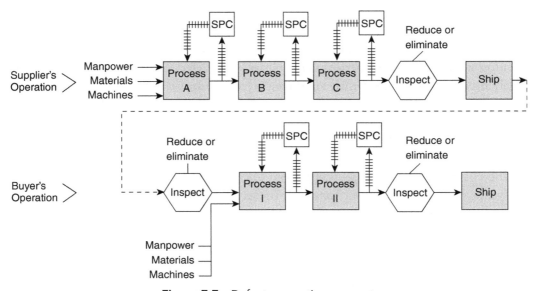

Figure 7.7 Defect prevention concept.

save on work allocation times, because the worker only needs to attend to the machine when it stops because of a problem. Jidohka forces resolution of the cause of a problem at the source when it happens and is still traceable. The fact that jidohka can shut down a production line highlights the importance of resolving the cause of the discrepancy, whereas defects discovered through traditional inspection are frequently not traced to the cause. As a result, traditional inspection procedures rarely reduce the need for same inspection in the future.

Jidohka is usually coupled with the use of poka-yoke, which could be called another form of inspection; however, most poka-yoke methods do not delay the product flow in the way that traditional inspection does because poka-yoke builds the inspection into the process. In the early 1960s, Shingo developed the concept of poka-yoke (or defect = 0), meaning mistake-proofing: source inspection is employed actively to identify process errors before they become defects; when an identification takes place, the process is stopped until the cause is determined and eliminated. Monitoring potential error sources takes place at every stage of the process, so that errors (leading to defects) are detected as soon as possible and corrected at the source, rather than at a later stage.

Needless to say, Ohno's and Shingo's techniques can be effective only if a certain (high) level of quality has already been built into the process as early as possible. This can be achieved through the use of appropriate statistical tools by the workers, who should be allowed the initiative to eliminate bottlenecks and interruptions, as well as through proper cooperation with the suppliers of the process components. Otherwise, jidohka or poka-yoke will cause a prohibitively large number of stoppages, and cost efficient production will never materialize.[43]

Critique

In summary, it is important that a buying organization develop some type of system that monitors a supplier's quality performance. Historically, defect detection systems have been widely used, but they are expensive and results frequently are not as effective as desired. In his classic book, *Quality is Free*, Philip Crosby summarizes the attitudes in American industry well in the following statement:[44]

> A prudent company makes certain that its products and services are delivered to the customer by a management system that does not condone rework, repair, waste, or nonconformance of any sort. These are expensive problems. They must not only be detected and resolved at the earliest moment, they must be prevented from occurring at all.

Once a defect prevention system is in place, benefits of improved quality control capability and reduced costs of quality flow to both the buying and supplying firms. Properly implemented, it should be a win-win situation that tends to improve cooperation and set the stage for a continuing relationship that is mutually beneficial.

Supplier Development

A *cooperatively* developed defect prevention system, including appropriate training of suppliers' personnel by the buying firm, creates a situation that should enhance a supplier's motivation to perform satisfactorily for the buying firm. The anticipated prospect of follow-on business in a longer-term continuing collaborative relationship normally provides a significant incentive for the supplier to perform well in the quality area. We mention supplier development here in order to stress its importance and refer the reader to Chapter 13 for further discussion of supplier development.

Certification

The concept of supplier certification has been practiced by a number of progressive firms for years. However, the increased emphasis on quality in the last two decades has sharpened the focus on the real values of a certification program. The certification concept recognizes the fact that a supplying and a buying firm's quality systems are two parts of a larger quality system, and that through integration of the two the total costs associated with quality can be reduced.

Certification agreements take many forms, ranging from a simple supplier's guarantee of quality to a formally negotiated document that specifies the responsibilities of both parties for specifications, process design inspection procedures, SPC applications and training, reporting and correction procedures, and so forth. Although certification programs vary widely from firm to firm, the general approach to supplier certification involves three steps:

1. Qualification
2. Education
3. The certification performance process

Qualification

Because of the mutual trust and dependence that exists in a certification relationship, a supplier is not considered for certification until the buying organization has had a fairly lengthy positive experience with the firm. In addition to quality and reliability performance, the supply manager again verifies the broader supplier characteristics that are important to the relationship—management philosophy, financial stability, R&D capability, shop organization and management, and manufacturing support capabilities, including supply management, etc.

The technical qualification requirements for certification typically are rigorous. They start with product and process design, tightened process capability studies, and a stringent quality capability survey to ensure mutual agreement on the quality system the supplier will employ. Applications for potential design of experiments and statistical process control are identified and the procedures are detailed. At this point in the process, the supply manager's quality personnel work with their counterparts in the supplier's organization to fine-tune the system and develop the procedures to be used in the final inspection.

Education

Two types of supplier education typically are required. The first deals with the buying organization's structure, people, mode of operation, and the resulting expectations for a certified supplier's performance. The second focuses on specific quality concepts and techniques that may be new to the supplier, but that are needed for successful operation—things such as various applications of SPC, unique inspection measurement techniques, the philosophy of Six Sigma or TQM, and so forth.

The extent to which education is required obviously varies from supplier to supplier. The important point, however, is that the buying firm must assume responsibility for ensuring that the education function is accomplished. In some cases, this education must continue up the supply chain to the supplier's suppliers.

The Certification Performance Process

To use an old operations expression, this is where "the rubber meets the road." The supplier must now demonstrate that it can meet the buying firm's requirements for certification.

After the manufacturing and assembly processes are stabilized and the quality assurance system is in place, the supplier's test period begins. Initially, both parties subject the supplier's output to 100% inspection. This process facilitates the identification and correction of unanticipated problems. When the predetermined quality level has been maintained for a specified period of time, full inspection is replaced by sampling inspection of declining severity as quality levels are maintained. At some point in the process, the supplier provides key evidence of implementing proactive quality management methods, such as SPC control charts, to the buying firm for continued analysis. Periodically the buying firm's quality expert visits the supplier's plant to ensure that control tests are being conducted appropriately. This process is continued for a period of perhaps six months to several years, depending on the specifics of a given operation, until the predetermined quality performance requirements have been fulfilled.

Once a supplier has gained certification, the buying firm's goal is to do very little, if any, inspection of incoming materials. When possible, material is delivered directly to the point of use after the required receiving activities. Periodically, however, the buying organization checks supplier performance in one of several ways: (1) by reviewing the supplier's critical manufacturing operations through their control charts and other statistical outputs, (2) by using a minimal sampling inspection program, (3) by reviewing test reports from the buying firm's laboratories, and (4) by periodic visits to the supplier's plant.

The concept of supplier certification fits very logically with supplier quality and collaborative arrangements, but there are potential problems:[45]

▲ Once a process has been "certified," it tends to be perceived as "unchangeable."

▲ The process of documentation and then certification can be long and resource consuming, such that once documentation does exist, companies don't want to commit the additional resources to improve it.

▲ Quite a few organizations have a team of full-time staff dedicated to maintaining certification documents and conducting internal compliance audits, but fewer or no people focused on actually improving the processes.

▲ Some companies do use their certification efforts to improve their processes, but such instances are relatively rare.

Despite these potential problems, firms that have used certification programs generally have experienced favorable results. Inspection costs typically are reduced, while quality levels usually remain high. Most suppliers take pride in being included on a customer's certification list. They are also aware that good performance places them in a favored position to receive additional business.

Summary

Quality management is a major component of supply management's supplier performance management responsibility. Quality failures lead directly to costly difficulties that reduce productivity, profit, and often market share. To preclude such losses, supply management should participate creatively in the corporate quality management program and in the firm's critical supplier quality efforts.

It is important to remember quality assurance programs and procedures are the same whether outsourcing overseas or using domestic suppliers. If your supplier is in China, it is critical to establish who monitors the quality programs and how it is verified that they are being followed.

To contribute most effectively to the organizational effort, supply management's role in the program should include (1) participation in the development of specifications;

(2) participation in the selection of appropriate quality control, inspection, and test requirements; (3) active involvement in analysis and development of proactive prevention measures at the suppliers facility; (4) the selection and motivation of qualified suppliers; and (5) the subsequent monitoring and nurturing of the ongoing relationships between the buying and supplying firms.

This chapter stresses the importance of assessing quality management prior to entering into long-term contracts with suppliers. All supply managers should have at least a rudimentary understanding of quality systems, tools of quality, process capability analysis, and SPC concepts. A world-class supply management firm views its quality system and the supplier's quality system as two parts of a single integrated system. The supply manager responsible for developing the relationship between the buying and selling firm is the middle person in this integrated system. He or she first must be able to cooperate with in-house quality people in designing and making operating decisions about the firm's quality system. Because the supply manager is the key communication link with suppliers, he or she must be able to deal with them effectively on a wide variety of quality issues.

Finally, the recent quality problems of Toyota regarding sudden and unexpected acceleration is an excellent example of the importance of testing the interaction of general systems when they are brought together into one system. In addition, the Toyota case history is a good lesson for the need to handle product recalls immediately and openly.

Endnotes

1. N. Logothetis, *Managing for Total Quality—from Deming to Taguchi to SPC*, The Manufacturing Practitioner Series (United Kingdom: Prentice Hall International, 1992). The authors highly recommend this book due to its conciseness, straightforward clarity, and global view.
2. David N. Burt and Richard L. Pinkerton, *Strategic Proactive Procurement* (New York: AMACOM, 1996), 185. A letter from Deming to Pinkerton, October 14, 1982.
3. Logothetis, op. cit.
4. M. Imai, *Kaizen: The Key to Japan's Competitive Success* (New York: Random House, 1986).
5. Logothetis, op. cit., 90.
6. Logothetis, op. cit., 62.
7. Logothetis, op. cit., 64.
8. While a majority of a firm's quality problems commonly can be attributed to purchased materials and subassemblies, the root cause of most of these problems is in the design and resulting specifications of the items and their production processes. Supply

management professionals have two areas of responsibility in the design/quality issue: They and invited suppliers must work cooperatively with the firm's design engineers to design appropriate quality characteristics into the materials to be purchased, and (2) this same group must ensure that suppliers design variability out of their production processes to the extent that is practical.

9. As quoted by Paul Moffat in "Quality Assurance," *The Purchasing Handbook* (New York: McGraw-Hill, Inc., 1992) 421.

10. As quoted by Greg Hutchins, *ISO 9000* (Essex Junction, VT: Oliver Wight-Publications, Inc., 1993) 4.

11. James F. Cali, *TQM for Purchasing Management* (New York: McGraw-Hill, 1993) 36.

12. James W. Martin, *Lean Six Sigma for Supply Chain Management: The 10 Step Solution Process* (New York: McGraw-Hill, 2007) 63–64. (See mhprofessional.com)

13. Logothetis, op. cit.

14. Peter S. Pande, Robert P. Neuman, and Roland R. Cavanagh, *The Six Sigma Way* (New York: McGraw-Hill, 2000) 77.

15. Ibid., 29.

16. Ibid., 37.

17. Ibid., 33.

18. Ibid.

19. Address to General Electric Company Annual Meeting, Cleveland, OH, April 21, 1999.

20. General Electric Company Annual Meeting, Charlotte, NC, April 23, 1997.

21. www.iso.ch, online website for the International Organization of Standardization, February 2008.

22. Keki R. Bhote, *World Class Quality* (New York: AMACOM, 1991). Also see the book review by Richard L. Pinkerton published in the *International Journal of Purchasing Materials Management,* Summer, (1993) 51–52.

23. A. V. Feigenbaum, *The Challenge of Total Quality Control, Industrial Quality Control,* May, 17–23, 1957.

24. J. M. Juran, *Quality Control Handbook* (New York: McGraw-Hill, 1951).

25. Many textbooks present figures for the COQ showing a plot where the optimum point is plotted somewhere midway between 0% and 100% good. This has confused many students over the years in believing that the optimal point is somewhere around 50% good.

26. Jack Campanella, Ed., *Principles of Quality Costs*, 2nd ed. (Milwaukee, WI: ASQC Quality Press, 1990).

27. G. Taguchi, and D. Clausing, "Robust Quality," *Harvard Business Review,* January-February, (1990): 65–75.

28. In practice, the natural capability range of the process frequently is *estimated* by taking the mean output quality value +3 standard deviations of the output values. In the case of normally distributed output values, this means that the process capability range includes 99.7% of the expected population values. In other words, there are 3 chances in 1000 that the process output will fall outside the estimated process capability range. Some firms with high precision requirements estimate the natural capability range of a process by using the mean output value ±4, ±5, or ±6 standard deviations.

29. The statistical central limit theorem states that the *means of small samples* tend to be normally distributed regardless of the type of distribution from which the individual sample values are taken.

30. Formerly the National Association of Purchasing Management. www.ism.ws, February 2008.

31. APICS stood for the American Production and Inventory Control Society. After the society changed its name, it decided to continue using its well-known acronym of APICS.

32. Steven Casper and Bob Hancke, *Global Quality Norms within National Production Regimes: ISO 9000 Standards in French and German Car Industries, Organization Studies* (Berlin: Walter De Gruyter and Company, 1999).

33. The American National Standards Institute (ANSI) is the U.S. member of the ISO.

34. Hutchins, op. cit., 3.

35. www.iso.org—Online website for the ISO, February 2008.

36. Robert E. Cole, *Learning from the Quality Movement* (Boston MA: Harvard Business School Publishing, 1998) 67.

37. James R. Evans and William M. Lindsay, *The Management and Control of Quality* (Mason, OH: Southwestern College Publishing, 1999) 143–144.

38. Logothetis, op. cit., 28.

39. For two classic discussions about the individual's responsibility for quality, see William Ouchi, "The Q-C Circle," *Theory Z* (Reading, MA: Addison-Wesley, 1981) 261–268; and Robert M. Smith, "Zero-Defects and You," *Management Services* January-February (1966): 35–38.

40. Industry uses a wide variety of statistical sampling plans. Two of the more commonly used sources are Dodge and Romig, *Sampling Inspection Tables-Single and Double Sampling* (New York: John Wiley & Sons, Inc., 1959); and Freeman, Friedman, Mosteller, and Wallis, *Sampling Inspection* (New York: McGraw-Hill Publishing Company, 1948). The Dodge and Romig tables are designed specifically to minimize total sampling. Sampling inspection, by Freeman et al., contains plans that are particularly useful in inspecting material coming from statistically controlled production processes. These plans tend to minimize the "consumer's risk" of accepting off-spec material.

41. In many types of inspection activities involving human operation or judgment, experience indicates that 100% inspection may detect only 80 to 95% of the defects present. Consequently, when quality is extremely critical, a second inspection operation may be required.

42. Figure 7.7 was developed from material in Gordon K. Constable, "Statistical Process Control and Purchasing," *Freedom of Choice* (Tempe, AZ: National Association of Purchasing Management, 1987) 15.

43. Logothetis, op. cit., 97–98.

44. Philip B. Crosby, *Quality is Free* (New York: Mentor New American Library, Times Mirror, 1979) 106.

45. Pande, Neuman, and Cavanagh, op. cit., 63.

Suggested Reading

Adams, Cary W., Praveen Gupta, and Charles E. Wilson. *Six Sigma Deployment*. Woburn, MA: Butterworth-Heinemann, 2003.

Akao, Yoji. *Quality Function Deployment: Integrating Customer Requirements into Product Designs*. New York: Productivity Press, 2004.

Basu, Ron, and J. Nevan Wright. *Quality Beyond Six Sigma*. Woburn, MA: Butterworth-Heinemann, 2003.

Brue, Greg. *Six Sigma for Managers*. New York: McGraw-Hill Companies, Inc, 2002.

Cali, James F. *TQM for Purchasing Management*. New York: McGraw-Hill Companies, Inc, 1993.

Crosby, Philip B. *Quality is Free: The Art of Making Quality Certain*. New York: McGraw-Hill, 1979.

Deming, W.E. *Out of the Crisis*. Cambridge, MA: MIT Center for Advance Engineering Study, 1986.

Fernandez, Richard R. *Total Quality in Purchasing and Supplier Management*. Boca Raton FL: St. Lucie Press, 1995.

Garvin, David A. *Managing Quality: The Strategic and Competitive Edge*. New York: The Free Press, 1988.

Greico, Peter Jr., Michael W. Gozzo, and Jerry W. Claunch. *Supplier Certification II: A Handbook for Achieving Excellence through Continuous Improvement*, 3rd ed. Plantsville, CT: PT Publications, 1992.

Harry, Mikel, and Richard Schroeder. *Six Sigma: The Breakthrough Management Strategy Revolutionizing the World's Top Corporations*. New York: Doubleday, 2000.

Juran, I. M. *Juran on Planning for Quality*. New York: The Free Press, 1988.

Keller, Paul A. *Six Sigma Demystified.* New York: McGraw Hill Professional, 2005.

Martin, James W. *Lean Six Sigma for Supply Chain Management: The 10-Step Solution Process.* New York: The McGraw Hill Companies, 2007.

Ruamsook, Kusumal, Dawn Russell, and Evelyn Thomchick. "U.S. Sourcing from Low-Cost Countries: A Comparative Analysis of Supplier Performance." *The Journal of Supply Chain Management* 43(4)(2007): 16–29.

CHAPTER 8

Purchasing Equipment

The Nuances of Capital Equipment Procurement

Nonrecurring Purchases

The purchase of a particular piece of capital equipment typically occurs no more than once every three to five years. For example, a supply manager recently purchased a unique high-temperature electric furnace for use in her company's research and development laboratory. Since the furnace is used only periodically for experimental work, it is very unlikely that another purchase of this kind of equipment will be made in the foreseeable future.

In contrast, a few industrial operations require the use of many identical machines in their production process. For example, in petroleum and chemical processing plants, the product is transported by pipeline throughout most of the production operation. This requires dozens, at times hundreds, of similar pumps which vary only in size and details of construction. To keep capital expenditures at a fairly uniform level from year to year and to minimize maintenance costs, pumps are often replaced on a continuing basis rather than all at once. Although relatively uncommon, this type of operating equipment purchase has some of the characteristics of conventional production purchasing.

The lead-time requirement is a unique feature of most equipment purchases. While some types of equipment are standard off-the-shelf products, many are not. Much production machinery and prime moving equipment is built, at least in part, to operate under specific conditions that are particular to each purchaser's operation. Consequently, manufacturing lead time for potential suppliers is usually a matter of months or perhaps years. The production of a large steam turbine generating unit, for example, may require negotiating and expediting work substantially different from that normally required in production procurement.

Nature and Size of Expenditure

An expenditure of company funds for capital equipment is an investment. If purchased wisely and operated efficiently, equipment generates profits for its owner. Because it affects the costs of production, the selection of major capital equipment should be a matter of significant concern to top management.

The purchase of most major equipment involves the expenditure of a substantial sum of money. The purchase price for a piece of equipment, however, frequently is overshadowed by other elements of cost. Since a machine is often used for many years, the cost of operation and maintenance during its lifetime may far exceed its initial cost. For example, downtime costs easily may exceed the equipment's purchase price. An auto assembler estimates its downtime cost as $26,000 *per minute*. Hence, the *total life cost* of a machine, relative to its productivity, frequently is the cost factor of primary importance.[1] Although estimating operating and maintenance costs that will be incurred in future years is not easy, such costs will be incurred and must be addressed when one is comparing the total cost of ownership of two or more items of equipment that will satisfy the firm's needs. Thus, post-sale technical service and the availability and cost of replacement parts may be critical supplier selection criteria.

The timing of many equipment purchases often presents a paradoxical situation. Typically, the general supply capabilities of equipment producers cannot be adjusted quickly to changes in demand. Thus, because most firms' equipment purchases are made infrequently and often can be postponed, producers of industrial capital goods frequently find themselves in a "feast or famine" type of business. When a potential purchaser's business is good, it needs additional production equipment as quickly as possible to satisfy customers' burgeoning demands. But because other purchasers are in the same situation, the buyer may find that equipment prices are rising in a market of short supply. Conversely, when a buyer's business is down and additional production equipment is not needed, equipment is in plentiful supply, often at reduced prices.

Finally, the installation cost may be as much as 30 to 50% of the total cost and involve multiple contractors for site preparation. The contract must cover all such details.

Building the Foundation

Identify the Need for a Procurement

At least five functional areas may identify a need for the acquisition of equipment: the using department, marketing, process engineering, supply management, and plant engineering (Figure 8.1). The using department may desire equipment which is more productive, that

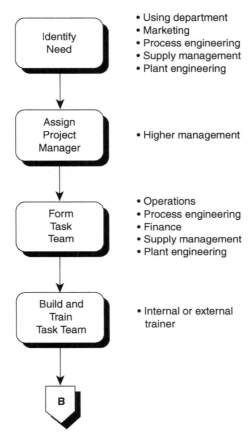

Figure 8.1 Capital equipment procurement
Phase 1: build the foundation.

is, requires less equipment or operator time (or both) per unit of throughput. Marketing may identify new products whose production processes require new equipment. Process engineering (or operations) is concerned with the equipment's ability to meet new and changing requirements. Supply management is responsible for monitoring threats and opportunities in the supply world. In the process, supply management may identify relevant new technology and new equipment. Additionally, supply management should be potential suppliers' primary point of contact. Supply management should be both a filter and a conduit for suppliers whose products may be of interest to other functions within the firm. Plant engineering may stimulate equipment procurement by identifying potential risks of downtime if an item of equipment is not replaced.

Project Management

If the equipment is critical on the basis of either cost or schedule, a project manager should be selected to drive the process. Ideally, that individual will champion and oversee the project to success. Good people, communication, leadership, and project management skills should be a requirement for the person who will fill this position.

Selection of an Equipment Sourcing Team

Depending on the criticality of the procurement, the following functional areas may be represented on the team responsible for obtaining the equipment: operations, process engineering, finance, supply management, and plant engineering.

Operations is responsible for identifying the required and desired operating characteristics of the equipment. (These characteristics are described in greater detail in the next section.) Process engineering, which is also known as manufacturing engineering, is concerned with the equipment's ability to meet current and likely future needs. The process engineer must balance two frequently conflicting forces: a very specialized piece of equipment may be the most productive one but may be incapable of being adapted to possible future production needs.

Finance's involvement is based on four primary interests in equipment purchases and leases. First, this department usually administers the firm's capital budget; therefore, it is concerned with the allocation of funds for the proposed purchase. If the budget contains a provision for such equipment, all is well; if it does not, the team will have to secure a budget authorization. Second, the finance department has the responsibility for deciding how to finance such purchases. Is enough cash available internally? Can a long-term loan be arranged? Will it be necessary to raise the money through a bond issue? For large purchases, the answers to these questions bear heavily on the final equipment selection decision. Third, the finance department should be involved in the economic analysis of alternative machines. Fourth, finance normally chairs the lease vs. buy analysis. In some firms, the finance department conducts the original analyses; in others, the analyses are done by engineering or supply management. In any case, the finance department normally is involved in these activities in connection with its capital budgeting responsibility.

Supply management plays many roles. It is the primary point of contact with potential suppliers and a conduit for the flow of information. Supply management ensures that the statement of work or specification that is developed to describe the firm's needs is sufficiently specific to protect the firm's interests and broad enough to ensure competition, assuming that competition is appropriate. Supply management normally guides the sourcing process and leads the negotiating team, and it also is responsible for the post-award activities.

Plant engineering is concerned with both current and future issues. Immediate considerations include physical issues such as size, foundation, and power requirements. Future concerns include reliability, maintainability, service support, and the availability of replacement and spare parts.

Build and Train the Team

The careful selection of the "right" representatives from the appropriate functional areas is an essential task. Unless the representatives have recent successful experience working as a team, an internal or external trainer should be called on to build and train the team.

Identify Objectives and Estimate Costs

Identifying Objectives

As is true with production requirements, some 80% of the costs associated with the procurement of equipment are built in during the requirements development stage! See Figure 8.2 for further explanation. Estimating the acquisition cost (purchase price, installation, spares, and training costs) and the total cost of ownership is always difficult. It is especially challenging at this early stage. However, it is strongly recommended that the team agree on both a target acquisition cost and a target total cost of ownership. Normally, the total cost of ownership is based on the present value (P.V.) of the anticipated stream of expenditures and downtime minus the P.V. of the item's estimated salvage value.

The desired operating and engineering characteristics are by far the most influential factors in selecting the supplier for a particular item of equipment. The user and appropriate engineering personnel must clearly establish the function the equipment is to perform and its design and operating capabilities. Operating characteristics include: the equipment's capacity, setup and run times, product yields, operator ease of use, and adaptability to be able to meet unforeseen requirements.

Closely related to the equipment's operating characteristics are its engineering features. *Ideally, these features will be compatible with the buying firm's existing equipment, process, and plant layout.* They also must comply with standards established by state and federal regulatory agencies such as the Occupational Safety and Health Administration (OSHA) and the Environmental Protection Agency (EPA). Some major engineering considerations are reliability (how long will it operate before requiring maintenance or replacement?), size and mounting dimensions, interface with other equipment, power and maintenance requirements, safety and OSHA requirements, and pollution and EPA requirements.

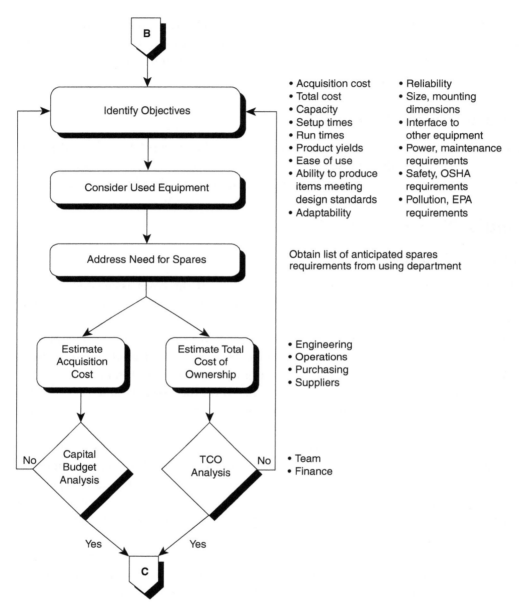

Figure 8.2 Capital equipment procurement Phase II: identify objectives and estimate costs.

Used Equipment

A buying firm is by no means restricted to the purchase of new capital equipment. Purchases of used machinery constitute an important percentage of total machinery sales.

Reasons for Purchasing Used Equipment

The firm may consider buying or leasing used equipment for several reasons. First, the cost of used machinery is substantially less than that of new equipment. Analysis of payback or return on investment may well reveal that a piece of used equipment is a better buy than a new machine. Even if this is not the case, a firm's financial position may dictate the purchase or lease of a lower-priced used machine. Second, used equipment frequently is more readily available than new equipment. In some situations, availability may override all other considerations.

A third and very common reason for the purchase or lease of used equipment is that used equipment may adequately satisfy the purchasing firm's need, in which case there is no point in acquiring new equipment. In cases in which operating requirements are not severe, a used machine in sound condition frequently provides economical service for many years. If equipment is needed for standby or peak-capacity operation, or for use on a short-lived project, more often than not used equipment can satisfy the need very well. The used equipment must not affect the manufacturing process negatively (i.e., the equipment is not a critical "technology"-dependent item).

The Used Equipment Market

Used equipment becomes available for purchase for a number of legitimate reasons. When a firm buys a new machine, it frequently disposes of its old one. Although the old machine may be obsolete relative to the original owner's needs, it is often completely adequate for the needs of many potential buyers. If significant changes are made in the previous owner's product design or production process, it may be advantageous to the original owner to purchase or lease more specialized production equipment. Finally, some used equipment becomes available because the owner lost a particular contract or has discontinued operation altogether.

Whatever the reason, a great deal of used equipment is available and commonly is purchased from one of four sources: (1) used equipment dealers, (2) directly from the owner, (3) brokers, and (4) auctions. In recent years, the majority of these purchases have been made from used equipment dealers who specialize in buying, overhauling, and marketing certain types of equipment. Dealers are usually located in large industrial areas, and, as a rule, they periodically advertise the major equipment available. Just remember, "as is, where is" means just that. There is no warranty of any kind (except for fraud—it isn't what was advertised) and the buyer has to come get it, moving the equipment at the buyer's expense.

Used Equipment Dealers

These dealers typically specialize in certain kinds of equipment and sell two types of machines—"reconditioned" and "rebuilt." Generally speaking, a reconditioned machine carries a minimal dealer warranty and sells for approximately 40 to 50% of the price of a similar new machine. The machine usually has been cleaned and painted, broken and severely worn parts have been replaced, and the machine has been tested under power. A rebuilt machine typically carries a more inclusive dealer warranty and sells for perhaps 50 to 70% of a new machine's price. A rebuilt machine usually has been completely dismantled and built up from the base. All worn and broken parts have been replaced, wearing surfaces have been reground and realigned, the machine has been reassembled to hold original tolerances, and it has been tested under power.

Sale by Owner

Some owners prefer to sell their used equipment directly to the next user because they think they can realize a higher price than by selling to a dealer. Some buying firms also prefer this arrangement. It permits them to see the machine in operation and learn something about its usage history before making the purchasing decision.

Brokers

A broker is an intermediary who brings buyers and sellers together but generally does not take title to the equipment sold. Brokers sometimes liquidate large segments of the equipment of a complete plant. Occasionally, an industrial supply house or a manufacturer's agent will act as a broker for a good customer by helping the firm dispose of an odd piece of equipment that has a limited sales market.

Auctions

Auction sales represent still another source of used equipment. Several types of auction firms are in operation. Some actually function as traders, buying equipment and selling from their own inventory. More common, however, are firms that simply provide the auction sale service. Their commission is usually somewhat less than a broker's commission. Generally speaking, buying at auction is somewhat more risky than the other supply sources because auctioned machines usually carry no warranty, and rarely is it possible to have the machine demonstrated. In some cases, however, machines can be purchased at auction via videotape or closed-circuit TV; this permits the buyer to see the machine operating in a distant plant.

Cautions in Purchasing Used Equipment

The age-old adage of *caveat emptor*—let the buyer beware—is particularly applicable when purchasing used equipment. It may be difficult to determine the true condition of a used machine and to estimate the type and length of service it will provide. For this reason, it is wise to have one supply professional specialize in used equipment. Moreover, it is virtually essential to enlist the cooperation of an experienced production or maintenance specialist in appraising used equipment. Send an experienced operator and maintenance mechanic to conduct the final inspection. If the company pilot says "don't buy this aircraft," don't buy it. It is always sound practice to check the reputation of a used equipment supplier and to shop around, inspecting several machines before making a purchase. Whenever possible, a machine should be observed under power through a complete operating cycle. Finally, a prospective buyer should determine the age of a machine. If not available in the seller's records, the age of a machine can be traced through the manufacturer by means of serial number identification. The combined knowledge of age and usage history is a key guide to the future performance of a used machine.

In preparing a purchase order or contract for used capital equipment, one must take care to include all essential data. In addition to an adequate description of the machine, an order should specify the accessories included, warranty provisions (if any), services to be performed before shipment, and financing as well as shipping arrangements. Generally, sellers do not provide service for used equipment after the purchase. All transportation, handling, installation, and start-up costs, as well as risk, usually are borne by the purchaser.

Spares

When is the optimal time to obtain prices on spares and service agreements? Obviously, when competition is present! Wise supply professionals obtain a list of anticipated spare parts and service requirements from the prospective user of the equipment so that they can solicit prices for these items when soliciting the price of the equipment itself. Otherwise, excessive prices frequently become the norm when the need for the spare part or service arises.

Estimating Acquisition Costs and the Total Cost of Ownership (TCO)

After the desired operating and engineering objectives have been identified, the team should develop both acquisition cost and TCO estimates. If the item to be purchased is a standard one, supply management will obtain estimates of the purchase price from its files or from potential suppliers. If customized or non-standard equipment is to be purchased, it is desirable for supply management to obtain informal estimates from potential suppliers. These estimates are compared with the budget authorization and are input to the total cost of ownership analysis. If these analyses indicate that either cost is likely to be excessive, the team must reexamine the list of desired objectives and make appropriate adjustments. Ideally, those analyses and adjustments are made before proceeding to the development of the appropriate specifications.

Develop Specifications and Initiate Sourcing, Pricing, and TCO Analysis

Develop Specifications

Normally, a performance specification is developed by the team. As was seen in Chapter 6, a performance specification describes the desired performance (160 units per minute, setup time of 5 minutes, tolerances, etc.) together with the required engineering features. *If* another type of specification is determined to be appropriate, one or more qualified potential suppliers may become involved in the process of developing the specification (see Figure 8.3).

One of the advantages of formally establishing a sourcing team is the fact that more *cooperative* action usually is generated in attacking the procurement. This can be extremely useful in the development of the equipment specifications. Quality/cost trade-offs are best addressed by a team. When specifications are nearing completion and requests for proposal are to be issued, a good supply management professional should function in the role of an informal auditor. Although technical requirements predominate, the supply management professional should make every effort to see that specifications are written as functionally as possible. Most equipment users are biased toward or against specific types of equipment. Every effort should be made to exclude personal biases from the specifications. The nature of many equipment requirements limits the number of possible suppliers. This number should not be reduced further by arbitrarily excluding certain potential suppliers on the grounds of personal prejudice. After development of the appropriate specification, sourcing, pricing, and negotiations are accomplished.

Sourcing

The sourcing of equipment suppliers involves the quantitative and qualitative analyses described in detail in Chapter 13. Briefly, the first step in equipment sourcing is the development of a request for proposal. Once proposals are received, they are evaluated for responsiveness. The supplier or suppliers that appear to be most attractive are identified. Some suppliers are more qualified in the "soft" or qualitative area than are others. The degree of qualification should be considered carefully by the team in deciding which machine to buy. The team must determine the level of a supplier's technical, production,

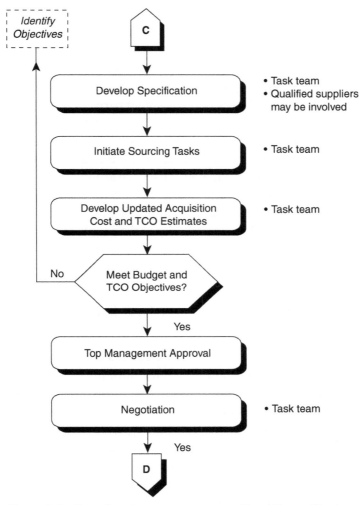

Figure 8.3 Capital equipment procurement Phase III: specifications, sourcing, pricing, and TCO analysis.

and commercial capabilities. The team must also assess the supplier's capability and willingness to provide any engineering service required during the installation and startup of the new equipment. This is an extremely important financial consideration when complex expensive equipment such as steam turbines and numerically controlled machine tools are involved. Closely related to this factor is the necessity of training operators. What service is the supplier willing to provide in this area? The reliability of a supplier in standing behind its guarantees is another important consideration. A combination of the supplier's history of satisfactory performance and financial viability must be addressed. Once the equipment is installed, unexpected problems beyond the purchasing firm's control sometimes add significantly to the total cost of a machine. Finally, what is the supplier's policy on providing spares and replacement parts? What will be the availability of obsolete parts when the purchased machine is superseded by a new model? The policy of one pump manufacturer, for example, is to produce a small stock of replacement parts for obsolete equipment once every six months. The semi-annual production policy of this manufacturer, combined with its low inventory levels, forces some customers to carry unreasonably large stocks of major replacement parts. The other costly alternative for the customer is to risk occasional breakdowns, which may leave a machine out of service as long as three or four months, waiting for the next run of parts. In practice, unfortunately, such considerations frequently play a minor role in the initial selection of equipment suppliers, only to assume major proportions at a later date. It is the responsibility of the supply management department to evaluate potential suppliers in light of these qualitative factors and to bring significant considerations before the evaluating group.

Develop Updated Acquisition Cost and TCO Estimates

When proposals are received, a supply management professional tabulates them and makes the necessary calculations so they can be interpreted on a comparative basis by the team responsible for the final recommendation. Because administration and control of such activities are clearly related to the capital budgeting function, the finance department frequently assumes responsibility for conducting a total profitability study. The authors' view, however, is that once management has selected the types of analyses to be used, the supply management department can perform the analyses more easily and effectively. Those analyses are a logical extension of the supply management department's proposal analysis activities. Clearly, the supply management professional is familiar with any proposal complications. Through his or her involvement in the preceding technical discussions, that professional should also understand any technical problems involved in developing estimates for maintenance and operating costs. Consequently, an individual with a good understanding of the total cost situation may effectively prepare, interpret, and present the complete package of price, cost, and profitability data for the group's consideration.

The team now has considerable information to update its TCO estimate including all likely acquisition costs and data on actual operating characteristics. As a result, a reasonably accurate TCO estimate can be developed. One of the most challenging issues confronting the team responsible for the selection of an expensive item of equipment is the possible conflict between the budget authorization and the total cost of ownership. The budget focuses on "now" costs (purchase price, transportation, installation, training,

Table 8.1   Total Cost of Ownership for Items X and Y

	X	Y
Acquisition cost	$1,000,000	$1,200,000
Present value of future costs for spares, maintenance, operator labor, downtime, etc.	$2,000,000	$1,300,000
Total	$3,000,000	$2,500,000

and initial spares); the total cost of ownership addresses both now and likely future costs. Compare the total cost of ownership for items X and Y in Table 8.1. If the budget authorizes the expenditure of $1,000,000, the sourcing team will tend to acquire X, incurring a likely $500,000 excess cost, based on the total cost of ownership for the two alternatives.

Meet Budget and TCO Objectives?

The team now ensures that the updated acquisition and TCO estimates are at or below its objectives. If they exceed either objective, the project is returned to the objectives phase for revision (see Figure 8.2).

Top Management Approval

If significant funds are involved, once the project satisfies budgetary and total cost considerations, it should be forwarded to top management for review and approval.

Negotiation

Once appropriate approvals have been received, the team proceeds to negotiate all the terms and conditions of a contract with the most attractive potential supplier as discussed in Chapter 16. During negotiations, the negotiating team may explore the advantages of leasing.

Leased Equipment

In addition to the possibility of purchasing new or used equipment to satisfy a firm's requirements, an equipment customer has a third alternative—that of *leasing* the equipment (see Figure 8.4). In recent years, leasing has become big business. It is now a $190 billion industry whose volume of business has doubled in the last five years. Approximately 20% of the new office and industrial equipment used in American business today is leased. Generally accepted

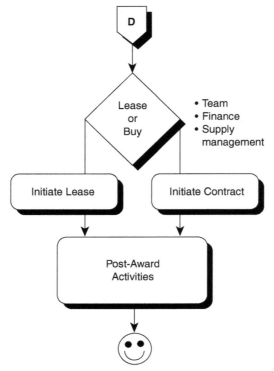

Figure 8.4 Capital equipment procurement
Phase IV: sourcing, lease/buy analysis, and
post-award activities.

reasons underlying this trend appear to be the heavy demand for capital in most firms, the cost of capital, and the increased flexibility that can be negotiated in the contract.

Types of Leases

If one looks at leases in terms of the basic purposes for which they are used, most fall into one of two categories: operating leases and financial leases.

As the name implies, an *operating lease* is used by most firms as a vehicle to facilitate business operations. The focus typically is on operating convenience and flexibility. Frequently, a firm has a temporary need for equipment to be used in the office or on a special production or maintenance job. The firm requires the use of the asset but is not interested in owning it. In some cases, the need stretches beyond the temporary period but the firm still is not interested in ownership and the risks and responsibilities that accompany it. Equipment obtained by means of an operating lease may fit such a firm's needs well. Most operating leases are short term, for a fixed period of time that is considerably less than the life of the equipment being leased. In many instances, an operating lease is used when the customer firm wants the freedom of being able to avail itself of new and unexpected technology.

A *financial lease*, in most cases, is used for a very different purpose. When operating equipment is obtained by means of a financial lease, the primary motivation for using this type of lease is to obtain financial leverage and related longer-term financial benefits.[2] Relatively speaking, a financial lease is a long-term lease that usually covers a time period just a bit shorter than the approximate life of the equipment being leased. Many financial leases cannnot be canceled. Several purists argue that such leases distort the firm's financial reports by reducing debt, which might otherwise be required to finance major purchases. Additionally, assets will be understated, resulting in a lower asset base when calculating the firm's return on assets.

Factors Favoring Leasing

Operating and Managerial Convenience

While there are different types of leasing organizations, most industrial leasing firms are called *full-service* lessors. This means that the leasing organization owns the equipment, has its own continuing source of financing, and is prepared to assume all the responsibilities of ownership for the lessee. Hence, the lessee has full use of the equipment and can concentrate on its regular business operations without having to worry about maintenance, special service, and other administrative tasks associated with equipment ownership. This can become a major benefit in the case of complex equipment requiring highly specialized technical support.

Operating Flexibility

With relatively short-term leases for selected pieces of operating equipment, a lessee is not locked into long-term commitments resulting from large capital investments. The lessee can maintain maximum flexibility in its operations to respond to changing business conditions and subsequent production requirements. It can use a leasing arrangement to meet temporary operating needs with relative ease, and in the same manner it can test new equipment prior to making a longer-term purchasing decision.

Obsolescence Protection

Leasing substantially reduces the risk of equipment obsolescence. In many businesses, particularly those using high-tech equipment, some machines become technologically obsolete in a very short period. In leasing things such as data processing equipment, for example, an arrangement usually can be made with the lessor to replace or upgrade the old equipment. This can be an extremely important consideration in a highly competitive industry.

Financial Leverage

A major advantage of leasing expensive equipment stems from the fact that a leasing decision typically replaces a large capital outlay with much smaller, regularly timed payments. This frees working capital for use in meeting expanded operating costs or for investment in other segments of the business operation. Leasing equipment reduces the upfront start up cost for new businesses.

Viewing the lease strictly as a financing mechanism provides a cash flow advantage over a conventional bank loan financing arrangement. Most equipment leases can be stretched over a longer payment period, thus reducing the relative size of monthly payments.

In the United States, the Financial Accounting Standards Board (FASB) requires that all financial leases be capitalized. This means that the present value of all the lease payments must be shown as a part of the firm's debt.

Income Tax Considerations

If a lease fulfills the Internal Revenue Service requirements for a "true lease" from an accounting point of view, lease payments are recorded as operating expenses. As long as lease payments exceed the value of allowable depreciation (if the asset were owned), an additional tax shield is provided for the lessee. Comparatively speaking, the amount of taxable income is reduced by the difference between the lease expense payments and allowable depreciation expenses. Therefore, if a lessee plans on obtaining this tax benefit, it is important to compare the proposed lease payments with the corresponding allowable depreciation figures (if the asset were owned) over the life of the lease to ensure that the anticipated positive relationship actually exists.

It is important to bear in mind that the IRS has established rather stringent guidelines to distinguish between a true lease and a lease that is really intended to be a disguised conditional sales contract. Whenever it appears that the actual intent of the parties is simply to use the lease as a financing arrangement for a subsequent purchase of the equipment, no incremental tax benefit is allowed.

The Tax Reform Act of 1986 has had a significant influence on the lease/buy decision for some firms. The 1986 tax law eliminated the investment tax credit and modified the allowable depreciation preferences, forcing many firms into an alternative minimum tax position. The combined effect of these two provisions has made it more costly, from a tax perspective, for some capital-intensive firms to buy capital equipment. Although the situation is different for each firm, experience to date indicates that the 1986 law has made leasing a more attractive alternative for many firms.

Factors Weighing Against Leasing

Cost

As a general rule, the primary disadvantage of leasing is its cost. "Margins" are typically higher on leases than interest rates are on direct loans. This is understandable, because in addition to covering financing charges, the lessor must also bear all the risks associated with ownership (including obsolescence and inflation risks). The typical financial lease runs for approximately three-quarters of the equipment's estimated useful life, and monthly fees total approximately 120 to 135% of the purchase price over the life of the lease contract.

Control

A second disadvantage arises from the fact that the lessor retains control of the equipment. This loss of control often places restrictions on the manner in which the equipment is operated; it also requires that the lessee give the lessor access to the equipment for inspection and maintenance. There are usually times when such control by the lessor creates inconveniences for the lessee. Closely related to this fact is the possibility that the lessee may have its purchasing prerogatives constrained with respect to the purchase of operating supplies for the leased equipment. A lessor normally wants to have the equipment operated with its own supplies. If use of other manufacturers' supplies could conceivably impair the machine's performance, the lessee usually agrees to such an arrangement. Frequently, higher prices are paid for such supplies than if they were purchased on the open market.

To Lease or to Buy?

Cost Comparison

It is imperative for the supply or finance professional to make a comparative analysis of the cost to lease and the cost to own. A discounted cash flow analysis of the two alternatives over the life of the lease is the most accurate and straightforward approach to use.

Procedurally, the same approach is used in this analysis as is used in making a life cycle cost analysis. Basically, all cost and savings factors for the lease alternative are identified and quantified—and then projected to appropriate future dates when they actually will be incurred. This produces a cost matrix over the life of the lease. All future costs are then discounted to their present values and subsequently summed to express the total cost in present value terms.[3] The same procedure is followed for the "buy" alternative. Total present value costs then can be compared directly to determine the additional true cost (or saving) associated with the leasing alternative.

The Decision

The lease-or-buy decision should be made just as is any other sound purchasing decision. First, the merits of the alternative items of equipment must be assessed relative to the buying firm's functional needs. The total cost of each then is considered in light of the preceding functional analysis. To these factors are added the relevant qualitative considerations that may vary among suppliers, markets, economic conditions, and so on. A decision then is made on the basis of the relative cost/benefit assessments. In the case of a lease-or-buy decision, this process can be summarized in the following four steps:

1. Determine the *operating* (including financing considerations) advantages and disadvantages of leasing and of owning. Input from operations, finance, and supply management is required.
2. Compare the two alternatives and answer the following question: From an operating point of view, is leasing the preferred alternative?
3. If leasing is preferable, calculate and compare the present *value costs* of the two alternatives.
4. Make the decision: Are the operating benefits of leasing worth the additional cost?

More often than not, the final decision will center on a determination of whether the extra cost entailed in leasing is justified by the avoidance of the major risks and responsibilities associated with ownership.[4]

Initiate Lease or Contract

The purchase order, contract, or lease agreement should be written with care, specifying the responsibility of both the buying and supplying firms for equipment performance and post-sale activities. Acceptance testing and inspection methods, acceptance timing, machine specifications and performance standards, and guarantee conditions should be addressed during negotiations and in the resulting contract. In the aviation industry, it is common for penalties to be paid by the supplier if performance standards are not met. For example, International Aero Engines when selling its new engine (V2500) agreed that if the fuel consumption was above specifications, IAE would pay a penalty to the engine user. Similarly, supplier responsibility for post-sale services pertaining to installation, start-up, operator training, maintenance checks, and replacement parts should be spelled out clearly so there is no question about what is to be furnished and at what price.

In the event that the purchase or lease involves a lengthy manufacturing period, a special follow-up and expediting program should be developed. This may call for periodic plant visits and in-process inspection of the work. Responsibility for monitoring this activity normally rests with supply management.

Post-Award Activities

We strongly recommend that the key individuals from the buying and supplying firms who are concerned with timely delivery, installation, training, and delivery of spares meet to ensure total understanding of each firm's and each function's responsibilities. Normally, supply management will assume responsibility for the day-to-day management of the contract and the relationship.

After a machine is purchased, the wise supply professional works closely with plant engineering in keeping and interpreting historical records (part by part) of machine performance. Data of this kind are valuable in making similar future analyses. Such information may indicate that reimbursement under the warranty is appropriate.

Summary

In most firms, equipment is not purchased frequently. Such purchases may represent important management decisions. Equipment purchases can be major investments that lead to the manufacture of more competitive products or the delivery of more competitive services, both of which increase sales in the marketplace, and lead to improved productivity.

The role of supply management is distinctly different in this type of buying activity from what it is in production buying. In the procurement of equipment, supply management personnel function in a creative capacity as facilitators, coordinators, contract administrators, and consultants to management. Specifications must be precise and complete, yet they must be written as functionally as possible. Economic analyses should utilize appropriate techniques, must be thorough, and must be based on data that are as accurate as possible. In many cases, supply management becomes the champion of total cost of ownership analysis. Supply management should be actively involved and optimally guide the team through the sourcing and negotiation activities. The contract must be precise and complete. There should be no doubt about installation and start-up responsibilities, performance requirements, test and inspection methods, related post-sale responsibilities, spare part support and warranties.

Endnotes

1. This type of analysis is also called *life cycle* costing. The term and the concept were originally developed and refined in military procurement. Subsequently, industry adopted the concept and now it is widely used in most industries. This topic is discussed in more detail shortly. The terms *total life cost* and *life cycle cost* are virtually identical to *total cost of ownership* described in Chapter 15.

2. For the financial analysis of leases, see R.A. Brealey and S.C. Myers, *Principles of Corporate Finance,* 6th ed. (New York: Irwin McGraw-Hill, 2000), Chap. 25.

3. The cost of capital for evaluating lease cash flows is the firms' after-tax cost of debt. See Chapter 25 of Brealey and Myers op. cit.

4. For a comprehensive analysis of the lease/buy question, see William L. Ferrara, James B. Thies, and N. W. Dirsmith, *The Lease-Purchase Decision* (New York: The National Association of Accountants, 1979.)

Suggested Reading

Auguston Field, Karen. "Calculating ROI: Hope for Best, Plan for Worst: By Examining a Proposed Investment's Financial Performance Over a Range of Conditions, You'll Know How Wrong You Can Be and Still Be Right." *Supply Chain Management Review* 6(2)(2002): 554.

Contino, Richard. *The Complete Equipment-Leasing Handbook.* New York: AMACOM Books, 2002.

Harding, Mary Lu. "Applying TCO to Capital Equipment." *NAPM InfoEdge* 5(3)(2000).

Menezes, Sanjit. "Calculating the Total Cost of Ownership." *Purchasing Today®* 12(5) (2001): 16.

Newan, Richard G. and Robert J. Semkins. *Capital Equipment Buying Handbook.* New York: AMACOM Books, 1998.

Woodside, Arch G., Timo Liukko, and Risto Vuori. "Organizational Buying of Capital Equipment Involving Persons Across Several Authority Levels," *Journal of Business & Industrial Marketing* 14(1)(1999): 30.

CHAPTER 9

Purchasing Services

Introduction

The procurement of services is an activity of increasing importance. Expenditures on services by commercial firms, not-for-profit organizations, and government increase each year. In some cases, services procurements represent more than 25% of the organization's expenditures. Purchased services play key roles in the successful operation of these organizations. In many instances, the impact of the services on the success of the organization's operation is far greater than the impact of the dollars spent. Services ranging from architectural engineering, promotion and advertising, and the development of software, to the maintenance and repair of production equipment are of critical importance to the operation of the organization. More mundane purchases such as cafeteria and janitorial operations affect the morale of all employees.

A tidal wave of "outsourcing" of services is taking place in America and abroad. At one level, services that are not at the core of the organization's competencies, such as management information systems, payroll, travel services, delivery services, and even the procurement of MRO supplies and services, are being outsourced to service providers. Those suppliers have the expertise and economies of scale that allow them to provide the services at the same or higher quality level than the purchasing firm and at a lower total cost. At another level, in economies that traditionally have been characterized by a high level of government participation in service delivery in the provision of health, transport, utilities, and municipal services, there has been large-scale outsourcing to private sector providers. This move to outsourcing adds a new dimension to procurement. Expenditures on such third-party service provision can reach 60 to 70% of total organizational expenditure, and require higher-level supply management skills than has been imagined.

Obtaining services is one of supply management's most challenging responsibilities. In no other area is there a more complex interdependency between the purchase description (statement of work), method of compensation, source selection, contract administration,

225

and a satisfied customer.[1] The purchasing of services frequently leads to relationships with suppliers that focus on trust rather than the transaction, and can "stretch" supply professionals beyond their traditional zones of comfort and competence.

Hidden Opportunities

Warren Norquist, former Vice President of International Materials Management at Polaroid, involved his staff in many of the following nontraditional procurements:

- ▲ Print ad production
- ▲ General consultants
- ▲ Computer consultants
- ▲ Computer network management
- ▲ Design of exterior of products
- ▲ Television ad production
- ▲ Outplacement agencies
- ▲ Training consultants
- ▲ Network TV time
- ▲ Market research
- ▲ Financial auditors
- ▲ Training courses
- ▲ Per diem help
- ▲ Placement agencies
- ▲ Technical consultants
- ▲ Spot TV and radio time
- ▲ Telephone customer service
- ▲ Annual reports
- ▲ Logistics and inventory control

Mr. Norquist's experience was that when qualified procurement personnel were involved in the planning and procurement of such services, savings of approximately 25% were achieved with equal or improved quality and service.[2] In addition, an increasing number of proactive supply management operations now purchase services such as utilities, disposal services, and insurance.

Stephen Sutton, Supply Manager for Ok Tedi Mining Limited (OTML) in Papua, New Guinea, talks of the increasing significance of services procurement in world-class organizations. OTML's spending on services now exceeds 50% of total expenditures. Sutton cites examples of outsourcing major areas of core on-site activity at OTML such as blasting (where a contractor now delivers a total drilling, charging, and explosives detonation service) and major equipment maintenance for which a contractor has total responsibility on site for activities that include buying and holding inventory, and the provision of the associated maintenance labor requirements. Such examples clearly indicate the shift in the manner in which leading-edge organizations operate in their supply markets today.[3]

The Statement of Work

As is true in the purchase of production requirements and capital equipment, the most critical prerequisite to a successful procurement of services is the development and documentation of the requirement—the statement of work (S.O.W.). As is true with production requirements and capital equipment, one of the keys to success is the involvement of qualified supply management personnel at this point in the procurement process. In fact, supply professionals know that many of their service customers lack training and experience in the development of service requirements or specifications. Accordingly, a supply professional can provide invaluable assistance during this phase of the procurement. In many instances, supply professionals invite two or three carefully prequalified potential contractors to aid in the development of the statement of work. Their early involvement helps the internal customer fully understand the organization's true needs. At the same time, the potential contractor gains insight into the nature and level of effort required.

The S.O.W. identifies what the contractor (supplier) is to accomplish. The clarity, accuracy, and completeness of the S.O.W and the way the effectiveness of delivery will be measured will determine to a large degree whether the objectives of the contract will be achieved. The S.O.W. clearly identifies first the primary objective and then the subordinate objectives so that both the buyer and seller know where and how to place their emphasis. Those responsible for developing the S.O.W. must ask themselves questions such as the following: Is timeliness, creativity, or artistic excellence the primary objective? How will customer satisfaction be measured?

One of the objectives of writing an S.O.W. is to gain understanding and an agreement with a contractor concerning the specific nature of the technical effort to be performed. Satisfactory performance under the contract is a direct function of the quality, clarity, and completeness of its statement of work, and of the contractor's understanding of the outcome required.

The S.O.W. also affects the administration of the contract. It defines the scope of the effort; that is, what the contractor is to do, what the buyer is to receive, and how satisfaction is to be measured. The manner in which the scope of work is defined governs the amount of direction that the supply professional can give during the contract's life.

A well-written S.O.W. enhances the contractor's performance in pursuit of S.O.W. objectives. Before writing it, those responsible must develop a thorough understanding of all the factors that will bear on the project and that are reflected in the S.O.W.

Four Formats for Statements of Work

The majority of formats fall into four basic types of S.O.W., plus a combination of these four types, called a "hybrid" S.O.W. The four S.O.W.s are:

▲ *Performance S.O.W.*—details everything wanted by the buyer. This S.O.W. is broken down into tasks describing the required outcome performance of the task.

▲ *Functional S.O.W.*—defines what the buyer is "trying to do," leaving the seller free to come up with the most efficient means to do so.

▲ *Design S.O.W.*—used mainly in the construction and manufacture of goods or equipment projects. S.O.W.s of this type require the inclusion of plans, blueprints, CAD designs, or specifications.

▲ *Level-of-Effort S.O.W.*—a specialized version of the performance S.O.W., generally used on research and development or studies contracts.[4]

Planning the Statement of Work

An S.O.W. frequently is described as a document that details a strategy for contractor and buyer accomplishment of the objectives of a project. However, before any strategy can be developed, certain basic questions must be answered and understood:

▲ What are the objectives of the project?

▲ Where did the objectives come from, who originated them, and why were they originated?

▲ What is the current status of the effort?

▲ Based on the current status, what are the risks associated with the achievement of the project objectives?

The planning phase of S.O.W. preparation is aimed at a thorough investigation of the why and what of the project. The following checklist will assist the program manager and the buyer in this determination.

▲ Identify the resource, schedule, and compensation constraints for the project.

▲ Identify all customer and contractor participation needed for the project, and define the extent and nature of their responsibilities. All customer support, such as customer-furnished equipment, materials, facilities, approvals, and so forth, should be specifically stated.

▲ Challenge the tasks identified, including sequencing and interrelationships of all required tasks. For example, on a janitorial services contract, should the contractor be required to wash (vs. completely erase) blackboards every evening or only once a week? On a landscape maintenance contract, should the contractor be required to furnish expendables such as fertilizer?

▲ Identify contractor delivery requirements at specified points in time; include details about the type and quantity of any deliverables.

▲ Identify specific technical data requirements such as plans, specifications, reports, and so on.

▲ Identify realistic desired or required service levels.[5]

O'Reilly, Garrison, and Khalil provide the following list of typical elements that may be in an S.O.W.:

- ▲ Description of the work
- ▲ Schedule
- ▲ Specifications and requirements
- ▲ Quality requirements
- ▲ Performance measurements
- ▲ Deliverables
- ▲ Delivery and performance schedule
- ▲ Service levels
- ▲ Changes and modifications
- ▲ Bonds
- ▲ Charges and costs
- ▲ Project management
- ▲ Reporting requirements
- ▲ Safety
- ▲ Supplier responsibilities
- ▲ Buyer responsibilities
- ▲ Work approvals
- ▲ Use of subcontractors
- ▲ Authorized personnel
- ▲ Exhibits, schedules, and attachments[6]

Writing the Statement of Work

As a result of a thorough planning effort, the individuals writing the S.O.W. should have determined the tasks and details that need to be included. These must now be documented. This is even more important as the Uniform Commercial Code (UCC) does not apply to services.

Writing a quality S.O.W. is not an easy task. The S.O.W. must maintain a delicate balance between protecting the customer's interests and encouraging the supplier's (contractor's) creativity during both proposal preparation and contract performance. For example, when purchasing janitorial services, some well-intended firms specify the number of personnel that the contractor must supply, in the rather questionable belief that this provision will guarantee satisfactory performance. But that action blocks the supplier's creativity and generally results in needlessly high prices. The use of a carefully developed S.O.W. that specifies the required performance and procedures for monitoring (inspecting) the contractor's performance allows the contractor to apply its experience and creativity—

usually at a significant savings. To complicate the task, those developing the S.O.W. must remember that it will be read and interpreted by customer and contractor personnel of widely varying experience and expertise.

The following issues merit special attention on a case-by-case basis. Required provisions may be in either the S.O.W. or special terms and conditions included in the request for proposal and the resulting contract.

▲ *A performance plan.* The contractor is required to develop a non-subjective, quantifiable blueprint for providing the services. Staffing, equipment, and supplies should all be identified. After developing the blueprint, the contractor must identify all required processes.

▲ *Quality monitoring system.* The contractor will be required to specify and implement fail-safe measures to minimize quality problems.

▲ *Personnel plan.* The contractor is required to develop and maintain recruiting and training programs acceptable to the customer.

▲ *Performance and payment bonds.* The contractor must provide performance and payment bonds equal to x% of the value of the contract amount.

▲ *Metrics.* When possible, performance objectives should be quantified.

▲ *Progress reviews.* If progress reviews appear to be appropriate, how, when, where, and by whom should be specified.[7]

Artificial Intelligence

While Executive Director of Purchasing at Pacific Bell, colleague Joe Yacura began the development of artificial intelligence applications used to develop specifications for S.O.W.s for both services and products. He carried this effort forward while Senior Vice President of Purchasing at American Express. His efforts have progressed to the point that an individual requiring a service to be purchased from an outside provider can go to a Web site that answers a number of detailed questions. The system algorithm and business rules then generate an S.O.W. together with a set of performance metrics and a total life cycle cost model.[8]

Tips on Writing an Effective Statement of Work

Experience indicates that a clause such as the following should be added to the Bidder's instructions section of the request for proposal (RFP) to ensure that the Bidder's proposal will be responsive to the mandatory requirements:

> This RFP describes the minimum content and general format for responding to our RFP. Your reply shall be submitted on the forms and in the formats requested

(or equivalent) with all questions answered in detail. Elaborate format and binders are neither necessary nor desired. Legibility, clarity, and coherence are more important. Your proposal should present information in the order requested in the RFP. It is mandatory that the Bidder use the same numbering format as used in this RFP so that responses correlate to the same paragraph in the RFP requirement. This will make your proposal more "evaluator friendly" to the evaluation team conducting the evaluation of the proposals.

Whenever a question is asked in the RFP subparagraphs or a requirement stated by the use of the phrase "The Bidder Shall," the Bidder is expected to answer these as fully and completely as possible. Failure to do so may deem your proposal as "nonresponsive" to that requirement.

Responsiveness will be measured by the Bidder's response to the requirements in each paragraph to the RFP. Merely "parroting" back the RFP requirements statement in the Bidder's proposal may deem the response as nonresponsive. The Bidder's response must demonstrate an understanding of the requirements. This might be done by providing what was asked for, or citing how the Bidder has achieved the requirements in its normal business practices (such as submitting samples of procedures, or award letters).[9]

To make sure that the S.O.W. accurately reflects what the contracting parties have agreed to, follow these suggestions:

▲ Be clear—use simple, direct language. Avoid ambiguity.

▲ Use active, not passive tenses. (The seller "shall conduct a test," as opposed to "a test should be conducted by the seller.") Active verbs assign responsibility more clearly than do passive verbs.

▲ Be precise, especially about task descriptions. The clarity of the S.O.W. affects the administration of the contract because it defines the scope of work to be performed. Work outside this scope will involve new procurement with probable increased costs.

▲ Spell out the buyer's obligations carefully. Don't infer or "back into" a work requirement.

▲ Limit abbreviations to those in common usage and spell them out in the first usage with the abbreviation in parenthesis. Provide a list of abbreviations and acronyms to be used at the beginning of the S.O.W.

▲ Include procedures. When immediate decisions cannot be made, it may be possible to include a procedure for making them (for example, "as approved by the purchaser," or "the seller shall submit a report each time a category B failure occurs").

▲ Do not over-specify or overstate. Depending upon the nature of the work and the type of contract, the ideal situation may be to specify results required or end times to be delivered and let the contractor propose the best method.

▲ Eliminate extraneous statements. If a statement has no practical value, it should not be in the S.O.W.

▲ Include all relevant reference documents.

▲ Don't mix general/background information, guidance and specific direction/ requirements.

▲ Don't sole source the work statement unless competition is not desired.

▲ Describe requirements in sufficient detail to assure clarity, not only for legal reasons, but for practical application, such as in closing loopholes.

▲ Be aware that contingent actions may have an impact on price as well as schedule.

▲ Provide a ceiling on the extent of services, or work out a procedure that will ensure adequate control, where appropriate (for example, a level of effort, pool of labor hours).

▲ Avoid incorporating extraneous material and requirements that may add unnecessary cost. (Data requirements are common examples of problems in this area.)

▲ Don't repeat detailed requirements or specifications that are already spelled out in applicable documents. Instead, incorporate them by reference.

▲ Explain the interrelationship between tasks, and how tasks are related to required results and deliverables.

▲ Identify all constraints and limitations.

▲ Include standards that will make performance measurement possible and meaningful.

▲ Be clear about phase requirements, if applicable, and the timing used to gauge work phases.

▲ Proofread for errors and omissions, as well as for format and information consistency.[10]

Selecting Service Contractors

Selecting the "right" source is much more of an art when purchasing services than when purchasing materials. Because of the complexity of many service procurements and the unexpected problems that tend to arise, usually it is prudent practice to select only established, reputable firms. Exceptions may be made occasionally in cases involving promising new suppliers who have not yet established a "reputation." Unless the potential supplier possesses some truly unique skill or reputation, competition typically is employed. In some service markets, however, experienced supply managers find that the competitive process is not completely effective because of the structure of the market. This issue is discussed later in the chapter.

When a large number of potential contractors is available and the dollars involved warrant the effort, the customer firm's sourcing team normally reduces this list to three

Table 9.1 Total Costs for the Construction Project

	Firm X	Firm Y
Construction cost	$10,000,000	$11,000,000
Design fee	739,200	660,000
Total cost	$10,739,200	$11,660,000

to five firms. Ideally, a weighted scorecard should be developed to facilitate the process. The team interviews prospective contractors' management, talks with previous customers, and checks out employees through random interviews. The supply manager then invites proposals only from the potential suppliers with which the buying firm would be comfortable doing business.

During the evaluation process, emphasis should be placed on the total cost and total benefits to the purchasing organization. Assume, for example, that two architect-engineering (A-E) firms are under consideration for the development of plans and specifications for a new building estimated to cost approximately $10 million. Firm X has a reputation of designing functional buildings whose costs are relatively low. Firm Y, on the other hand, has a reputation of designing more elaborate and aesthetically more attractive buildings whose costs tend to run about 10% more than X's. For the sake of illustration, however, assume that firm X's professional fees tend to run about 12% more than Y's. Table 9.1 illustrates these cost differentials and shows the overriding influence of construction costs in the complete analysis. Hence, in this case, the contractor's design fee is a relatively minor item in the total cost package.

In addition to the traditional concerns about a prospective contractor's financial strength, management capability, experience, and reputation, the area of technical capabilities requires special analysis. An article in *Purchasing World* identifies the following issues that should be addressed when selecting a contractor for computer maintenance. This list of issues is introduced simply as an example of the depth of analysis required when selecting a contractor for this specialized service.

▲ Will the contractor maintain all the equipment in your computer installation?
▲ Can the contractor quickly correct the problem?
▲ How close to your facilities is the contractor's field engineering office? Does the contractor specialize in your type of equipment?
▲ Does the contractor have a prescribed schedule of service calls?
▲ Does the contractor have troubleshooting escalation procedures, skilled field engineers, and ready availability of spare parts?

▲ If there is any possibility of having to move the computer equipment, does the contractor have proven successful experience moving computers?

▲ Does the contractor offer equipment brokerage?

▲ Does the contractor have the technical ability to make low-cost modifications to your equipment? If so, can the firm support the resulting system?

▲ Will the contractor service refurbished equipment?

▲ Does the contractor have high hiring standards, require appropriate training, and equip field service personnel with appropriate tools and equipment?

▲ Does the contractor supply maintenance documentation?

▲ Will the contractor develop custom products for your special needs?

▲ Is the contractor flexible in meeting your specific requirements?[11]

The selection of suppliers for repair services depends on the situation. The best way to cope with emergency services is to anticipate them. Vehicles, office machines, and plant equipment break down. Sewer lines get clogged. In many cases, it is possible, and certainly desirable, to establish the source and the price or dollar rates for such services before the emergency occurs. In purchasing transportation services, consistent on-time pickup and delivery, equipment availability, and service to particular locations typically are more important than price.[12]

Competitive prices should be solicited every two or three years for recurring services. This action tends to avoid complacency and helps to maintain realistic pricing. More frequent changes in contractors often cause too many service disruptions.

Tips from a Professional

Barbara Stone-Newton, Purchasing Manager for the State of North Carolina, provides the following tips:

▲ *Partner with users* Be sure that you understand their needs, goals, and constraints. You will be the bridge between them and potential service providers. If users are not already familiar with your organization's procurement procedures (and possible options), use this time to go over them.

▲ *Learn from the past* For ongoing services, review the prior specifications and any comments in the contract administration file. If there were questions during the previous procurement, try to incorporate answers into your new request for proposal (RFP) or invitation for bid (IFB).

▲ *Update specifications* Service standards and environment often evolve over the contract term. Generic specs might leave out important features that users have come to expect and thus create problems. For example, the current contractor may price for today's level of service (assuming rightly or wrongly that you want that to continue) while others submit costs for exactly what is in the specs. Be equally cautious about overspecifying because that can increase costs or limit competition.

▲ *Minimize assumptions* Be specific about deliverables, schedule, performance measures, and similar expectations. The "fudge factor" is a particular concern in service procurements.

▲ *Encourage questions* Provide some mechanism for questions—a deadline for written inquiries, a site visit, or a pre-proposal conference. Issue a summary addendum formalizing substantive questions and answers so that all potential suppliers have the same information.

▲ *Facilitate comparison* Make it easy for prospective suppliers to provide everything you need to evaluate responses. Include a checklist, outline, or similar section detailing what information is to be included in the responses. A "fill-in-the-blank" format is great for simple procurements.

▲ *Plan evaluation* Outline the evaluation process and any weighting or scoring method to be used.

▲ *Reduce surprises* Include contractual terms and conditions in the solicitation document. This counters the mind-set of "we'll fix it with the contract" and streamlines your award process.

▲ *Check yourself* When you think the RFP or IFB is ready to go, put it aside (at least for a few hours), and then take a fresh look. Read it from the suppliers' viewpoint. Does it contain everything they will need to offer a competitive response? Finally, look forward to contract administration. Are the deliverables, standards, etc. defined clearly so that both parties can measure performance? If the answers are positive, you probably have a good solicitation document.[13]

The Ideal Services Supplier

The ideal services supplier listens to what users complain about most and then designs service products that supply the market's missing ingredients. Satisfaction is built into service products rather than added as an afterthought. Employees of ideal services providers are given every conceivable form of automation to help them deliver a consistently satisfactory service product. The ideal services supplier invests to increase both employee productivity and customer satisfaction.

If an ideal services supplier or contractor is not available, the purchasing firm should consider the development of a long-term relationship with a supplier willing and able to grow into an ideal provider.

Pricing Service Contracts

Procurement authority Louis J. DeRose writes that "the competitive process is not truly efficient in services markets. It is constrained by three forces and factors of supply:

▲ One of the strongest factors influencing competition and prices—a continuing or cumulative supply—is absent.

▲ Interchangeable services generally are not available due to the personal effort and involvement of the supplier.

▲ The supply of services is more easily restricted or restrained than is the supply of commodities or products.

It is for these reasons, DeRose writes, "that buyers must negotiate service agreements."[14] Yet, there are some situations where competitive bidding is an effective method of determining both source and price. A janitorial services contract for which competition is intense is an example. Again, the supply professional's judgment plays a key role: Are all of the conditions required for the use of competitive bidding (as discussed in Chapter 13) satisfied? Are time and qualified resources available to prepare for and conduct negotiations? Are supply management's internal customers prepared to play a constructive role in professional negotiations? In most instances, negotiation often results in better pricing and the supply of a more satisfactory service.

Too frequently, the pricing of service contracts is not tailored to motivate the supplier to satisfy the organization's principal objective. Once the primary requirement (artistic excellence, timeliness, low cost, and so on) is identified, the supply professional must ensure that the resulting contract motivates the supplier to meet this need. When conditions require, the contract should reward good service and penalize poor service.

Professional Services

Architect-engineering firms, lawyers, consultants, and educational specialists are representative of the individuals and firms that provide professional services. Supply professionals pay particular attention to the relationship between the price mechanism (e.g., firm fixed price, cost plus incentive fee, fixed price with award fee, and so on) and the contractor's motivation on critical professional services contracts. For example, fixed price contracts reward suppliers for their cost control. Every dollar that the supplier's costs are reduced results in a dollar of additional profit for the supplier.

Assume that you were selecting an individual or a firm to prepare a fairly complex personal income tax return for a gross income of $125,000. Firm C advertises that it will prepare any tax return, regardless of its difficulty, for a guaranteed maximum of $200. Firm D offers rates of $75 per hour. Discussion with a representative of Firm D indicates that approximately five hours will be required to prepare your return. If forgone tax savings are considered as a cost, which firm is more likely to provide the service at the lowest total cost? The use of Firm D will cost an estimated $375 ($75 × 5 hours), or $175 more than Firm C. Assuming that C and D had similar hourly costs, it is likely that D would spend about two hours more preparing the tax return. Most individuals with a $125,000 income would pay the extra $175 in hope of offsetting the outlay with a larger tax saving.

Cost-type contracts should be considered when there is significant uncertainty about the amount of effort that will be required or insufficient time to develop a realistic S.O.W. Obviously, the dollar amount involved must warrant the administrative cost and effort involved. For smaller dollar amounts, a time and materials or labor-hour contract should be considered to avoid contingency pricing. Such contracts require close monitoring to ensure that the specified labor skill is furnished and that the hours being billed are in fact required.

Administratively, it may be impractical to use anything except a fixed price contract or an hourly rate price for relatively small professional services contracts. Even on larger dollar amounts, the supplier's reputation may allow the use of a fixed price contract. However, supply managers should be aware of the potential effect of the pricing mechanism on the contractor's performance.

Technical Services

Technical services include such things as:

- Research and development
- Software development
- Machine repairs
- Printing services
- Payroll services
- Mailroom services maintenance
- Elevator maintenance services
- Pest control
- Energy management
- Accounting and bookkeeping services
- Advertising and promotion
- Heating and air-conditioning
- Copyroom and message

R&D services normally are purchased through one of two methods of compensation: a fixed price for a level of effort (e.g., 50 days) or a cost plus fixed or award fee. Software development lends itself to cost plus award fee contracts. This approach rewards excellent performance and punishes poor performance while ensuring the contractor that its costs will be reimbursed and at least a minimum fee will be received.

In a competitive market, once a good S.O.W. is available for services such as printing, promotional services, and the development of technical manuals, competition should be employed to select the source and determine the price, using a fixed price contract.

Operating Services

Janitorial, security, landscaping, and cafeteria operations are typical operating services. Experience has shown that obtaining effective performance of such services can be very challenging for contract administrators. Accordingly, the compensation scheme should reward the supplier for good service and penalize it for poor service. (The use of a fixed price award fee scheme, as described in Chapter 17, may be appropriate.) That approach to pricing greatly aids in the administration of the contract and frequently results in a far higher level of customer satisfaction.

Insurance, plant and equipment maintenance, and anticipated emergency services should be sourced and priced through the use of competition of carefully prequalified suppliers. Unanticipated emergency repairs normally are purchased on a "not-to-exceed" time and materials basis (as described in Chapter 17).

Third Party Contracts

Contracts for the provision of a service to a third party may result in a nonlinear supply chain and create ambiguities in contract relationships, for example, between the supplier and the end user. While a contract may exist between an organization and its customer, and between an organization and its supplier, it does not necessarily follow that a contract will exist between the supplier and the customer unless explicitly provided for in the contract.

An example is a nationwide automobile insurer that puts in place across all regions of the country arrangements with suppliers for the provision of windshields for its customers. The question of whether a contract exists between the windshield supplier and the customer of the insurance company probably will be incidental if the relationship between the insurance company and its supplier is well managed. But if this is not the case, the relationship between the insurer and the customer may be at risk.

So, Your Services Contract Is About to Expire

The requirement for many services continues beyond the duration of the service contract. If a collaborative, mutually beneficial relationship has been established, many members of the purchasing firm may want to extend the contract. Additional pressure to extend the contract results from switching costs. We estimate that it may cost as much as 5% of the face value of some services contracts to switch to a new supplier.

But what should the price be for such an extension? If we enjoy an open book relationship, we could study the present supplier's costs and use this information as the basis of a contract extension. A more objective and more scientific basis for determining the price of a contract extension is to apply the "experience" or "learning curve" as discussed in Chapter 16. The supplier, its site management, and its direct labor should learn how to do things more efficiently the more times they do a task or activity. Supply management should work with the supplier to develop a realistic estimate of what performance "should cost" based on past and future learning. While the mechanics of such an approach are beyond this chapter, we emphasize that our readers should be sensitive to some of the subtleties of extending services contracts.

Contract Administration

The four keys to successful service contract administration are (1) a sound S.O.W., (2) selection of the "right" source, (3) a fair and reasonable price, and (4) aggressive management of the contract. The administration of many service contracts can be a very challenging responsibility. The supply professional needs to monitor and have a realistic degree of control over the supplier's performance. Crucial to success in this area is the timely availability of accurate data, including the contractor's plan for performance and the contractor's actual progress. The supply professional must manage the relationship to ensure success.

Services Purchases and the Internet

Clearly, the Internet is becoming increasingly important in the purchase of services. Firms are employing electronic requests for proposals and receiving proposals electronically. Electronic collaboration both within a firm and with potential suppliers is becoming increasingly common. The Internet allows purchasing firms to obtain increased competition and lower prices for some services. When used properly the Internet has the potential of reducing total cost of ownership, reducing order processing costs, compressing the sourcing cycle time, and improving the flow of information required to manage the resulting contract. Electronic marketplaces can provide a directory of services suppliers and frequently can provide the role of matchmaker.

Leading firms use the prospect of incorporating the Internet as a stimulus to optimize their services supply chains. Reengineering is frequently appropriate. One Fortune 500 company reengineered its temporary labor service process prior to going digital. The documented savings was in the millions of dollars. This and a host of similar experiences cause us to encourage firms to reengineer to optimize the process before developing (or acquiring) the enabling e-solution.

There are some services that lend themselves to reverse or "e-auctions." The critical aspect of this sourcing technique is the complete S.O.W. to ensure comparing apples to apples.[15]

Construction Services

The purchase of new facilities is a commitment for the future. Quality, productivity of the new plant or office, the time required to affect the purchase, and cost all must be considered. Aesthetic requirements, time requirements, and the availability of highly qualified designers and builders all will tend to influence the selection of a purchase method.

There are five common methods of purchasing construction; however, it is unlikely that any one of those methods consistently will be the proper choice for all building requirements. Figure 9.1 provides a graphic presentation of the various steps involved in each method, from start to completion of a construction project.

Conventional Method

This is the most frequently employed approach to buying building construction in the United States. With this approach, design of the required facility is done by architects and engineers without the involvement of a builder. The design of the facility is completed before potential suppliers are requested to submit bids. Two separate organizations are responsible: one for the design work and one for the construction phase of the project.

Many architects are not noted for being cost-conscious. Nevertheless, the cost factor can be controlled in several ways. A common approach involves employment, on a consulting basis, of a "cost-control architect" who is concerned solely with cost reduction. Naturally, the general architect typically does not appreciate having his or her work reviewed or "second guessed." However, use of a consulting architect undeniably tends to make the primary architect more cost-conscious.

If a consulting architect is not used because of the general architect's sensitivity, alternative methods of cost control are available to the supply professional. For example, for major interior furnishings, the architect or interior decorator can be required to specify three manufacturers' products that can be purchased through competitive bids (any one of the specified products being satisfactory). The actual purchasing can be done by the customer's organization, the architect, or the interior decorator. The three-bid requirement ensures competitive pricing. As an added bonus, this practice helps eliminate conflicts of interest, unreasonable personal bias, and the specifying of low-volume, proprietary items.

Design and Build, Firm Agreed Price Method

This approach could be described as construction with gratuitous design. The owner determines the basic facility requirements, such as size, temperature, electrical, mechanical, and so on. Those requirements become the basis of a performance specification. That specification is furnished to carefully pre-qualified builders who, with their prospective

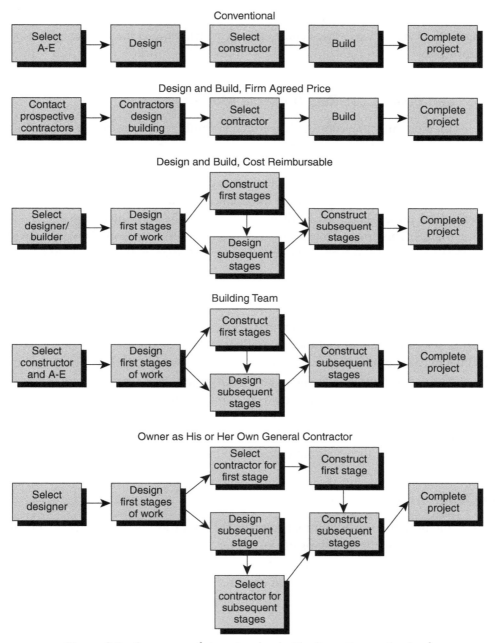

Figure 9.1 Sequence of steps involved with alternative methods of purchasing construction.

subcontractors, prepare a bid package consisting of a design and price proposal. The purchasing firm awards a fixed-price contract for construction to the builder whose bid is most attractive.

Design and Build, Cost-Reimbursable Method

With this method, only one contract is awarded for both design and construction. Design is accomplished by architects and engineers employed by the general contractor. Thus, the builder has ample opportunity to influence the design of the required facility. With this approach, construction of a work element (excavation, structural work, and so on) proceeds when the design of the element has been completed. It is not necessary to wait to design the total project because one firm is responsible for both the design and the construction phases. This approach is particularly useful when a structure is required within a very short time and the design and build, firm agreed price method is not applicable.

Building Team

With this approach, the owner retains both a designer and a builder concurrently. In contrast to the conventional method, the builder is retained during the design phase and is expected to contribute information on costs, procedures, and time requirements to the designer. As the A-E completes the plans and specifications for a work element, the builder either accomplishes the work with its own crews or obtains prices from several specialists and awards the work to the qualified subcontractor that makes the best offer (price, time, and quality considered). As with the other methods, the general contractor oversees and integrates the efforts of the subcontractors.

The Owner as a Contractor

With this method, the owner contracts directly for the various work elements (including design) and performs the functions of integrating and controlling that otherwise would be accomplished by a general contractor. Since purchase orders and contracts are awarded on a work element basis, it is possible for construction to proceed prior to completion of the total design phase.

Findings

Research conducted by one of the authors on these five methods shows that the conventional method is, by far, the most costly approach to purchasing construction. Savings of approximately 25% were found to result when the design and build, firm agreed price

Table 9.2 Purchase Price of a Hypothetical Building under Five Compensation Methods

Method	A&E Fee	Construction Contract	Total	Time (months)
Conventional	$740,000	$12,300,000	$13,040,000	16
D&B (firm agreed price)	—	9,750,000	9,750,000	11.5
D&B (cost reimbursable)	—	11,830,000	11,830,000	12
Building team	670,000	11,160,000	11,830,000	12
Owner as his or her own general contractor	600,000	11,800,000	12,400,000	15.5

method was used rather than the conventional method. Savings of 9% resulted when the design and build, cost-reimbursable or the building team method were used in lieu of the conventional method. Savings of about 5% resulted when the owner acted as his or her own general contractor (see Table 9.2).

The amount of time from first contacting the designer or builder until completion of the facility frequently is as important or more important than the price paid. Availability of the required facilities varies significantly with the methods used. On a typical 130,000-square-foot manufacturing plant, 16 months were required with the conventional method; 11.5 months with the design and build, firm agreed price method; 12 months with both the design and build, cost-reimbursable method and the building team method; and 15.5 months when the owner acted as the general contractor.

Supply professionals know that selection of the most appropriate method of purchasing plant facilities can significantly reduce the cost and time required to purchase new facilities. Their early involvement in such projects is a key to saving both time and money.

Construction Purchasing Entails Unique Problems

Construction purchasing is a highly specialized field. Of particular importance is the fact that proper financial, legal, and planning actions must be undertaken to prevent possible losses. For example, a purchasing organization sometimes discovers that after completion of a new building a mechanic's lien is filed against the firm. In such cases, the organization has typically paid the general contractor, but the general contractor has not paid its subcontractors. Under the law, if a "general" does not pay its "subs," the owner of the building is financially responsible.

The proper financial and legal steps that must be taken differ among states and municipalities; hence, the specific steps to be taken must be determined individually in each case. In general, though, the first step that must be taken when selecting the supplier is to carefully analyze the financial status of all prospective contractors. Next, the supply professional should consider the desirability of utilizing such protective devices as bid, performance, and payment bonds,[16] liquidated damages contract clauses, and the development of construction cost estimates by the organization's own engineering personnel (perhaps with the assistance of a specialized consultant). Finally, legal protection is achieved when the organization properly files all required completion and related reports, at the appropriate times, at the appropriate courthouses.

Construction Insurance

Because construction is a high-risk business, the insurance a purchasing organization requires the contractor to carry is very important. All construction contracts should stipulate specific insurance responsibilities. For example, a contract clause might require the contractor, at its expense, to maintain in effect during performance of the work certain types and minimum amounts of insurance coverage, with insurers satisfactory to the customer.

In addition, the contract should specify that prior to the performance of any work, the contractor must provide certificates of insurance as evidence that the required insurance is actually in force and that it cannot be canceled without 10 days' written notice to the customer organization. Failure to require any of the foregoing insurance provisions could be very costly. Other unique legal nuances in construction include the concept of Promissory Estoppel. A good example of this legal principle is a sub-contractor cannot "at will" revoke a quote offer, which is part of a prime contractor's proposal, prior to a reasonable time period. For example, a plumbing company offers to do the plumbing for a general contractor and it knows the general contractor has used this offer as part of the general contract proposal. The plumbing company must allow a reasonable time to lapse to allow the general contractor's proposal to be accepted or rejected, that is, the offer from the plumber is "irrevocable" until notified by the general contractor.[17]

Performance Contracting

There are many opportunities to correlate the design and construction fees to the "performance" of the resulting facility. Throughput and productivity are obvious examples. Energy conservation is an especially "hot" example today, although it has been around since the 1970s. An article in *Facilities Design and Management* cites an example of how performance contracting saves the facility owner money while reducing energy demands. When the owner of one of the largest high-rise office/retail complexes in the United State—comprising two 51-story buildings and an underground retail center for a total of 1.56 million square feet—undertook a major facilities upgrade to reduce operating costs and retain Class A tenants, it used a performance contract to execute the project. Not only did the performance contract provide turnkey delivery for design, construction, and maintenance, it also allowed the owner to finance the project off its balance sheet, secured against operations costs and installed equipment value, which would have been difficult to arrange under a traditional design/construction contract.

The project included the design and installation of a direct digital controlled energy management control system; variable frequency drives; variable air volume terminal units; fire- and life-safety control upgrades; utility control and monitoring systems; a fiber optic Ethernet network with multiple workstations; and the development of custom software drivers to third-party terminal unit controllers. After completion of the project, operating savings were 10% overall, more than double the owner's target.[18]

The Special Case of Contracting Consultants

While all the criteria for buying services as previously noted in this chapter apply, there are some unique factors to consider when determining which consultant to hire. First, there are several different types of consultants:

1. *Big management consulting firms:* Most of you know their rather famous names and some of them specialize in purchasing and supply management.
   Advantages: In depth staff resources who are well educated, experienced, and trained; a large data bank of case histories and examples; corporate experience; and knowledge of best practices.
   Disadvantages: High overhead that translates into high per hour fees, a possibility that the experienced senior partner who sold you the contract will be replaced by a young MBA to perform the work; their recommendations often require additional services such as follow-up training or installation of new systems, that is, you become dependent upon the consulting firm to actually operate the new program.

2. *Solo expert and small firms:* These are usually talented individuals with direct work experience who have decided to launch out on their own. Often named "John Doe & Associates," they have loose affiliations with others in similar professions. As they are small, they are usually very specialized as to their business expertise. Quite often, they have extensive experience gained from previous employment in the large consulting firms.
   Advantages: Extensive, hands-on work experience in specific functional areas; much lower overhead due to modest offices (many work out of their homes). They outsource many functions to keep the number of employees and equipment to a minimum.
   Disadvantages: They have limited resources, which hinders their flexibility. Many are lone rangers and cannot quickly accept a new contract. They may not be up-to-date

on the latest, best practices. For example, are they simply selling an IT package that does not fit your needs?

3. *Academics:* Many individuals teaching business at local universities or colleges have had experience in the business world and have complimented this knowledge with research and academic theory. They can offer a valuable blend of the latest theories and best practices, tempered by the knowledge of real-world application.

    Advantages: As they do not depend on your business for making a living, their response to your needs should be purely objective. Often your case will be viewed with great interest from the perspective of documenting and publishing the findings. As this requires your permission, it can be an effective bargaining chip in negotiating fees. Through their work with students, academic faculty are often potentially ideal as trainers. They tend to be knowledgeable of leading edge processes, practices, technology, and techniques.

    Disadvantages: Some have too much theory and not enough practical work experience. A few tend to have ego issues that hinder their ability to establish a connection with the rank and file employees. As lecturing is their business, it can interfere with their ability to listen, which is critical in the business of consulting.

Qualifying the Consultants

Conduct a thorough check on their credentials, in particular the list of degrees and previous employment, former clients, and publications. Verify this data and phone key references. Make sure you have a complete resume and remember, people will tell you more on the phone than in any form of writing. Watch for statements like "attended" x university, which almost always indicates failure to graduate. However, work history may be much more important than a degree.

Perhaps the key decision factor is cultural fit and comfort level with the individual. Are you comfortable with this person? Do you have confidence in him or her? There has to be some personality fit. Have a long dinner with the individual to see if there is the correct chemistry.

Finally, the issue of contracts with an hourly fee plus expenses vs. fees for the entire project with progressive payments must be resolved. We prefer the total project costs plus expenses contract, which divides the risk on an equal basis. Remember to prepare a thorough S.O.W. regarding precise deliverables.

Summary

The procurement of services is one of supply management's most interesting and challenging assignments. Large sums of money are involved. Of equal or greater importance, successful operation of the organization is affected by the effectiveness with which key services are purchased. Supply management frequently must assume a far more active role in all phases of a service procurement than when purchasing materials.

Endnotes

1. This material is based, in part, on David N. Burt, Warren E. Norquist, and J. Anklesaria, *Zero Base Pricing™: Achieving World Class Competitiveness through Reduced All-in-Cost* (Chicago, IL: Probus Publishing, 1990).
2. Burt, Norquist, and Anklesaria, op. cit., 177.
3. Personal interview with Laurie LeFevre, October 14, 2000.
4. Janet Sickinger, "Writing a Complete and Effective Statement of Work," *InfoEdge*, Nov. 1997, 3.
5. A desired service level might call for response within three hours 97% of the time.
6. Peter O'Reilly, David H. Garrison, and Frediric Khalil, "Purchasing Professional Services," *InfoEdge*, May, 2001, 8.
7. Sickinger, op. cit., 2.
8. Mr. Yacura currently is CEO of Ridgewood Development Corporation.
9. Sickinger, op. cit., 9.
10. Sickinger, op. cit., 11.
11. "How to Choose a Computer Maintenance Service," *Purchasing World* August (1987): 73.
12. James R. Stock and Paul H. Finszer, "The Industrial Purchase Division for Professional Services," *Journal of Business Research* February (1987): 3.
13. Barbara Stone-Newton, quoted in "Services Purchases in the Public Sector" by Julie Roberts, *Purchasing Today* February (2001): 57.
14. Louis J. DeRose, "Not by Bids Alone," *Purchasing World* November (1985): 46.
15. Bryan Robinson, "How to E-Auction the Intangible," *Inside Supply Management* July (2005):10–11.
16. Some organizations require performance bonds on all construction contracts, at times mistakenly believing that performance bonds per se assure quality performance.
17. Richard A. Mann and Barry S. Roberts, *Smith & Roberson's Business Law*, 13th ed. (Mason, OH: West, Thomson/South-Western, 2006) 176, 214, 259–260.
18. Trevor Foster, "Performance Contracting Can Yield Significant Returns," *Facilities Design & Management* January (2000): 34.

Suggested Reading

Axelsson, Björn, and Finn Wynstra. *Buying Business Services.* West Sussex, England: John Wiley & Sons, 2002.

Duffy, Roberta J., and Anna E. Flynn. "Services Purchases: Not Your Typical Grind." *Inside Supply Management*® 14(9)(2003): 28.

Hatfield, Jo Ellen. "Purchasing Services on the Internet: Is There a Fit?" *Inside Supply Management*® 13(5)(2002): 20.

O'Reilly, Peter, David H. Garrison, and Frederic Khalil. "Introduction to Purchasing Services." *NAPM InfoEdge* 6(3)(2001).

Roberts, Julie S. "Service Purchases in the Public Sector." *Purchasing Today*® 12(2)(2001): 53.

Smeltzer, Larry A., and Jeffrey A. Ogden. "Purchasing Professionals' Perceived Differences between Purchasing Materials and Purchasing Services." *Journal of Supply Chain Management* 38(1)(2002): 54.

CHAPTER 10

Production and Inventory Control

PART I: THE FUNDAMENTALS OF PRODUCTION PLANNING

The objective of the production planning and control function is to coordinate the use of a firm's resources and synchronize the work of all individuals concerned with production to meet required completion dates, at the lowest total cost, consistent with desired quality.

Historically, all firms conducted their production planning and control activities manually, with the use of a variety of Gantt charts and specialized visual scheduling/control boards. Today, most firms utilize some type of computer-based system to perform essentially the same types of activities in a more comprehensive, semiautomatic manner. Regardless of the specific operating system used, an effective production planning and control operation must accomplish five general activities:

1. Preliminary planning
2. Aggregate scheduling
3. Detailed production scheduling
4. Release and dispatching of orders
5. Progress surveillance and correction

Preliminary Planning   After the initial product design and process design work are completed by the respective engineering groups, the preliminary planning work begins. The product's engineering bill of materials is restructured for compatibility with the firm's planning system. *For the specific product (or special job),* analysts then determine the specific material requirements, standard labor and machine requirements, and tooling requirements. In most intermittent manufacturing operations, one or more work-flow routings through the shop are determined.

Aggregate Scheduling   The next step in the process is scheduling—first aggregate scheduling and then detailed production scheduling. As orders and forecasts are generated, they are matched against and coordinated with the facility's overall capability. Aggregate scheduling is simply a first-pass, broad-brush determination that shop capacity—equipment and people—and required materials probably can be made available through careful, detailed scheduling work.

Detailed Production Scheduling   The ensuing step consists of the detailed production scheduling work. The aggregate planning work is broken down into specific product models and configurations, and for each one the detailed manufacturing steps are scheduled into specific work centers or on specific machines. Start and completion dates for each operation or set of related operations are assigned, indicating the desired production priorities. Specific material and tooling requirements for the job are determined, as is the specific shop routing for the job.

Release and Dispatching of Orders   The work completed up to this point in the process has led to the development of an operating plan. When the order is released to the shop, the plan becomes operational and the order is dispatched from one operating unit to the next until it is complete. One of the key functions in this activity is to review the production priorities established in the prior scheduling activity. Any desired changes can be made at this point.

When the order is released, it is accompanied by a packet of paperwork and instructions. The packet typically contains such things as the engineering drawings and perhaps the bill of materials, tooling and material requisitions or computer entry instructions, a routing sheet, and detailed operations instructions for the production people. It may also contain instructions for charging labor and moving the job from one operation to the next.

Progress Surveillance and Correction   The last step in the process is the control function. Progress at each stage of the operation is monitored and fed back to the shop dispatcher and the production scheduler, who compare actual performance with the plan. Significant deviations from schedule typically require some type of corrective action: rerouting, rescheduling, the use of overtime work, and so on. These decisions, if fairly routine in nature, typically are made by production planning and control personnel. If a serious trade-off of resources or priorities is involved, marketing and manufacturing personnel may also enter the decision-making process.

Modern Production Planning Systems

The preceding discussion sketched the fundamental activities that must be accomplished in planning and controlling production operations effectively. In practice, manufacturing firms conduct those activities in a variety of ways, some with great detail and precision and others with less sophistication. As competition in the marketplace has become increasingly keen, however, firms have been forced to meet higher performance standards and

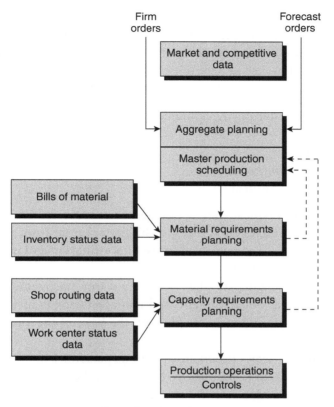

Figure 10.1 A basic flowchart for a modern production planning system.

do that cost-effectively. This economic reality, coupled with the availability of relatively inexpensive computing capability, has spawned a new era in production planning. Within a few years, a progressive firm without a sophisticated, computer-based production planning and control system will be a rarity.

Although numerous different systems are evolving, both as custom-designed and as standard commercial software packages, most utilize essentially the same basic elements. Figure 10.1 portrays, in flowchart form, the basic operating elements found in most current computer-based production planning systems.

Aggregate Planning and Master Scheduling

The development of a viable aggregate plan and a coordinated master production schedule is the starting point for the use of a detailed computer-based planning system.

The *aggregate plan* is based on the expected receipt of a certain number of orders for a specific family of products during the planning period. For the near term, a number of firm orders typically are in hand. As the planners peer further into the future, they use various forecasting techniques to determine an approximate aggregate demand for the product family. The most commonly used forecasting approaches are the following:

▲ Bottom-up analysis, utilizing the opinion, judgment, and market surveys of field sales personnel
▲ Time series analysis
▲ Exponential smoothing techniques
▲ Regression and correlation analysis

Forecasting activities typically are conducted or coordinated by a specialized staff group and generally are handled as a responsibility separate from the computerized planning system activities. In a majority of firms, forecasting demand is the responsibility of the marketing/sales department.

Development of the aggregate plan is usually a top management responsibility. General management, sales management, and manufacturing management personnel jointly develop the initial version of the plan, on the basis of the known and expected order data.

With the assistance of senior production planning personnel, in most firms the plan is developed for a period of 6 to 12 months. To be effective, the schedule must cover a time span that exceeds the cumulative lead-time of the finished product. The plan must be firmed up for a reasonable period because overall production volume cannot be changed abruptly without incurring significant unplanned costs. Every production volume utilizes a particular mix of labor, materials, and equipment. When the output rate is changed, a new optimal mix must be achieved by readjusting the usage rate of the various resources. In the longer term, this can be done by replanning the variables: the employment level, the use of overtime, the use of subcontracting, and the variation of inventory levels. In the short run, it usually is difficult to do efficiently.

The *master production schedule (MPS)* is developed directly from the aggregate plan and is the instrument that drives the firm's entire production system. The aggregate plan establishes an overall level of operations that balances the plant's capability with external sales demand. The master schedule translates the aggregate plan into specific numbers of specific products to be produced in identified time periods.

The relationship between the master schedule and the aggregate plan is shown in Figure 10.2. In this hypothetical illustration, an appliance manufacturer's aggregate plan for refrigerator production is shown at the top of the figure. (When this plan is added to the firm's aggregate plans for ranges, washing machines, and dryers, the firm's total

Aggregate Plan for Refrigerators

Month	Jan.	Feb.	March	April	May	June	July	Aug.	Sept.	Oct.	Nov.	Dec.
Number of refrigerators	500	500	600	700	700	800	800	700	600	500	400	400

Master Production Schedule for Refrigerators

Month	Jan.	Feb.	March	April	May	June	July	Aug.	Sept.	Oct.	Nov.	Dec.
Business Model												
▲ Standard	—	100	—	200	—	200	—	150	—	100	—	100
▲ Heavy Duty	—	100	—	100	—	100	—	100	—	50	—	50
Model A												
▲ Standard	200	100	300	150	300	200	350	200	250	150	150	100
▲ Deluxe	50	—	50	—	100	—	100	—	50	—	50	—
▲ Executive	—	100	—	150	—	200	—	150	—	100	—	100
Model B												
▲ Standard	200	100	200	100	200	100	250	100	250	100	150	50
▲ Deluxe	50	—	50	—	100	—	100	—	50	—	50	—

Figure 10.2 Illustrative relationship of the aggregate plan and the master production schedule.

production capacity will have been utilized for the year.) Note that the master production schedule breaks down the aggregate production of refrigerators into the production of seven specific models, time-phased by quantity per month for the planning period of a year. For example, the 500 units planned in January consist of 200 Model A Standards, 50 Model A Deluxe units, 200 Model B Standards, and 50 Model B Deluxe units. The master schedule for the rest of the year is constructed in the same manner.

The next step in the development of the master schedule is to evaluate its feasibility by means of simulation, checking the availability and balance of required materials and capacity resources. If bottlenecks or imbalances are encountered, the schedule is modified by trial and error until an acceptable arrangement is found. Many computerized planning systems have this simulation capability built in, as will be discussed shortly.

Once an acceptable schedule has been determined, its outputs—the volume and timing of the production of specific products—become the inputs required for the subsequent detailed computer planning work that drives the production, inventory, and purchasing operations.

Before moving on to this part of the discussion, however, it is necessary to say just a word about timing and modification of the original plan. The time interval used in master scheduling obviously varies from firm to firm; it depends on the types of products produced, the volume of production, and lead times of the materials used. However, weekly periods probably are used most commonly, followed by biweekly and monthly intervals.

Thus, within the time frame of a 6- to 12-month aggregate plan, the master schedule typically is updated weekly to reflect changing sales demands and perhaps internal problems that require rescheduling. If the system is to work effectively, it is essential that the schedule be updated regularly. After each update, many firms follow the policy of holding the next four weeks' schedule *firm*, providing only modest flexibility for adjustment in the schedule for weeks 5 through 8, and providing considerable flexibility for change in the schedule for weeks 9 through 12.[1]

Material Requirements Planning

Material requirements planning (MRP) is a technique used to determine the quantity and timing requirements of dependent demand materials used in the manufacturing operation. The materials can be purchased externally or produced in-house. The important characteristic is that their use is directly dependent on the scheduled production of a larger component or finished product, hence the term "dependent demand." For example, a refrigerator door is a dependent demand item in the production of a refrigerator.

The MRP and the capacity requirements planning (CRP) segments of the production planning system are the responsibility of the production planning and control group. In practice, the actual number crunching and paperwork generation usually is accomplished by computer.

Production planning personnel are responsible for structuring and formatting the product bills of material that eventually are contained in computer memory. The same group is responsible for setting up the part and component inventory status records (perpetual inventory records), which also are computerized. In addition to the current inventory balance, an inventory status record typically contains the timing and size of all open (scheduled) orders for the item, the lead time, safety stock levels, and any other information used for planning purposes. When this preliminary planning work is completed, material requirements for a specific time period can be generated.

Although details of the software operation may vary, generally speaking, the MRP segment works as follows. It takes the master production schedule output for a particular product and calculates precisely the specific part and component requirements for that product during the given period of operation. This is done by "exploding" the product bill of material (listing separately the quantity of each part required to make the product) and extending these requirements for the number of units to be produced. Since a part often is used in more than one finished product, the process is repeated for all products. Then all products' requirements for a given part can be summed to obtain the total requirement for the part during the stated period of operation.

For illustrative purposes, let us return to the refrigerator example in Figure 10.2. Assume that all refrigerator models produced except Model B Deluxe use one standard 1-horsepower electric motor. Because of an additional freezing compartment, Model B Deluxe uses two standard 1-horsepower motors. For the month of January, MRP system would determine that the manufacturing operation requires 550 standard 1-horsepower electric motors.

Returning now to our general discussion of the MRP processing activity, after a part's requirements for the operating period are calculated, the computer compares those requirements with the inventory balance, considering open orders scheduled for receipt, to determine whether a new order needs to be placed.

The output of the MRP system, then, can be the following items:

1. Current order releases to purchasing (or to previously selected suppliers), with due date requirements.
2. Planned order releases to purchasing for ensuing periods (considering inventory balance, scheduled requirements, and lead-time requirements).
3. Current and planned order releases for in-house production, with completion date requirements.
4. Feedback to the master production scheduler in case operating changes or supplier performance has produced material availability problems.
5. With revised output from the master production schedule, the MRP system will replan and schedule the material requirements.

A more detailed illustration of an MRP system's operating logic is provided in the following section.

MRP Logic and Format[2]

As they are utilized today, most versions of MRP act as information processing systems that seek to develop and maintain a set of orders that support the production plan, while simultaneously maintaining inventories within the production system at reasonably low levels. Orders within an MRP system fall into two categories: (1) open orders that have been released but have not arrived yet and (2) planned orders that are developed in anticipation of future releases. As was noted previously, each category can contain both purchase orders and shop orders.

The processing logic of MRP centers on the development of a materials planning record for each item. Figure 10.3 illustrates a typical planning record based on the data contained in the master production schedule, bill of materials, and inventory record file shown at the top. The top row of each record shows the "gross requirements," or anticipated usage of the item projected into the future. The second row shows all

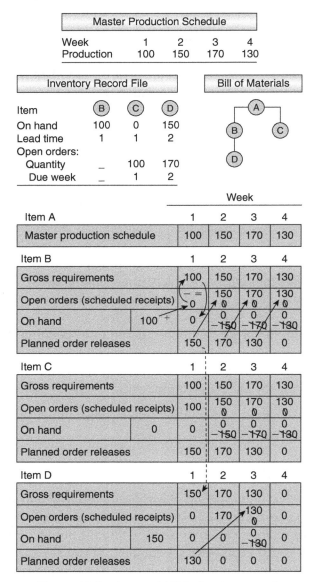

Figure 10.3 A typical MRP planning record.

current "open orders (scheduled receipts)," with each order assigned to the time period in which it is expected to arrive. The "on hand" row projects the inventory balance into the future. This is calculated by determining the impact of gross requirements (planned withdrawals) and open orders (planned receipts) on the inventory balance.

The current inventory balance is shown to the left of period 1. The bottom row shows the "planned order releases" for the item, with each order assigned to the time period when it should be released.

The development of the planning record is based on three fundamental concepts, which form the essence of the MRP-based approach to materials planning and control.

▲ Dependent demand
▲ Inventory/open order netting
▲ Time phasing

Dependent demand takes the multistage product into account in the planning for individual items. Clearly, decisions to acquire purchased materials should be based on anticipated production plans. Those decisions include both quantity and timing considerations. Dependent demand logic is used to calculate the gross requirements for each planning record. This projected usage includes the planned production of all other products that require the item being planned.

The inventory/open order netting concept is used to develop the "on hand" balance row of the planning record. Efficient use of inventory implies that current stocks should be largely depleted before the acquisition of additional inventory. The netting process accomplishes this by allocating current inventory and open orders to the earliest requirements. When the on-hand balance falls below zero, additional stock must be ordered. This process not only signals the need to plan an order but also determines when the order should arrive. This "need date" becomes the due date for the order.

Time phasing utilizes lead-time information and need dates. The "planned order releases" row of the planning record shows time-phased orders whose placement dates are offset from the need dates of the orders by the lead time of the item. Each order, if released in the time period designated by the planned order row, should arrive exactly at the time it is needed by a following production stage.

Utilization of the three basic concepts can be observed in the simplified illustration of Figure 10.3. The bill of materials diagram shows that item B is a fabricated part that is manufactured using D as a raw material and that B is used as a component of A. The dependent demand concept is seen in the gross requirements data; the planned usage of B matches the planned production of A, in both quantity and timing.

The netting process calculations determine the need for additional orders to be planned. For item B, in week 2, 150 units are needed. Since the on-hand quantity of 100 is used to meet week 1 requirements, an order release for 150 units is shown in week 1. Time phasing maintains the timing difference between the planned order and its need date; this is the item's lead time. In this case, the lead time is one week. A similar process is used to develop the remaining orders in the "planned order releases" row. For item D, for example, the process is repeated, the only difference being that D is dependent on B.

MRP not only plans for each order, but also allows the replanning of orders. Replanning is generally necessary when the status of an item is changed as a result of new information. An example of a status-changing event could be notice of a late delivery. Using item D as an illustration, an open order currently is planned to arrive in week 2. If this order is known to have an altered delivery date of week 3, the current plan should be adjusted (replanned) to reflect this new information. The impact of this change can be traced to every other item by reversing the processing logic and working from bottom to top rather than from top to bottom.

The impact of the shortage also can be reduced by adjusting the orders for the other items. Take item B, for example. The planned order release in week 2 is no longer feasible and should be shifted into week 3. The requirement that was to be covered by this planned order no longer can be supported. It also should be shifted into the next week, thus affecting the master production schedule.

Thus, the MRP system generates a complete set of planned orders for all manufactured parts and purchased materials based on the information inputs. Clearly, the validity of the plan produced by the system is dependent on both accurate and timely lead-time information from purchasing personnel. At the same time, if system planning is done far enough in advance (and rescheduling activity is kept to a minimum), the advance knowledge about specific material requirements certainly can facilitate planning and conduct of the buying activities.

It is important to understand that all MRP systems rely on the following:

1. *A very accurate master schedule*, which is based on the sales forecast. This means production planners must interact constantly with the marketing department, scheduling starts with the organization's customers, not in the operations department.
2. *Absolute accuracy of all inventory counts*, including on-hand, on order, and encumbered (ear-marked). For example, sloppy inventory counting, including failure to adjust for shrinkage, will result in inaccurate records being "read" by the computer.
3. *Up to date bills of materials*, which reflect all engineering and quantity changes. Many MRP systems have failed because specification revisions failed to reach all members of the supply chain.

Capacity Requirements Planning

The next step in the production planning process is capacity requirements planning (CRP). The function of the CRP segment of the process is to convert the shop orders produced by the MRP system into scheduled workloads for the various factory work centers. In addition to the MRP system output, two other sets of data inputs are required to do this: (1) shop routing data and (2) work center status data. These inputs for CRP operation are analogous to the bill of materials and inventory status inputs for MRP operation.

In an intermittent production operation, the manufacture of each product or component requires that a series of specific machine or human operations be performed on the item as it progresses toward completion. Those required operations define the "route" the item must travel through the manufacturing facility. Sometimes more than one routing sequence is possible. In any case, one of the required preliminary planning activities is to develop one or more shop routing plans for each product and component produced in-house. In addition to the physical routing plans, standard processing time requirements must be determined for each operation in the sequence and included in the shop routing data file. In a continuous manufacturing operation, these activities usually are simplified by the design and layout of the production facility.

The work center status data file maintains a perpetual record of the capacity—equipment and human—that is available (and committed) in each of the factory's work centers. Capacity typically is measured in standard man or machine hours per time period.

With shop routing data and work center status data files loaded in computer memory, the CRP system is ready for operation. Recall that during the development of the master production schedule, a preliminary analysis of work center capacity was done before the schedule was firmed up. While this "rough cut" produced approximate data accurate enough for overall scheduling purposes, it is not precise enough for the detailed work-loading job at hand. Therefore, with the MRP-generated current and planned shop order releases now known with certainty for a specific time period, a second pass is made. The CRP program first obtains the necessary routing and timing data for each scheduled order and then checks the appropriate work center status files to determine if the required capacity is available. Frequently the proposed plan does not mesh satisfactorily with the availability and timing of capacity existing in the required work centers. In this case, the CRP activity becomes an iterative process, and replanning continues until realistic work center loads are developed. The variables that can be manipulated by the system and the planner to achieve a reasonable balance typically include the following:

- ▲ Alternative routings
- ▲ Personnel reallocation
- ▲ Use of overtime
- ▲ Inventory-level variations
- ▲ Use of alternative tooling
- ▲ Use of subcontracting—outsourcing

Occasionally, an impossible scheduling situation is encountered. In this case, the system communicates to the scheduler the need for selected capacity modification or for a revision of the master schedule. Most of the time, however, a reasonable fit can be achieved. The normal output of the system in these cases is:

1. Verification of the planned orders from the MRP system
2. Work center load reports that reflect the priorities established by the MRP system

This information subsequently is used in the final stage of the planning and control process.

Control of Production Activities

According to a time-honored adage, "the proof of the pudding is in the eating." The validity and usefulness of the detailed planning done to this point in the process will now be seen as it is applied to production operations in the shop.

The output of the MRP and CRP systems is transmitted to the manufacturing organization in the form of order releases and a dispatch list. Referring again to Figure 10.3, the planned order releases shown on line 4 of the planning record, when released as the date moves into the current period, officially become the open orders (scheduled receipts) shown on line 2 of the planning record. Before releasing an order, the planner must make a final check to ensure that the priority sequencing of the order is still valid, capacity is still available, and materials are available. If all factors are not "go," release may be delayed rather than having the order held up after it is started in the shop.

When an order is released, typically it is accompanied by a packet of materials and instructions required to complete the job. Included are such things as engineering drawings, bills of materials, route sheets, move tickets, materials requisitions, and labor charge forms. Depending on the extent to which the entire system is computerized, some of these functions may be handled through the use of online terminals.

The dispatch list, containing a series of order releases, is prepared by the planner and may cover a time period ranging from a day to a week. It goes to the appropriate work center foreman, who schedules his machines and people in accordance with the due dates specified for each order on the list.

If daily dispatch lists are used, the foreman generally has little discretion in scheduling the jobs in the work center. In contrast, if the list covers several days of work or more, the foreman has an increasing amount of latitude in his detailed scheduling activities. This provides a better opportunity to maximize the operating efficiency of the work center. Perceptive planning and scheduling by the foreman usually can minimize setup and material move costs and maximize the utilization of equipment and labor. Consequently, it is important that planners coordinate the development of their dispatch lists closely with the appropriate foremen in an effort to optimize both planning control and shop efficiency. Team-type spirit and effort usually produce the best results.

As noted earlier, control is an essential element if the entire process is to work effectively. In most operations this is accomplished at two levels. First, as a job progresses, status and related information are fed back to the planner from either the operator or the foreman on the job. The information reported typically includes order status, anticipated delays, materials shortages, and rework and scrap data. As appropriate, such feedback occurs either daily or when a job is started and completed. Some firms report only on an exception basis. In continuous manufacturing operations, such reporting typically is done less frequently at predetermined checkpoints in the process. In any case, such information is used by the planner to determine whether replanning or other corrective action is necessary to meet the firm's sales commitments.

In addition to order status types of control reports, most systems require one or more types of capacity control reports. The most commonly used one is an input/output report, typically developed weekly. This type of report usually shows the hours of work planned for a particular work center, the number of hours actually worked on the planned jobs, and the difference between the two. Significant deviations from plan can produce obvious problems in the CRP and subsequent scheduling activities. Hence, in most operations, this type of control is essential.

Reporting and monitoring methodology varies among firms. Some are totally computerized, with computer graphics output, while others use a combination of computer and manual communication and charting techniques. The use of Gantt charts and schedule/control boards is still fairly common.

In most organizations, control is also exercised at a second level: personal control right on the shop floor. One or more dispatcher/expediters, usually assigned to the production planning and control group, spend most of their time on the floor visually following jobs through the manufacturing operation. This individual's job is threefold: (1) to ensure the integrity of the job priority plan; that is, make sure the routing and scheduling instructions for jobs are implemented reasonably; (2) to ensure reasonable capacity control; that is, ensure the hours scheduled in the various work centers are actually being worked; and (3) to help the operating people solve unexpected planning and scheduling difficulties. As problems arise, the dispatcher/expediter works with the various foremen in helping to resolve them. This person is the planning group's representative on the shop floor. Within reason, he or she has the authority to suggest certain micro planning and scheduling changes that may be able to resolve a problem directly and expeditiously.

To summarize, the production planning activities all come together on the shop floor to initiate and control the production operations. Overall responsibility for *control* of operations usually is vested with the production planning and control manager or the materials manager. In firms without a materials management department, the manufacturing manager is sometimes responsible. The control function usually has both a centralized and a decentralized component. Order release, dispatching, and formal status control are the responsibility of the centralized production planning and control group. Decentralized informal control on the shop floor is the joint responsibility of dispatcher/expediter personnel and the line foremen responsible for the actual production operations. Viewed in another sense, it is the responsibility of the foremen to do the micro work center scheduling and to run each production operation so that planned completion dates are met. It is the responsibility of production planning and control to keep work flowing through the shop at a steady rate, focusing always on order priority control and capacity control.

Two Management Considerations

It is appropriate at this point to reiterate two important concepts that may have gotten buried in the details of our discussion of the total planning system.

The first concerns the *multilevel nature of the operation of the production planning system.* A review of Figure 10.1 will reveal that for the most part, the aggregate planning and the master scheduling activities are top-management and staff responsibilities. In comparison, most activities associated with the MRP and CRP activities are primarily the responsibility of production planning and control personnel. Finally, the control of production operations themselves is a joint responsibility of production planning, control personnel, and supervisory operating personnel. For a system to function effectively, the coordinated efforts of all three groups are required.

The second concept focuses on the *dynamic nature of the total production planning system.* Although time periods vary for different organizations, a majority of firms utilizing a comprehensive computer-based planning system work from an aggregate plan structured for the coming year. The master schedule subsequently covers the same year's period, but typically it is further delineated by month and by week. Material and capacity requirements are also structured in weekly "time buckets." To maintain a current plan that correctly reflects changing sales demands and internal scheduling and capacity constraints, the entire operation typically is replanned on a weekly basis. Therefore, the firm, semifirm, and flexible portions of the total operating schedule simply drop the week just past and encompass one new week as replanning occurs each week. Some systems utilize a technique in which only selected portions of the operation that are influenced by changing conditions are replanned; this approach has some obvious logistical advantages. Regardless of the specific technique used, however, the important point is that the dynamic nature of the planning system keeps it current on at least a weekly basis. From a practical point of view, this is the feature that makes such a system so valuable to a large complex firm operating in a competitive environment.

Evolution of MRP and MRP II Systems

As innovations are accepted and refined in the business world, inconsistencies in the use of terminology and variations of the concept inevitably emerge with the passage of time. Such has been the case with MRP and its subsequent derivatives. The following paragraphs describe briefly the evolution of MRP and its progeny, MRP II.

The computer-based material requirements planning technique found its first significant industrial usage in the early 1970s. Although it became known as MRP, for a number of years it was used primarily to generate orders for parts and materials that related to a specific demand schedule. Later, users found that with some refinement it could be used as a scheduling technique. It could be used to feed back schedule change data and subsequently reschedule existing orders to maintain valid material and shop order dates. Hence, it became a much more valuable tool.

However, from a production planning point of view, MRP still left something to be desired because it was unable to deal with the capacity variable. Before long, though, necessity proved to be the mother of invention, because soon a capacity requirements planning module was developed and linked to the original MRP module. With further development of the master production schedule concept, in many firms the bulk of the planning activities shown in Figure 10.1 were integrated into a single planning and scheduling package. Today, with the exception of the aggregate planning and the production operation controls segments, this entire integrated package is identified as a *closed-loop MRP system*. Thus, the concept and the terminology have both changed with time.

The last step, to date, in this evolutionary process is an expanded system known as *MRP II—manufacturing resource planning*. This system simply adds two new capabilities to a closed-loop MRP system. The most significant addition is the financial interface. This module provides the ability to convert operating production plans into financial terms, so the data can be used for financial planning and control purposes of a more general management nature. Related to this feature, the second addition provides a simulation capability that makes it possible for management to do more extensive alternative planning work in developing the marketing and business plans. This can be done by asking "what if" types of questions—that is, by modifying an operating variable and receiving a systemwide response to the proposed operating change.

As MRP and MRP II systems developed in different firms, they often included somewhat different levels of capability. For example, some MRP systems did not originally include comprehensive CRP capability and did not function fully as a closed-loop system. When MRP II capability subsequently was developed, it included these features that previously were missing. As a result of this situation, coupled with the fact that MRP II is more comprehensive in nature, many users see MRP II as the "umbrella" system, with MRP as a major component of that system.[3]

Impact on Purchasing and Supply

Sooner or later, most manufacturing firms will use some type of MRP-based system as a central component of their production planning system. If present experience is a reasonable indicator of the future, purchasing operations will be affected in the following important ways:[4]

1. Expanded use of the buyer-planner or the supplier scheduler concept.
2. Expanded use of contract buying.
3. Necessity of greater supplier flexibility and reliability.
4. Development of closer relationships with suppliers, including more partnering arrangements.
5. Increased accuracy and timeliness of materials records.
6. Direct interface between the buyer's and supplier's MRP system.
7. The buying organization must share its forecast with suppliers.

Buyer-Planner and Supplier Scheduler Concepts The nature of an MRP operation places the planner in close, continuing contact with material requirements and their frequently changing schedules. Typically, the planner has a more sensitive feel than does the buyer for the probable usage pattern of most materials. Consequently, to improve efficiency of the planning-buying activity as well as communications with suppliers, many firms have used one of several organizational schemes that utilize the planner as the supplier contact person for day-to-day material flow activities.

The buyer-planner concept is one commonly used approach. In essence, the buyer's job and the planner's job are combined into a single job done by one individual. This person obviously handles a smaller number of items than originally were handled by either the buyer or the planner. The buyer-planner is responsible for determining material requirements, developing material schedules, making order quantity determinations, issuing all material releases to suppliers, and handling all of the activities associated with the buying function. Thus, in this integrated role, the buyer-planner maintains close contact with various supplier personnel.

Another popular approach is simply to assign to the planner the responsibility for dealing directly with suppliers in releasing and following up materials orders. In this arrangement, the buyer handles all the normal purchasing responsibilities except requirements releases against existing contracts. The planner handles this latter function and becomes the buying firm's supplier contact on all day-to-day material scheduling matters. Most firms refer to this arrangement as the *supplier scheduler concept*.

In a recent survey of present MRP users, researchers found that approximately 55% of the firms utilize the supplier scheduler concept and that 30% use the buyer-planner organizational arrangement. Only 15% of the firms utilize the traditional pattern of operation.[5] It seems clear that continued expansion of the use of these two concepts will accompany MRP development in the future.

Contract Buying Because an MRP system requires the placement of frequent orders for relatively small quantities of materials, it obviously would be inefficient, if not impossible, to make a new buy for every weekly requirement. The alternative, of course, is to place annual or longer-term contracts with suppliers for the required materials and then simply issue a telephone or an MRP schedule release against the contract as the production operation requires.

Not only is this buying approach required in an MRP-scheduled operation, as a general rule it is excellent buying practice. It permits more careful purchasing planning and more thorough market and supplier research, and it needs to be done only once every year or two for each material. In addition, such contracts usually produce attractive pricing arrangements and improved supplier relations.

Supplier Flexibility and Reliability Because of the weekly updating of most MRP systems, coupled with the frequent rescheduling that sometimes occurs, a supplier has to be more than reasonably flexible. Even if a supplier has the buyer's MRP schedule with weekly or biweekly requirements for the next two months, the irregularity of demand and the short notice given on schedule changes present a difficult operating situation for most suppliers. Resolution of the potential problems requires careful cooperative planning and usually some compromises by both parties.

It is obvious that supplier reliability is a must. The buying firm typically carries some inventory, but not as much as in the traditional operating situation because one of the objectives of the system is to reduce inventory levels. Hence, there is much less cushion in the system to handle the problems of late deliveries and off-spec materials.

The bottom line of these two stringent operating requirements is that supplier selection is a critical, yet more difficult, task.

Closer Relationships with Suppliers The use of contract buying and the need for unusual supplier flexibility and reliability create an operating situation in which the buyer-supplier relationship must be closer and more cooperative than it might normally be. As discussed in Chapters 13 and 14 on sourcing, this type of operating situation requires the ultimate in coordination, cooperation, and teamwork. A mutual understanding of each other's operations and problems is essential in achieving this type of effectiveness. It literally is an informal partnership operation, and it must turn out to be a win-win deal.

The buyer-planner or the supplier scheduler must stay in close touch with the supplier's counterpart on a week-to-week basis as far as scheduling and delivery matters are concerned. Also, the buyer (or buyer-planner) must handle the broader issues of the relationship with appropriate supplier sales and technical personnel on a regular and timely basis.

As one reviews the MRP segment of the production planning system, it is readily apparent that the accuracy of the system will be no better than the accuracy of the data used in its calculations. If the system is to work effectively, records such as specifications, bills of materials, supplier lead times, receiving reports, and inventory balances must be as close to 100% accurate as possible.

Just-in-Time Production Planning

Although originally pioneered by Henry Ford, the JIT manufacturing concept has been refined and developed over the past several decades in Japanese industry. The purpose of this recent concerted effort was to improve quality and reduce costs to help Japanese businesses become more competitive in world markets for selected product lines. The resounding success of the Japanese effort prompted a growing number of U.S. firms to develop and implement modified versions of the system in this country.

The JIT concept is considered by many to be a technique used for reducing inventories, but in reality, it is much more. The complete JIT concept is an operations management philosophy whose dual objectives are to reduce waste and to increase productivity. However, operationally, the basic theme of the JIT concept is that *inventory is evil*. Inventory is considered to be undesirable for three reasons:

1. It hides quality problems.
2. It hides production inefficiencies and productivity problems.
3. It adds unnecessary costs to the production operation—carrying costs of approximately 25 to 35% of the inventory value per year.

Inventories of production materials permit suppliers' quality deficiencies to be covered up, and in-process inventories permit off-spec work in-house to be given less attention than it should. This occurs because the unacceptable items can be replaced with good items from inventory while those that are unacceptable are being reworked. The same rationale applies to schedule slippages caused by inefficiencies in the workplace and in the system itself. The end result, claim JIT proponents, has been a tendency among U.S. managers and their employees to accept mediocre, second-rate work as the norm.

Hence, in an effective JIT application the operating policy is to minimize production inventories and work-in-process inventories by providing each work center with just the quantity of materials and components needed to do a specific job at the exact time they are needed. In an ideal situation, each unit of output would be produced just as it was needed at the succeeding work station. In reality, this ideal is not achieved, but within reason, it is a viable objective in many organizations. Practically speaking, the result is a reasonably continuous flow of small-lot production. At the supply end of the operation, materials that are procured in a JIT mode are delivered frequently by suppliers in fairly small quantities. Deliveries may range from twice a day to once a week.

Consequently, throughout the system, with only minimal inventories on hand to cover for poor-quality materials and workmanship, the focus is on consistently high-quality material and in-process work. Without that, the system breaks down.

To summarize, the basic operating plan is to gear production and final assembly as closely to sales demand as possible. Individual production operations are also geared more closely together. This is accomplished either by means of a product-type layout of equipment and work centers[6] or by means of a material-pull, "Kanban" type of material movement system[7] in a process-oriented layout. Finally, the firm's total production operation, through its purchasing activities, is geared as closely as practical to key suppliers' production operations. Thus, the characteristic of small-lot flow can be traced through the entire system from a supplier's plant, through the buyer's plant, and out into the finished goods distribution system. The actuating element is sales demand, which *pulls* the various stages of in-process work and materials through the complete system.

A JIT Illustration

Figure 10.4 provides a flow diagram of the major production operations in an electronic instrument manufacturing plant both before and after the firm implemented a JIT system. The top portion of the figure shows the original operation. After incoming material was received, counted, and logged into the system, it went through a standard visual receiving inspection operation where potential quality problems were detected and perhaps submitted to quality assurance for further detailed inspection. The next step was to prepackage the materials, parts, and components that subsequently would be used in putting together a particular subassembly. This was done for each subassembly produced to facilitate stock picking for the later assembly work. Most production inventories thus were stored in this subassembly kit form. After subassembly operations occurred, subassembly units were inventoried until they were used in a product's final assembly operation. Approximately one week's finished goods inventory for most products was maintained at the plant.

The lower portion of Figure 10.4 shows the dramatic change that occurred as a result of JIT implementation. The receiving inspection and quality control technical inspection operations for purchased materials were completely eliminated. *The responsibility for incoming quality was placed with the purchasing department and delegated to each supplier organization.* This required a reasonable amount of supplier education.

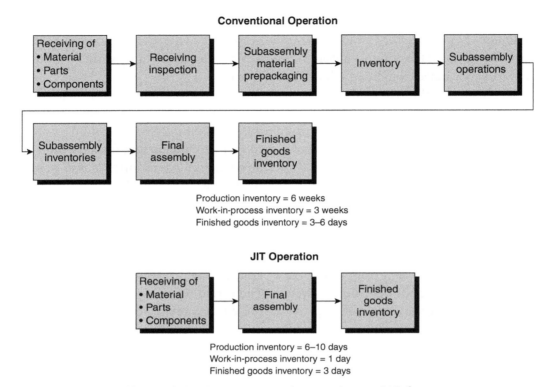

Figure 10.4 How JIT was implemented in one U.S. firm.

In most cases, purchasing and quality control (QC) worked with suppliers to develop and install statistical process control (SPC) systems in their manufacturing operations. SPC control charts were then required to be submitted with each shipment of delivered material.

The next major change occurred on the shop floor. The facility originally utilized a specialized process-type layout, similar to a large job shop. This was revised to achieve a modified product-type layout. Although the firm produced approximately half a dozen different product lines, there was enough similarity between products to permit the use of several product-flow types of facilities arrangements. This layout permitted the use of an open-type storage system adjacent to the production operations so that incoming materials were delivered directly to the point of use in the shop.

Production scheduling subsequently was based completely on units of finished product rather than on the production of subassemblies. This made it practical to eliminate the subassembly prepackaging and storage activities as well as the subassembly operations themselves. The firm's closed-loop MRP system was still used to generate requirements and overall schedules, but it was necessary to "smooth" the master schedule to facilitate the reasonably continuous, small-lot production. Consequently, the total shop now resembles a continuous manufacturing operation in contrast to its previous job shop character.

As shown underneath each of the two flow diagrams, inventory levels were reduced greatly. Production inventory was decreased approximately 70%, and in-process inventories were dropped from about a 15-day supply to a 1-day supply. Finished goods inventory was reduced by about 40%. Hence, the total float figure declined from approximately 50 days to 12 days, and the firm reports that quality problems have declined noticeably.

In summary, the following elements tend to characterize most successful JIT operations.[8]

1. The JIT concept is most applicable to manufacturing operations that produce a relatively small number of different products in at least a quasi-continuous environment.
2. Product demand must be reasonably predictable, and requirements must be generated accurately. A closed-loop MRP system can be used to do this, but typically the master production schedule must be smoothed on a daily basis.
3. Statistical process control typically is used in both the buyer's and suppliers' organizations to ensure tight control of material and production quality. This is vital to the functioning of the low-float, small-volume, relatively smooth-flowing operation.
4. Production operation setup requirements must be able to be reduced to relatively short times. Most firms target for tool changes and equipment setups of less than 10 minutes. Without this capability, small-batch and smooth-flow production of different models or different products cannot be accomplished efficiently.
5. Purchasing must be able to reduce materials replenishment lead times. This usually is accomplished by reducing the four major elements of lead time—internal paperwork and ordering time, supplier queue and manufacturing time, transportation time requirements, and incoming receiving and inspection requirements.
6. Successful JIT operation suppliers must be able to be flexible to meet the buying firm's stringent, short-fused material requirements—and they must be reliable to the nth degree.
7. As previously noted, the buying organization must share its forecasts with suppliers with the necessary lead times. This rolling forecast must have specific locked-in build dates for the suppliers. One major manufacturer calls this procedure schedule stabilization. Such interface obviously demands a direct computer link between the buyer and supplier's production control personnel. It is imperative to ensure that forecast revisions are sent to all impacted suppliers and internal buying company employees.

These considerations lead logically to a discussion of the impact of a JIT system on the purchasing operation.

JIT's Impact on Purchasing and Supply

Purchasing and supply plays a key role in any JIT operation. *Whether a JIT production system works depends on how well purchasing does its job in selecting and managing suppliers.* Obviously, it is not practical to procure all materials on a JIT basis. Most successful JIT firms buy from 5 to 10% of their individual materials—those that account for 60 to 75% of the firm's materials expenditures and those that are space-intensive—in a JIT mode. This keeps the administrative part of the job manageable.

Finding reliable suppliers that are willing to comply with a JIT buyer's stringent requirements typically is not an easy task. Consequently, purchasing usually utilizes three basic strategies:

▲ A specific plan to reduce the number of suppliers utilized, using single and dual sources in many cases
▲ Extensive use of long-term contracting
▲ Share forecasts

Evaluation and qualification of JIT suppliers clearly is a critical and time-consuming activity.

The nature of a JIT purchasing operation requires, and in fact usually creates, a closer, more cooperative relationship between the buying and supplying firms. Hence, from a practical point of view a reduced supplier base is a necessity, and a longer-term contract is the primary incentive that attracts a supplier to consider the arrangement. Only with knowledge of the buyer's long-term requirements schedule can a supplier schedule production and size inventories so that replenishment lead time can be reduced, while simultaneously providing both flexible and reliable service.

As was noted earlier, the basic objective of the "partnering" relationship is to reduce costs, improve efficiency, and increase profitability for *both* organizations. The development of scheduling guidelines and parameters, the implementation of SPC quality systems, and the conduct of value analysis work on the purchased items must be done jointly in a team-type environment. To assist in all of these activities, the buyer often makes greater use of *performance specifications* to encourage the supplier to exercise as much creativity as possible.[9]

Reducing the delivering carrier's transportation time is also an important objective. Consequently, suppliers located near the buyer's operation may offer a distinct advantage. The most important strategy in this element of the equation, however, is to work out a longer-term contractual JIT arrangement with a small number of selected carriers. This type of transportation service can be purchased in the same manner material is purchased from a JIT supplier. This topic is discussed further in the appendix to this chapter.

A final impact of JIT is seen in the form of a shift in the workload within the purchasing and supply department. The buyer's job now involves more responsibility for contract administration and supplier management than it did previously. The tight delivery schedules, the emphasis on control of quality and performance, and the joint resolution of problems with suppliers require this. At the same time, the nature of the JIT buying operation now requires less routine, nitty-gritty buying work. In effect, the JIT buyer's job tends to require a broader range of professional and managerial skills than typically was the case previously.

ERP Systems

Finally, one must remember that enterprise resource planning (ERP) systems are macro data input software programs for all of the firm's information. ERP systems include production planning and inventory control programs but go well beyond their scope.

The benefits of ERP systems include cycle time reduction of key business processes, better financial management, and facility linkage to e-commerce systems; they also help make knowledge and information systems explicit. When the ERP system is well designed, there is complete transparency throughout the enterprise and across the supply chain. However, critics of ERP give many negatives, including long implementation periods, inflexibility, excessive hierachical organizations, obsolete technology, and high cost.[10] However, there are many successful ERP implementations, and all of the major ERP packages now support XML.[11]

In addition, the ERP software firms are obviously dominated by information technology personnel who often write unrealistic rules such as "no back orders" or "no partial shipments." There must be professional purchasing input to avoid such mistakes. Do not let any software package dictate an unrealistic or ineffective procedure contrary to supply chain best practices.

Radio Frequency Identification System (RFID)

Radio frequency identification systems are an improvement to bar coding and involve using inexpensive, thin, integrated circuitry fixed or embedded stickers or labels to electronically send a tracking number with product identification. This allows for faster scanning and overcomes many bar code reading problems caused by dirt or poor positioning. As one might expect, Walmart was an early adopter.[12]

Summary

The interface between procurement and production planning is an extremely important one. Production planning decisions influence the parameters within which purchasing does its work. At the same time, the effectiveness with which purchasing does its job directly influences the success of the production planning system.

In recent years, with the aid of computerized systems, production planning has evolved into a highly specialized and sophisticated activity. Closed-loop MRP systems, MRP II systems, and JIT systems all significantly affect the design and implementation of a firm's purchasing and supply systems. To a great extent, purchasing strategies used in prior years must be modified, and in some cases replaced with new approaches to create the supply environment required to support and sustain these evolving planning systems.

In this dynamic environment it is imperative that the procurement and production planning functions be developed in close coordination because of the interdependencies and that operationally they be coordinated effectively on a day-to-day basis.

PART II: THE FUNCTIONS OF INVENTORIES

The preceding discussion of the JIT concept highlighted the disadvantages inventories can bring to a manufacturing operation. In many circumstances, however, inventories have some redeeming values—they are not all bad. The trick is to obtain the best of both worlds at a reasonable cost.

Generally, inventories make possible smooth and efficient operation of a manufacturing organization by *decoupling individual segments* of the total operation. *Purchased-part* inventories permit the activities of purchasing and supply personnel to be planned and conducted somewhat independently of shop production operations. By the same token, those inventories allow additional flexibility for suppliers in planning, producing, and delivering an order for a specific part.

Inventories of *parts and components produced in-house* in an intermittent operation decouple the many individual machines and production processes from various subassembly and assembly activities. This typically enables management to plan production runs in individual production areas in a manner that utilizes labor and equipment considerably more efficiently than would be the case if all were tied directly to the final assembly line. In addition, *finished goods* inventories perform the function of decoupling the total production process from distribution demands, allowing the development of similar efficiencies of production on a broader scale. These inventories also help *balance the firm's supply with the market forces of demand.*

Thus, well-planned and effectively controlled inventories can contribute to the effective operation of a firm and to a firm's profit. The basic challenge is to determine the inventory level that works most effectively with the operating system or systems existing within the organization and that realistically is the most feasible in dealing with given suppliers and material markets.[13]

If the inventory issue is viewed through the eyes of various operating department managers, an interesting situation appears. The *marketing manager* tends to favor larger inventory stocks to assure rapid assembly and delivery of a wide range of finished product models. This capability obviously can be used as an effective sales tool. The *production manager* is inclined to go along with the marketing manager, but for a very different reason. He or she argues for higher inventory levels because they allow more flexibility in daily planning; unforeseen problems in producing a component can be mitigated if productive efforts can be transferred easily to another component for which the required raw materials are on hand. Likewise, a reasonable inventory of the required items *ensures against production shutdowns* caused by delivery problems, supplier problems, and stockouts, thus avoiding the incurrence of high production downtime costs. These arguments reveal the flip side of the arguments favoring a JIT system.

The *financial officers* of the firm, in contrast, argue convincingly in favor of very low inventory levels. They point out that the company's need for funds usually exceeds availability and that reduced inventories free sorely needed working capital for other uses. They note also that total indirect inventory carrying costs drop proportionately with the inventory level. The *purchasing and supply manager* is the final participant; he or she is concerned with the size and frequency of individual orders. Purchasing often favors a policy of placing fewer and larger orders. Unless contractual arrangements with routine delivery-release systems can be worked out with suppliers, fewer and larger orders usually increase the total inventory level. At the same time, however, they tend to minimize operating problems with suppliers, and in some cases they may reduce unit material prices. Large-volume buying also permits more *efficient utilization of buying personnel* and more effective advance planning for major activities such as market studies and supplier investigations. Thus, it is clear that each departmental executive supports his or her position with legitimate justification.

Concepts and techniques useful in analyzing the inventory issue in order to arrive at sound policy decisions are the focal point of our investigation in this chapter.

Definition of Inventories

Although inventories are classified in many ways, the following classification is convenient for use in further discussion of the topic:

1. *Production inventories*   Raw materials, parts, and components which enter the firm's product in the production process. These may consist of two general types: (1) special

items manufactured to company specifications and (2) standard industrial items purchased "off the shelf."

2. *MRO inventories*   Maintenance, repair, and operating supplies which are consumed in the production process but which do not become part of the product (e.g., lubricating oil, soap, machine repair parts).

3. *In-process inventories*   Semifinished products found at various stages of the production operation.

4. *Finished goods inventories*   Completed products ready for shipment.

In most manufacturing companies, production and MRO inventories together represent the major segment of total inventory investment.

Let us now place the discussion in perspective. Inventory planning occurs at several levels in an organization and covers various time spans. The concern in this chapter is the planning and control of production and MRO inventories in a short-run situation, involving weekly, monthly, and in some cases quarterly or yearly decisions. Hence, the discussion assumes that the longer-range activities of sales forecasting, product modification, aggregate planning, and master scheduling have been completed. The investigation begins at this point in the planning cycle.

Inventory Analysis

The inventory of a typical industrial firm includes as many as 5000 to 50,000 different items. Initial planning and subsequent control of such an inventory is accomplished on the basis of knowledge about *each* of the individual items and the finished products of which each one is a part. Consequently, the starting point for sound inventory management is the development of a complete inventory catalog, followed by a thorough ABC analysis.

Inventory Catalog

After all inventory items have been completely described, identified by manufacturer's part number, cross-indexed by user's identification number if necessary, and classified generically for indexing purposes, some form of inventory catalog, usually computerized, should be prepared for use by all personnel.[14] Careful preparation and maintenance of such a catalog pays two important dividends.

An inventory catalog serves first as a medium of communication. It enables personnel in many different departments to perform their jobs more effectively. A design engineer, for example, may have a choice between two standard parts in an experimental design; an inventory catalog quickly indicates whether either part is carried in inventory and may be available immediately for use in the experimental work. Suppose the item in question is to be used in large quantities on the production line. If one of the alternative parts is a stock item and the other is not, the engineer knows immediately that procurement time and cost probably will be lower for the part that already is being purchased and used elsewhere in the plant.

As a further example, envision a mechanic who has just removed a faulty bearing from a major production machine that has broken down. Upon examination, the mechanic finds the manufacturer's name and part number stamped on the edge of the bearing. Unfortunately, though, the storeroom clerk cannot help because that particular bearing is not carried in stock. If the mechanic or a supervisor consults the inventory catalog, however, he or she may well find that a satisfactory substitute bearing, carried under another manufacturer's part number, is in stock. Proper cross-indexing in the inventory catalog can inform users about common interchangeable parts, a typical situation with many MRO supplies.

A second significant benefit produced by an inventory catalog accrues to the inventory control operation itself. This benefit takes the form of more complete and correct records through the reduction of duplicate records for identical parts. A purchasing department often buys the same part from several different suppliers, under various manufacturers' part numbers. Unless control requirements dictate otherwise, identical parts from all suppliers should be consolidated on one inventory record. A simple situation? Perhaps, but one is amazed to find in highly reputable companies many similar cases in which two or more inventory records bear different numbers for the same part. A carefully constructed catalog significantly reduces the possibility of such problems.

ABC Analysis—The 80–20 Concept

As soon as an inventory is identified and described, the manager must determine the importance and the dollar value of each individual inventory item. This calls for a study of each item in terms of its price or cost, usage (demand), and lead time, as well as specific procurement or technical problems. Without the data provided by such a study, an inventory manager normally does not have enough information to determine the best allocation of departmental effort and expense to the tasks of controlling thousands of inventory items.

A study of several hundred medium-sized West Coast manufacturing firms conducted by the authors reveals the data illustrated in Figure 10.5. This figure shows that in the typical firm a small percentage of the total number of items carried in inventory constitutes the bulk of the total dollars invested in inventory. In the study cited, 10% of the inventory items account for approximately 75% of the investment, and only a quarter of the items make up approximately 90% of the total investment. The remaining 75% of the items constitute roughly only 10% of the inventory investment. While these figures

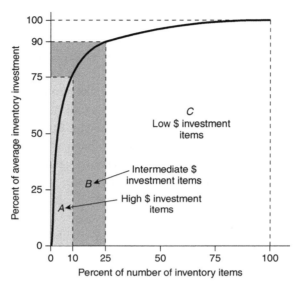

Figure 10.5 Graphic analysis of production and MRO inventories.

vary somewhat from one firm to another, the magnitude of variation usually is not great. Several similar studies in large corporations have produced strikingly similar results, leading to what some firms call the *80–20 phenomenon* (20% of the items account for 80% of total inventory investment). Historically, some firms have termed this phenomenon the *Pareto principle,* based on the law of "the vital few and the trivial many," developed by the Italian economist Vilfredo Pareto around the turn of the twentieth century.

In different companies this type of analysis is known by several different names, *ABC analysis* or *Pareto analysis* being the most common. In practice, such an analysis can be made on the basis of either the average inventory investment in each item or the annual dollar usage of each item. The analysis is easy to conduct once inventory has been identified properly and usage records have been maintained for a complete operating cycle. First, *all items are simply ranked in order of their average inventory investment* (or dollar usage). The total of these values (average inventory investment or annual usage) for all inventory items is then computed. The value of each item is next expressed as a percentage of the total. By going down the list and successively cumulating the individual percentages for each item, one can determine which items make up the first 75% of inventory investment, the first 90%, and so on. If it is convenient to use the three arbitrary classifications noted above, they can be labeled A, B, and C, respectively, and each inventory item becomes an A, B, or C item.

The value of such an analysis to management is clear. It provides a sound basis on which to allocate funds and time of personnel with respect to procurement management and the refinement of control over the individual inventory items. Obviously, no supply manager wants to spend as much time and effort managing the items that make up 20% of the investment as is spent on those making up the remaining 80%. In this sense, management may take several forms. It may involve minimizing acquisition cost, maximizing service and reliability, minimizing inventory investment, minimizing indirect costs associated with inventory, or utilizing personnel effectively. The concept clearly permeates a number of departmental operations: purchasing, production control, stores, and accounting, for example.

In practice, a never-ending problem is that of adequately planning for handling the thousands of low-value C items. In many cases, availability and reliability for these items are just as important as they are for the A and B items. Even with good purchasing planning, because of the sheer number of C items, low-value nuisance purchases frequently require more time than should be allotted to them. Consequently, they reduce the amount of time available to purchasing and supply personnel for supplier studies, value analysis, and other creative work involving high-value *A* and *B* items.

A problem that has grown out of the discussion in the preceding paragraph focuses on another potential dimension of the *ABC* classification. Some firms have observed that in addition to the varying dollar magnitude (or turnover) represented by each material in the three categories, the *criticalness* of each material to the firm's operation also varies and is important from the standpoint of managerial control.

To provide additional guidance for supply managers, James A. G. Krupp, director of corporate materials for Echlin Inc., suggests that each material might also be classified according to its service or operating importance on a three-point scale: 1—critical, 2—medium, and 3—noncritical. Thus, a less important A material would be designated an A-3 item, while a critical C material would be identified as a C-1 item. Depending on the circumstances in a specific situation, it is possible that the C-1 item might require more stringent management attention than the A-3 item. In any case, some firms have adopted this two-digit classification system to provide additional managerial guidance.[15]

An effective inventory management system must help resolve the problems identified in the preceding paragraphs.

Dependent and Independent Demand

As discussed in the MRP section in this chapter, to do the job well, an inventory manager needs one additional bit of information about each of the items in inventory—is the demand (usage) for the item "dependent" or "independent"?

An item is said to exhibit *dependent demand* characteristics when its use is directly dependent on the scheduled production of a larger component or parent product of which the item is a part. Hence, in a plant producing automobile engines the demand for engine block castings is a dependent demand; once the production schedule for a group of engines is established, the planner knows with certainty that one block will be required for each engine. Conversely, the demand for cutting oil used by the machines on the line cannot be calculated accurately from the production schedule and bills of materials; thus, cutting oil is said to have an *independent demand*. Generally speaking, in an assembly or fabrication-type operation, most production inventory items will have a dependent demand, while MRO and similarly used items will have an independent demand.

Although the distinction seems relatively simple, it is important for the inventory manager to know whether an item exhibits a dependent or an independent demand. Certain inventory control systems function more effectively with one type of item than with another.

Costs Associated with Inventories

From a managerial point of view, two basic categories of costs are associated with inventories: (1) inventory carrying costs and (2) inventory acquisition costs. These plus a related variable cost are discussed in the following paragraphs.

Carrying Costs

Carrying material in inventory is expensive. Before the relatively recent periods of higher interest rates, a number of studies determined that the *annual cost* of carrying a production inventory averaged approximately 25% of the value of the inventory. The escalating and volatile cost of money in recent years, however, has increased the typical firm's annual inventory carrying cost to a figure between 25 and 35% of the value of the inventory. Five major elements make up these costs in the following manner:

1. Opportunity cost of invested funds	12–20%
2. Insurance costs	2–4%
3. Property taxes	1–3%
4. Storage costs	1–3%
5. Obsolescence and deterioration	4–10%
Total carrying costs	20–40%

Let us briefly examine these carrying costs.

1. *Opportunity cost of invested funds*   When a firm purchases $50,000 worth of a production material and keeps it in inventory, it simply has this much less cash to spend for other purposes. Money invested in productive equipment or in external securities earns a return for the company. Conceptually, then, it is logical for the firm to charge all money invested in inventory an amount equal to that it could earn if invested elsewhere in the company. This is the "opportunity cost" associated with inventory investment.

2. *Insurance costs* Most firms insure their assets against possible loss from fire and other forms of damage. An extra $50,000 worth of inventory represents an additional asset on which insurance premiums must be paid.

3. *Property taxes* As with insurance, property taxes are levied on the assessed value of a firm's assets; the greater the inventory value, the greater the asset value, and consequently the higher the firm's tax bill.

4. *Storage costs* The warehouse in which a firm stores its inventory is depreciated a certain number of dollars per year over the length of its life. One may say, then, that the cost of warehouse space is a certain number of dollars per cubic foot per year. Conceptually this cost can be charged against inventory occupying the space.

5. *Obsolescence and deterioration* In most inventory operations, a certain percentage of the stock spoils, is damaged, is pilfered, or eventually becomes obsolete, a situation often called inventory shrinkage. No matter how diligently warehouse managers guard against these occurrences, a certain number always take place. With new products being introduced at an increasing rate, the probability of obsolescence is increased accordingly. Consequently, the larger the inventory, typically the greater the absolute loss from this source.

Generally, this group of carrying costs rises and falls nearly proportionately with the rise and fall of the inventory level. Further, the inventory level is directly related to the quantity in which the ordered material is delivered. When the complete order is shipped at one time, the larger the order quantity, the higher the *average* inventory level during the period covered by the order. Hence, the costs of carrying inventory vary nearly directly with the size of the delivery quantity. This relationship is illustrated by the CC curve in Figure 10-6.

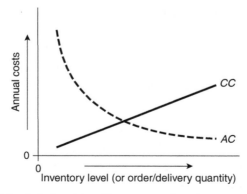

Figure 10.6 Relationship of inventory-related costs to inventory-level (*AC* = acquisitions costs; *CC* = carrying costs).

If a firm has estimated its approximate inventory carrying cost as a percentage of inventory value, the *annual* inventory carrying costs that would be generated by delivery quantities of various sizes can be calculated as follows:

$$(\text{Carrying cost per year}) = (\text{average inventory value}) \times \begin{pmatrix} \text{inventory carrying cost as} \\ \text{a \% of inventory value} \end{pmatrix}$$

$$\begin{pmatrix} \text{Carrying cost} \\ \text{per year} \end{pmatrix} = \begin{pmatrix} \text{average inventory} \\ \text{in units} \end{pmatrix} \times \begin{pmatrix} \text{material unit} \\ \text{cost} \end{pmatrix} \times \begin{pmatrix} \text{inventory carrying} \\ \text{cost as a \% of} \\ \text{inventory value} \end{pmatrix}$$

$$CC = \frac{Q}{2} \times C \times I$$

where CC = carrying cost per year for the material in question

Q = order or delivery quantity for the material, in units[16]

C = delivered unit cost of the material

I = inventory carrying cost for the material, expressed as a percentage of inventory value

Acquisition Costs

When one looks at inventory costs in another light, a different set of indirect materials cost factors emerges. These factors all contribute to the cost of generating, processing, and handling an order, along with its related paperwork. Examples of these costs are listed next and can be thought of as inventory acquisition costs.

1. A *certain portion of wages and operating expenses* of such departments as purchasing and supply, production control, receiving, inspection, stores, and accounts payable—those departments whose personnel devote time to the generation and handling of the order
2. *The cost of supplies* such as engineering drawings, envelopes, stationery, and forms for purchasing, production control, receiving, accounting, and so forth
3. *The cost of services* such as computer time, telephone, fax machines, and postage expended in procuring material

When considering this group of acquisition costs, observe that they behave quite differently from carrying costs. Acquisition costs are not related to inventory size per se; rather, they are a function of the number of orders placed or deliveries received during a particular period of time.

One simplified example will illustrate this point. Suppose a buyer in the purchasing and supply department receives a requisition for a special fabricated part used in the manufacture of one of the firm's products. Assume further that the part has been purchased before and that price quotations from three or four shops are on file. The buyer first reviews the current inventory situation and probably checks with production control to see if any significant changes are anticipated in future production. Drawings and specifications of the part are then reviewed to refresh his or her memory regarding required tooling and other technical details of the purchase. Next, the buyer reviews the quotations to determine why the order was placed with supplier A last time. Before deciding if supplier A should again receive the order, the buyer must also review supplier performance data. Finally, the buyer decides which supplier should receive the order and subsequently inquires about the firm's current shop loads and any other matters that have arisen during the investigation. It is entirely possible that a negotiation session may also be required.

In total, the buyer's investigation may require anywhere from one hour to several days. The total cost of the buyer's time to the company will be the same whether the purchase order is written for 20 parts or 200 parts. This process may result in the development of a term contract with the supplier, in which case the buyer's effort is spread over all deliveries of the item during the life of the contract. If this is not the case, however, the next time the buyer receives another requisition for this part, he or she will go through somewhat the same process, generating almost the same indirect cost for the company.

The largest segment of the acquisition cost element is made up of these types of indirect labor and overhead costs, generated in purchasing and in the other departments that subsequently become involved in handling some activity associated with the purchase. The cost of supplies and services consumed in the placement and handling of an order typically varies directly with the number of orders placed. While these costs are significant, they are considerably less so than the human and related overhead cost figures just discussed. The variable acquisition cost per order varies widely among firms, depending on the specific cost inclusions and type of material. Today the range appears to run from approximately $50 to $125 per order.

If a firm experiences a certain annual usage of an item, the number of orders placed during the year will decline as the individual order quantity increases, thus generating lower *annual* acquisition costs. The experience of numerous firms over the years reveals that this relationship is not linear, but follows the approximate contour of the *AC* curve shown in Figure 10.6.

If a firm's cost accounting department can estimate its approximate acquisition cost *per order,* the *annual* acquisition costs that will be generated by order quantities of various sizes can be calculated as follows:

$$(\text{Acquisition cost per year}) = \left(\begin{array}{c} \text{number of orders} \\ \text{placed per year} \end{array}\right) \times \left(\begin{array}{c} \text{acquisition cost} \\ \text{per order} \end{array}\right)$$

$$AC = \frac{U}{Q} \times A$$

where AC = acquisition cost per year for the material in question

$\quad U$ = expected annual usage of the material, in units

$\quad Q$ = order or delivery quantity for the material, in units[17]

$\quad A$ = acquisition cost per order or per delivery for the material

Economic Order Quantity (EOQ) Concept

If one has to make decisions about managing an inventory, it is useful to understand the behavior of the inventory-related cost factors discussed previously. These factors often help a manager determine which items should or should not be carried in inventory, what inventory levels should be carried for specific items, and what order quantities are appropriate for particular items.

The latter part of this chapter discusses several types of systems that can be used in managing an inventory. In each case, one of the short-term operating questions that must be answered is: How much of the item should be ordered? Among the factors that often enter this decision process is a concept known as EOQ—the notion of an economic order quantity. As its name suggests, this concept holds that the appropriate quantity to order may be the one that tends to minimize all the costs associated with the order: carrying costs, acquisition costs, and the cost of the material itself.

Concentrating for the moment on the first two costs, Figure 10.6 shows clearly that as the order or delivery quantity increases, carrying costs rise—and at the same time acquisition costs decrease. To see the total picture more clearly, if carrying costs and acquisition costs are added together over the order quantity range shown on the graph, the total indirect materials cost curve, *TC,* is produced. This transformation is shown in Figure 10.7. The economic order quantity concept simply says that the sum of all the indirect costs associated with inventory will be minimized on an annual basis if the material for which the graph is drawn is ordered (or delivered) consistently in the quantity that corresponds with the low point on the *TC* curve. This is the *economic order quantity.*

Note that the low point on the total cost curve coincides with the point at which the carrying cost curve intersects the acquisition cost curve. This makes it easy to develop the

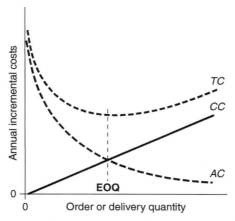

Figure 10.7 Graphic representation of the EOQ concept (*AC* = incremental acquisition costs; *CC* = incremental carrying costs; *TC* = total incremental costs.)

basic formula that can always be used to calculate a material's *basic* EOQ. Recall the two simple cost formulas developed for annual carrying costs and annual acquisition costs. Those forumulas can be used to develop the EOQ formula.[18]

The EOQ occurs when

Annual carrying cost = annual acquisition cost[19]

$$CC = AC$$

$$\frac{QCI}{2} = \frac{UA}{Q}$$

Solving for Q:

$$Q^2CI = 2UA$$

$$Q = \sqrt{\frac{2UA}{CI}}$$

This formula is the fundamental mathematical representation of the EOQ concept. It can be modified to accommodate numerous special conditions, but in practice it probably finds its most effective application in this form.

Professor Daniel Jones, who has researched various lot sizing concepts, says that the EOQ concept can be used in conjunction with a variety of inventory management systems, including JIT. He writes: "When the EOQ model is properly employed, there is little difference between lot sizes based on the JIT model and the EOQ model." He points out that all relevant incremental costs must be included when using the EOQ model. This is perhaps an obvious observation, but it is one that he finds is violated frequently in practice.[20]

Thus, despite some criticisms, the EOQ concept continues to be a versatile and useful tool if it is applied properly.

Incremental Costs and Stability

Note that the vertical axis in Figure 10.7 is labeled "annual *incremental* costs." It is appropriate at this point to emphasize the fact that the costs that are relevant in making an EOQ analysis are incremental costs. *Incremental costs are those costs that actually change as a result of a particular operating decision.* For example, if the decision to issue more purchase orders during the year actually increases supply and service costs, these are incremental costs. If it requires the addition of a buyer or clerical person to handle the load, the additional payroll costs are incremental costs. *Incremental costs are either variable costs or opportunity costs* that represent a forgone opportunity to utilize an asset in some other productive way.

By their nature, most of the inventory carrying costs discussed here are incremental costs and are reasonably stable. The distinction is less clear when dealing with inventory acquisition costs. Judgment usually is required in estimating the portion of the human effort that represents a legitimate opportunity cost. In any case, EOQ produces valid results only when the I and A cost factors are largely incremental, and when the usage and unit cost elements, as well as I and A, are reasonably stable over the operating period.

Material Prices and Quantity Discounts

In the discussion of EOQ analysis to this point, it has been assumed that material prices and transportation costs are constant factors for the range of order quantities considered. In practice, some situations occur in which the delivered unit cost of a material decreases significantly if a slightly larger quantity than the originally computed EOQ is purchased. Quantity discounts, freight rate schedules, and perhaps anticipated price increases may create such situations. These additional variables can also be included in the basic formula, but from a practical point of view they usually can be handled more easily with a separate simple calculation.

If one uses Q computed with the basic formula, such alternative quantity decisions can be made quickly and accurately. If one simply compares annual material cost savings resulting from the purchase of the additional quantity with the additional inventory carrying costs occasioned by the increased purchase, the most economical decision quickly becomes evident. With a limited amount of practice, a buyer can determine in a matter of seconds whether material cost savings exceed carrying costs for the additional inventory.

Other Uses of the EOQ Concept

A final word should be said about *general* usage of the EOQ concept. Even though this discussion has been set entirely in the purchasing environment, the EOQ concept logically has broader application as well.

A very common situation in which the concept is used is the determination of economic production lot sizes in a manufacturing operation. Consider the formula a moment, and look at the individual factors in light of both the purchasing operations and the production operations. When the formula is converted for production use, the annual usage and carrying cost factors are the same as they were in the purchasing application. The unit cost factor, however, is no longer delivered price; instead, it consists of direct labor and materials and production overhead costs. Production acquisition cost is similar to purchasing acquisition cost except that production setup cost replaces most of the purchasing and related departmental wage and operating costs.

Buyers also should consider (from a supplier's point of view) the formula in this form when determining lot sizes on term contract purchase orders going to various types of job shop suppliers. A supplier's costs and subsequent product price obviously are influenced by the size and frequency of such orders.

The Weaknesses of the EOQ Formula

The EOQ Formula has been explained in detail more for its philosophical merit than its actual use. In other words, the logic, themes, and concepts are all proper elements to consider. In practice, many experts have questioned the use and accuracy of the formula for the following reasons:[21]

▲ It is not fully compatible with MRP and JIT (despite the case presented by Professor Jones).

▲ All demand inputs are yearly *estimates* so the output is just an estimate.

▲ Only incremental costs are appropriate, that is, the cost of the *next* setup, the cost of the *next* purchase order preparation, the cost of the *next* stored unit (warehouse cost).

▲ Most users of this technique erroneously use "average costs," which destroys the logic. Until you need to build a new warehouse, there is no increase in "sunk" warehouse cost as inventory increases.

▲ For inventory that moves from location to location, one set of costs does not apply.

▲ The carrying costs usually are badly overstated, based on broad "industry averages" and often on pure estimates just short of guesses.

▲ Purchase ordering costs for long-term contracts are one-shot expenses. Release costs, especially by computer, would be or are fractions of a cent or, for a fax, perhaps $1. Repeat purchase orders for the same material would be nominal.

▲ Few firms have accurate studies of carrying cost to hold inventory beyond the cost of capital of the material. This cost could never be much more then the going commercial loan rates and the average rate of return of the firm. Thus, figures of 25 to 35% are suspect.

▲ The opportunity cost aspect of EOQ rests on gigantic assumptions such as storage costs. Most warehouses are on the books at sunk costs with zero alternative use in the short term.

▲ EOQ assumes constant demand for the next year, which hardly ever is the case.

We agree with James Gardner, a leading author and consultant in materials management, when he says

In the same vein, suppose there is excess storage capacity from time to time. The accounting department will tell you to include all available space into carrying costs because if you are not using the space, it could be rolled out. They are dealing in pure fantasy. To whom are you going to rent it? How would you go about finding someone who would inconvenience himself by renting out a few feet of your floor space on a temporary basis at your stated rate when he could get all the space he needs at a public warehouse and probably save money in the bargain.[22]

Types of Inventory Control Systems

To this point in the discussion, we have considered background concepts that are useful in formulating fundamental aspects of the plans to manage inventories. The discussion now turns to the specific operating systems that can be used. Generally speaking, four types of inventory control systems are in use: (1) the cyclical or fixed order interval system, (2) the JIT approach, (3) the MRP-type system, and (4) the order point or fixed order quantity system. Each system monitors and controls inventory levels. Each system, on the basis of its unique characteristics, provides the inventory manager with information that helps answer the two basic questions of *when* to order and *how much* to order.

Cyclical or Fixed Order Interval System

The cyclical system, or fixed order interval system as it is sometimes called, is the oldest and simplest of the systems now found in use. Years ago, when most businesses were small and uncomplicated, this control system was used in all types of operations: manufacturing, service, wholesale, and retail. With the exception of one variation called flow control, the cyclical system is not widely used today except in smaller and medium-size operations.

Operationally, the system works like this. It is *a time-based* operation which involves *scheduled periodic reviews* of the stock level of all inventory items. Looking at it in a manufacturing setting, when the stock level of a specific item is not sufficient to sustain the production operation until the next scheduled review, an order is placed to replenish the supply. The frequency of reviews is based on judgment and varies with the degree of control desired by management; A items might be reviewed weekly (or more often), B items monthly or bimonthly, and C items quarterly or semiannually.

Stock levels can be monitored by physical inspection, visual review of perpetual inventory record cards, or by automatic computer surveillance. In most operations, a perpetual inventory record is maintained,[23] either by computer or manually, except in simple flow-controlled shops (which will be examined shortly). Physical stock counts are required once or twice a year to reconcile actual values with book values in all systems utilizing perpetual inventory records.

The first operating question—*when to order*—is answered or controlled by the review dates established by the inventory manager. If material usage has remained reasonably stable, an order (or a release against an order) is usually placed each time the item is reviewed.

The order date decision also is affected by the quantity previously ordered, so let us consider the second question: How much should be ordered? The quantity to be ordered generally is determined by three factors: (1) the number of days between reviews, (2) the anticipated daily usage during the cycle period, and (3) the quantity actually on hand and on order at the time of the review. One of the primary reasons this system is used is to control high-value items closely and to maintain a relatively low investment in inventory. Hence, the order quantity typically is the quantity required to cover only the ensuing period, with allowance for order lead time. Occasionally a two- or three-period supply is ordered, but not as a rule.

Consequently, as its name implies, the system works in cyclical fashion, with an order typically placed at each review date for a quantity large enough to cover the ensuing cycle plus the order lead time. A small safety stock generally is carried, based on the observed lead time variability. Inventory levels and tightness of control are thus determined by the establishment of the period of the cycle. High-value A and B items typically are placed on short cycles and C items are on longer cycles.

This system can be used with both dependent demand and independent demand materials. It works most effectively in an organization that has a continuous operations function, manufacturing or service, in which demand is fairly stable and can be predicted with reasonable accuracy. Additionally, it is probably the most efficient system to use for independent demand items that experience irregular or seasonal demand and for any items whose purchases must be planned months in advance because of infrequent supplier production schedules. In these cases, it tends to keep inventory levels lower than would be possible with the other applicable systems. When used for materials with these characteristics, however, the system must be augmented with a minimum balance figure that signals the need for an early reorder in the case of a sharp usage increase.

To conclude, the cyclical system finds its greatest usage in organizations that have large numbers of independent demand items to control, or in relatively simple process operations where dependent item demand can be projected easily from the production schedule. When used for dependent demand items in an intermittent manufacturing operation, it becomes difficult to determine cycle period demands if many products are involved or if individual items are used in several different products. The bill of materials explosion and time-phasing capabilities of an MRP system must all be handled manually in the cyclical system. As product complexity increases, this becomes a virtual impossibility, so historical demand data tend to become the basis for order quantity determination—and this soon leads to unreasonably high inventory levels because of the uncertainties associated with near-term demand. For these reasons, MRP *systems have replaced most cyclical systems in intermittent manufacturing operations.*

Flow Control System

The flow control method of managing inventories is a special variation of the cyclical system. This method is applicable in continuous manufacturing operations that produce the same basic product in large quantities day after day. Most materials used in such an operation are purchased on term contracts and scheduled for daily or weekly delivery throughout the term. The production cycle is often a day or less in duration, and in effect material flows through the plant in continuous streams. Inventory floats consequently can be kept quite low, thus requiring a minimum investment in production inventory.

In this type of operation, an open stores system is used for most production materials, and the individual items are stored on the line near the point of use. Stores personnel visually review the level of all material stocks daily and report any imbalances to the purchasing or production control department. Changes in production schedules must be relayed immediately to buyers so that delivery schedules can be revised accordingly.

The Just-in-Time (JIT) Approach

The just-in-time concept was explored in detail earlier in this chapter. It was pointed out that in total, JIT is an operating management philosophy. With reference to that philosophy, a number of specific operating techniques have been developed for manufacturing operations, for production planning, and inventory management. Those dealing with inventory management are the products of the JIT decisions made in the manufacturing and planning areas.

The operating concept of the system is to gear factory output tightly to distribution demand for finished goods, gear individual feeder production units tightly together, and gear the supply of production inventories tightly to the manufacturing demand schedule. This means that all inventories in the system, including production inventories, are maintained at absolutely minimal levels.

It should be emphasized at the outset, however, that as a practical matter, most firms utilize the JIT concept for no more than 5 to 10% of the materials handled by the purchasing and supply activity, regardless of the extent of the commitment in the manufacturing operation. This means that the production inventory items handled in the JIT inventory system are primarily high-value A items. All these items are purchased on a long-term contractual basis, with small-volume deliveries scheduled as frequently as once or twice a day or as infrequently as once or twice a week.

If one observed a JIT operation strictly from an inventory point of view, it would look very much like a flow control operation, with material flowing into and through the plant operation in continuous streams. In fact, it functions much as a flow control operation, only more tightly and stringently controlled. From strictly an inventory point of view, the systems have almost identical objectives. Many JIT materials are delivered directly to the production operation and are stored close to the point of use; others are handled in a conventional closed stores operation.

From a practical point of view, a JIT inventory system in its purest sense is workable only in continuous manufacturing and processing operations or in intermittent operations that produce a small number of standard products and, because of that fact, are similar to continuous operations. Most, if not all, of the materials handled are dependent demand items.

How are the *when* and *how much* questions answered? As was discussed in the preceding section, the buyer and the supplier work together closely on the matters of delivery volumes and scheduling. The buying firm's production schedule drives the entire process. The detailed production schedule typically is firmed up for one or two weeks at a time, and in more general terms for a month or so ahead. Specific daily requirements for JIT materials can be determined from this schedule and are relayed directly to the contracting supplier. The exact size and frequency of each delivery are worked out jointly in an attempt to minimize the buyer's incremental inventory-related costs and, at the same time, maintain an efficient and practical operation for the supplier. As a general rule, the buyer does not identify a specific safety stock component in the firm's inventory figure. Depending on the material, the buying firm typically works on an inventory of several days' to a week's supply. As is the case in a flow control system, stores personnel on the shop floor visually monitor stock levels at least daily and communicate potential overage or shortage problems to the appropriate buyer.

To this point, our discussion has covered only the 5 to 10% of a firm's production materials that typically are handled by its JIT purchasing and inventory management system. What about the remaining 90 to 95% of items? As a rule, they are handled in the more conventional manner with one of the other standard systems: an MRP or an order point system.

Material Requirements Planning (MRP) System

The section on production planning in this chapter detailed how closed-loop MRP systems function as complete production planning and control systems. The MRP module is an integral part of such a system. Through its bill of materials explosion and aggregation process, this element of the system generates on a weekly basis the projected materials requirements for all the finished products included in a firm's updated master production schedule for the coming two- to three-month period.

After taking the projected gross requirements for a specific material during the planning period, the MRP module calculates the net requirements by subtracting on-hand inventory and any scheduled receipts of the item as production is scheduled to progress through the planning period. This produces a "time-phased" purchase order requirement to be released at a calculated future date. The reader may wish to review this logic by referring to Figure 10.3.

The inventory that is carried in the system is a function of three factors: (1) the quantity purchased when each order is placed, (2) the purchase lead time specified by the buyer, and (3) any safety stock that is carried routinely. The objective of time-phasing the order point is to keep the inventory as close to zero as is practical until the material actually is needed for production. Consequently, using an MRP system, the average inventory levels of most materials are relatively low over the long term.

In the case of some materials, no safety stock is carried. In other cases, a one- to two-week supply may be carried as a hedge against uncertainties such as possible fluctuations in demand, variations in supplier lead-time requirements, or anticipated scrap or reject rates. Variations in supplier lead-time requirements also may be covered by simply extending the lead-time figure used in calculating the order release date; in this case, safety stock would be reduced correspondingly. These safety stock and lead-time hedge values typically are determined judgmentally on the basis of past experience with specific materials and suppliers.

The *when to order* question then, is answered by the logic of the system. Deciding *how much to order* is in part a judgment call. The most common approach, as was the case in the cyclical system, is to order the quantity required during the planning period—the "lot for lot" approach. This method typically tends to minimize the inventory in the system. At times, however, the lot for lot approach may produce an order quantity that is too small to be economical. Because of high acquisition costs or production setup costs, order size may have to be larger. In this case, the EOQ or a related least-cost calculation is frequently used to obtain a more appropriate order quantity figure. A number of other decision rules are sometimes used, but those just mentioned appear to be the most common.

The MRP system is designed for use with dependent demand items, that is, production materials. The only way it can handle an independent demand item is by tying such an item's use into a product bill of materials. For production tools and certain other MRO supplies, it is sometimes possible to do this in an approximate way by estimation. However, the system's most important use by far is with dependent demand materials in an intermittent manufacturing operation. An MRP system can be adapted for use in a continuous or processing-type operation, but it does not fit such operations well and usually it offers few significant advantages over the other types of systems.

Order Point or Fixed Order Quantity System

The order point system, historically known as the fixed order quantity system, is another inventory control system that has been used for years in this country by both manufacturing and nonmanufacturing organizations. The system recognizes the fact that each item has its own unique optimum order quantity, and it is therefore based on *order point* and *order quantity* factors rather than on the time factor.

Operation of an order point system requires two things for each inventory item:

1. The *predetermination of an order point,* so that when the stock level on hand drops to the order point, the item is automatically "flagged" for reorder purposes. The order point is computed so that estimated usage of the item during the order lead-time period will cause the actual stock level to fall to a planned minimum level by the time the new order is received. Receipt of the new order then increases the stock level to a preplanned maximum figure.
2. The *predetermination of a fixed quantity to be ordered* each time the supply of the item is replenished. This determination typically is based on considerations of price, rate of usage, and other pertinent production and administrative factors.

The automatic feature of the system is achieved by maintaining a perpetual inventory record for each item. The computer, or an inventory clerk in the case of a manual system, continues to post all material issues until the balance of an item falls to its order point. At this point the system notifies the appropriate buyer, who replenishes the stock in a quantity that takes the inventory to its planned maximum level. During the course of operation, the ongoing inventory level is thus maintained between the planned minimum and maximum values.

The predetermined order point, then, tells the buyer *when* to order. In most organizations the order point is determined in the following manner: First, basic operating data about demand and lead time must be obtained. Next, a decision must be made about the desired service level. For most materials, most firms aim for 100%; that is, they do not want to run out of stock before the new order arrives. At this point in the discussion, the process can be described most easily with the use of a simple illustration. Suppose the following data have been determined for a given inventory item:

Purchasing lead time = 1 week (very stable; little chance of variation)

Material usage = 50 units per week, with ±10% variation over the long run

Therefore,

Maximum usage during lead time = 55 units

Average usage during lead time = 50 units

Minimum usage during lead time = 45 units

Figure 10.8 shows a simplified, or idealistic, inventory movement pattern for the material in question, with the usage rate constant at the maximum level of 55 units per week. If the buyer does not want to run out of stock, at what inventory level should the new order be placed? If lead time is known to be one week, and the *maximum* usage has been determined to be 55 units per week, the new order clearly should be placed when the stock level falls to 55 units. Under these conditions, the new order will arrive just when the stock level reaches zero. Thus, the order point is 55 units.

What happens when the usage rate runs around 50 units per week, as it does much of the time? The inventory movement pattern shown in Figure 10.8 then becomes the

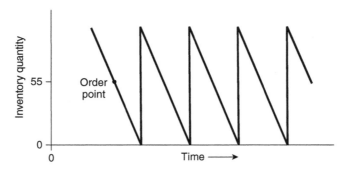

Figure 10.8 Illustrative simplified inventory movement pattern for a given material, with a maximum usage rate.

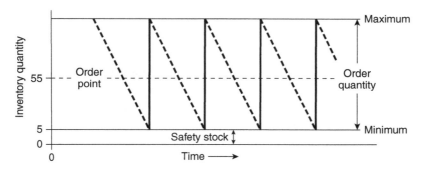

Figure 10.9 Movement pattern in Figure 10.8 with an average usage rate, showing safety stock determination.

dashed sawtooth pattern shown in Figure 10.9. With an order point of 55 units, as long as the *average* usage rate of 50 units per week prevails, the new order will arrive when 5 units (55 − 50 = 5) are still left in stock. This brings us to the definition of safety stock. In an order point system, set up as just described, *safety stock* is normally defined as *the maximum lead-time usage minus the average lead-time usage.* In this case, then, the order point is 55 units and the basic safety stock is 5 units.

In operation, over a period of time, this means that the low point of the inventory sawtooth pattern will occasionally fall to 0 as the new order arrives (55 − 55 = 0); and when usage is at its lightest, the low point of the sawtooth will be as high as 10 units when the new order arrives (55 − 45 = 10). Most of the time, the low point of the sawtooth will fluctuate between these two extremes, with occurrences concentrated around the safety stock value of 5, which is also defined as the *theoretical* planned minimum. Figure 10.10 depicts this situation hypothetically.

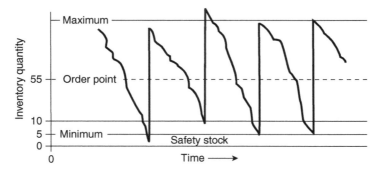

Figure 10.10 Typical inventory movement pattern for a reasonably stable material, with a fixed order quantity.

Now that the question of *when* to order has been answered, how does the buyer determine *how much* to order? Any number of decision rules can be used, but the most common approach is to determine the EOQ value. A fixed order quantity system is a natural for EOQ application. The EOQ value can be calculated automatically almost instantaneously with a computer-based order point program. In a manual system, tables of EOQ values can be precalculated and prepared for the near-instantaneous use of inventory clerks. Nomographs and preprogrammed hand-held calculators can also be used easily and quickly. More general decision rules sometimes used by experienced managers are to order a one- to four-week supply of A items, a one- to two-month supply of B items, and a four- to six-month supply of C items.

The major advantages of an order point system are (1) unlike the cyclical and the JIT approaches, the EOQ concept (if the data inputs are accurate) can be applied easily in this system, so each material can be procured in the most economical quantity; (2) this preplanned approach utilizes the time and efforts of people efficiently—purchasing and inventory control personnel automatically devote attention to a given item *only when the item requires attention;* and (3) within limits, control can be exerted easily to maintain inventory investment at a target level simply by varying the planned maximum and minimum values.

On the other side of the coin, the system also has some serious limitations.

▲ The most serious problem in using an order point system stems from the fact that it works on the basis of historical rather than actual demand data. Therefore, the order point cannot be time-phased to correspond with actual demand requirements. This means that over the long term, an order point system almost always produces a higher average inventory level than would a comparable MRP system.

▲ The system functions correctly only if each of the materials exhibits reasonably stable usage and lead-time characteristics. When these factors change significantly, a new order point and a new order quantity must be determined if the system is to fulfill its objectives. Consequently, although it can be adapted, the system becomes costly and cumbersome to operate effectively when applied to materials with highly unstable demand and lead-time patterns.

Bearing in mind the advantages and disadvantages discussed previously, it is appropriate to note that an order point system can be used equally well with both independent and dependent demand items. In a *dependent demand* manufacturing environment, a time-phased MRP system is far superior to an order point system because of its use of actual demand data and its ability to maintain average inventories at significantly lower levels. However, if it is not practical to use MRP for low-value C items, an order point system can be used very effectively, and the increased inventory levels will affect total costs very little. In managing *independent demand* inventories, such as retail, a well-designed order point system has no peer; it works extremely well.

As pointed out earlier, an order point system is versatile: It can be used in any type of operation from manufacturing to service. For school districts, hospitals, banks, retail, and numerous other service-oriented institutions, virtually all inventory items exhibit independent demand characteristics. As one would suspect, order point systems are used in a vast majority of these types of operations.

Two-Bin System

A variation of the basic order point system is found in the operation of the simple two-bin system. The distinguishing feature of this system is the absence of a perpetual inventory record. In practice, the stock is physically separated into two bins, or containers. The lower bin contains a quantity of stock equal to the order point figure. This, typically, is just enough stock (or slightly more) to last from the date a new order is placed until the incoming material is received in inventory. The upper bin contains a quantity of stock equal to the difference between the maximum and the order point figures. At the outset, stock is used from the upper bin; when this supply is depleted, it signals the clerk that the order point has been reached. At this point an order is placed, and material is used from the lower bin until the new stock is received. Upon receipt of the new order, the proper quantities of material are again placed in the two bins. This method demonstrates very simply the fundamental concept that underlies the basic order point system.

The two-bin method is widely used in all types of operations in handling low-value hardware and supplies whose usage is not recorded on a perpetual record. The major advantage of the method is the obvious reduction of clerical work. Issues do not have to be posted to determine the proper reorder time. Receipts, however, are usually posted to reveal significant changes in usage or lead time. A possible disadvantage of the system in some cases is the requirement of additional storage facilities and perhaps some practical difficulty in keeping the two stocks properly separated.

To generalize, it should be pointed out that none of the systems or their adaptations discussed in this chapter are mutually exclusive. Several or all of them may be used advantageously for different materials in a single firm.

Supplier Managed Inventories

Many firms have outsourced some of their inventory management. Large industrial distribution companies like Granger are contracted to replenish stock of MRO supplies under long-term contracts. Granger issues monthly inventory usage reports and, using reorder points established by the buying company, they restock the material as necessary;

this results in significant savings. Other applications of this concept include tools, office supplies, and other support materials.

Summary

For the last five years or so, production planning and inventory management activities have been in a state of change. New and evolving systems have spawned new and increasingly sophisticated planning and control capabilities. In some cases, the more traditional inventory control systems have been replaced for certain types of applications, in other cases they have not. Table 10.1 draws together in summary form the most significant applications features of systems in use today.

It is becoming increasingly difficult for most firms to control inventories effectively. One major reason for this is the expansion of product lines and models. A second key reason is that more and more components going into the typical firm's products are being *purchased* as fabricated parts rather than being produced from basic materials in the firm's own shops. In many firms, this means that the number of inventory items to be managed is constantly growing. Because of the increasingly technical nature of materials today, the number of dollars that may be invested in inventory is growing at an even faster rate than the number of items.

In its daily operation, inventory control should be largely a series of clerical or computerized activities, carried on within a carefully defined and controlled framework. The routinization of the daily operation, however, often camouflages the importance of sound management in this area.

At this point in the discussion, there should be little doubt that the basic responsibility for inventory control should lie with top management. The effects of poor inventory management, unfortunately, are not directly visible on the operating statement as a composite cost of inventory management. Nevertheless, in most organizations these indirect costs, dispersed and hidden throughout the operating statement, can have a significant impact on profit. For this reason, top management should carefully formulate and periodically review the basic policies and operating plans that constitute the framework within which the daily inventory control operation functions.

Appendix: The Fundamentals of Inbound Transportation[24]

Buyers and other supply personnel are heavily involved in determining how the materials are delivered to the internal customers. This appendix gives detailed checklists to help achieve the lowest possible freight costs and how to work with the transportation department. It also defines the multitude shipping terms. Attention to details is very

Table 10.1 Comparison of Characteristics and Applications of Inventory Control Systems

Type of System/Characteristics	Order Point	Cyclical	MRP	Flow Control	JIT
Maintains low inventory levels	Fair	Fair/good	Good/excellent	Excellent	Excellent
Application to items, type of demand	All types ◂ Particularly good for independent demand	All types	Primarily dependent demand	Primarily dependent demand	Dependent demand
Application to type of operations	◂ All types of manufacturing operations ◂ Particularly good for service operations	◂ All manufacturing operations ◂ Service operations	Primarily intermittent manufacturing operations, with great product variety	Continuous manufacturing operations, with little product variety	Continuous manufacturing, with moderate product variety
Demand data used	Historical	◂ Actual in simple operations ◂ Historical in complex operations	Actual	Actual	Actual
Time-phased order point	No	◂ Yes in simple operations ◂ No in complex operations	Yes	Yes	Yes
Computer required?	Optional	Optional	Yes	Optional	Optional
Bill of materials explosion/ aggregation capability	No	No	Yes	Yes	Yes
Administrative effort required	Minimal	Moderate	Moderate	Heavy	Heavy

important. For example, incorrect classifications of freight can double or more the transportation cost.

Transportation as a Strategic Objective

Many purchasing and supply managers are responsible for buying inbound transportation services for their organizations. Yet, many practitioners may not feel fully "at home" in this complex field, especially with recent deregulations and reregulations.

Mastering this area can add value to the organization and to the purchasing and supply management functions because transportation, which can be very costly, can have a significant impact on the bottomline. For one thing, if the supplier makes transportation decisions, there is no guarantee that the choices made will be the most cost-effective for your organization. Profits can be improved by:

▲ Improving carrier services
▲ Reducing inventory
▲ Advancing the use of JIT
▲ Reducing delivery costs
▲ Cutting lead times

It is purchasing and supply management's responsibility to team with other functions in the organization to plan and negotiate the most cost-effective transportation method and align it with the organization's service requirements. This goal can be a challenge. It can be met with a thorough analysis and assessment of material requirements and by knowing how to negotiate and partner with, when appropriate, selected carriers.

Benefits of a Transportation Program

When purchasing/supply management takes control of, or contributes to, transportation decision making, it should develop into a program that is managed as carefully as any other within the function. The benefits of such a program can add value to the organization through:

▲ Lower delivered costs coupled with improved delivery service
▲ Identifying key areas of opportunity for improving the delivery of goods and services at the lowest total costs
▲ Control of the business terms of sale to best determine the routing, cash flow, and risk management associated with delivering goods to end users
▲ Recognizing strategic opportunities to interface with other functional units within and between the organizations of the purchasers, suppliers, and transportation carriers to achieve inbound supply chain management initiatives

▲ Identifying options to manage and control the inbound routing of freight, including the outsourcing of third-party logistics services

Understand the Impact of Transportation on Inventory

The time it takes to manufacture and deliver a product is frequently the basis of supplier performance measurement. Freight carriers who most often supply consistent, timely service can help reduce manufacturing/delivery time, which in turn helps an organization to reduce its inventory and plan production using tighter schedules. Carriers who provide effective service can also contribute to JIT efforts, provided they can work reliably within the precise schedules required.

Checklist: How to Evaluate a Carrier

Selecting a carrier is actually a standard sourcing issue, which includes a thorough investigation of specific factors. The following points should enter into the evaluation of a carrier to help determine reliability, flexibility, consistency, and financial stability and strength:

▲ Assets and debts
▲ Revenues and expense categories
▲ Operating ratio—operating expenses divided by revenues
▲ Method of financing capital improvement and new equipment
▲ Billing process/methods
▲ Billing accuracy
▲ Pricing and contract terms
▲ On time delivery including JIT capability
▲ Use of electronic data interchange (EDI) and other information technologies
▲ Damage and claims record (including time to correct)
▲ Length of time in business
▲ Number of employees in the organization
▲ Equipment—type, variety, condition, age
▲ Availability of additional/backup equipment
▲ Tracing and tracking abilities
▲ Terminals and warehouse services
▲ Safety programs
▲ Training and management of drivers
▲ Insurance for bodily injury, property damage, cargo, and catastrophic risks
▲ General reputation with other shippers
▲ Intermodal capability

▲ Hazardous materials programs
▲ Accident/safety record
▲ Citations record
▲ Reputation with other shippers
▲ Services—such as in-transit options and auditing
▲ Routes and geographical area served
▲ Outsourcing logistics services capability
▲ Freight forwarding services
▲ State and federal registrations

Know How to Select the Best Carriers

When purchasing/supply management manages or works with the transportation function, the selection of a carrier is in its hands, and is specified on the purchase order (called the routing). The routing is also provided via a routing letter or in a routing guide to the supplier, and to all parties who have a need to know this information, such as the receiving or traffic departments.

Carrier selection involves determining first, the appropriate method, and second, the specific carrier that can provide the best service to meet delivery requirements. These factors should be considered:

▲ Door-to-door or door-to-point of use (such as the shop floor) costs
▲ Door-to-point of use transit time
▲ Financial stability of the carrier
▲ Willingness to negotiate rate changes
▲ Carrier reliability and consistency
▲ Frequency of service
▲ Scheduling flexibility
▲ Capabilities of operating personnel
▲ Shipment expediting and tracing
▲ Claims processing and prevention
▲ Special equipment or services available

In the current deregulated environment, customer service has become an important attribute of carrier selection. Nonetheless, there must be complete understanding between purchasing and the carrier regarding the service level required and the criteria for measuring performance as defined by the factors listed previously. Carrier billing charges are often not what is actually paid because factors in pricing structures, such as discounts, and negotiated terms can alter such charges significantly.

Most commercial transportation is via rail, pipeline, motor carriage, air, or water. About 75% of the agricultural products and manufactured goods tonnage is transported by truck, which is usually faster, more flexible, and more versatile than rail.

Note: Suppliers should be asked for their recommendations because they may have other customers in your area with whom you can combine loads. There also may be opportunities for backhauling, where trucks carrying your product to your customers can then carry material from your suppliers back to your facility.

The following are guidelines for choosing the mode of transportation, which is the first step in the traffic-transportation decision.

Motor Carriers: Motor carriers transport goods Less Than Truckload (LTL) or Truckload (TL). To determine which is appropriate, purchasers should consider the weight and volume (cube density) of the shipment. Rates are significantly lower for TL and among TL carriers can vary as much as 10 to 20%. Freight forwarders take advantage of long-haul TL rates by consolidating small shipments from LTL shippers into TL shipments.

Rail Carriers: Rail is generally used for bulky and heavy goods shipped in large volumes, such as chemicals, coal, petroleum, and lumber, and for goods that have low per unit value. Rail shipments generally encompass goods traveling long distances, usually beyond a one-day truck trip (about 450 to 500 miles). There are fewer rail carriers today than there have been in the past, and they generally operate on contract rates (indicating a long-term need for rail services), although tariff rates have not been completely abandoned. Rail rates tend to be cheaper than those of motor carriers, but when compared to motor carriers, rail does have disadvantages, including

▲ Less flexible routes
▲ More time consuming
▲ Diverting freight from one destination to another is not as easy
▲ Obtaining goods in transit may be inconvenient
▲ Requires more secure blocking and bracing to help prevent damage

Water Carriers: Except for pipeline, the cheapest way of moving freight is generally water-inland—inland barge carriers, great lakes carriers, and coastal/intercoastal and ocean carriers. Water is best for shipping high-bulk, low-value commodities; basic raw materials, including coal, iron ore, and petroleum; and agricultural products, such as grains. Water carriers charge on the basis of weight or volume, depending on which yields the greater revenue for the carrier. The disadvantage of shipping by water is that transport is slow. This is due to infrequent sailing schedules, the lack of water routes in many commercial areas, and the limited speed of marine vessels, although ships are getting faster.

Air Carriers: Air, while the most expensive, is the fastest transportation service. It is generally used for high-value, low-density items. Frequency and reliability are high, but disadvantages include:

▲ Time lost to terminal delays
▲ Lack of direct, door-to-door capability
▲ High transport charges

Yet, air transport offers trade-off possibilities for the purchaser. Aside from the common use of special small package services, such as UPS, RPS, and FedEx, air can be cheaper than truck or other modes if one looks at total landed costs as opposed to freight charges alone. Packaging and handling costs can be much less than truck and rail. In addition, many organizations have discovered they can substantially reduce inventories by shipping via air.

Intermodal Carriers: As the name implies, intermodal carriers combine modes of shipping goods, such as trailer or container on a flatcar (motor-rail), and landbridge (rail-ocean). Intermodal transportation is common for shipments arriving from foreign destinations. It should also be considered as an alternative to motor carriage. Service can be very cost-effective, especially where rail carriers have a heavy volume of traffic to keep costs low.

Dealing with Hazardous Materials

Hazardous materials are defined as those dangerous goods which pose a risk to health, safety, and property. They include:

▲ Explosives
▲ Flammables
▲ Poisons
▲ Corrosives
▲ Radioactive and nuclear materials

Hazardous materials must be packaged, labeled, handled, and transported according to stringent regulation from several agencies. Current U.S. regulations appear in the Code of Federal Regulations, Title 49, Parts 171-178. However, the Hazardous Materials Transportation Uniform Safety Act of 1990 is the primary law covering the transportation of hazardous materials. International shipments must comply with docket HM-181, where the term dangerous goods is used interchangeably with hazardous materials.

Make sure you, the shipper, and the carrier comply with all rules as they apply to hazardous materials. Never assume your shipment is exempt, and make sure the carrier you select is not only competent to move these goods, but has a record of doing so safely.

Know Carrier Classifications

There are four types of general transportation carriers:

General Purpose: These carriers handle goods, as the name implies, without specificity. However, some will provide a unique service designed to meet the purchasing organization's specific requirements. This can include common carrier or contract carrier services. A common carrier holds itself out to service the general public, but only for the types of freight for which it has the capability and capacity to transport, such as food, chemicals, or machinery. A contract carrier operates only under negotiated contractual agreements, which are generally reviewed every one to two years.

Carriers of High-Risk, Dangerous Bulk Commodities: Dangerous cargos that are transported in bulk include flammable liquids, certain wastes, nuclear materials, and other hazardous materials. Due to the sensitive nature of transporting these items, rates are not a priority in selecting a carrier. Purchasers should focus instead on building a fixed-price relationship based on mutual trust, shared risk, and reward.

Specialized Carriers: Specialized carriers are those which have developed services based on certain requirements. For example, refrigerated trucks for carrying food items, cooled and air-ride trucks for electronic equipment, and trucks to carry oversized items, such as rigging equipment. Selection of this type of carrier depends on the basis of competitive advantage balanced against other costs.

Commodity-Type Carriers: This category includes carriers who offer less specialized services, and includes most LTL common carriers. Since these suppliers are generally interchangeable, rates are important in the selection process.

Common Shipping Terms

The following terms are the most commonly used, but note there are several variations.

▲ Free on Board (FOB) Purchaser's Plant Freight Prepaid, also called a "destination" sale. Title is with seller until delivery. The seller must:

1. Place goods on or in transport vehicles
2. Secure receipted bill of lading
3. Pay all transportation charges to destination
4. Be responsible for filing cargo loss-claims with carrier

The purchaser must:

1. Move the goods after arrival

2. Be responsible for loss or damage after arrival
3. Pay any detention or demurrage charges and storage fees

▲ FOB Seller's Plant, Freight Collect, also called an "origin" contract. Title passes to the purchaser when the shipment is turned over to the carrier. FOB seller's plant is generally the preferred method because it allows the purchaser to control the transportation. The seller must:

1. Place goods on or in transport vehicles
2. Secure receipted bill of lading
3. Be responsible for cargo loss or damage until goods have been placed in or on transport vehicle at point of origin and a clean bill of lading has been furnished

The purchaser must:

1. Provide for movement after goods are on board
2. Pay all transportation charges to destination
3. Be responsible for filing loss or damage claims
4. Pay any detention or demurrage charges and storage fees

▲ FOB Named Point of Origin Freight Allowed. This is the same as FOB seller's plant except the shipper agrees to reimburse the purchaser for the freight charges for a selected mode, such as rail. Buyers often prefer truck transport and opt to pay for truck service. Title passes to purchaser when the shipment is turned over to carrier. The seller must:

1. Place goods on or in cars or trucks
2. Secure receipted bill of lading
3. Be responsible for loss or damage until goods have been placed in or on transport vehicles at point of origin and clean bill of lading as been furnished by the carrier

The purchaser must:

1. Provide for the movement of goods after they are on-board
2. Be responsible for loss/damage-claims while in transit
3. Pay all transportation charges to destination
4. Pay any detention or demurrage charges and storage fees

For international terms, see "A Guide to Incoterms."

Be Familiar with Shipping Terms

To simply indicate "ship the cheapest or best way" on a purchase order is an abdication of any freight analysis. It is critical to know shipping terms, especially because they indicate who owns what and when they own it. The point at which a purchaser takes title determines who selects the carrier, who pays the carrier, who negotiates with the carrier, and who files any necessary claims. Freight costs are part of the product costs and, ultimately, the purchaser will pay all costs unless the freight costs are negotiated to include a discount. Under a delivered, freight prepaid by seller, FOB purchaser's plant, or warehouse agreement, the freight costs are "hidden" or averaged somewhere in the price.

Transportation Documentation: Domestic and International

The most common documentation used in transportation is as follows:

▲ *Bill of Lading or Air Bill*—certifies the transfer of goods from shipper to carrier and carrier to recipient. It is the basis for proof of delivery as well as the shipment contract for transportation (contract for cartage). The bill of lading may be incorporated in a master contract.
▲ *Manifest*—consolidates multiple shipments from the shipper to the carrier for single or multiple recipients, and lists the contents of all orders on one document.
▲ *Delivery Receipt*—a document signed by the recipient showing the shipment has arrived in good order and is not damaged. The bill of lading also can be used for this purpose.

International shipments can require the following:

▲ *Letter of Credit (LC)*—letter from purchaser's bank saying funds have been set aside to pay for the shipment. The LC specifies exact terms and conditions for shipment and must coincide with the terms and conditions of the purchase order.
▲ *Commercial Invoice*—standard invoice indicating all charges and the value of the shipment, used for customs purposes.
▲ *Shipper's Letter of Instruction (SLI)*—letter to a freight forwarder with a confirmation to all parties, including the purchaser, of the terms of purchase and responsibilities for risks of transport.
▲ *Export Declaration*—content and value of the shipment prepared for the U.S. Department of Commerce
▲ *Certificate of Origin*—document states the country of origin of the goods shipped.
▲ *Packing List*—contains a list of contents of all cartons in shipment. Domestic bills of lading are usually included to cover motor transport to and from ports.
▲ *Airway or Ocean Bill of Lading*—same as packing list, but for air and sea shipments.
▲ *Dock Receipt*—proof of delivery to dockside outbound freight terminal.
▲ *Independent Inspection Certificate*—in those cases where an independent inspection takes place (quality control or calibration check at the shipper's facility before crating) a certificate of inspection is provided. This must be stated in terms of the letter of credit.

▲ *Import License*—many countries require a basic license or special permit to bring specific goods in.

▲ *Arrival Notice*—arrival of the vessel, identifies shipment, number of packages, weight, and when free time transpires.

▲ *Consumption Entry Form*—U.S. Customs form showing origin, value, description of goods, and duty estimated.

▲ *Immediate Delivery Entry Form*—expedites clearance, allows up to 10 days to pay duty and processing consumption entry ID.

▲ *Immediate Transportation Entry Form*—allows goods to be moved from dockside to inland destination on a bonded carrier without payment of duties and finalization of entry forms.

▲ *Carriers Certificate and Release Order*—advises customs of the details of the shipment, its ownership, port of lading; carrier certifies ownership.

▲ *Delivery Order*—consignee or customs broker order (authority) to release shipment to inland carrier.

Understand Freight Rates

The common freight rates are:

1. Contract rates, such as point-to-point specific rates for a given material shipped in large quantities, that is, TL on a continuing basis
2. Common carrier tariffs rates, which are offered by the carrier in a price list, and are usually lower than common carrier rates

The actual freight rate paid by the shipper requires analysis; use a spreadsheet application for comparison and options. As a rule, negotiated rates should be based on cost, value, and competitive factors.

Freight equalization is sometimes allowed for uniform or like commodities and raw materials such as steel. For example, basing point systems charge all freight from one mill shipping point to foster competition. For example, if the actual CWT is $0.75 per hundred pounds for a distant supplier compared to the close supplier rate of $0.50 per hundred pounds, the shipper can pay the freight bill at $0.75 and subtract the $0.25 from the total commercial invoice.

Consider a Third-Party Provider

Freight forwarders provide transportation service to organizations purchasing goods in LTL, airplane load, ship load, or carload lots. Forwarders consolidate small shipments from several clients into unit loads and schedule the shipment as a large load at lower rates.

Freight forwarders have carrier responsibility, move freight on their master bills of lading or waybills, and generally do not own line-haul transportation equipment.

Long-haul and pickup and delivery owner/operators service many of these forms of carriers as independent contractors but are not themselves a registered carrier providing services at a price comparable to that of a carrier.

Other third-party arrangements include:

▲ Transportation property brokers, which arrange freight for motor carriers but are not themselves a carrier with cargo liability
▲ Private logistics service providers who may partner with your organization
▲ Small package/courier service carriers, which simplify and reduce costs for small shipment service
▲ All-service logistical organizations with whom a purchaser-shipper can outsource the entire transportation function

The practice of outsourcing the transportation function is becoming more popular as purchasers look at the total landed cost vs. a single freight rate. These providers offer a wide range of integrated services including transportation, and warehousing and distribution functions. They can perform trade-off analyses, conduct commodity rate investigations, perform auditing, handle intermodal needs, provide pool car arrangements, and tender and track shipments.

Forging a Long-Term Contract

Services defined in a long-term contract go beyond the common carrier services offered to the general public. The driving force in such contracts is to identify the distinctive needs, then document those performance agreements with the attendant legal charges. The principles apply for all physical modes of carriage. The following is a review of types of services that lend themselves to long-term agreements between shippers and carriers.

1. Volume of services over a period of time. Contracts for a volume of shipments during a specified term. This factor offsets the services strictly on a transactional, shipment-by-shipment basis.
2. Continuous move, or round-robin rates between two points. The trip includes two or more shipments for which even trip billing may occur.
3. Monthly agreements as to average weight agreements per shipment. Most often, shipment charges are assessed on scaled weights or on a metered basis. If the shipper and carrier agree with the weights determined on a computed basis, then actual assessment of shipment charges can be computed on an agreed-upon billing cycle, such as weekly

or monthly, and weights monitored or checked on a sample basis for average weight. Individual freight bills do not need to be generated, only batched in a summary form on a weekly or monthly basis.

4. Debit/credit loading and unloading detention demurrage charges on a periodic basis. Instead of assessing detention/demurrage charges on a per shipment basis, an average amount of time can be agreed upon, such as two hours for truckload traffic, and then debits and credits accumulated over a periodic basis, such as a month. Specifically, if a shipper or receiver loads or unloads in 45 minutes vs. the 2 hour standard, then the shipper gets a debit in his account. If the shipper or receiver takes 4 hours to load or unload, then they get a credit in their account—a liability to pay. At the end of the month, the shipper and carrier tally up to see who pays.

5. Performance-based contracts with perks and penalties for on-time delivery. If a carrier delivers service above the agreed upon standard, such as 95% on-time pickup and/or delivery, then the carrier is entitled to a perk—a bonus. However, if the carrier does not meet the standard, then the carrier can be penalized for the poorer service. Monthly billing is conducive to this type of arrangement.

6. Performance of value-added services including light duty assembly, and JIT delivery with the possible use of returnable containers to a receiver for a supplier. With so many carriers developing contract logistics services beyond carriage, contracts make sense in defining the business and legal responsibilities of the parties involved. Specifically, legal party responsibilities include that of a supplier, a carrier, a warehouser, and possibly a light assembly supplier agent. This situation exemplifies the need for a contract or contracts among the supplier, the third-party agent, carrier, and receiver for providing defined duties, the nature of the involved legal parties, the charges for performing the duties, and how billing and payment procedures are to be performed.

7. Performance of services involving high degrees of information technology. Many evolving relationships among shippers, carriers, receivers, and third-parties involve information technologies, such as bar coding, EDI, radio frequency transponders, image processing systems, GPS/GIS positioning systems, and radio frequency tracking systems that go beyond offered common carrier services. These services relate to shipment tracking and diversion services, billing and invoicing, and payment processes.

Be Savvy About Contracting

Many carriers and shippers adopt carrier rules tariffs into long-term contracts. Unless a review process with formal notice and addendum procedure is incorporated into the contracts, shippers may be subjecting themselves to rules, additional charges, and rate increases of which they are not aware. Carefully review contracts as to how rules tariffs and classification systems are referenced. Other clauses to include are damaged freight provisions, salvage rights, a statement that the carrier is subjected to common carrier liability, and any unique needs such as dedicated equipment for the shipper.

Note: The statute of limitations for overcharge claims for common carriers is now six months, which is a compelling reason to use contract carriage and to pre-audit freight bills (see next two sections). Finally, make sure freight discounts are part of your contract or purchase order, and that the purchasing organization receives this discount if paying assessed freight charges on commercial invoices. Also, include an arbitration or dispute resolution procedure requirement in the event of disputes.

The first wave of transportation deregulation came with the Air Cargo Act of 1977, followed by the Staggers Rail Act of 1980, the Motor Carrier Act of 1980, and the Shipping Act of 1984. These Acts substantially reduced the rules of entry, price discounting, route offerings, service offerings, and the role of the Interstate Commerce Commission (ICC—which became defunct) in regulating the air, truck, rail, and water carriers involved in interstate commerce. These acts provided freedom and ease to negotiate rates, expand private contracting, and offer other services.

However, these acts also brought problems and instability. For example, the ensuing increased competition in motor trucking brought about discounting practices which, in turn, led to bankruptcy. After several Supreme Court cases (in particular the Maislin and Transcon cases), Congress passed the following amendments:

1. *The Negotiated Rates Act of 1993 (NRA)*   Effective December 3, 1993, this Act prohibits motor common and contract carriers from providing a tariff rate reduction to the nonpayer of freight charges except under certain rules of disclosure to the party responsible for paying (or his agent), when indicated on the freight document and for performance of a particular service.

2. *The Transportation Industry Regulatory Reform Act of 1994 (TIRRA)*   Requires motor carriers to provide, on request of the shipper, a written or electronic copy of the rate, classification, rules, practices, and any other information affecting the rate. This requirement is also known as "verified pricing documentation." This act also eliminated the (then) ICC rules for motor carrier contracting.

3. *The Federal Aviation Administration Act of 1994*   An aviation bill rider, this Act takes the states out of the intrastate economic regulation of motor carriers. It carries the provisions of the Motor Carrier Act of 1980 down to the state level, that is, states cannot regulate intrastate motor carriers, unless the motor carriers in the state petition the state to exercise regulation.

The latest legislation is the ICC Termination Act (ICCTA) of 1995.

Understand "Common Law" Price Lists and Rules Tariffs

Carriers (motor and rail) can still include rules tariffs in their "common law" price lists, which allow for amendments without the knowledge or consent of the shipper. For example, references to classification tariffs mean that the *shipper* is subject to the rules related to uniform bill of lading forms, liability limits, package specifications, released rates, and other rules and charges in the National Motor Freight Classification (NMFC) for trucks, the Uniform Freight Classification (UFC) for rail, and loading rules published in the Association of American Railroads (AAR) loading pamphlets.

These rules tariffs include provisions for surcharges, various value-added ancillary services for which the carrier assesses additional charges. Many carriers, instead of relying upon the NMFC, are now publishing their own rules tariffs. The new law also enables carriers and shippers to develop their own unique bills of lading rather than using the uniform bills of lading incorporated in the NMFC or in the Uniform Commercial Code of states.

Know Where Contract Carriage Fits in the ICCTA

Two precepts of the ICC Termination Act (ICCTA) are that the federal government has jurisdiction to oversee (as opposed to regulate) U.S. transportation policy, and that carriers provide *non-discriminatory* service at reasonable charges. Some attorneys and practitioners believe that under the policies of ICCTA, all carriers are now common carriers with the right to develop long-term contracts (contract carriage). However, under deregulation these carriers are not required to file tariffs with any agency. The question arises as to how shippers, receivers, and third-parties will know if they are being discriminated against by other service/price offerings? How can one use contract carriage to legally discriminate among shippers? Contracts are a legal means as long as distinctive services are identified in the contracts. The answers to these questions may lie in the courts. Parties who claim to be injured by discriminatory services and prices may only be able to know if and when court cases force discovery. If this is the case, further legislation may be needed.

The ICC Termination Act of 1995 (ICCTA)

Effective January 1, 1996, the Interstate Commerce Commission (ICC) was abolished, and limited functions were transferred to the newly created Surface Transportation Board (the Board), within the Department of Transportation (DOT). ICCTA is a massive revision of the Interstate Commerce Act. The major changes relevant to purchasing include:

▲ Elimination of railroad tariff filings (but rail carriers must publish rates, including a 20-day notice of rate increases, even though the Act does not specify where the rates are to be published).

A Guide to Incoterms

International shipping terms and rules are called "INCOTERMS." Here is a listing of the most important INCOTERMS.

	General Definition	**Specific Incoterm**
Group E	Goods are made available at the supplier's named site. Buyer assumes all responsibility for transportation, cost, and risk beyond this point.	**EXW** Ex works (supplier's site)
Group F	Supplier delivers goods to specified carrier location. Buyer assumes all costs and risks associated with the main carriage and beyond.	**FCA** Free carrier at a named place. Supplier clears goods for export. **FAS** Free alongside ship (on dock). Buyer clears goods for export. **FOB** Free on board (on ship). Supplier clears goods for export.
Group C	Supplier contracts for main carriage and assumes all costs to the destination country. Buyer assumes all risks during main carriage, plus cost of delivery from destination dock to buyer's site.	**CFR** Cost and freight to the port of destination. Supplier clears goods for export. **CIF** Cost, insurance, and freight to the port of destination. Same responsibilities as CFR, except supplier is responsible for insurance during main carriage. **CPT** Carriage paid to the named destination. Supplier clears goods to export. **CIP** Carriage and insurance paid to the named destination. Same responsibilities as CPT, except supplier covers insurance during main carriage.
Group D	Supplier assumes all risks and costs to the specified destination.	**DAF** Delivered at frontier (country border) at a specified location, but before the customs border of the adjoining country. Supplier clears goods for export. **DES** Delivered ex ship at specified port of destination. Goods are available aboard ship, uncleared for export. **DEQ** Delivered ex quay (dock) at the named port of destination. The duty can be paid or unpaid, depending on the agreement. **DDU** Delivered duty unpaid at the specified place of destination. **DDP** Delivered duty paid at the specified place of destination. Supplier bears all risks and costs, including duties, taxes, and other charges of delivering the goods, cleared for import.

Source: Donald W. Dobler and David N. Burt, *Purchasing and Supply Management: Text and Cases*, 6th ed. (New York: McGraw-Hill, 1996) 574. Reprinted by permission of McGraw-Hill, Inc.

▲ Limitation of the authority of the Board to set minimum rail rates, and investigate and suspend new rail rates except to prevent irreparable harm. The Board also cannot initiate rail rate investigations but can respond to complaints.

▲ Use of stand-alone cost (SAC) methods to determine rail rate reasonableness (and other methods when SAC preparation is too costly).

▲ Repeal of the Elkins Act against rail rebates, along with the commodities clause and special regulations for transportation of recyclable or recycled materials.

▲ Retention of the motor carrier, water carrier, freight forwarder, and broker provisions of TIRRA of 1994, which eliminated much of the tariff filing and rate regulation for interstate transportation.

▲ Retention and some expansion of the provisions of the Federal Aviation Administration Reauthorization Act of 1994 and the Negotiated Rates Act of 1993 (NRA).

▲ Inclusion of domestic port-to-port water and intermodal motor-water carriage regulations under the Board from the Federal Maritime Commission.

▲ Requirement that collective rate-making agreements be reviewed every three years and tested for public interest.

▲ Modification of the 180-day time limit to file claims to include 180 days from the start of rebillings by carriers.

▲ Substitution of market-determined rates to prevail instead of "rate reasonableness" and "tariff requirements," which now apply only to household goods.

▲ Simplified licensing procedure for new carriers, freight forwarders, and brokers, who register with the Federal Highway Administration (FHWA).

▲ Eliminates the distinctions between common and contract carriers.

▲ Authority for transportation carriers to offer line-haul and various accessorial, value-added services either under common carriage for the general public or under contracts for specific services.

▲ Broadens the term "transportation" to include packing and unpacking activities.

▲ Removes the states from any obligation to determine and regulate tariffs.

The following are the basic steps needed to establish and maintain an effective transportation program.

Step One: Take Control

The importance of gaining control of inbound transportation is twofold.

1. The supplier determines these costs, and not always to the purchaser's benefit.
2. Within the organization, purchasing/supply management has the best opportunity to work with suppliers to reduce transportation costs.

Purchase orders must state that the buying organization is paying for inbound transportation because it is also essential that senior management commit to purchasing and supply management's participation in transportation supply chain policy. This is necessary if purchasing/supply management is to take control of transportation, which would otherwise be determined by the supplier. In most cases, it is a benefit to the purchaser to specify on the purchase order *free on board (FOB) origin, freight collect, or FOB shipping point, freight collect.*

Step Two: Coordinate Efforts with Suppliers

Even though your supplier will no longer determine the specifics of inbound transportation, supplier cooperation is vital to a successful program. Your supplier may have contacts that can prove valuable to you, and will also be required to help coordinate the movement of goods from its facility to yours.

Step Three: Form a Team

In large organizations, as well as small ones, the team approach is the most effective in establishing a capable transportation program. This is because cooperation and coordination is necessary from several functions within the organization, including:

▲ Transportation/traffic
▲ Manufacturing/operations
▲ Production planning and inventory management
▲ Warehousing and receiving
▲ Product and supply quality assurance
▲ Accounting

Purchasing/supply management should assume team leadership, especially since the success of the program depends on the ability of purchasing/supply management to gain the cooperation of suppliers whose commitment must be long-term.

Step Four: Analyze Requirements

Several factors should be considered preparatory to selecting a carrier. These include:

▲ Types of goods to be transported—size, weight, density, packaging, and specific physical characteristics
▲ Mode of transportation—motor, rail, air, water, or combinations of these
▲ Shipping point and volume "traffic lanes"
▲ Delivery requirements—time needed or required
▲ Organizational requirements
▲ Carrier capabilities

Step Five: Outsource if Appropriate

Analyzing requirements may lead to a decision to outsource one or more of the inbound transportation functions. This may be especially true if purchasing/supply management wishes to control inbound transportation but does not have the time or necessary staff to support the program.

Step Six: Analyze Opportunities

Any number of creative methods can be used to maximize inbound transportation. Here are some ways to manage the process proactively:

▲ Switch from FOB Delivered, Freight Prepaid to FOB Origin, or Freight Collect, which can be a cost saving if actual freight is less than the average freight charges built into the delivered price of the goods.
▲ Use customer pickups to build leverage from outbound freight delivery—backhauled inbound freight can be negotiated to your advantage.
▲ Negotiate the right of routing for better control over loading and unloading schedules.
▲ Check shipment weights—always determine the correct weight for shipments.
▲ Review supplier source locations.
▲ Consolidate where possible.

Step Seven: Investigate Tariffs

Under deregulation, the transportation industry has become very competitive. Carriers of the same mode now may compete with one another, as well as with carriers in a different mode.

Understand the rate structures of each mode of transportation and the rates offered by carriers in each mode. Do not assume that the rate/tariff quoted is valid until you are presented with a copy of the actual tariff showing the date stamp and effective date of the tariff. If a lower negotiated rate is used before the effective date, the carrier may later file an undercharge claim and the purchasing organization would be legally obliged to pay it. Tariffs are available in electronic format on floppy disks or through computer transfer. Purchasers and supply managers should always keep a printed copy of any tariff transactions on file.

Transportation Contract Checklist

The following items should be included when contracting transportation services: Be sure to check the glossary for any unfamiliar terms.

▲ Responsibility for cargo loss and damage—defines the basis of liability and the rule of mitigating damages and claims filing

▲ Disclosure of goods—determines the value of goods shipped (without this notice, the carrier can avoid paying for any damages claimed)

▲ Routing mode

▲ Method of transportation—eliminates any possibility of inappropriate methods or equipment

▲ Responsibility for equipment specification—designates which party is responsible for all equipment

▲ Volume requirements—defines the minimum amount of goods required of the shipper and the minimum of equipment to be provided by the carrier

▲ Rate escalation/de-escalation—specifies changes in rates, including escalation and de-escalation; defines what indexes the rates are tied to, how rates can or will change, when, under what circumstances, and with what notice

▲ Terms of transportation charges—includes all descriptive detail including who pays for what services

▲ Billing and payment—specifies documents to be used, what constitutes a bill, billing error rectification, credit extended, rights of lien, and all other such matters

▲ Force Majeure

▲ Title of goods—defines when title passes and who is entitled to make a claim

▲ Applicable law—specifies which court system will have jurisdiction over dispute resolution

▲ Assignability—specifically, will it be allowed

Step Eight: Evaluate and Select a Carrier

Selecting a carrier encompasses standard sourcing methods. A thorough investigation of candidates should focus on items such as financial strength, billing procedures, pricing and contract terms, equipment upkeep, personnel training, delivery record, damage and claims record, safety programs and safety record, insurance, reputation among other shippers, service capabilities, geographical area served, and routing. The selection process should also take into account the use of a third-party provider, such as a freight forwarder or transportation broker.

Step Nine: Negotiate Price

Purchasing and supply management's task is to negotiate with the carrier for the lowest class of rates available for the purchased goods. Transportation consultants are sometimes employed for this purpose, especially if an organization is small or does not have a transportation or traffic manager on staff. Class rate negotiation is important because it lays the platform for all further negotiations about price.

Once class rates have been established, negotiate for the best discount you can obtain. It is the combination of the class rate and the discount that eventually determines the charge that will appear on the invoice. This is calculated once the goods have been delivered and the weight or volume of the order has been established. Charges are ultimately based on:

▲ Origin and destination points
▲ Volume or weight
▲ Packaging
▲ Specific characteristics such as perishability
▲ Handling

Step Ten: Negotiate a Contract

For other than one-time deliveries, contracts are essential for carriers who will provide long-term services to your organization.

Note: Contracts are also necessary for one-time deliveries in which a special service will be provided, such as the use of oversized equipment.

As with any other supplier, it is important to develop and partner carriers with whom a relationship of mutual trust can be built. Contracts should ultimately aim for the best service at the most competitive price. A transportation consultant can also be helpful when negotiating such a contract.

Caveat: A carrier will sometimes try to insert their rules tariffs into a contract. Do not allow this to happen. Rules tariffs should be incorporated into the carrier's bill of lading. Rules tariffs contain items such as late payment penalties and liability limitations which, if cited in the contract, require the shipper to surrender certain rights. When rules tariffs are part of the bill of lading, the purchasing organization retains its right to challenge any claims made by the carrier.

Step Eleven: Expediting and Tracing Shipments

Expediting means the shipper wants or needs faster than normal service-delivery time. This request must be made prior to shipment and may incur extra charges. Expediting should be used only when necessary, not only due to the cost, but because the requests will be ignored if they become habitual. Expediting should never be a substitute for production planning.

Tracing systems are used to locate a shipment while it is in the possession of the carrier, and can take place even while the shipment is in transit. To trace a shipment, the following information is needed:

▲ Date tendered to the carrier
▲ Shipper/origin
▲ Destination
▲ Freight bill or pro number
▲ Description of the goods

- ▲ Recipient of the goods
- ▲ Date shipped
- ▲ Appropriate documents
- ▲ Truck or car number

Most tracing systems are automated and utilize electronic/satellite tracking, which literally follows the shipment at any point in the route.

For tracking and tracing individual boxes, pallets, cases, etc. aside from the well-known bar code system, many companies have used the radio frequency identification system (RFID). RFID is an improvement to bar coding and involves using inexpensive, thin integrated circuitry fixed or embedded stickers or labels to electronically send a tracking number with product identification. This allows faster scanning and overcomes many bar code reading problems caused by dirt or poor positioning. As one might expect, WalMart was an early adopter.[25]

Step Twelve: Pre-Audit Freight Invoices

Shippers have the right to pre- and post-audit their freight bills within a six-month period. Pre-auditing is the prudent course of action, for it can save a great deal of money. Pre-auditing is particularly important in international shipping (export or import) as tariff and custom duties are based on how the goods are classified. Many independent rate consultants figure that the average shipper overpays by 15 to 20% in areas that include:

- ▲ Incorrect classifications
- ▲ Duplicate payments
- ▲ Mistakes in rate-invoice calculations, such as incorrect mileage or weights
- ▲ Incorrect FOB terms
- ▲ Incomplete documentation
- ▲ Wrong routing
- ▲ Inclusion of carrier rules tariff for late payment penalties, falsely claimed liability limitations, exemption from special damages, and other such "hidden charges"
- ▲ Poor delivery/missed arrival deadlines
- ▲ Wrong mode
- ▲ Incorrect packaging
- ▲ Wrong source
- ▲ Miscellaneous extra charges (such as demurrage) not justifiable and any other deviation from the intended contract

If your organization does not have in-house resources, a special auditing firm will provide pre-auditing services at a 50% commission on savings on net recoverable overcharges.

Caveat: Freight bills should never be paid without first being audited.

Step Thirteen: File Claims When Necessary

Inevitably, even with the best carriers, claims will be filed for over, short, or wrong goods shipped by the supplier, or goods damaged while in transit. Claims are filed when either the supplier or carrier fails in their efforts to correct the situation. Sometimes it is difficult to assign culpability. For example, material delivered short can be the result of a mistake by either party. The receiving function, which should be part of your team, is vital in this phase of the transportation program. Receiving must verify that all shipments arrive as specified.

The rules governing the filing of claims are founded in law, and must be strictly followed. Certain government and international rules also apply. In addition, claims rules are found in the carrier's tariffs or in the bill of lading.

Step Fourteen: Evaluate Carrier Performance

Standard performance measures should be applied to carrier performance. Specific parameters to be measured include:

▲ Transit time between facilities
▲ Equipment utilized
▲ Billing accuracy
▲ Claims occurrence
▲ Pickup and delivery performance
▲ Rate negotiation service
▲ Sales representative information
▲ Technology or innovations offered
▲ Follow-up performance

Many carriers supply a monthly service report to customers containing shipment and delivery date information. If such a report is not normally issued by a carrier, ask for one. Measuring carrier performance should not only serve as a means of monitoring performance, but should also be a part of continuous improvement operations.

Glossary

The following are common transportation terms. Not all will have been discussed in text, due to space limitations. These definitions are from the NAPM Glossary of Key Purchasing Terms.

Air waybill—The document used for the shipment of air freight providing information such as commodities shipped, shipping instructions, and shipping costs.

Assignment—A transference of a property right or title to another party. In shipping, it is commonly used with a bill of lading, transferring rights, title, and interest to a named party.

Bill of lading (B/L)—A carrier's contract and receipt for goods transported from one place to another, delivered to a designated person. Types of bills of lading include:

- A certified bill of lading, provided when a consular officer endorses an ocean bill of lading, specifying that the shipment meets the requirements of the country of importation.
- A clean (clear) bill of lading, or a carrier-receipted bill of lading, provided when a shipment is deemed in good condition with no apparent loss or damage. This means the bill of lading contains no exceptions and is a readily negotiable instrument.
- An export bill of lading, issued by an inland carrier to contract the movement of goods from an interior point of origin to a foreign destination.
- A government bill of lading, supplied by the U.S. government for shipment of government-owned property or goods within the government.
- A negotiable (order) bill of lading, consigned directly to the order of a party, usually a shipper or a bank, whose endorsement is required to transfer the title of the goods. The bill of lading must be surrendered to the carrier before the goods are released.
- A nonnegotiable (straight) bill of lading, consigned directly to the consignee and not negotiable. The goods usually are delivered without surrender of the bill of lading.
- An ocean bill of lading, issued by an ocean carrier for marine transport of goods.
- An order-notify bill of lading, similar to a negotiable bill of lading, except that it contains an additional clause directing that a specified third party, usually at the port of destination, be notified upon arrival of the goods. However, this party does not take title.
- A short-form bill of lading, unlike the straight bill of lading, only refers to the contract terms but does not include them.
- A through bill of lading, issued by a shipping company or its agent, covering more than one mode of transportation.

Carload (C/L)—The weight and/or volume necessary to qualify for a rail carload rate, or a rail car loaded to its capacity. The carload quantity referred to on a carload rate has nothing to do with the actual quantity required to fill the rail car, but is the minimum weight specified to qualify for a lower-class rate.

Cartage—A charge made for the hauling and transferring of goods, usually on a short haul basis; drayage. Also, the physical movement of the goods.

Class freight rate—A rate resulting from a classification rating of the freight. While commodity rates are available only on selected commodities, a class rate can be found for almost all commodities.

Commodity freight rate—A rate for a specific commodity, moving between specified points, sometimes in a certain direction, and typically for a specific minimum quantity.

Common carrier—A common law carrier holds itself out to serve all customers, but carries only the types of freight for which it is registered. The most accepted characteristic of a common transportation carrier is the availability of nondiscriminatory service to anyone seeking such transportation.

Consignee—The person or organization to whom a shipper directs the carrier to deliver goods, generally the purchaser of the goods.

Consignor—The shipper of a transportation movement.

Consolidation—Combining less-than-truckload or less-than-carload shipments from various facilities for transport as one larger shipment, typically at a lower freight rate.

Containerization—Using large, sealed, standard-size containers primarily for intermodal and international shipping. The containers can be transloaded between rail, motor, and water carriers to reduce transit time, theft, packaging requirements, damage, and costs. "Piggyback" often is used to designate containers on flat cars (COFC) and trailers on flat cars (TOFC), with "fishyback" used for water transfer. Roll on-Roll off (RoRo) is used when equipment or containers can be driven directly onto the water carrier.

Contract carrier—A contract carrier, regardless of mode, provides transportation and/or related services according to a contractual agreement. Tariff rates do not apply; contract rates are generally lower than regular common carrier rates.

Cost and freight (CFR)—The supplier quotes a price for the goods being sold, including the cost of transportation to a specified destination; it is used most commonly in international shipping.

Cost, insurance, and freight (CIF)—A sales practice in international trade whereby the supplier quotes a price that includes the cost of the material, freight charges to a destination point, and marine insurance en route.

Declared value—The practice of stating the value of goods being transported on the shipping document, often to achieve a lower freight rate or to obtain insurance.

Demurrage—A fee charged by a carrier against a consignee, consignor, or other responsible party to compensate for the detention of the carrier's equipment in excess of allowable free time for loading, unloading, reconsigning, or stopping in transit. The term is also used by suppliers of material delivered in a variety of returnable containers, such as gas cylinders.

Door-to-door—The through transportation of a shipment from the consignor directly to the consignee.

Expedite—To contact a supplier or carrier to speed up delivery of an inbound shipment.

Free along side (FAS) vessel—The supplier agrees to deliver the goods in proper condition along side the vessel, with the buyer assuming all subsequent risks and expenses after delivery to the pier.

Free on board (FOB)—In domestic trade (when the term is used with no further explanation), FOB means delivery of the goods with all charges paid aboard the carrier's equipment without cost to the buyer. Modified FOB terms include:

- ▲ FOB destination, freight collect means that title passes from the supplier to the buyer at the destination point, and that the freight charges are the responsibility of the purchaser. (The supplier owns the goods in transit and is responsible for filing loss and damage claims against the carrier, but the purchaser pays and bears the freight charges and files any overcharge claims.)
- ▲ FOB destination, freight prepaid means that title passes from the supplier to the buyer at the destination point, and that the freight charges are paid by the supplier. (The supplier pays and bears the freight charges, owns the goods in transit, and may file claims for overcharges, loss, and damage.)
- ▲ FOB destination, freight prepaid and charged means that the title passes at the destination point, and that the freight charges are paid by the supplier and added to the invoice. (The supplier pays the freight charges, owns the goods in transit, and files all claims for overcharges, loss, and damages. The purchaser bears the freight charges.)
- ▲ FOB origin, freight allowed means that the purchaser obtains title where the shipment originates and is responsible for all claims against the carrier, but that the supplier pays the freight charges.
- ▲ FOB origin, freight collect means that title passes to the buyer at the point of origin; the buyer pays the freight charges. (The buyer owns the goods in transit, and files all claims against the carrier.)
- ▲ FOB origin, freight prepaid and charged means that title passes to the buyer at the point of origin, and that the freight charges are paid by the supplier and then collected from the purchaser by adding the amount of the freight charges to the invoice. (The supplier pays the freight charges and files claims for overcharges. The purchaser bears the freight charges, owns the goods in transit, and files claims for loss and damage with the carrier.)

Freight bill—The carrier's invoice for transportation charges applicable to a shipment.

Freight bill audit—A critical review of freight bills to determine classification, rating, or extension either by a third party or an inside auditor.

Freight claim—A claim against a carrier due to loss or damage to goods transported by that carrier; or for erroneous rates and weights in assessment of freight charges.

Freight forwarder (domestic)—A carrier that collects small shipments and consolidates them into larger shipments for delivery to the consignee.

Intermodal freight shipments—Transportation shipments involving more than one mode, for example, rail-motor, motor-air, or rail-water.

In-transit privileges—Special privileges which give rail shippers (buyers and sellers) the right to stop a shipment en route, unload it, perform certain processing operations on the material, reload the processed material, and continue the shipment at the original rate plus a modest additional charge.

Inventory in transit—Physical inventory en route aboard a carrier. The term also indicates the capital costs of materials, parts, and finished goods en route aboard a carrier. This cost is commonly computed by multiplying the opportunity cost rate by the value of the inventory, then by the percentage of time (annualized) the goods are en route, plus the cost of the material itself.

Knocked down (KD)—Disassembled goods for the purpose of reducing the cube space of the shipment for transportation and storage.

Less than carload (LTC)—A shipment which is less than the amount required to be eligible for carload rates.

Less than truckload (LTL)—A shipment which is less than the amount required to be eligible for truckload rates.

Loading allowance—A reduced rate or refund offered to shippers and/or consignees who load and/or unload the shipment.

Manifest—A list of items shipped, plus related details (often a copy of the freight bill).

Over, short, and damage report—Issued by a freight agent indicating any discrepancies between the bill of lading and freight on hand.

Packaging list—An itemization of package contents prepared by the shipper.

Pickup and delivery (PU & D)—Transport from the shipper's dock to the consignee's dock.

Point of origin—The location where a transportation company receives the shipment.

Prepaid freight—Transportation charges which are paid by the shipper at the point of shipment.

Private carrier—A carrier that owns or leases vehicles and provides transportation services for the firm that owns it.

Released value rate—A transportation rate based on a reduced value of the shipment which, in turn, limits the carrier's liability to a lesser amount.

Routing—Determination of how a shipment will move from the point of origin to the destination, including selection of carriers and geographic routes.

Shipping release—A form used by the purchaser to specify shipping instructions for goods purchased for delivery at a future date.

Stopover privilege—An arrangement whereby a shipment can be stopped at stations en route to take advantage of transit privileges, such as complete loading or partial unloading.

Store door delivery—Delivery to the consignee's receiving platform by motor vehicle or rail car.

Transshipping—Transferring goods from one transportation line to another.

Uniform freight classification—A listing of commodities showing the assigned class rate for determining rail freight rates, together with governing rules and regulations.

Waybill—A document prepared at a shipment's point of origin showing point of origin, destination, route, consignor, consignee, description of the shipment, and the amount charged for transportation.

Weight, gross—Total combined weight of the article, container, and packing material.

Weight, net—Weight of the contents of a container or the cargo of a vehicle.

Weight, tare—Combined weight of an empty container and packing materials.

Facing the Future

Taking control of transportation represents a challenge and an opportunity for the purchaser or supply manager to add significant value to the organization. With deregulation there is intense competition for the movement of goods, with truck, rail, and water carriers competing for tonnage, and with carriers in the same mode vying for business. It behooves purchasers of transportation not only to become familiar with the fundamentals of transportation, but also to look for ways to creatively use price and service options to advantage.

Practitioners should also keep abreast of developments in the oversight of the transportation industry. Degregulation and reregulation may not be over. Expect cases to come to court which may challenge the ICC Termination Act. Any decisions handed down may prompt new legislation and regulations.

Bibliography

Armstrong's Guide to Third Party Logistics Providers and Dedicated Contract Carriers. Stoughton, WI: Transportation Education Specialists, 1995.

Augello, W. J. "Negotiating with Carriers in Light of New Laws." *NAPM Insights,* February (1995).

Farris II, M. Theodore. "Are You a Victim of Off-Bill Discounting?" *Purchasing Today*™, March (1996).

Hommrich, Thomas P. "Logistics: The Road Less Traveled?" *Purchasing Today*™, April (1996).

Marien, Edward J. "Overcoming Claims as Losing Propositions." *Transportation & Distribution,* May (1996).

Marien, Edward J. "Contract Carriage with Less Regulation." *Transportation & Distribution,* July (1996).

Marien, Edward J. "Making Sense of Freight Terms of Sale." *Transportation & Distribution,* September (1996).

Marien, Edward J. "Making Logistics Count in Your Company." *Transportation & Distribution,* November (1996).

Pohlig, Helen M. "Supreme Court Rules for Shipper in Undercharge Case." *NAPM Insights,* April (1995).

Endnotes

1. Some firms use time frames of five or six weeks rather than four weeks.

2. Based on Daniel J. Bragg and Chan K. Hahn, "Material Requirements Planning and Purchasing," *Journal of Purchasing and Materials Management* Summer (1982): 18–20. Used with permission of the authors.

3. For an interesting discussion of MRP II development, see Troy Juliar, "Completing the Mix: Materials Management and MRP II," *Purchasing Management* February (1987): 6–11.

4. An MRP-type system also generates output that can be utilized easily and effectively as input for an electronic data interchange (EDI) system. Consequently, as EDI operations become more widely used, MRP systems will tend to facilitate their use.

5. J.E. Schorr and T.F. Wallace, High Performance Purchasing: Manufacturing Resouce Planning for the Purchasing Professional," Wight Publications Inc., Essex Junction, VT (1986): 150.

6. A *product-type layout* is one in which the various types of equipment required to make a given product are arranged adjacent to each other so that operations occur sequentially, in line, from one end of the manufacturing process to the other. An automobile

assembly plant is a good example of product layout. A process-type layout groups similar types of equipment together, without regard for the flow of the product being manufactured—for example, all lathes are grouped together, all boring machines are grouped together, and so on. A traditional job shop exemplifies this type of layout.

7. A *Kanban material production and movement system* is simply one in which no more than approximately one hour's supply of material is produced in one run in each work center—and the next production run does not occur until the material is called for by the succeeding operation. It is this feature of the operation that produces the *small batch* and *pull-flow* characteristics of the system.

8. The following articles provide good in-depth analyses of JIT applications: Chan K. Hahn, Peter A. Pinto, and Daniel J. Bragg, "Just-in-Time Production and Purchasing," *Journal of Purchasing and Materials Management* Fall (1983): 2–15; C. H. St. John and K. C. Heriot, "Small Suppliers and JIT Purchasing," *International Journal of Purchasing and Materials Management* Winter (1993): 11–16; P. A. Dion, P. M. Banting, S. Picard, and D. L. Blenkhorn, "JIT Implementation: A Growth Opportunity for Purchasing," *International Journal of Purchasing and Materials Management* Fall (1992): 32–38.

9. For a detailed discussion of the buyer-seller relationship, see C. R. O'Neal, "The Buyer-Seller Linkage in a Just-in-Time Environment," *Journal of Purchasing and Materials Management* Spring (1989): 34–40.

10. Thomas H. Davenport, *Mission Critical—Realizing the Promise of Enterprise Systems* (Boston: The Harvard Business School Press, 2000): 6–7.

11. Ibid., 16–19. See also Matthew Friedman and Marlene Blanshay, *Understanding B2B* (Chicago, Dearborn Trade, 2001).

12. Donald F. Blumberg, *Introduction to Management of Reverse Logistics and Closed Loop Supply Chain Processes* (Boca Raton, FL: CRC Press, Taylor & Francis, 2005) 84, 184–185.

13. For an interesting discussion of the JIT system and the value of inventories, see George Newman, "As Just-in-Time Goes By," *Across the Board* October (1993): 7–8.

14. Most firms have a company-wide materials standardization program. The materials included in the inventory catalog should all have been accepted as company "standards." This topic is discussed further in Chapter 6.

15. For a complete discussion of this interesting approach, see James A. G. Krupp, "Are ABC Codes an Obsolete Technology?" APICS—*The Performance Advantage* April (1994): 34–35.

16. When the entire order is delivered in one shipment, Q for the order and Q for the delivery are the same number. When the order is delivered in several shipments, Q per delivery is smaller than Q for the order. In this case, Q for the delivery should be used in the formulas to calculate carrying cost.

17. When the entire order is delivered in one shipment, the Q value and the A value for the order are the same as for the delivery. However, in the case of a term contract, Q and

A values should be calculated on the basis of each delivery. Thus, purchasing expenses incurred in generating the order and administering the contract should be spread over the deliveries so that *A* equals the acquisition cost per delivery and *Q* equals the delivery quantity. This approach produces the most useful *A* value, and it is also consistent with the approach used in calculating inventory carrying costs.

18. A more straightforward mathematical solution can be obtained using differential calculus: (1) Write the equation for the total cost curve; (2) differentiate the equation; (3) find the minimum value of the function by setting the derivative equal to 0 and solving for *Q*.

19. See footnotes 13 and 14 for use of the EOQ formula in the case of term contracts with multiple deliveries.

20. Daniel Jones in "Don't Let JIT Overrule EOQ," *Supplier Selection and Management Report* June (1991): 10.

21. David N. Burt, and Richard L. Pinkerton, *A Purchasing Manager's Guide to Strategic Proactive Procurement* (New York: Amacom, The American Management Association, 1996): 90–91.

22. James A. Gardner, *Common Sense Manufacturing: Becoming a Top Value Competitor* (Burr Ridge, IL: Business One Irwin, 1992): 24.

23. A perpetual inventory record for a material controlled in a closed stores system is maintained simply by posting receipts from invoices or receiving reports and disbursements from stores material requisitions or similar withdrawal authorizations. In most firms today, perpetual records are computer-based.

24. By Richard L. Pinkerton, Ph.D., C.P.M., and Edward J. Marien, Ph.D. Published as the April 1997 issue of *INFO EDGE*, National Association of Purchasing Management (NAPM), now the Institute for Supply Management (ISM), Tempe, AZ. © Institute for Supply Mangement™. All rights reserved. Reprinted, with minor editing, with permission from the publisher, the Institute for Supply Management. The authors of this book want to thank Joseph V. Shannon, President of PO$E, a transportation consulting firm in Cleveland, OH. Mr. Shannon verified on March 27, 2008, that this appendix is "still current, useful and educational today, 11 years later."

25. Donald F. Blumberg, *Introduction to Management of Reverse Logistics and Closed Loop Supply Chain Processes* (Boca Raton, FL: CRC Press, Taylor & Francis, 2005): 84, 184–185.

Suggested Reading

American Production and Inventory Control Society (APICS) Newsletters and other material. www.apics.org.

Ballou, Ronald H. *Business Logistics/Supply Chain Management*, 5th ed. Upper Saddle River, NJ: Prentice Hall, 2004.

Blumberg, Donald F. *Introduction to Management of Reverse Logistics and Closed Look Supply Chain Processes*. Boca Raton, FL: CRC Press, Taylor & Francis, 2005, 84, 184–185.

Bowersox, D.J., D.J. Closs, and M.B. Cooper. *Supply Chain Logistics Management*. New York: McGraw-Hill/Irwin, 2002.

Bragg, Steven M. *Inventory Best Practices*. Hoboken, NJ: John Wiley & Sons, Inc., 2004.

Burton, Terrence, and Steven Boeder. *Lean Extended Enterprise: Moving Beyond the Four Walls to Value Stream Excellence*. Fort Lauderdale, FL: J. Ross Publishing, 2003.

Chase, Richard B., F. Robert Jacobs, and Nicholas J. Aquilano. *Operations Management for Competitive Advantage*, 10th ed. New York: McGraw Hill/Irwin, 2004.

Coyle, John J., Edward J. Bardi, and Robert A. Novack. *Transportation*, 6th ed. Cincinnati, OH: South-Western College Publication, 2006.

Emmett, Stewart. *Excellence in Warehouse Management: How to Minimise Costs and Maximise Value*. West Sussex, England: John Wiley & Sons, Inc., 2005.

Esty, Daniel C., and Andrew S. Winston. *(New) Green to Gold: How Smart Companies Use Environmental Strategy to Innovate, Create Value, and Build Competitive Advantage*. New Haven, CT: Yale University Press, 2006.

Frazelle, Edward H. *World-Class Warehousing and Material Handling*. New York: McGraw-Hill, 2001.

Gross, John M., and Kenneth R. McInnis. *Kanban Made Simple: Demystifying and Applying Toyota's Legendary Manufacturing Process*. New York: AMACOM, 2003.

Hanna, Mark D., and W. Rocky Newman. *Integrated Operations Management: A Supply Chain Perspective*, 2nd ed. Mason, OH: Thomson Advantage Books, SouthWestern/Thomson Learning, 2007.

Incoterms. 2000. New York: International Chamber of Commerce (ICC) Publishing, Inc., 2000.

Krajewski, Lee J., and Larry P. Ritzman. *Operations Management*, 6th ed. Upper Saddle River, NJ: Prentice Hall, 2002.

Langley, C.J., G.R. Allen, and G.R. Tyndall. *Third-Party Logistics Study: Results and Findings of the 2001 Sixth Annual Study Cap Gemini Ernst & Young*. Tempe, AZ: Institute for Supply Management, 2001.

Muller, Max. *Essentials of Inventory Management*. New York: AMACOM, 2003.

Oliver Wight International. *The Oliver Wight Class A Checklist for Business Excellence*, 6th ed. Hoboken, NJ: John Wiley & Sons, Inc., 2005.

Seifert, Richard. *Collaborative Planning, Forecasting, and Replenishment: How to Create a Supply Chain Advantage*. New York: AMACOM, 2003.

Stock, James R., and Douglas M. Lambert. *Strategic Logistics Management*, 4th ed. New York: McGraw-Hill, 2001.

Wincel, Jeffrey P. *Lean Supply Chain Management: A Handbook for Strategic Procurement*. New York: Productivity Press, 2003.

CHAPTER 11

Demand Management and Logistics

The Key to Supply Chain Management

Supply management is the key to supply chain management (SCM). While many functional areas lay claim to the emerging field of SCM, supply management has the strongest claim. The argument is simple. Since supply professionals are responsible for developing and managing the contractual, behavioral interrelationships between supply chain members, supply management is the critical function responsible for managing the supply chain.

The issue of how to harness the power of SCM is creating debate in upper management boardrooms and academic classrooms across the world. Firms need to come to terms with how they are going to improve their competitiveness in the future through SCM. Competition is not just firm vs. firm, but chain vs. chain (or network vs. network.)

The SCM Triangle

As companies gear-up for this chain vs. chain struggle, the need to manage the supply chain effectively is becoming increasingly critical. We believe that successful strategic SCM consists of three critical components: world-class supply management, demand management, and logistics management.

Evolution to Strategic SCM

Currently, at least three stages of SCM exist: (1) SCM is the management of the internal supply chain, (2) SCM is supplier-focused, and (3) SCM is the management of a network of enterprises, which includes the customer as well as suppliers. Most firms invest their resources sequentially. Generally, firms evolve from the first stage through the last over the course of several years of effort to improve SCM. World-class SCM encompasses all three stages operating in parallel.

Internal Supply Chain Focus

In this stage, the first priority of a business enterprise is to integrate and optimize its own operations before making any attempt to extend supply chain rationalization to external organizations. The internal customers need to acknowledge the presence of the supply management organization when they are implementing the product development and planning process.[1] Therefore, SCM may be defined as the integration of previously separate operations such as marketing, engineering, and operations, within a business enterprise.

Historically, initial attempts at internal functional integration followed the materials management concept. Materials management was used to integrate activities such as purchasing, inventory management, material control, stores, warehousing, materials handling, inspection, receiving, and shipping. Further internal integration was achieved by linking information systems between sales order entry, production planning and control, and distribution. The focus on MRP/MRP II planning and control systems, along with the alignment of customer demand and supplier response, led to further functional integration.[2] In the 1990s, enterprise resource planning (ERP) systems were developed to complete internal information system integration and automate many activities.

A majority of firms attempting to engage in SCM are still preoccupied with the internal integration of functional activities and material and information flows. The real potential of SCM can be realized only after external integration of customers, key suppliers, and information flows are attained.[3]

Supplier Focus

The supplier focus stage of SCM emphasizes the importance of fostering long-term, collaborative relationships with key suppliers. Collaboration with suppliers has been the focus of much of the supply management and contract management literature in recent years.[4] The thrust of this thinking is based on a belief in the elimination of adversarial relations between buyer and seller and the formation of long-term relationships with fewer suppliers to gain the benefits of economy of scale, reliability, and quality.

Often overlooked in the supplier-focus stage of SCM are the potential benefits derived from expanding the role of key suppliers in the supply chain. The expertise of key suppliers can provide technical and supportive assistance at a nominal cost when secured through purchased services.[5]

Network Management Focus

The network management focus stage of SCM recognizes the complex nature of business relationships. The drivers of this perspective emphasize outsourcing of nonstrategic functions and processes. The high level of outsourcing often results in greater physical separation of operations and functional areas. In such an environment, information technology becomes critical to maintain communications.

A network view helps one visualize the true relationships between supply chain entities. The network view aids in developing an appreciation of the range of significantly different products, processes, markets, geographical markets, and time that are concurrently present in any supply chain.[6] The network perspective properly focuses upon a reality wherein multiple supply chains exist within the same network.

Strategic Demand Management

A major driver of the recession of 2001–2002 was the proliferation of errors in forecasts, which were not identified, analyzed, and acted upon until billions of dollars worth of inventories had accumulated in some supply chains. The primary reason for the inaction on the part of supply chains that were caught with excessive inventory levels was the absence of demand management. Demand management is used to estimate, control, smooth, coordinate, balance and influence the demand and supply for a firm's products and services, in an effort to reduce total costs for the firm and its supply chain. Demand management recognizes that forecasts are developed at several points throughout an organization. Demand management does not develop forecasts. Rather, it accepts forecasts from other functions and updates them on the basis of actual, real-time demand. Demand management's need to reconcile forecast errors with the actual order rate of an enterprise is one of the most overlooked potentials in the successful management of inventory levels, customer satisfaction, staffing strategies, and facilities expansion or contraction.

In addition, demand management works with the supply side to adjust the inflow of materials and products. Demand management is "a collaborative process that involves accurately determining how much product needs to be produced (the demand) at each level of the supply chain through the end customer."[7] It is responsible for creating a smooth master production schedule and for smoothing production after schedules have already been released to internal production and external suppliers. Smoothing requires that demand managers recognize that demand management is a process that requires the utmost in coordination and communication between the responsible parties. Demand can change daily, weekly, and monthly. Demand managers must have contingency plans developed with supply chain members to allow modification of short-term schedules when necessary. Demand management also balances the total costs of not meeting demand against the total costs of adding additional resources required to meet demand.

Figure 11.1 shows the primary information flows required for demand management. No single figure can completely capture the full complexity of demand management activities and information flows between a single buying firm and a single supplying firm.

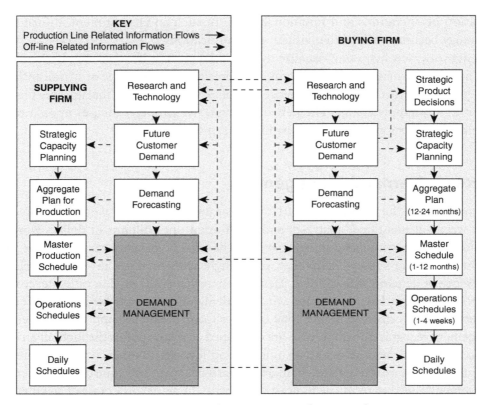

Figure 11.1 Demand management information flows.

The complexity increases exponentially as the "real world" situation of multiple suppliers and buyers is included.

Figure 11.1 assumes that all firms have dedicated demand managers who are actively involved in demand decisions throughout the life cycle of the product or service. Demand managers are involved with product and service design teams in the early stages of design. Early involvement in design provides a critical opportunity for demand management to convey the "voice of the customer" on need issues and provide proactive contributions on strategic product mix decisions. Subsequent to design, demand managers must work tightly with internal marketing and production and supply managers to ensure that production planning at the strategic capacity and aggregate planning levels is communicated to and reconciled with suppliers.

At the demand planning level, demand management work with production and supply managers to reconcile the many problems inherent in the master schedule, which is based on the forecast and actual orders. The basic objectives of demand management are to create close collaboration among all trading partners and to unify the demand and supply requirements of the business.[8] Effective demand management takes tactical and strategic corrective actions as required to bring forecast demand in line with actual demand from the marketplace. Tactical level corrective actions require adjustment to the master schedule. At a strategic level, integrated operational planning "links three or more supply chains working together to determine the end-users needs and then taking action up and down the chain to optimize operations."[9] If these adjustments to the schedule are not effectively managed, they can result in drastic fluctuations producing a bullwhip effect throughout the supply chain.

The Bullwhip Effect

Failure to estimate demand accurately and share information among supply chain entities can result in bloated inventory levels due to a cumulative effect of poor information cascading up through a supply chain. Poor demand data forces the supplying firm to either carry additional inventory or increase lead times to account for the uncertainty. Either way, inventory levels in the supply chain are increased. If lead times are increased, then the buyer (based on conventional reorder point calculations) will increase order quantities. The supplier will interpret the increase in the order quantity as increased customer demand. The supplier then will need to take action to increase capacity to meet the fictional trend. To add more irony, just as the supplier has added capacity to meet the increase in demand, demand falls off because the buying firm has excessive stock available. The supplier then will need to reduce its capacity through firing, selling assets, or some other approach. The problem of fictional or "phantom" demand has been termed the bullwhip effect in SCM. The following example presents another way that the bullwhip effect can occur.

Bullwhip Effect Example[10]

An extreme example of this is the behavior of an individual employed as a sales representative at a large tobacco firm in Richmond, Virginia. Every evening he would telephone 10 tobacco retailers out of several hundred to obtain reports on the day's sales of his firm's cigarettes. Each of the retailers believed that the frequency of these queries was an indication of likely increased demand for cigarette products, and they all increased their orders with the tobacco firm. The sales representative noted the increase in retail orders, and thus modified his own sales forecast input to the factory.

The manufacturing manager at the cigarette factory observed the upsurge in retail demand in the form of actual orders as well as his sales representative's modification (upward) of the forecast. He planned the addition of a third shift operation for the next month, informed supply management to order more materials, and asked human resources to hire additional workers for the next month. To meet the immediate upturn in demand, he authorized the use of overtime and expedited raw tobacco deliveries to fill the increased orders from the retailers. The cigarette firm had increased its capacity and its orders with its tobacco producers. The firm's tobacco growers concluded that the increase in orders from the factory reflected an increased market demand for cigarettes. Accordingly, they made plans to expand planting acreage for the next season, and purchase additional harvesting equipment and hire additional casual labor.

However, toward the middle of the following month, tobacco retailers noticed that their stocks of cigarettes were not moving from the shelves as they had expected. In fact, sales were declining because of ongoing antismoking campaigns and the increase in the cost of tobacco products resulting from increases in tobacco product taxes. Swamped with unsold inventories of cigarettes, the retailers called the sales representative and cancelled virtually all outstanding orders. They maintained that they now had many weeks of inventory on the shelves and that they must work off the inventory before placing additional orders. The tobacco firm's sales manager reacted quickly by slashing the sales forecast. The manufacturing manager reacted to the change in the sales forecast and the cancellation of retail orders by eliminating the third and second shifts, and furloughing all of the newly hired workers, as well as eliminating all overtime. Supply management cancelled its orders for raw tobacco with the growers, who in turn cancelled their orders for new harvesters, labor, and so on. Everyone involved in this scenario wondered, "What went wrong?"

Evolution of Strategic Demand

Forecasting has often been viewed as being minimally more accurate than a glance into a crystal ball. Historically, many organizations have employed manual or visual systems of order replenishment. Such approaches generally fall under the category of reorder point replenishment (ROP) techniques. Later, calculations on the effects of set up, holding and carrying costs in addition to purchased and manufactured lot size quantities were considered and resulted in replenishment calculations known as statistical reorder point techniques. With the introduction of the computer into the manufacturing and distribution environment, material requirements and distribution requirements planning (MRP and DRP) were able to schedule orders for products against lead times and bills of materials. The resultant effect was a drastic decrease in inventory at all levels (finished goods, work-in-process, components, and raw materials) held in the production system. It is important to remember that wise forecasters combine many different forecasts using multiple quantitative and judgment methods and they keep adjusting for new developments.[11]

Further integrative refinements (known familiarly as manufacturing resources planning and distribution resources planning—MRP II and DRP II) of computer-based information systems provided capabilities to manage production capacity as well as the demand for materials. These planning and information systems enabled business enterprises to effectively gauge the financial impact of various inventory-customer service-capacity decisions, and to run "what-if" simulations of various material and capacity scenarios, without the risk of inventory, labor, equipment, or facilities commitment.

Later, the many influences of planning techniques such as Just-in-Time (JIT) manufacturing and its requirements for lean operations, total quality, continuous process improvement, and the elimination of waste in all forms further defined the need for many functional disciplines to communicate and collaborate in ways never before envisioned.

Most recently, many businesses have purchased and installed enterprise information systems. The potential of such enterprise systems lies in the seamless integration of the many databases typically found in any firm. Thus equipped, the prototypical planning and control information systems applied to production and distribution have become known as the step beyond MRPII—enterprise resource planning (ERP).

While highly integrated enterprise planning systems and the application of e-Commerce, e-Procurement, and business-to-business (B2B) capabilities have manifested themselves in recent years, by themselves they do not address the fundamental questions posed by effective demand management. For instance, thorny issues often arise when trade-off issues are considered. When the following questions arise as a firm examines the levels of customer service in terms of inventory on-hand, there is often no clear solution.

- ▲ What is enough inventory?
- ▲ What is too much inventory?
- ▲ What are the cost implications?
- ▲ What are the effects on customer service levels?
- ▲ What short- and long-range capacity management decisions must be made to address demand (i.e., overtime for line workers or outsource to contract manufacturers)?

With traditional manual replenishment techniques, and later disaggregated legacy information systems, the answers to these vital questions were most often not available. There were simply not enough resources on hand (both people and machines) to answer such questions easily or quickly. With the advent of enterprise information systems, and the emergence of the transparent and electronic seamless transfer of planning information between businesses, the true potential of sharing planning information throughout the supply chain may finally be realized.

One of the authors worked as a Gantt analyst in a large manufacturing company. He soon discovered there were three different forecasts in the company; the sales forecast, the production control scheduling forecast, and the materials-purchasing forecast. The way they resolved the question of which numbers to follow was by establishing a joint forecasting committee to massage the conflicting numbers into one master plant forecast.

Forecasting Demand

The most important output of demand management for a particular product or service is an accurate forecast of customer demand. What is a forecast? In terms of SCM, a forecast is an estimate of future demand. In other words, it is a calculated guess or estimate about the future demand for a firm's products and services under conditions of uncertainty. Forecasts fall into two categories: quantitative and qualitative.

Quantitative methods require mathematical analysis of historical data. Common mathematical approaches based on historical data are regression analysis, moving averages, and exponential smoothing. A frequently favored forecasting method by managers who have little knowledge of forecasting techniques is the naïve method, where the last period's historical value becomes the forecast for the next period. However, historical data may not be complete or available and, obviously, the past may not be repeated in the same numerical form.

Qualitative forecasts are created subjectively, using estimates from sources such as market surveys, in-depth interviews, and experts. When historical data are available, qualitative forecasting usually is employed to verify or adjust quantitative forecasting methods. In some cases, historical data are not available and qualitative forecasting is the only alternative.

In most organizations, forecasting starts in the marketing department or with an economic forecasting staff. In other words, we take our cue from the sales forecast. For example, this forecast is the source for the master schedule in MRP and ERP programs.

Why are forecasts necessary in SCM? The primary reason is that lead times exist for production, distribution, and services. If lead times were zero, then demand could be met as it arose. Since lead times do exist, supply chains must operate based on forecasts.

Forecasting Fundamentals

1. Accurate forecasts are based on expected results from the execution of strategic and tactical operational activities. They are based on assumptions of the internal and external corporate environments within the particular industry. These internal factors include labor productivity, facility and equipment production rates, and other factors under control of the firm. The external factors include competitive forces, regulatory controls, technology changes, world economic conditions and stability, climate and weather, supplier performance, and other future events not under control of the firm. If these environments change, and they most certainly will by varying degrees, then the plans must change and so must the forecast. This means we must constantly monitor the environment in which we operate.[12] Usually, the marketing department is the key player in this area with input from all the stakeholders. The stability of prices and supply

availability for our critical materials and components would be a major input from purchasing. Does your purchasing department have this input in both the planning and budgeting functions?

2. As previously noted in the proceeding chapters, share the forecast with the key players in the channel of distribution (the supply chain) and obtain their input. As mentioned previously, make sure you achieve one forecast by collaboration; often a cross-function team is necessary to resolve the different estimates.

   For consumer high-velocity manufacturers sending products through various channels of distribution to point-of-sale retailers, a total supply network system called collaborative planning, forecasting, and replenishment (CPFR) helps to ensure high pipeline efficiency.[13] Using CPFR, both long- and short-term information regarding point-of-sale data, forecasting, shipping, production plans, and order generation is jointly planned by key trading partners using Web-based collaboration software, creating a glass pipeline where all relevant information is shared in real time.[14] The obvious users are Walmart, Safeway, Ace Hardware, Procter & Gamble, and other similar global giants.

3. Contrary to popular belief, quantitative forecasting is widely used as advanced computers and software have made those applications easier to use and more user-friendly. The Institute of Business Forecasting (IBF) offers seminars, workshops, and networking (see www.ibf.org). Software such as Forecast X help take the mystery out of the use of various forecasting tools such as moving averages and exponential smoothing, seasonal indices, multiple regression, time series, box-Jenkins, and event modeling.[15]

4. Use multiple forecasting methods and then arrive at a range of possible outcomes. Have the final review of the numbers done by one or more experts on analysis, with a deep understanding of the various markets and the external and internal environments within the industry on a global scale. Make sure everybody is reading market intelligence reports such as "Global Business Trends" and the "ISM Report on Business" in each issue of *Inside Supply Management* (www.ism.ws). These market experts can identify flaws in the numbers, that is, regardless of the math used, the numbers can be distorted by such elements as using an average with large ± spikes in the individual number inputs.

5. Remember most statistical techniques use correlation data, not experimental design methods, to determine cause and effect. They also assume the past trend will continue, which may or may not be true. We must brainstorm what might/could change in the future.

6. Many forecasts are merely educated guesses by managers not trained in the game of forecasting. They simply state, "We will produce 10% more of X next year," with no backup analysis except their "gut feeling." If no one in your organization has forecasting training, one possible solution is to seek help from your local college. Find

out who is teaching marketing research, statistics, and hopefully forecasting. Most of these instructors welcome the opportunity to have a real class project and the costs are nominal. Of course, you may have to hire the professor as a consultant.

Hopefully, you will be able to get help in-house from marketing or industrial engineering departments, and the large firms usually have at least one professional forecaster. Better forecasting will reduce the danger of stockouts, expediting, excessive inventory, low inventory turnover, late or partial shipments, and lower operating costs while increasing your customer satisfaction levels. The more time and skill we apply to forecasting, the better the results in our total operation, that is, it is a major step to reduce risk of failure. As the late and great professor of forecasting at the University of Wisconsin, Dr. Harry Dean Wolfe usually said at the end of his course, "remember, don't try to find the 10th decimal point when you don't know the whole number."

Planning with Time Fences

Many companies reduce uncertainty in demand by establishing time fences. A time fence reflects management decisions regarding production and supplier commitments about change allowed to the scheduling of materials and capacity elements. Figure 11.2 presents three typical approaches to time fences: frozen, slushy, and fluid.

Typically, the demand time fence is for a short period, such as the current production month. The demand time fence establishes planning rules, which if broken, may prove very disruptive and costly to a firm. The most common rule is that production schedules

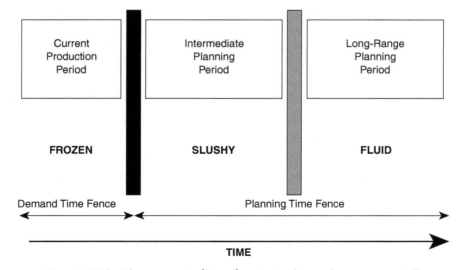

Figure 11.2 The concept of time fencing in demand management.[16]

that are within the demand time fence cannot be changed, even if the demand forecast is changed. Thus, schedules in the demand time fence are often said to be "frozen," that is, made very difficult or even impossible to change without top management approval.

Within the "slushy" time fence, many commitments for a firm's material, capacity, capital equipment, and related financial resources can be made on an advisory basis. The intermediate period is usually two to six months into the future. Some actions cannot be delayed until the frozen period. For example, long lead-time materials may need to be ordered.

"Fluid" time fences acknowledge that greater uncertainty exists as forecasts are extended further out in the planning horizon. Commitments for materials, capacity, capital equipment, and related financial resources are highly subject to change. The chief advantage of planning during the "fluid" time fence for materials, capacity, capital equipment, and finances is to provide as much forward visibility as possible. In SCM, such forward visibility about the changing nature of demand is of critical importance for suppliers so that they may effectively manage their own resources. Buying firms that communicate planned orders on a continuous basis with their supply chain members will achieve a distinct advantage over those firms that do not share their plans.

Implications for Supply Management

Since a primary role of demand management is in coordinating demand between the buying and supplying firm, demand management must work effectively with supply management and the supply chain management. Demand management requires open, honest, trusting, and collaborative relationships between customers and suppliers throughout the supply chain. This point underscores the critical importance of trust and relationship management in effective SCM.

The greatest potential—and challenge—for effective supply and demand management is to achieve collaborative planning, forecasting, and replenishment driven by actual customer demand. Changes in demand patterns suggest that the formal planning evolutions once largely confined to processes and functions within a firm must be shared with all supply chain members, particularly downstream customers.

Demand management requires a shift from a focus on component and commodity level planning to one of strategic product and subassembly level planning. Demand and supply managers must become involved in strategic planning if they are to be relevant in the management of the supply chain.

Strategic Logistics Management

Logistics management forms the third side of the SCM triangle. Logistics professionals play an important role in the success of SCM in the management of transportation, storage, and warehousing activities. Unfortunately, many companies define logistics as synonymous with the term SCM, thus ignoring the contributions and roles of supply management and demand management.

Logistics Defined

Logistics management deals with the handling, movement, and storage activities within the supply chain, beginning with suppliers and ending with the customer. One of the best-selling books in logistics states:

> Logistics is the part of the supply chain process that plans, implements, and controls "the efficient, effective flow and storage of raw materials, in-process inventory, finished goods, services, and related information from point of origin to point of consumption (including inbound, outbound, internal and external movements) for the purpose of conforming to customer requirements."[17]

The Role of Logistics in Supply Chain Management

As the competitive context of business continues to change, logistics activities and processes must be integrated into the strategic-level thinking and planning. "From procurement, to inventory management, inbound supplies and materials, and outbound distribution, logistics is the performance that links them together physically."[18] Because of the critical role of logistics, much of the focus on logistics has been on reduction of cycle times in logistics activities. Cycle-time reduction and the elimination of waste in logistics processes have had a direct correlation to the enhancement of customer satisfaction. Customer needs vary, and firms can tailor logistics systems to serve them better—and more profitably.

Logistics and supply management will realize their greatest gains in efficiency and effectiveness through collaboration. Only then can both traditionally separate disciplines achieve a world-class supply chain. The major professional organization in logistics is the Council of Supply Chain Management Professionals in Chicago, Illinois.[19]

Eliminate the Warehouse, If Possible

Sun Microsystems (now owned by Oracle Corporation), in Santa Clara, California, has developed what it calls a "one-touch supply chain," which eliminates the distribution warehouse component of its supply chain.[20] Under the old system, the supplier (in Asia) built and boxed the server board and shipped it to an external manufacturing hub and then to a third-party integration-box configuration center, which sent the product to a distribution hub, which sent it to a channel and reseller for shipment to the final customer. Under the new system, the board and box go directly to the customer for standard products and to a geographic configuration center for custom orders. Sun now ships more than 50% of the orders directly to customers from suppliers, and that has reduced logistics costs by 20% and finished goods inventory by as much as 40% and increased customer predictability performance by more than 10%. Interestingly, Sun had to fool the ERP system into believing the material was on-site in order to ship and invoice the order. The lesson here is never be a slave to an ERP system.

Summary

For many in the field of SCM, the future holds substantive and far-reaching changes. These changes are largely driven by the following trends:

- ▲ Institutionalization of the SCM perspective
- ▲ Increasing emphasis on supply chain relationships
- ▲ Increasing emphasis on the long-term view
- ▲ Use of information technology to enhance supply chain communications
- ▲ Use of information technology to foster rapid decision making
- ▲ An increasing focus that looks "outward" toward the intricacies of supplier and customer relations
- ▲ The emergence of the supply management professional as a "manager and facilitator of relationships" vs. an information broker whose attention is defined by commodity knowledge

A thorough grasp of the continually evolving perspectives of SCM is a necessary component of the skill sets of any proactive supply management professional. Such perspectives recognize the continuing need to integrate the many competencies traditionally resident in other functional disciplines, particularly demand management and logistics. Further, successful supply chain optimization depends on the coordination of cross-functional competencies in cross-enterprise teams. The supply management professional is the logical "team leader" for such initiatives.

Endnotes

1. Peter E. O'Reilly, "Attaining a World Class Supply Management Organization through Strategic Initiatives" presented at the 89th Annual International Supply Management Conference, Philadelphia, PA, (April 2004).
2. Martin Christopher, *Logistics and Supply Chain Management: Strategies for Reducing Cost and Improving Service* (London: Financial Times–Pitman Publishing, 1998): 18–19.

3. Christine Harland, "Supply Chain Management." *Blackwell Encyclopedic Dictionary of Operations Management*, ed. Nigel Slack (Oxford, England: Blackwell Publishers, Ltd., 1997): 213–215.

4. Richard L. Pinkerton, "The Evolution of Purchasing to Supply Chain Management," *Business Briefing: European Purchasing and Logistics Strategies* (London: World Markets Research Centre, 1999): 16–28 and reprinted in *The Purchasing and Supply Yearbook, 2000 Editorial*, ed. John Q. Woods, and National Association of Purchasing Management, New York: McGraw-Hill.

5. O'Reilly, op. cit.

6. Harland, 213–215.

7. Steve Miller, "21st Century Logistics: Harnessing the Demand Chain" (presented at the 86th Annual International Conference Proceedings–2001. Institute for Supply Management (ISM) Orlando, FL).

8. John Yuva, "Demand Management: Creating Balance Through Collaboration," *Inside Supply Management* 15(3)(2004): 20.

9. Miller, op. cit.

10. The problems identified with the bullwhip effect and false demand were first identified as "The Phoney Backlog," by the late Oliver Wight in the late 1970s. See Oliver W. Wight, (*MRP II: Unlocking America's Productivity Potential* (Williston, VT: Oliver Wight Publications, Ltd. 1981): 33–38.

11. Philip Kotler, *Marketing Management, Analysis, Planning, Implementation, and Control*, 10th ed. (Upper Saddle River, NJ: Prentice Hall, 2000).

12. See Richard L. Pinkerton, "Industry Analysis for Materials Management: NAICS Industry Classifications," *ISM Material Management Newsletter*, ed. Sheila D. Petcavage, 1(4)(2005): 12–15.

13. Dirk Seifert, *Collaborative Planning, Forecasting and Replenishment: How to Create a Supply Chain Advantage* (New York: AMACOM, The American Management Association, 2003), From the front cover jacket inside based on original research at the Harvard Business School. This book includes 38 case histories plus the 9 critical steps to successfully implanting CPFR.

14. Ibid.

15. J. Holton Wilson, Barry Keating, and John Galt Souttons, Inc., *Business Forecasting with Forecast X™* (Burr Ridge, IL: McGraw-Hill/Irwin, 2009): 2–3.

16. Thomas E. Vollmann, William L. Berry, and D. Clay Whybark, *Manufacturing Planning and Control Systems* (New York: MgGraw-Hill, 1997).

17. J.J. Coyle, Bardi, E.J., and R.A. Novak, *Transportation*, 6th ed. (Thompson South-Western, 2006).

18. Roberta J. Duffy, "Logistics: A Custom Link," *Purchasing Today* 10(3)(1999): 22.

19. Founded in 1963 as the National Council of Physical Distribution Management, the council changed its name in 1985 to the Council of Logistics Management, and again in 2005 to the Council of Supply Chain Management Professionals.
20. John Yuva, "A One-Touch Supply Chain: How Sun Microsystems Streamlined Its Inventory Process," *Inside Supply Management*, December (2007): 25–27.

Suggested Reading

Ballou, Ronald H. *Business Logistics/Supply Chain Management*, 5th ed. Upper Saddle River, NJ: Pearson/Prentice Hall, 2004.

Bechtel, Christian, and Jayanth Jayaram. "Supply Chain Management: A Strategic Perspective." *International Journal of Logistics Management* 8(1)(1997): 15–34.

Bender, Paul. "Going Around the World Makes the (Business)World Go 'Round." *Inside Supply Management* 17(2)(2006): 34.

Bowersox, Donald, David Loss, and M. Bixby Cooper. *Supply Chain Logistics Management* New York: McGraw-Hill/Irwin, 2005.

Bruning, Bob, and Rich Hoole. "Adding Value through Demand Management." *Inside Supply Management*, 16(6)(2005): 8.

Cavinato, Joseph L. "The Latest Trends in Logistics" *Purchasing Today* 11(5)(2000): 76.

Chase, Robert B., F. Robert Jacobs, and Nicholas J. Aquilano. *Operations Management for Competitive Advantage*, 10th ed. New York: McGraw Hill, 2004.

Chopra, Sunil, and Peter Meindl. *Supply Chain Management: Strategy, Planning, and Operation* Upper Saddle River, NJ: Prentice Hall, 2004.

Cohen, Shoshanah, and Joseph Roussel. *Strategic Supply Chain Management: The Five Disciplines for Top Performance.* Chicago, IL: McGraw-Hill, 2005.

Cohen, Shoshanah. "Forging the Chain." *American Executive* May (2005): 28–30.

Cook, Thomas A. *Global Sourcing Logistics: How to Manage Risk and Gain Competitive Advantage in a Worldwide Marketplace.* New York: AMACOM, 2007.

Cooper, M. C., and M. Ellram. "Characteristics of Supply Chain Management and the Implications for Purchasing and Logistics Strategy. *International Journal of Logistics Management* 4(2)(1993): 13–24.

Coyle, J.J., E.J. Bardi, and R.A. Novak, *Transportation* 6th ed. Mason, OH: Thompson South-Western, 2006.

Emmett, Stuart. *Supply Chain in 90 Minutes (Integrating the Flow of Goods and Information from Supply to Demand.* Gloucestershire, UK: Management Books, 2005.

Hoole, Rick. "Demand Management—Optimizing the Demand-Side Drivers of Total Cost of Ownership." Paper presented at the 89th Annual International Supply Management Conference, The Institute for Supply Management (ISM) Tempe, AZ, April 2004.

Krajewski, Lee J., Larry P. Ritzman, and Manoj K. Malhotra. *Operations Management: Process and Value Chains*, 8th ed. Englewood Cliffs, NJ: Prentice Hall, 2007.

Lambert, Douglas, M. *Supply Chain Management: Processes, Partnerships, Performance*, 2nd ed. Sarasota, FL: Supply Chain Management Institute, 2006.

Lambert, D. M., and R. Burduroglo. "Measuring and Selling the Value of Logistics." *International Journal of Logistics Management* 11(1)(2000): 1–17.

Penton Media's, Supply Chain Group, New York.

Penton Media's, *Logistics Today Magazine* and *Supply Chain Technology News*. See www.penton.com or call 212-204-4200.

Seifert, Dirk. *Collaborative Planning, Forecasting, and Replenishment: How to Create a Supply Chain*. New York: AMACOM, 2003.

Tracy, John J. Jr. "You Can See Eye to Eye." *Purchasing Today* 10(3)(1999): 12.

Wight, Oliver. *The Executive's Guide to Successful MRP II (Oliver Wight Manufacturing)*. New York: John Wiley & Sons, 1993.

Van Hoek, Remko I. "The Purchasing and Control of Supplementary Third-Party Logistics." *The Journal of Supply Chain Management* 36(4)(2000): 14.

Yuva, John. "Collaborative Logistics: Building a United Network." *Inside Supply Management* 13(5)(2002): 42.

Yuva, John. "Demand Management: Creating Balance Through Collaboration." *Inside Supply Management*, March (2004): 20–27.

CHAPTER 12

Outsourcing

Outsourcing: A Growth Industry

The use of outsourcing by both manufacturing and service industries is increasing rapidly. An increasing number of business functions are being outsourced. To meet competitive challenges, corporations are outsourcing to highly specialized firms that can use their expertise to increase the efficiency of the outsourced function. The increase in outsourcing has resulted in lower staffing levels, reduced costs, and more flexibility. Outsourcing is more than a means of cutting costs; it provides an opportunity to achieve innovation. Managers are transforming from traditional roles into brokers or facilitators of outsourced activities. The "make/buy" decision continues to be one of the key strategic issues and options confronting the purchasing function.[1]

Strategic Issues

The starting point most firms use in conducting the strategic outsourcing analysis is to identify the major strengths of the firm and then build on them. Senior management needs to ask, "What is it we really do better than most firms?"

A firm's competitive advantage is often defined as cost leverage, product differentiation, or focus. It is important to perform a competitive analysis before initiating the outsourcing analysis. A competitive analysis will provide a report of the firm's strategic position relative to the market, industry, and its competitors.

Core Competencies

Do our strengths lie in certain design skills, unique production skills and equipment, or different types of people skills? A thorough investigation of these types of questions is what many people today call identifying the firm's existing core competencies. The next step in the process is to look at the current and expected future environment in which the firm operates, including the competition, the governmental regulatory climate, and the changing characteristics of sales and supply markets. Subsequently, the bottom line question management must answer is, "Precisely what business do we really want to be in to maximize the use of our core competencies as we proceed into the future?"

Once a clear answer to this question has been formulated, the supply manager must identify expected competency requirements necessary for future operations. Competency requirements are then compared with existing core competencies to determine which ones need to be refined and which ones need to be supplemented with related competencies that must be developed to create a competitive advantage. Two researchers place these ideas in sharp focus when they say, "Senior managers must conceive of their companies as a portfolio of core competencies rather than just as a portfolio of businesses and products."[2] The products and the nature of the business flow from the core competencies.

In considering what to make and what to buy, the decisions should nurture and exploit the firm's core competencies. The items or services that should be made or done in-house are those that require capabilities that are closely linked with the core competencies and are mutually reinforcing, as opposed to those that can be separated. This is the fundamental strategic consideration that guides the original make-or-buy decisions that ultimately shape the character of the firm.[3]

Supplier Dominance

Chris Lonsdale at the Centre for Business Strategy and Procurement at the University of Birmingham in the United Kingdom observes that a majority of the problems outsourcing firms have experienced can be traced to suppliers who exploit the leverage they gain through the relationship. Lonsdale writes,

> The significance of asset specificity for outsourcing is that if activities that require "transaction-specific investments" are outsourced, the firm will find itself locked into its supplier, as it will not want to write off those investments by revisiting the market. This lock-in can then be exploited by the supplier, by renegotiating the terms of the contract or insisting on different terms next time around. This post-contractual lock-in (dependency) will cause the power relation between the two parties to change—the situation can become one of supplier dominance.[4]

We share the concerns of Lonsdale and his colleague, Andrew Cox, with the potential for suppliers to exercise the power they gain through such relationships. However, we believe that a carefully crafted and managed alliance will prevent such problems.

The Creation of Strategic Vulnerabilities

Michael E. Porter, author of the landmark book *Competitive Strategy*, inserts a timely cautionary note by observing that

> when you outsource something, you tend to make it more generic. You tend to loose control over it. You tend to pass a lot of the technology, particularly on the manufacturing or service delivery side, to your suppliers. That creates strategic vulnerabilities and also tends to commoditize your product. You're sourcing from people who also supply your competitors.[5]

The Dangers of Vertical Integration

If a decision to "make" or "in-source" results in vertical integration, the critical connection between output and rewards is broken. Cost and responsiveness both suffer. Vertical integration frequently results in a loss of flexibility and responsiveness.

Horizontal Integration

A general trend in competitive strategy is emerging in which all noncritical activities are outsourced to achieve significant cost advantage. Known as critical dependencies, these activities are the same in other businesses and are not a unique part of the firm's product. For this approach to be successful, the minimum resources and value add activities that are key to supporting the firm's core competencies must be identified and defined. All other activities (called noncore activities) are potential candidates for outsourcing.

Companies are outsourcing a large range of services, including customer service, warehousing, training, and travel. Horizontal integration, often referred to as a "virtual corporation," involves outsourcing nearly everything except a few core activities. Companies that separate intellectual activities from the resulting processes and then outsource those processes create virtual corporations. Cisco Systems is an ideal example of this type of organization. The company outsources most of its manufacturing, order fulfillment, and distribution. New product development is outsourced to small companies that frequently are acquired by Cisco Systems. This company has developed a competitive advantage over rivals and established a standard for the virtual corporation. The company is able to provide its products faster and at a lower cost than its competitors due to its flexibility.

New Product Development and Outsourcing

The option to make-or-buy is first presented at the beginning of a product's life cycle, during new product development. The process of designing a new or modified product often is accompanied by new technologies, minimal information, and a high level of uncertainty. It is important that supply managers conduct an analysis that is based on the availability of resources in terms of timeliness and optimal cost. Extensive supplier market research should be conducted on new technologies and innovations. The firm's technological core competencies must be identified and defined. Often suppliers will develop technology and new innovations that are beyond the reach of the firm's core competencies. To take advantage of new technology, many manufacturers outsource development to suppliers. By utilizing suppliers during the development of new and modified products, manufacturers are gaining a competitive advantage and developing dependencies. To maintain a technological competitive advantage, manufacturers need suppliers to continue developing innovative designs.

A great deal of project planning is involved in the make-or-buy decision for new product development. Although outsourcing is considered a long-term activity, it can be considered in the context of finite projects. A project's strict deadlines may force a firm to make the decision to outsource. Coordination of the development activities of selected suppliers may be necessary depending on the number of suppliers involved and the type of development required. The supply manager will need to coordinate horizontal suppliers producing parts that will affect the product as a whole. Vertical coordination is necessary for suppliers interacting on different tiers when technological collaboration is required.[6]

Lean Manufacturing

The term "Lean manufacturing" is a macro term used to describe all the methods, techniques, procedures, and tools to eliminate waste and "do it right the first time." In this chapter, we apply the term to the question of outsourcing.

If one steps back to assess the current situation in American industry, it is clear that the concept of Lean manufacturing is widely embraced for competitive purposes. A rule of thumb used by some firms is to outsource subsystems and components unless they fall into one of the following three categories:

1. An item that is critical to the success of the product, including customer perceptions of important product attributes
2. An item that requires specialized design and manufacturing skills or equipment and the number of capable and reliable suppliers is extremely limited
3. An item that fits well within the firm's core competencies or within those the firm must develop to fulfill future plans

Components or subsystems that fit into one of these categories are considered strategic and are produced in-house if possible. The analytical procedure used in making these decisions is straightforward and is shown in Figure 12.1.

Deal first with subsystems of the product:

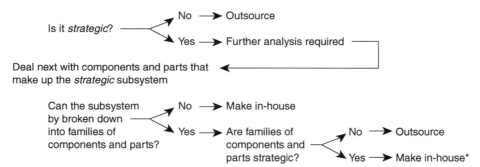

*In some cases, in the short run it is not possible to make such an item in-house. This may be due to budget constraints, capability problems, capacity limitations, in-house technical skills, equipment needs, and so on. In these cases, until the problem is resolved, the item must be outsourced under a carefully crafted and managed contract.

Figure 12.1 Analyzing strategic outsourcing decisions.

If the analysis to this point indicates that a make decision is desirable from a strategic point of view, before the decision is made several additional factors must be analyzed. These practical considerations focus on a comparison of the firm's present situation with that of potential suppliers with respect to the matters of design, manufacturing, and quality capabilities. Similarly, relative costs and volume requirements also need to be compared and evaluated as supplementary information to be used in conjunction with the strategic analysis in reaching a final decision.

This, then, is the approach used at the strategic level of analysis to determine whether a firm should make an item or outsource it. These are the crucial decisions that to a great extent shape the destiny of the firm.

Let us turn now to make-or-buy decisions that are made at the operating level.

Tactical Decisions

After the strategic make and outsourcing decisions are finalized and as operations progress, a number of situations inevitably arise that require additional make-or-buy analyses at something less than a strategic level. Unsatisfactory supplier performance in the case of some outsourced items, cost considerations, changing sales demands, restricted manufacturing capacity, and the modification of an existing product are just a few of the operating factors that generate these needs. As a general rule, from a make perspective these tactical make-or-buy situations involve items for which the firm already has most of the necessary production resources. Small investments in tooling, minor equipment, or a few additional personnel usually are all that would be needed to do the job in-house. Consequently, these investigations tend to be driven by operating considerations of efficiency, control of quality and reliability, cost, capacity utilization, and so on.

In any case, the make-or-buy possibility requiring only a small expenditure of funds in the event of a make decision is the type most commonly encountered by supply managers. A decision of this type usually does affect a firm's resource allocation plans; however, its effect on the firm's future is minimal compared with a decision requiring a major capital investment. Although the decision requiring a nominal expenditure of funds does not require direct top-management participation, it does require coordinated study by several operating departments—perhaps using a team approach. Top management's responsibility is to develop an operating procedure that provides for the pooling and analysis of information from all departments affected by the decision. In other words, management should ensure that the decision is made only after all relevant inputs have been evaluated.

Factors Influencing Make-or-Buy Decisions

Two factors stand out above all others when considering the make-or-buy question at the tactical level: total cost of ownership and availability of production capacity. A good make-or-buy decision, nevertheless, requires the evaluation of many less tangible factors in addition to these two basic factors. The following considerations influence firms to make or to buy the items used in their finished products or their operations.

Considerations that favor making:

1. Cost considerations (less expensive to make the part)
2. Desire to integrate plant operations
3. Productive use of excess plant capacity to help absorb fixed overhead
4. Need to exert direct control over production and quality
5. Design secrecy required
6. Unreliable suppliers
7. Desire to maintain a stable work force (in periods of declining sales)

Considerations that favor buying:

1. Limited production facilities
2. Cost considerations (less expensive to buy the part)
3. Small-volume requirements
4. Suppliers' research and specialized know-how
5. Desire to maintain a stable work force (in periods of rising sales)
6. Desire to maintain a multiple-source policy
7. Indirect managerial control considerations
8. Procurement and inventory considerations

Cost Considerations

In some cases, cost considerations indicate that a part should be made in-house; in others, they dictate that it should be purchased externally. Cost is obviously important, yet no other factor is subject to more varied interpretation and to greater misunderstanding. A make-or-buy cost analysis involves a determination of the cost to make an item and a comparison of that cost with the cost to buy the item. The following checklist provides a summary of the major elements that should be included in a make-or-buy cost estimate.

To make:

1. Delivered purchased material costs
2. Direct labor costs[7]
3. Any follow-on costs stemming from quality and related problems
4. Incremental inventory carrying costs
5. Incremental factory overhead costs
6. Incremental managerial costs
7. Incremental purchasing costs
8. Incremental costs of capital

To buy:

1. Purchase price of the part
2. Transportation costs
3. Receiving and inspection costs
4. Incremental purchasing costs
5. Any follow-on costs related to quality or service

To see the comparative cost picture clearly, the analyst must evaluate these costs carefully, considering the effects of time and capacity utilization in the user's plant.

The Time Factor

Costs can be figured on either a short-term or a long-term basis. Short-term calculations tend to focus on direct measurable costs. Therefore, they frequently understate tooling costs and overlook indirect materials costs such as those incurred in storage, purchasing, inspection, and similar activities. Also, a short-term cost analysis fails to consider the likely future changes in the relative costs of labor, materials, transportation, and so on. Consequently, in comparing the costs to make and to buy, the long-term view is the correct one. Cost figures must include all relevant costs, direct and indirect, and they must reflect the effect of anticipated cost changes.

Since it is difficult to predict future cost levels, estimated average cost figures for the total period in question are generally used. Even though an estimate of future costs cannot be completely accurate, the following example illustrates its value.

Suppose the user of a stamped part develops permanent excess capacity in its general-purpose press department. The firm subsequently decides to make the stamped part that previously had been purchased from a specialized metalworking firm. Because this enables the firm to reactivate several unused presses, the additional cost to make the item is less than the cost to buy it. However, the user finds that the labor segment of its total cost is much higher than the labor segment of the automated supplier's cost. If labor costs continue to rise more rapidly than the other costs of production, the user's cost advantage in making the part may soon disappear. Therefore, an estimate of future cost behavior can prevent a make decision that might prove unprofitable in the future.

Another factor that should be considered is the need for time-based competition. Many companies, especially high-tech firms, compete by reducing the amount of time necessary to produce or complete activities. The ability to reduce cycle time is a key strategy for gaining competitive advantage. Time is a critical issue in developing new products and bringing them to market. Making the decision to outsource research and development, the manufacturing, or distribution of a product will affect the time required to bring new technology to market. Make-or-buy decisions have a strong impact on reducing cycle time in all aspects of a business.

The Capacity Factor

When the cost to make a part is calculated, determining the relevant overhead costs poses a difficult problem. The root of the problem lies in the user's capacity utilization factor. As in most managerial cost analyses, the costs relevant to a make-or-buy decision are the incremental costs. In this case, incremental costs are those costs that would not be incurred if the part were purchased outside. The overhead problem centers on the fact that the incremental overhead costs vary from time to time, depending on the extent to which production facilities are utilized by existing products.

For example, assume that an automobile engine manufacturer currently buys its piston pins from a distant machine shop. For various reasons, the engine producer now decides that it wants to make the piston pins in its own shop. Investigation reveals that the machine shop is loaded to capacity with existing work and will remain in that condition throughout the foreseeable future. If the firm decides to make its own piston pins, it will have to either purchase additional machining equipment or free existing equipment by subcontracting to an outside supplier a part currently made in-house. In this situation, the incremental factory overhead cost figure should include the variable overhead caused by the production of piston pins, plus the full portion of fixed overhead allocable to the piston pin operation.[8]

Now assume that the same engine manufacturer wants to make its own piston pins and that it has enough excess capacity to make the pins in its machine shop with existing equipment. Investigation shows that the excess capacity will exist for at least the next two or three years. What are its incremental overhead costs to make the pins in this situation? Only the variable overhead caused by production of the piston pins! In this case, fixed overhead represents sunk costs which continue to accumulate whether or not piston pins are produced. The total machine shop building continues to generate depreciation charges. Heat, light, and janitorial services are still furnished to the total machine shop area. Also, property taxes for the machine shop remain the same regardless of the number of machines productively employed. The firm incurs these same fixed costs regardless of the make-or-buy action it decides on. Such costs, under conditions of idle capacity, are not incremental costs and, for purposes of the make-or-buy decision, must be omitted from computation of the cost to make a new part.

The concept can be observed from a slightly different point of view in the graphic representation of Figure 12.2. Note that a 12.5% increase in production volume (an increase from 80% to 90% of capacity) can be achieved by a total cost increase of only 10%. This favorable situation results simply because the 12.5% increase in production is accomplished by activating unused capacity. Fixed overhead costs are incurred irrespective of the decision to make piston pins by utilizing unused capacity.

Finally, consider a third common situation in which the same engine manufacturer wants to make its own piston pins. Investigation in this case reveals that enough excess capacity currently exists in the machine shop to permit production of the piston pins. However, management expects a gradual increase in business during the next several years, which will eliminate all excess capacity by the end of the second year. How should the make-or-buy decision be approached in this particular case?

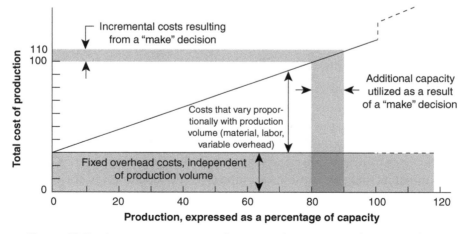

Figure 12.2 A representative case illustrating the incremental costs resulting from a make decision when operating at 80% of production capacity.

As always, the starting point of the analysis is an estimate of the costs to make vs. the costs to buy. For the first one to two years, the cost to make piston pins will not include fixed overhead because excess capacity exists. Beyond two years, however, the cost to make must include fixed overhead; if piston pins are not made, increased production of some other part will, in the normal course of business, carry its full share of fixed overhead. One alternative is to consider the make-or-buy decision separately for each of the two time periods. While the analysis may indicate that it is profitable to make the pins in both cases, it will probably reveal that it is profitable to make the pins only for the first two years and to buy them beyond that date. In this case, several qualitative factors must be investigated to determine the practical feasibility of a split course of action. If the split course of action does not appear feasible, a second alternative is to compute a weighted average cost to make the pins during both time periods. The cost data can then be used in considering the total make-or-buy question.

In practice, an infinite number of situations exist between the two extremes of excess capacity and full capacity. There is no simple, absolutely accurate solution to any of these problems. Each situation must be analyzed in its own dynamic context.

In summary, an analyst should be guided by several basic ideas. First, incremental costs are virtually always the costs relevant to the managerial decision-making process. Second, the determination of a realistic cost to make an item requires a realistic estimate of the future conditions of capacity. When capacity can be utilized by existing business or by alternative new projects, incremental overhead costs to make a new item must reflect total overhead costs. During any period when this condition does not exist, the incremental overhead to make consists only of variable overhead. When conditions of capacity normally fluctuate frequently between partial load and full load, it is likely that the make-or-buy decision for a new project of an extended period will turn largely on considerations other than comparative cost.

Precautions in Developing Costs

If a firm decides to buy a part that it has made in the past, it must exercise particular care in interpreting the quotations it receives from potential suppliers. Some suppliers may prepare the quotation carelessly, with the mistaken idea that the user does not really intend to buy the part. Other suppliers may bid unrealistically low in an attempt to induce the user to discontinue making the part in favor of buying it. Once the user has discontinued its make operation, resumption of the operation in the future may be costly. Thus, the user may be at the mercy of the supplier if the supplier later chooses to increase the price.[9] It is essential that the supply management professional carefully evaluate the reliability of all quotations in his or her attempt to determine a realistic estimate of the total cost to buy the part.

In estimating the cost to make a part, an analyst must be sure that the firm possesses adequate equipment and technical know-how to do the job. Moreover, in an industry where technological change occurs rapidly, a firm can find its equipment and know-how competitively outmoded in a few short years. Thus, the factor of obsolescence should also be given adequate consideration in determining the ultimate costs of equipment and personnel training.

The proper equipment to make an item sometimes may be easier to acquire than the properly skilled labor. Large-volume requirements, complex skill requirements, and unique geographic locations can precipitate shortages of adequately skilled labor. In preparing cost-to-make estimates, the local labor situation must be evaluated. If it is necessary to import adequately skilled personnel, total labor costs can exceed initial estimates substantially.

In the case of a make decision, it is equally important to investigate the availability and price stability of required raw materials. Large users of particular materials generally find the availability and price structure of those materials much more favorable than do small unspecialized users. Wise analysts ensure that their estimates for raw material are realistic.

Finally, in estimating the cost to make a part for the first time, the analyst must also investigate several practical production matters. The first deals with a partnering arrangement or a long-term requirements contract may be used to help control price increases such as the cost of unacceptable production work. What is the expected rate of rejected and spoiled parts? Equally important, what learning curve can the production department reasonably expect to apply? Answers to these questions may vary substantially, depending on the complexity of the job and the type of workers and equipment available. The resulting influence on the make-buy cost comparison can be considerable, however, and realistic answers should be sought.

Control of Production or Quality

Consider now some of the factors other than costs that influence make-or-buy decisions. Two conditions weigh heavily in some firms' decisions to make a particular part: control of production and control of quality.

Production Requirements

The need for close control of production operations is particularly acute in some firms. A company whose sales demand is subject to extreme short-run fluctuations finds that its production department must operate on unusually tight time schedules. This kind of company often produces a small inventory of parts used in several different products.

However, it produces to individual customer order the parts unique to a particular product or customer specification. Sales fluctuations for products that use unique parts therefore influence the planning and scheduling of numerous assembly and subassembly operations as well as single-part production operations. Efficient conduct of assembly operations depends on the firm's ability to obtain the unique unstocked parts on short notice.

Most suppliers that serve a number of customers normally cannot tool up and fit an order for a unique part into existing schedules on a moment's notice unless it is operating under some type of JIT or partnering arrangement. If a user cannot tolerate suppliers' lead-time requirements, its only other major alternative is to control the part production operations itself. Thus, by making the item, the user acquires the needed control. It then is possible to revise job priorities, reassign operators and machines to specific jobs, and require overtime work as conditions demand.

Some firms also choose to make certain critical parts to assure continuity of supply of those parts to succeeding production operations. This type of integration guards against production shutdowns caused by supplier labor problems, local transportation strikes, and miscellaneous supplier service problems. These are particularly important considerations when dealing with parts that feed an automated production operation whose downtime is tremendously expensive. If such action reduces the risk of a production stoppage, it may well justify the incurrence of extra materials costs.

Quality Requirements

Unique quality requirements often represent a second condition requiring control of part production operations. Certain parts in technical products are occasionally quite difficult to manufacture. Compounding this difficulty at times is an unusually exacting quality specification the part must meet. In certain technological fields or in particular geographical areas, a user may find that the uniqueness of the task results in unsatisfactory performance by an outside supplier. Some companies find that their own firm is in a better position to do an acceptable production job than are external suppliers.

A user normally understands more completely than an outside supplier the operational intricacies connected with the use of the part. Therefore, if the using firm makes the part itself, there can be greater coordination between the assembly operation and the part production operation. Conducting both operations under one roof likewise eliminates many communications problems that can arise between a supply management professional and supplier whose operations are geographically separated. Finally, large users often possess technological resources superior to those of smaller suppliers. Such resources may be needed in solving new technical problems in production.

For example, one producer of hydraulic systems makes a practice of subcontracting production of some of the valves used in its systems. The production of one particular subcontracted valve involved difficult interior machining operations as well as tight quality requirements. Of the supplier's first four shipments, the systems manufacturer rejected 80% of the valves for failure to meet quality specifications. During the ensuing months, the systems manufacturer worked closely with the subcontractor in an attempt to solve the quality problem. With the passage of time, however, it became clear that the systems manufacturer was contributing considerably more to solution of the problem than was the supplier. Eventually, the systems firm decided to make the valve. Although production of the valve still remained a difficult task, the systems manufacturer was able to develop the techniques necessary to produce a valve of acceptable quality, with a greatly reduced reject percentage.

The American airline industry has learned that outsourcing of critical operations such as aircraft maintenance can be very dangerous, causing major investigations by the Federal Aviation Administration (FAA). Outsourcing may require more quality control supervision than internal opertions.

Business Process Outsourcing

As outsourcing has gained popularity, the opportunity to outsource business processes has increased greatly. Lisa Ellram and Arnold Maltz have described business process outsourcing as the transfer of responsibility to a third-party of activities, which used to be performed internally.[10] Many consulting firms are competing to be external providers of these services because of the "commoditization" of generic services. Companies that formerly outsourced only information technology (IT) now outsource many business processes that are not core competencies. Advertising, maintenance, auditing, travel, and human resources are functional services commonly outsourced. In addition, companies are beginning to outsource major business systems including logistics, real estate, and software systems development. Companies are realizing more than cost reduction by outsourcing these activities; they are able to take advantage of innovative, specialized suppliers. James Brian Quinn has stated that "proper outsourcing of entire business processes can speed and amplify major innovative changes."[11]

The success experienced by outsourcing business processes has encouraged many firms to outsource entire operational functions. Firms have been outsourcing supply management without receiving much improvement in cost reduction or innovation. Often internal purchasing has proven to be more effective than using a third-party supplier for supply management. Firms that use programs such as JIT or VMI (vendor managed inventory) have the most success outsourcing the activities associated with those programs. MRO purchasing is another activity that is often outsourced successfully.

The high-tech industry may benefit the most by outsourcing supply chain management activities. The need for faster cycle times and flexibility was the reason for Toshiba's decision to combine operations with a supply management provider. Toshiba outsourced responsibility for supplier management, logistics, manufacturing, testing, and order processing. Yasuo Morimoto, president and CEO of Toshiba Semiconductor Co., explained, "To be successful in this extremely competitive environment, it is absolutely essential to have the most efficient and flexible method to service the customer, wherever the customer need arises."[12] The business model Cisco Systems established relies on the outsourcing of many supply management activities to strategic partners.

While the outsourcing of some supply activities is beneficial, the need to perform critical sourcing activities internally has become apparent. Loss of control over performance or cost is important to consider when deciding on outsourcing supply management activities. A third party may not be aggressive in seeking improvements. Supply management skills are strategic, hard to duplicate, lead to success in multiple business units, and can lead to dominance over competitors.[13]

Design Secrecy Required

Although their number is small, a few firms make particular parts primarily because they want to keep secret certain aspects of the part's design or manufacture. The secrecy justification for making an item can be found in highly competitive industries where style and cost play unusually important roles. Also, a firm is more likely to make a key part for which patents do not provide effective protection against commercial copying.

If design secrecy is really important, however, a firm may have nearly as much difficulty maintaining secrecy when it makes a part as it does when a supplier makes it. In either case, a large number of individuals must be taken into the firm's confidence, and once information leaks to a competitor, very little can be done about it. Nevertheless, a firm usually can control security measures more easily and directly in its own plant. In either case, however, the element of trust is extremely important. World-class firms work hard to create an atmosphere of trust for both internal and external activities.

Technology Risk and Maturity

Technology life cycles are an important factor in the make-or-buy decision. Technology that is changing continually is not mature and has short life cycles. It may be too risky to make parts that use this type of technology internally. The investment in capital equipment would not be reasonable. Outsourcing technology that is changing rapidly places the risk on the supplier. Such suppliers are less likely to invest in production of a part that uses mature technology. The life cycle of mature technology is reasonably stable and long term. Supply management professionals closely survey expected changes in technology. It is important to obtain insight into technology life cycles of potential materials and processes on a continuous basis.

Outsourcing is not a viable option for products with specifications that fluctuate as a result of continually developing technology or technology that is truly new. For example, Cisco Systems has decided to make many of the products in the demanding industry of optical networks. Cisco Systems is forced to integrate and perform product development internally to be competitive in this industry. Fiber optic technology is still in a fluctuating stage, which makes it difficult to define the specifications required to outsource development and manufacturing successfully.[14] Technology life cycles are a function of customer demand and technological advancements.

Unreliable Suppliers

Some firms decide to make specific parts because their experience has shown that the reliability record of available suppliers falls below the required level. The likelihood of encountering such a situation 30 years ago was infinitely greater than it is now. Today, competition in most industries is so keen that grossly unreliable performers do not survive the competitive struggle. With one major exception, unreliable delivery or unpredictable service is confined largely to isolated cases in new, highly specialized lines of business in which competition has not yet become established. Those businesses usually are characterized by low sales volumes, the requirement of highly specialized production equipment, or the unique possession of new technological capabilities.

The one major exception mentioned above is the case of the firm that purchases only an insignificant fraction of a specific supplier's total volume of business. Even the most reputable suppliers are forced, at times, to shortchange very small accounts to give significant attention to their major accounts. Regardless of the reasonableness of the cause, however, consistently unreliable performance by a supplier is sufficient grounds for switching suppliers or possibly reconsidering the original make-buy decision.

Suppliers' Specialized Knowledge and Research

A primary reason underlying most decisions to buy a part rather than make it is the user's desire to take advantage of the specialized abilities and research efforts of various suppliers.

Lest the preceding discussion of make decisions distort the total procurement picture, bear in mind that the typical American manufacturing firm spends more than 50% of its sales dollar for purchases from external suppliers. Modern industry is highly specialized. No ordinary firm, regardless of size, can hope to possess adequate facilities and technical know-how to make a majority of its production part requirements efficiently. Large corporations spend millions of dollars on product and process research each year. The fruits of this research and the ensuing technical know-how are available to customers in the form of highly developed and refined parts and component products. A firm that considers foregoing these benefits in favor of making an item should, before making its final decision, exam carefully the long-range values that accrue from industrial specialization.

Small-Volume Requirements

When a firm uses only a small quantity of a particular item, it usually decides to buy the item. The typical firm strives to concentrate its production efforts in areas where it is most efficient and in areas it finds most profitable. The work of designing, tooling, planning, and setting up for the production of a new part is time-consuming and costly. These fixed costs are recovered more easily from long production runs than from short ones. Consequently, more often than not, the small-volume user searches for a potential supplier that specializes in production of the part and can produce it in large quantities economically. Such specialty suppliers can sell to a large number of users in almost any desired quantity at relatively low prices.

Small-volume production of unique, nonstandard parts also may be unattractive to external suppliers. Every supplier is obligated to concentrate first on its high-volume, high-profit accounts. Thus, cases may develop in which a user is virtually forced to make a highly nonstandard part it uses in small quantities. Generally, however, as a part tends toward a more common and finally a standard configuration, the tendency to buy increases proportionally.

Limited Facilities

Another reason for buying rather than making certain parts is the physical limitation imposed by the user's production facilities. A firm with limited facilities typically attempts to utilize them as fully as possible on its most profitable production work. It then depends on external suppliers for the balance of its requirements. Thus, during peak periods a firm may purchase a substantial portion of its total requirements because of loaded production facilities, and during slack periods, as internal production capacity opens up, its purchases may decrease markedly.

Work Force Stability

Closely related to the matter of facilities is the factor of work force stability. A fluctuating production level forces a firm to face the continual problem of contracting and expanding its work force to keep in step with production demands. Significant continuing fluctuation, moreover, negatively affects the quality of workers such a firm is able to employ. The less stable an operation is, the more difficult it becomes to retain a competent work force.

At the time when a firm sizes the various parts of its production operation, many make-buy decisions are made. One factor that often bears heavily on the decision is the firm's desire to develop an interested, responsible group of workers with a high degree of company loyalty. Awareness that stable employment helps attain this objective sometimes prompts a firm to undersize its production facility by a slight margin. Its plan is to maintain as stable an internal production operation as possible and to buy requirements in excess of its capacity from external suppliers. This policy is most effective in firms whose products require a considerable amount of general-purpose equipment in the manufacturing operation. Equipment, as well as personnel, that can perform a variety of different jobs provide the internal flexibility required to consolidate or split work among various production areas as business fluctuates. This capability is necessary for the successful implementation of such a policy. As business increases, it is not feasible to place small orders for a large number of different parts with outside suppliers. It is much more profitable to farm out large orders for a small number of parts.

Firms that solve their work force problems in this way frequently create problems in the purchasing area. External suppliers, in effect, are used as buffers to absorb the shocks of production fluctuations. This action transfers many of the problems associated with production fluctuations from the user to the supplier. The supplier's ability to absorb these production shocks is therefore an important consideration. In some instances, it may be able to absorb them reasonably well; in others, it may not.[15] In all cases, however, the supplier prefers, as does the user, to maintain a stable production operation. Consequently, many suppliers are not interested in the customer who buys only its peak requirements. The question naturally arises, "Will a supplier ever be motivated to perform well for a purchaser who uses the supplier only for surplus work?" This question should be considered carefully before a "buy" decision of this type is made.

A third choice is available for maintaining a stable work force. A company may choose leased labor to alleviate the need to maintain a steady number of employees during fluctuations in production. External temporary labor support is utilized instead of permanent employees or completely outsourcing the task or function. Leased labor can run the gauntlet from clerical to highly skilled professionals. Quality is maintained by selecting leased employees who are competent in the areas required. The supply management professional must be aware of all aspects of the leased-labor business. Agencies providing leased labor should be analyzed and selected with great care. Payment terms, temporary to permanent employment options, and functional variations should be determined prior to implementing a plan to use leased labor.

Multiple-Source Policy

Some firms occasionally make and buy the same nonstandard part. This policy is followed for the explicit purpose of having available a reliable and experienced second source of supply. Firms that adopt a make-and-buy policy recognize that they may not always be able to meet their internal production schedules for certain parts. In an emergency, an experienced outside source is usually willing to increase its delivery of the part in question on a temporary basis until the situation is under control.

Managerial Control Considerations

Companies occasionally buy and make the same part for the purpose of developing managerial control data. Some firms use outside suppliers' cost and quality performance as a check on their own internal production efficiency. If internal costs for a particular part rise above a supplier's cost, the user knows that somewhere in its own production system some element of cost is probably out of line. An investigation frequently uncovers one or more problems, some of which may extend to other production areas. These consequent improvements may exhibit a compounding effect as they reach into other operations where inefficiencies might otherwise have gone undetected.

Procurement and Inventory Considerations

A buy decision produces several significant benefits in the management of supply and inventory activities. For supply management, a buy decision typically means that it has fewer items to buy and fewer suppliers to deal with. Usually, although not always, when a component is made in-house a number of different materials or parts must be purchased outside to support the make operation. A corresponding buy decision usually involves only one or two suppliers and a relative reduction in the associated buying, paperwork, and follow-up activities. The same relative reduction in workload occurs in receiving, inspection, stores, and inventory management groups. Typically, inventory investment is also reduced.

Netsourcing

The Internet has enabled companies to manage supply more efficiently and effectively. A supply management professional can now locate and research new suppliers by accessing information on the Internet. The Internet has become an open market for electronic business transactions. Many tactical supply management activities can be replaced by using electronic forms and direct connections with suppliers. Intranets have become a way to provide services to personnel by connecting employees directly with preferred providers. The Web offers an aggregation of common business tools allowing more efficient management of many business processes. The Internet can replace many different services and functions including human resources, accounts payable and receivable, and document storage systems. The online business revolution requires companies to develop infrastructures and Web sites quickly.

The Volatile Nature of the Make-or-Buy Decision

Although make-or-buy investigations normally begin with a cost analysis, various qualitative factors frequently foretell more far-reaching consequences than does the cost analysis. Therefore, a correctly approached make-or-buy decision considers the probable *composite effect* of all factors on the firm's total operation.

A thorough investigation is complicated considerably by the dynamics and uncertainties of business activity. Certain factors can have very different implications for a make-or-buy decision at different points in time and under different operating conditions. As has been pointed out, changing costs can turn a good decision into a bad one in a very short period of time. In addition, future costs, complicated by numerous demand and capacity interrelationships, are influenced substantially by such variable factors as technological innovation and customer demand. The availability of expansion capital also influences make-or-buy decisions. An "easy money" policy, a liberal depreciation policy, or liberal government taxing policies tend to encourage make decisions. The opposite policies promote buy decisions. These federal policies fluctuate with economic and political conditions.

The tendency toward favoring make decisions to stabilize production and work force fluctuation is usually greater in small firms than in large ones. In some small shops, the loss of just a few orders results in the temporary layoff of a sizable percentage of the work force until additional orders can be obtained. Generally, larger organizations do not have such severe problems because their fluctuations in production volume relative to total capacity are smaller. As large firms adopt compensation plans that move toward a guaranteed wage structure, however, they too will feel a similar pressure to favor make decisions.

To summarize, beware of rigid formulas and rules of thumb that claim to produce easy make-or-buy decisions. The make-or-buy question is influenced by a multitude of diverse factors that are in a constant state of change. Under such conditions, few easy decisions turn out well in both the short and the long run. Moreover, the relevant factors vary immensely from one firm to another. For these reasons, every company should evaluate the effectiveness of its past decisions periodically to generate information helpful in guiding future courses of make-buy action.

Insourcing

Core competencies change. Thus, periodic reevaluation of outsourcing decisions is a strategic necessity. Although it is often expensive and difficult to bring an activity back in-house, changes in core activities, technology, or strategy may require a company to reverse a make-or-buy decision. Continued involvement is always necessary when a "buy" decision has been implemented. Managing the supplier relationship for an outsourced activity will allow a company to make adjustments when problems arise. Change management has been recognized as a needed skill for handling the transition of such activities. Implementing the decision to outsource or bring an activity back in-house involves a formal transition process that affects the organizations of both the firm and the supplier.

Dangers of Outsourcing

Loss of Control

Entrusting an entire process to an external provider may cause loss of control and skills that result in over-dependency. A firm may lose key information resources without continuous and active management of the outsourcing contract. Information that is required to manage the business and future growth effectively will not be communicated. Inadequate involvement is often the cause of this problem. Supply professionals must be good at supplier management for an outsourcing program to work. The airline industry has outsourced significant percentages of aircraft maintenance causing great concern to the FAA over major maintenance problems.

Loss of Client Focus

The goals and objectives of the selected external provider may differ from the firm's goals. Eventually the provider will lose touch with the firm's business plan and strategy. The provider may cause a conflict of interest if similar outsourced functions are performed for other organizations. Key resources from one firm may be used to support other clients. These activities affect the timely and successful performance of the outsourced function.

Lack of Clarity

Failure to articulate clearly the responsibilities of the selected external provider is a major concern. A formal service level agreement (SLA) contract must be developed prior to commencing outsourced services. Without a clear and agreed upon contract, the outsource provider can cause costly and disruptive disputes with claims of "out of scope" work. A formal agreement avoids the extra charges for every change or request that such providers can assess. The agreement should specify how changes and requests would be processed with a mechanism or formula for pricing such changes.

Lack of Cost Control

Many outsourcing decisions are made in an effort to lower overall costs. Changes in company objectives and rising prices can take costs beyond estimates made for the initial analysis. If the outsourcing contract does not include long-term pricing with appropriate incentives, then the outsource provider will not be motivated to control costs or maintain quality. This problem is impacted by an inflexible contract. The buying company should specify in the outsourcing contract the degree to which it desires to manage costs and the appropriate mechanics. It can be very difficult to disengage from a poorly structured outsourcing contract. Inflexible conditions limit the ability to support changing business strategies and objectives. If a company is unable to manage an outsourcing contract appropriately, it may need to in-source in order to regain control of those activities. Bringing a function or service back in-house is expensive and the organization usually lacks employees with the required skill sets.

Ineffective Management

The selected external provider may not perform the outsourced function better than the client organization can. Many companies have outsourced a function without carefully pre-qualifying the provider based on efficiency, effectiveness, and total capabilities. Careless outsourcing ultimately costs more than keeping a function in-house. Without proper analysis and consideration, a function may be outsourced for the wrong reasons. Managers may decide to outsource due to a problem they are experiencing with a function in-house only to find that the selected provider cannot solve the problem. For example, management may believe that the internal IT department is not sensitive to the needs of the users. An external provider is not necessarily going to be more sensitive to user needs. This type of problem can be avoided by developing a precise statement of work, carefully researching and pre-qualifying providers, and utilizing a clear service level agreement.

Loss of Confidentiality

Outsourcing often means off-loading sensitive functions involving proprietary corporate data. Concerns about loss of controls of these functions and the protection of the underlying information are valid, particularly in an environment where hacking is a blood sport. This raises several issues:

▲ Is the outsourcing service provider's system secure against external and internal threats?
▲ What kind of security monitoring is in place?
▲ How will the service provider prevent intentional and unintentional disclosure of private data?

▲ What may the service provider do with data it has received from your company?

▲ How will your company know if sensitive data has leaked from the service provider's system?

Service providers must answer all these questions satisfactorily, and executives charged with outsourcing IT functions should validate all claims before retaining any service provider.[16]

Double Outsourcing

The practice of double outsourcing is essentially the subcontracting of an outsourcing contract. This type of arrangement can backfire if the client does not manage the agreements with both companies. Double outsourcing is common with functions that are often outsourced, such as IT. The external provider may not have the necessary technical skills to perform all of the outsourced work and will subcontract to fulfill the contract requirements. If the selected external provider is considering the use of a third party, then the supply manager should select and manage the subcontractor. Problems occur when the provider utilizes a subcontractor without involving the client. This leaves the client firm with little control over problems caused by the outsource provider's subcontractor.[17]

Administration of Make-or-Buy Activities

It is not difficult to find otherwise well-managed firms in which many tactical make-or-buy decisions are inadvertently delegated to an operating person in inventory control or production control. It should be apparent that this is a poor practice. In the first place, such a person normally does not have adequate information with which to make an intelligent decision from a companywide point of view. Second, even if adequate information were available, this type of person typically lacks the breadth of experience to evaluate fully the significance of the information and the resultant decision.

Chief Resource Officer

Strategic outsourcing is an emerging trend within the business world. A new management role is being developed as a result of this trend: the Chief Resource Officer (CRO).[18] The CRO manages and initiates outsourcing for direct support to the company's bottom line. Outsourcing deals are becoming an increasingly large amount of corporate expenditures. Both the overall risks and rewards are becoming greater with increased outsourcing activity. The need for active management and skilled leadership is increasing. The CRO, essentially the director of external resources, is responsible for all outsourcing relationships, ensuring they live up to expectations. Companies that are embracing strategic outsourcing are beginning to realize the need for a dedicated team of individuals to oversee outsourcing activities. John Chiazza, Chief Information Officer at Kodak, advocates establishing relationship management groups to "research potential outsourcers, negotiate terms, and address other matters of policy…individual relationship managers interact with the outsourcing providers on a day to day basis."[19] These outsourcing professionals possess three essential skills: negotiation, communication, and project management.

Framework for Outsourcing

Many frameworks have been developed for successful outsourcing. All of them involve information gathering, developing a strategic roadmap, and a decision flowchart.[20] Developing a "transition back" plan is an important step that is often ignored. It is less costly to have a plan in place in the event that the outsourced function needs to be brought back in-house. It will be easier to transfer the function back in-house or to another provider if such a contingency was planned for initially. In addition to providing an operational framework within which make-or-buy alternatives are investigated, a review system is necessary. The system should provide procedures for three important additional activities: (1) the entry of projects into the study system, (2) the maintenance of essential records, and (3) a periodic audit of important decisions.

Procedures should be established as part of a firm's product development program, compelling high-value and strategically oriented parts in new products to enter the make-or-buy analysis process. In some cases, this analysis can be effectively integrated with pre-production value engineering investigations. Similarly, existing production parts should be subjected to a systematic review which searches for borderline make-or-buy items warranting careful study.

Regardless of the source of entry, all make-or-buy investigations should be classified as "major" or "minor," based on the value and strategic nature of the part. In one firm, all items involving expenditures under $25,000 are classed as minor. Subsequent studies of minor items involve personnel from production, supply management, quality control, and occasionally design engineering. Major items entail expenditures over $25,000, and additionally involve personnel from finance, marketing, and other production areas.

Summary records are essential to full utilization of the data developed in make-or-buy investigations. The record should be designed to serve as a useful future reference. A brief discussion of all factors pertinent to the decision that was made should be included, as well as the primary reasons for the decision. Assumptions about future conditions should be stated. An accurate summary of cost data should always be included. Records of this type provide the information required when a firm is forced to make quick decisions about subcontracting work under peak operating conditions, or about bringing work back into the shop when business slumps. Accurate records can mean the difference between a profitable decision based on facts and a hopeful decision based on intuition and hunches. Finally, investigation records provide the basic data for post-decision audits.

Executive Level Involvement

In most cases, make-or-buy decisions should be made or at least reviewed at the executive level. The decision maker must be able to view such decisions with a broad company-wide perspective. Many progressive firms use a team or committee approach to analyze make-or-buy alternatives. The important point to keep in mind is that all departments that can contribute to the decision, or that are affected by it, should have some voice in making it. A team or committee accomplishes this directly. In other cases, a formal mechanism must be established which facilitates, and perhaps requires, all interested departments to submit relevant data and suggestions to the decision maker. Moreover, to ensure thoroughness and consistency, the system must detail the cost computation procedures to be used and assign cost investigations to specific operating groups.

The Special Disadvantages of Outsourcing Business Processes

While many of the negatives involved in make-or-buy decisions have been discussed in this chapter, there are special risks involved in business process outsourcing, the moving of "back office" functions such as accounting to third-party providers. "Many companies overweight the cost factor in their original analysis and underweight (or ignore) the other relevant variables."[21] The other variables include the quality of delivered services, risk to branding, loss of in-house knowledge and skills necessary to evaluate the service effort, connections to the other business functions, and documentation of the process for the hand-off to the supplier. Several firms have had to bring back telephone service centers at the insistence of their customers. In all buy and outsourcing decisions, there is some, and at times serious, loss of control.

Summary

If one takes a broad view of the American industrial scene over the past several decades, three characteristics stand out clearly: (1) Firms are becoming more aware of the strategic dimension of the make-or-buy decision—management is more proactive in identifying and exploiting the firm's core competencies as organizations adopt Lean manufacturing strategies; (2) most manufacturing firms have become much more specialized as technology has advanced—in the words of researchers Peters and Waterman, they "stick to their knitting";[22] and (3) the cost of materials, expressed as a percentage of total product cost, has continued to increase in many industries. These three factors lead to the inevitable conclusion that, in the aggregate, American firms are buying more and making less. Planned or unplanned, the trend continues to develop.

Yet at the managerial level, many successful firms have not handled the recurring make-or-buy issue in a well-organized, systematic manner. Instead, many have elected to deal with specific cases on an ad hoc basis as they arise.

This situation is understandable yet ironic. In earlier years when the cost-price squeeze was less severe for many firms, poor decisions in this area did not affect earnings dramatically. Yet, in the aggregate, make-or-buy decisions do significantly affect a firm's ability to utilize its resources in an optimal manner. Past practices are changing. Three forces will continue to stimulate this change:

▲ *Pressures on profit margins* are severe, and will continue to increase. Resources must be utilized more effectively.
▲ *Firms continue to become more highly specialized* in products and production technology, producing greater cost differentials between making and buying for many users.
▲ *Computer modeling capability* is becoming commonplace; make-or-buy evaluation and control systems can be developed and handled quasi-automatically with this capability.

Just as materials management organizations, MRP systems, and JIT systems have developed over the past several decades, implementation systems for recurring make-or-buy analysis are being developed.[23]

Endnotes

1. Lisa Ellram and Arnold Maltz, "Outsourcing Supply Management," *Journal of Supply Chain Management,* Spring, 35(2) (1999): 4–17.
2. C. K. Prahalad, "Core Competence Revisited," *Enterprise* October (1993): 20.
3. Robert B. Hays and Gary Pisano, "Beyond World Class: The New Manufacturing Strategy," *Harvard Business Review* 72, No. 1 (January-February 1994), pp. 77–84.
4. Chris Lonsdale, "Locked-In to Supplier Dominance: On the Dangers of Asset Specificity for the Outsourcing Decision," *Journal of Supply Chain Management,* May (2001).
5. Michael E. Porter quoted in an article by John A. Byrne, "Caught in the Net," *Business Week,* August 27, 2001, 35.

6. Björn Axelsson, Finn Wynstra, and Arjan van Weele, "Purchasing Involvement in Product Development: A Framework," *European Journal of Purchasing & Supply Management,* 5(3/4)(1999): 129–141.

7. It is assumed that all inspection costs associated with the "make" operation are included in the direct labor costs.

8. If the piston pin production replaces the production of another part, the piston pin operation should carry the same absolute amount of fixed overhead that was carried by the part replaced. If new equipment is purchased to produce piston pins, the piston pin operation should be charged with the additional fixed overhead arising from acquisition of the new equipment.

9. A partnering arrangement or a long-term requirements contract may be used to help control such price increases.

10. Ellram and Maltz, op.cit.

11. James Brian Quinn, "Outsourcing Innovation: The New Engine of Growth," *Sloan Management Review,* July 1, 2000, 13–28.

12. "Toshiba and Kingston Establish New Supply Chain Management Model," *Business Wire,* February 29, 2000.

13. Ellram and Maltz, op. cit.

14. Clayton M. Christensen, "Limits of the New Corporation," *Business Week,* August 28, 2000, 180–181.

15. Two factors largely determine a supplier's ability to absorb fluctuating order requirements from a user. They are:
    1. The similarity of the work involved in producing a particular user's requirement and in producing other customers' orders. The more similar the requirements are, the less is the expense of special setup work for a particular user.
    2. The extent to which other customers' orders offset the peaks and valleys in the supplier's production operation. The more stable the supplier's total production operation is, the less disturbing is the effect of an occasional fluctuating account.

16. Scott J. Nathan, "Reducing the Risk of Outsourcing," *Supply Strategy,* May/June (2001): 20.

17. "Getting Into Outsourcing," *NAPM InfoEdge,* January (1998): 1–16.

18. Frank Casale, "The Rise of The Chief Resource Officer," *Business Briefing: European Purchasing & Logistics Strategies,* July (1999): 81–82.

19. Scott Leibs, Special Report on Outsourcing: How You Slice It," *CFO Magazine,* February (2001): 81–86.

20. For more information see Eric Sislian, and Ahmet Satir, "Strategic Sourcing: A Framework and a Case Study," *Journal of Supply Chain Management,* Summer (2000).

21. Steve Matthesen, "It's Not Just about Cost," *Inside Supply Management,* March (2006): 12–14. Matthesen is Vice President and Global Leader for Supply Chain, Boston Consulting Group (BCG), in Los Angeles.
22. T. J. Peters and R. H. Waterman, *In Search of Excellence,* New York: Harper & Row Publishers (1982): 292–305.
23. Readers interested in a detailed examination of the make-or-buy issue should review the classical study conducted some years ago by J. W. Culliton, Make or Buy? Research Study 27, Graduate School of Business Administration, Harvard University, Boston, MA, 1942, 4th reprint, 1956.

Suggested Reading

Berry, John. *Offshoring Opportunities: Strategies and Tactics for Global Competitiveness.* Hoboken, NJ: John Wiley & Sons, Inc., 2006.

Carter, Joseph R., William J. Markham, and Robert M. Monczka. "Procurement Outsourcing: Right for you?" *Supply Chain Management Review* 11(4)(2007): 26.

Corbet, Michael F. *The Outsourcing Revolution: Why It Makes Sense and How to Do It Right.* Chicago, IL: Dearborn Trade Publishing, 2004.

Cohen, Linda, and Allie Young. *Multisourcing Moving Beyond Outsourcing to Achieve Growth and Agility.* Boston, MA: Harvard Business School Press, 2006.

Duening, Thomas N., and Rick L. Click. *Essentials of Business Process Outsourcing.* Hoboken, NJ: John Wiley & Sons, Inc., 2005.

Engardio, Pete, Michael Amdt, and Dean Foust. "The Future of Outsourcing." *Business Week–Special Report–Outsourcing,* January 30, 2006.

Fine, C. H. and D. E. Whitney. "Is the Make-Buy Decision Process a Core Competence?" MIT Center for Technology, Policy, and Industrial Development, Cambridge, MA, 1996.

Gietzmann, M. B. "Make or Buy Revisited: Some Lessons from Japan." *Management Accounting* 73(1)(1995): 24–25.

Greaver, Maurice F. *Strategic Outsourcing: A Structured Approach to Outsourcing Decisions and Initiatives.* New York: AMACOM, 1999.

Halvey, John K., and Barbara Murphy Melby. *Business Process Outsourcing: Process, Strategies, and Contracts,* 2nd ed. Hoboken, NJ: John Wiley & Sons, Inc., 2007.

McIvor, R. T., and P. K. Humphreys. "A Case-Based Reasoning Approach to the Make or Buy Decision." *Integrated Manufacturing Systems* 11(5)(2000): 265–310.

Parker, David W., and Katie A. Russell. "Outsourcing and Inter/Intra Supply Chain Dynamics: Strategic Management Issues." *Journal of Supply Chain Management* 40 (4) (2004): 56.

Perrons, R. K., M. G. Richards, and K. Platts. "What the Hare Can Teach the Tortoise about Make-Buy Strategies for Radical Innovations." *Management Decision* 43(5) (2005): 670–690.

Perrons, R. K. "The Long-Term Irrelevance of Trust in Outsourcing Decisions for New Technologies: Insights from an Agent-Based Model." *Supply Chain Forum: An International Journal* 6(2)(2005): 30–37.

Perrons, R. K., and K. Platts. "Make-Buy Decisions in the Face of Technological Change: Does Industry Clockspeed Matter?" *International Journal of Management & Enterprise Development* 2(1)(2005): 1–11.

Quinn, J. B. "Outsourcing Innovation: The New Engine of Growth," *Sloan Management Review* 41(2000): 13–28.

Tompkins, James A., Steven W. Simonson, Bruce W. Tompkins, and Brian E. Upchurch. "Creating an Outsourcing Relationship: Successful Outsourcing Depends as Much on the Kind of Relationship Developed as It Does on the Details of the Operational Execution....". *Supply Chain Management Review* 10(2)(2006): 52.

CHAPTER 13

Sourcing

The Strategic Sourcing Plan

Most veteran purchasing managers state, "Selecting the best supplier in the sourcing process is the most important purchasing decision. If you have a weak supplier, all the supply problems of poor quality, price, and delivery will surface and the contract will not solve these problems."

Development of a strategic sourcing plan is driven by the recognition that tactical sourcing will not succeed in developing a supply base that will yield the benefits[1] of collaborative relationships and alliances. Development of collaborative relationships and eventually world-class supply management requires concerted strategic planning. This chapter presents a generic road map that details how supply management can develop a strategic sourcing plan that will enable supply management to discover, evaluate, select, develop, and manage a viable supplier base. The "road map" is presented in Figure 13.1.

Before pursuing the development of supplier base for any material or service, the buying firm must first determine whether the material or service should be outsourced at all. Chapter 12 addresses this issue in detail. World-class firms conduct a strategic analysis of what their core competencies are. They analyze the skills and processes that form the basis of their success and competitive advantage. As Prahalad and Hamel say, "Core competencies are the wellspring of new business development."[2] If an item or service represents a core competency or supports or interfaces with such a competency, then the source of supply should be the firm itself.

Also prior to developing a strategic sourcing plan, company-wide support and financial backing must be given by the chief executive officer. The strategic sourcing plan is doomed to fail without company-wide support from the various departments and financial backing to allow supply management time to be spent on essential planning activities.

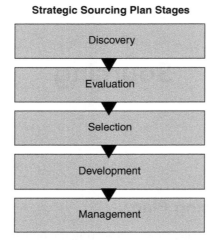

Figure 13.1 Strategic sourcing plan stages.

Discovering Potential Suppliers

Before the information age and globalization of markets, the discovery process for potential suppliers was greatly limited. Today, suppliers throughout the globe can be found by simply typing several key words into a Web search engine or letting suppliers find your company through a variety of posting methods. While the opportunities to source via the Web are amazing, supply managers should not ignore other sources of information. Many leading supply management organizations maintain existing information about present and potential suppliers in soft and hard copy media. The following information sources should prove helpful to a supply manager in establishing a robust list of potential suppliers.

▲ *Supplier Web Sites*   Most suppliers have Web sites that provide detailed information about their products and services. The sites usually are registered with search engine providers such as Lycos or Yahoo! Search engines enable supply managers to type in key words such as "third-party logistics" and receive a "hit list" of sites that contain the key words. The Web has become one of the primary ways of discovering new sources of supply.

▲ *Supplier Information Files*   Supply management departments should keep supplier information files on past and present suppliers that include the name of each supplier, a list of materials available from the supplier, the supplier's delivery history, the supplier's quality record, the supplier's overall desirability, and general information concerning the supplier's plant and management. In addition to a departmental file, supply managers usually maintain a personal supplier file for their own use. Supplier information files are important because many supply management operations are repetitive; hence, it would be poor management indeed if supply managers spent time repeatedly recapturing information that was once available to them but had been needlessly lost. Web-enabled centralized databases make maintenance and dissemination of supplier information files a relatively easy task compared to the hardcopy alternative.

▲ *Supplier Catalogs* Because catalogs are a commonly used source of supplier information, many supply management departments maintain a hardcopy catalog library. The alternative gaining in popularity is the use of electronic catalogs. Regardless of the storage medium, users may examine the catalogs to locate the materials they need. The firm's supply managers also use catalogs to determine potential sources of supply and, on occasion, to estimate prices and total cost of ownership.

▲ *Trade Registers and Directories* *Thomas' Register of American Manufacturers* is typical of several widely known trade registers and directories. These registers are available in book, searchable compact disk, and Web site formats. They contain information on the addresses, number of branches, and affiliations of all leading manufacturers. Financial standings of firms are also frequently given. The registers are indexed by commodity, manufacturer, and trade name or trademark description of the item. Kompass Publications in Europe provides similar information for European firms.[3]

▲ *Trade Journals* Trade journals are another excellent source for obtaining information about possible suppliers. Advertisements in trade journals are often a supply manager's first contact with potential suppliers and their products. For example, a supply manager in the aircraft industry would routinely read *Aviation Week*. Many journals have moved or are moving to the Web and compact disk where archived editions are easily searched for specific sourcing needs.

▲ *Phone Directories* Before the 1990s, this source of information was of limited value to industrial supply managers because local telephone books list only local companies and searching through volumes of telephone books was a cumbersome activity. Today, phone directories are available online and on searchable compact disks that are available for purchase in computer retail stores. Phone directories can serve as a useful starting point if other sources have proved fruitless or if local sources are desired.

▲ *Filing of Mailing Pieces* Many mail advertisements are worth saving. These should be given a file number, dated, and indexed by the name and number of each publication. When supply managers seek a new source, they can then refer to the index and review the appropriate brochures and booklets. Some supply management departments ask prospective suppliers to complete a simple form giving basic information about themselves and their products. This information, which includes company name, address, officers, local representatives, and principal products, is kept in a set of looseleaf notebooks or preferably in a searchable database or spreadsheet. By referring to these standardized data, a supply manager can obtain immediate, current information about potential new sources.

▲ *Sales Personnel*   Sales personnel are excellent sources of information about suppliers and materials. Not only are they usually well informed about the capabilities and features of their own products, but they are also familiar with similar and competitive products as well. By the very nature of their specialized knowledge, salespeople can often suggest new applications for their products that will eliminate the search for new suppliers. From their contacts with many companies, salespeople learn much about many products and services, and all this information is available to the alert, receptive supply manager. This is a key reason why sales personnel should always be treated courteously and given ample time to make their sales presentations. To deny them this opportunity is to risk the loss of valuable information, including information concerning new and reliable sources of supply.

▲ *Trade Shows*   Regional and national trade shows are still another way supply managers learn about possible sources of supply. Trade shows provide an excellent opportunity for supply managers to see various new products and modifications of old products. They also offer supply managers an opportunity to compare concurrently similar products of different manufacturers. Regional trade shows are sponsored periodically by many manufacturers, distributors, and trade organizations. Information about trade shows is usually sent to all interested supply management and technical personnel in the area. If not, then a quick search of the Web will usually yield information about upcoming shows.

▲ *Company Personnel*   Personnel from other departments in a supply manager's firm often can provide supply management with helpful information about prospective suppliers. Through their associations in professional organizations, civic associations, and social groups, these employees often learn about outstanding suppliers. Scientific, technical, and research personnel who use sophisticated materials or services always have many valuable suggestions to make regarding possible sources of supply. From their attendance at conventions and trade exhibits, and from their discussions with associates, these personnel are particularly well informed regarding new products, new methods, and new manufacturers.

▲ *Other Supply Management Departments*   Supply management departments in other firms can be helpful sources of information regarding suppliers. Information exchanged among individuals from these departments can be mutually beneficial for all participating companies; therefore, this source of information should be actively developed.

▲ *Professional Organizations*   Local supply management associations, such as the local affiliates of the Institute for Supply Management (formerly NAPM), the Purchasing Management Association of Canada, the National Institute of Governmental Purchasing, and the National Association of Educational Buyers, publish a list of their members. One of the basic objectives of a supply management association is that its members help each other in every possible way. Accordingly, members usually will do everything possible to help fellow members locate and evaluate new sources of supply.

The list of sources for suppliers left out the possibility that an existing supplier could be developed as a new supplier. Jim Wehrman, Assistant Vice President of Purchasing at Honda America Manufacturing, points out "top level supply chain management effort must focus heavily on the development of current suppliers, and must devote significant resources to strengthening current suppliers and improving their capabilities."[4] An existing supplier may have the capability to fulfill new sourcing needs, which, if selected, strengthens the existing relationship.

Evaluating Potential Suppliers

After developing a comprehensive list of potential suppliers, the supply manager's next step is to evaluate each prospective supplier individually. The type of evaluation required to determine supplier capability varies with the nature, criticality, complexity, and dollar value of the purchase to be made. The evaluation also varies with the supply manager's or sourcing team's knowledge of the firms being considered for the order.

In some cases, an evaluation is uneccessary. For many uncomplicated, low-dollar-value purchases, an examination of basic information readily available, such as a mailing or Web site, is sufficient. For complex, high-dollar-value, and perhaps critical purchases, additional evaluation steps are necessary. Marc Ensign, Director of Strategic Sourcing for Honeywell IAC, provides the following guidelines for whether an assessment is necessary.

▲ Is the supplier strategically important? If the supplier provides a product or access to a future product that is critical to the buying firm's success, take the time.

▲ Is the product or service being procured considered strategic? If yes, then take the time to perform the evaluation.

▲ Are there other short-term alternatives available? If supply management can modify the request (with the concurrence of the internal customer) to allow another product, service, or supplier to be quickly substituted, then they can reduce the thoroughness of the evaluation.[5]

Steps for complex, high-dollar-value, and other critical purchases can include surveys, financial condition analysis, third-party evaluators, evaluation conferences, plant visits, and selected capability analyses. Usually surveys and an analysis of the financial condition come first. Companies that have positive survey results and good financial standing still may require facility visits. As necessary, visits are followed by even more detailed analyses of the most promising suppliers' management, quality, capacity, service, JIT and IT capabilities. All of the approaches and analyses given here are by no means an exhaustive list, but they do provide a starting point for supply managers in evaluating potential suppliers. The approaches and analyses are discussed in greater detail in the following.

▲ *Supplier Surveys*   A survey should provide sufficient knowledge of the supplier to make a decision to include or exclude the firm from further consideration. A survey is based on a series of questions that often cover the following areas: principal officers and titles, bank references, credit references, annual history of sales and profit for the past five years, a referral list of customers, number of employees, space currently occupied, expansion plans (including sources of funds), an indication of the use of design of experiments (DOE), current production defect rate for similar products, number of inspectors used, quality methods adopted, and a list of all equipment and tools that would be used to manufacture, test, and inspect the purchase in question. Appendix A at the end of this chapter provides a sample supplier survey that can be used to develop other surveys.

▲ *Financial Condition Analysis*   Preliminary investigation of a potential supplier's financial condition often can avoid the expense of further study. A qualified supply manager or professional from the finance department conducts these investigations. A review of financial statements and credit ratings can reveal whether a supplier is clearly *incapable* of performing satisfactorily. Financial stability is essential for suppliers to assure continuity of supply and reliability of product quality. Imagine the difficulty of getting: (1) a financially weak supplier to maintain quality, (2) a supplier who does not have sufficient working capital to settle an expensive claim, or (3) a financially unsound supplier incapable of working overtime to meet a promised delivery date. For additional information on financial ratio calculations, refer to Appendix B at the end of this chapter.

▲ *Third-Party Evaluators*   Independent third-party firms can be hired to conduct many of the analyses given in this section. For example, a Dun and Bradstreet Information Services supplier evaluation typically contains sections describing the company's address, size, organizational structure, officers, financial condition, bankruptcies, suits, liens, newsworthy events, performance vs. industry competitors, operations, facilities, subsidiaries, corporate relations, minority ownership, payments, and other public record information. As the databases, accumulation procedures, and query procedures of third-party evaluators improves over time, the use of this source of information in evaluating suppliers will continue to grow.

▲ *Evaluation Conference*   For an extremely critical purchase, a supplier evaluation conference frequently is held at the supply manager's plant to discuss the purchase. From such a discussion, it is usually easy to differentiate among those suppliers who understand the complexities of the purchase and those who do not. By eliminating those who do not, the search for the right supplier is narrowed further.

▲ *Facility Visits* By visiting a supplier's facility, the sourcing team can obtain firsthand information concerning the adequacy of the firm's technological, manufacturing, or distribution capabilities, and its management's technical know-how and orientation. Depending on the importance of the visit, the company may send representatives from only supply management and engineering; or it may also include some combination of representation from these functions and finance, operations, quality assurance, marketing, and industrial relations. For example, engineering's task may be to review and assess the technological capability of the potential supplier. Occasionally, top management may also participate in the visit and the evaluation. When the concurrent approach to the design of new products is utilized, appropriate members of the cross-functional team conduct the visit and evaluation.

▲ *Quality Capability Analysis* The potential supplier's quality capability is a critical factor to examine. If the prospective supplier's process capability is less than the buying firm's incoming quality requirements, the supplier typically is not worthy of further investigation. An obvious exception is the case in which no supplier possesses the required process capability. In this case, the two firms will have to work together to improve the supplier's process capability. An analysis of the quality capability should also include investigating upper management's philosophy toward quality, the quality department (if one exists), and the firm's abilities with quality assurance techniques and DOE. These critical issues are discussed in detail in Chapter 7. According to Forker, Ruch and Hershauer, "Managers in customer firms should not underestimate the importance of a supplier's top management and its quality department in shaping and implementing quality improvement efforts. Top management support has been shown repeatedly to be paramount to the success of a quality improvement program."[6]

▲ *Capacity Capability Analysis* Ensuring continuity of supply is one of the most fundamental objectives of supply management. A supply manager is rarely noticed if materials arrive on time; but if materials are late such that expensive operations grind to a halt, the supply manager can become infamous and perhaps unemployed. Unscrupulous suppliers will often promise that they can meet future demands when in fact they do not have the capacity. Some supplier's salespeople do not have a solid understanding of their firm's manufacturing capacity and the demands on that finite capacity. In 2000, one large telecommunications company did not investigate its second largest supplier's capacity prior to entering into a "partnership length and quantity" contract for the critical components in DSL kits. The supplier did not have the capacity to meet the buying firm's demands. The buying firm's parts and components from other suppliers arrived, but the parts could not be kitted because all parts were not available.

▲ *Management Capability Analysis*   Evaluating an organization's management style and compatibility usually requires several visits to the potential supplier's facilities. A quick way to draw conclusions that will often hold true is to evaluate the management capability of the firm by evaluating the sales representative, facility grounds, and even the parking lot.[7] A properly trained sales representative knows his or her product thoroughly, understands the buying firm's requirements, gives useful suggestions to the supply manager and appropriate members of the buying team, commits the company to specific delivery promises, and follows through on all orders. This type of sales representative indicates that the supplier's firm is directed and managed by responsible and enterprising executives. Even a parking lot analysis can yield information about a potential supplier's management. Few cars and poorly maintained landscaping, structures, and pavement can be indicators of problems. A well-maintained and managed firm seldom experiences the instability that results from continual labor problems and always strives to reduce its costs. This type of company can be a good supplier. In addition, the supplier's purchasing expertise is a major factor in cost control.

▲ *Service Capability Analysis*   "Service" is a term that varies in meaning, depending on the nature of the product being purchased. Specifically, good service always means delivering on time, treating special orders specially, filling back orders promptly, settling disputes quickly and fairly, and informing supply managers in advance of impending price changes or developing shortages. In some situations, it means exceptional post-sale service. Service also can include actions such as stocking spare parts for immediate delivery, extending suitable credit arrangements, or warranting the purchased item's quality and performance to a degree beyond that normally required. In the aggregate, good service means that a supplier will take every reasonable action to ensure the smooth flow of purchased materials between the supplying and buying firms.

▲ *Flexibility Capability Analysis*   One issue that emerged from the stock market turmoil of 2001 was the importance of supplier flexibility to adjust production volumes with short notice and remove inventory out of the chain. Flexibility is achieved through methods encapsulated in the JIT philosophy, also known as Lean strategy. According to a 10 year forecast for the 2000s, "Lean supply chains will be a competitive strategy."[8] When properly implemented, a JIT system results in the following *supply chain* benefits: reduced inventory, increased quality, reduced lead time, reduced scrap and rework, and reduced equipment downtime.[9] JIT requires a high degree of integration of the customer's and supplier's operations. The inevitable changes in a customer's production plans and schedules affect the supplier's schedules. Experience has demonstrated that dependable, single-source collaborative relationships are virtually essential if the required level of integration is to result. A firm that is considering the adoption of JIT manufacturing must focus on its suppliers' abilities and willingness to meet the stringent quality and schedule demands imposed by the system. The sourcing team must carefully investigate a potential supplier's capability as a JIT manufacturer.[10]

▲ *Lead Time*   Although one might suspect lead time is the same for all suppliers for the same products, it is not. The components of supplier lead-time are:
▼ Order processing time
▼ Production control scheduling time
▼ Set-up time
▼ Manufacturing and assembly time; including ordering and inventory of materials and purchasing sub-components
▼ Inspection time
▼ Packaging and shipping time
▼ Receiving and inspecting time
▼ Routing to user work-stations time

There are many variables involved in the accumulated lead times quoted by suppliers and they constantly change depending on supply and demand factors and the internal efficiencies of the particular supplier.

▲ *Information Technology Capability Analysis*   Information sharing is a key enabler of effective supply chain management. Information sharing does not require technology, but technology is increasingly being used as the "vehicle of use." As reported in the *European Journal of Purchasing and Supply Management*, "Without a doubt, competitive advantage accrues to those who effectively adapt information technology to better disseminate information within the supply chains. In a number of industries, the ability to link electronically has become a right of entry and a prerequisite just to be considered as a potential supply chain partner."[11] What type of analysis is required depends greatly on the buying firm's technology capability.

The analyses given above should not necessarily be limited to potential first-tier suppliers. Today's supply chains have multiple tiers which may be critical to the success of the procurement. In a recent 10-year forecast on the future of supply management, several noted academicians stated, "Determination of first-, second-, and possibly third-tier suppliers will become more critical to supply chain dominant companies in the future."[12]

▲ *Other Considerations*   Other factors to consider include evaluating a supplier's green manufacturing programs (covered in this chapter) and risk management (covered in the Chapter 22 appendix).

A common approach to summarizing the analyses given above or to conduct them on an individual basis is a weighted-factor analysis discussed next.

Selecting Suppliers

After one or more potential suppliers have passed the evaluation process, the selection process must begin. The supply manager or the sourcing team will now invite potential suppliers to submit bids or proposals. A decision must be made whether to use competitive bidding or negotiation (or a combination of the two) as the basis for source selection.

Bidding vs. Negotiation

When competitive bidding is used by private industry, requests for bids are traditionally sent to three to eight potential suppliers depending on the dollar size and complexity of the purchase. Requests for bids ask suppliers to quote the price at which they will perform in accordance with the terms and conditions of the resulting contract, should they be the successful bidder. The traditional bidding process is usually one pass. Government supply managers generally are not able to restrict the number of bidders to only eight. Rather, all suppliers wishing to bid are permitted to do so (for large purchases of standard commodities, the numbers are literally in the hundreds). Under competitive bidding, industrial supply managers generally, *but not always,* award the order to the lowest bidder. By law, government supply managers are routinely required to award the order to the lowest bidder, provided the lowest bidder is deemed qualified to perform the contract. It is prudent to create a qualified bidders list (QBL) to help prevent awarding the contract to a low bidder who cannot fulfill the contract requirements.

Prerequisites to Bidding

The proper use of competitive bidding is dictated by five criteria. When all five criteria prevail, competitive bidding is an efficient method of source selection and pricing. The criteria are:

1. The dollar value of the specific purchase must be large enough to justify the expense, to both buying and selling firms, that accompanies this method of source selection and pricing.
2. The specifications of the item or service to be purchased must be explicitly clear to both the buying and selling firms. In addition, the seller must know from actual previous experience, or be able to estimate accurately from similar past experience, the cost of producing the item or rendering the service.
3. The market must consist of an adequate number of sellers.
4. The sellers that make up the market must be technically qualified and *actively want* the contract—and, therefore, be willing to price competitively to get it.

5. The time available must be sufficient for using this method of pricing—suppliers competing for large contracts must be allowed time to obtain and evaluate bids from their subcontractors before they can calculate their best price. Thirty days is not an uncommon time; however, the increasing use of online bidding using the Web is forcing compression of bid preparation time.

Conditions Demanding Negotiation

In addition to satisfying the preceding five prerequisites, four other conditions should *not* be present when employing competitive bidding as the means of source selection:

1. Situations in which it is impossible to estimate costs with a high degree of certainty. Such situations frequently are present with high-technology requirements, with items requiring a long time to develop and produce, and under conditions of economic uncertainty.
2. Situations in which price is not the only important variable. For example, quality, schedule, and service may well be negotiable variables of significant importance.
3. Situations in which the buying firm anticipates a need to make changes in the specification or some other aspect of the purchase contract.
4. Situations in which special tooling or setup costs are major factors. The allocation of such costs and title to the special tooling are issues best resolved through negotiation.

If these nine conditions are satisfied, competitive bidding usually will result in the lowest price and is the most efficient method of source selection. To ensure that the lowest prices are obtained, the competing firms must be assured that the firm submitting the low bid will receive the award. If the buying firm gains a reputation for negotiating with the lowest bidders *after* bids are opened, then future bidders will tend *not* to offer their best prices initially, believing that they may do better in any subsequent negotiations. They will adopt a strategy of submitting a bid low enough to allow them to be included in any negotiations. But their initial bid will not be as low as when they are confident that the award will be made to the low bidder without further negotiation.[13] When any of the prerequisites to the use of competitive bidding are not satisfied, the *negotiation process* should be employed to select sources and to arrive at a price.

In his now famous lectures in Japan, W. E. Deming extolled that organizations should "end the practice of awarding business on the basis of price tag alone."[14] Several progressive supply management professionals offer two additional arguments that favor the use of negotiation over competitive bidding for critical procurements:

1. The negotiation process is far more likely to lead to a complete understanding of all issues of the procurement. This improved understanding greatly reduces subsequent quality and schedule problems.

2. Competitive bidding tends to put great pressure on suppliers to reduce their costs in order to be able to bid a low price. This cost pressure may result in sacrifices in product quality, development efforts, and other vital services.

Normally the competitive bidding system itself, when all the prerequisite criteria prevail, evaluates the many pricing factors bearing on the purchase being made quite accurately. These factors include determinants such as supplier production efficiency, willingness of the seller to price this particular contract at a low profit level, the financial effect on the seller of shortages of capital or excesses of inventories, errors in the seller's sales forecast, and competitive conditions in general.

Reverse Auctions

In contrast to standard competitive bidding, reverse auctions use the Internet for online "real-time" interaction where the number of bids submitted is limited primarily by the time provided by the buyer for the process. In addition, the number of bidders in reverse auctions can be very large. Reverse auctions are often used to help drive prices down on products and services that are believed to be above the true market value. The results are reported to be an average of 15% savings.[15] However, as with competitive bidding, reverse auctions are not appropriate for all situations. When discussing reverse auctions with managers in industry, they urge caution in opening the bidding process to unqualified suppliers.[16] Suppliers must be pre-qualified for reliable completion of the contract. The reverse bid process can have adverse affects on long-term supplier relationships. Amid the frenzy of this bid process, suppliers can inadvertently bid below their costs. While this may be appealing for the short term, economics dictate this cannot be sustained for the long run. Reverse auctions, as well as some bidding situations, also assume all the participating suppliers and their products/services are identical and equal—very risky assumptions.

Two-Step Bidding/Negotiation

On occasion, large, technically oriented firms and the federal government use a modified type of competitive bidding called "two-step bidding." This method of source selection and pricing is used in situations where *inadequate specifications* preclude the initial use of traditional competitive bidding. In the first step, bids are requested only for technical proposals, without any prices. Bidders are requested to set forth in their proposals the technical details describing how they would produce the required materials, products, or services. After these technical bids are evaluated and it is determined which proposals are technically satisfactory, the second step follows.

In the second step, requests for bids are sent only to those sellers who submitted acceptable technical proposals in the first step. These sellers now compete for the business on a price basis, as they would in any routine, competitive-bidding situation. The price is determined in either of two ways: (1) Award may be based solely on the lowest price received from those competing, or (2) the price proposals for the accepted technical approaches may be used as the beginning point for *negotiations*. It is important that the supply manager specify *at the outset* which of the two procedures will be used.

The Solicitation

Once a decision has been made whether to use competitive bidding or negotiation as the means of selecting the source, an *invitation for bids (IFB)* or a *request for proposal (RFP)* is prepared. The IFB or RFP normally consists of a purchase description of the item or service required, information on quantities, required delivery schedules, special terms and conditions, and standard terms and conditions. The legal implications of these documents and processes are discussed in Chapter 17.

When an RFP is used in anticipation of cost negotiations with one or more suppliers, the supply manager should request appropriate cost data in support of the price proposal. The supply manager must also obtain the right of access to the supplier's cost records that are required to support the reasonableness of the proposal. The cost data and the right of access must be established during the RFP phase of the procurement, at a time potential suppliers believe that there is active competition for the job.

Weighted Factor Analysis

In many instances, one prospective supplier is obviously superior to its competition—and selection is a very simple matter. Unfortunately though, the choice is not always so clear. In these cases, a numerical weighted-factor rating system can greatly facililtate the decision process.

A weighted-factor system calls for four activities: (1) the development of the factors that serve as the selection criteria and the weight these carry in the decision-making process, (2) the development of sub-factors or performance factors within the broader selection criteria and the weighting of these factors, (3) the development of a scoring factor to evaluate potential suppliers, and (4) the scoring or evaluating of the supplier.

The first activity, identification of the key factors to be considered in the selection decision, along with their respective weights, and the second activity of sub-factors and weights, typically are accomplished by a committee of individuals involved in the purchase. Factors to consider are specific to the particular buy and could include technical, financial, managerial, quality, or capability factors. Weights are assigned according to the importance of each factor. If technology outweighs cost in a particular buy, technical factors would be given more weight.

The sub-factors indicated activities under each main factor or category. As shown in Table 13.1, sub-factors of the technical factor would include understanding the problem, the technical approach to the product, the level of technology in the production facilitites, the operators's technical capabilities, and the maintenance requirements. Weights for

Table 13.1   An Illustration of the Weighted-Factor Rating Approach

Factors	Weights and Subweights	Score[a] 0–10 (A)	Weighted Score (Supplier A)	Score 0–10 (B)	Weighted Score (Supplier B)
Technical	40				
Knowledge of problem	10	9	9	7	7
Approach	20	9	18	8	16
Production facilities	5	8	4	6	3
Operators	3	6	1.8	6	1.8
Maintenance	2	10	2	10	2
			Total 34.8		Total 29.8
Delivery	20				
Requested date	10	10	10	8	8
Lead time	10	10	10	8	8
			Total 20		Total 16
Price	20				
Cost structure available	10	8	8	10	10
Value analysis efforts	10	8	8	10	10
			Total 16		Total 20
Managerial capability	10				
Labor relations	5	10	5	8	4
Financial strength	5	9	4.5	8	4
			Total 9.5		Total 8
Quality	10				
Quality control processes	5	9	4.5	8	4
Acceptable defect rate	5	9	4.5	8	4
			Total 9		Total 8
RATING TOTALS			89.3		81.8

[a]Scoring: A score of 9/10 possible is 90% of available points. Ninety percent of 10 available points of the subweight for "Knowledge of problem" earns Supplier A 9 points. Supplier B received a score of 7/10. or 70% of the 10 available points in the subfactor "Knowledge of problem." Supplier B cams 7 points.

these sub-factors would result from assigning a portion of the total weight for the overall factor or category.

Step 3 requires the developing of a numerical rating system with which to evaluate each supplier on each factor and sub-factor. Generally, a scale to assess these issues would be a 5- or 10- point scale, clearly defined to reduce subjectivity in the process. Clearly defined would include a defined rating of 0 to 5 where 0 may be "not able to conform" to 5 being "supplier meets all needs of customer." The better defined the factors and rating system, the more optimum the decision outcome.

The last step requires the assignment of numerical ratings for each of the competing firms. These assessments are based on the collective judgements of the evaluators after studying all the data and information provided by the potential suppliers, as well as that obtained in field investigations.

In effect, a weighted-factor rating system breaks a complex problem down into its key components and permits analysis of each component individually. The approach is widely used in practice and generally leads to a fair and reasonably objective result.

Responsibility for Source Selection

While supply management normally has the ultimate responsibility for selecting the "right" source, the process is handled in many ways. Procedurally, the simplest approach is when a single *supply manager* conducts the analysis and makes the selection. A second common approach calls for the use of a *cross-functional team* consisting of representatives of supply management, design engineering, operations, quality, finance, etc. The third common approach is the use of a commodity team. Commodity teams are created to source and manage a group of similar components. Commodity teams frequently consist of supply managers, materials engineers, and production planners. Larger commodity teams include a commodity manager (normally from supply management) and representatives of materials, design and manufacturing engineering, quality, and finance. Commodity teams are essentially a type of cross-functional team. The principle difference between them is that commodity teams tend to be fairly permanent, while cross-functional teams tend to be one-time assignments.

Developing Suppliers

Most suppliers have awakened to the awareness that their ability to be competitive on cost, quality, timeliness, and service is dependent on their suppliers' abilities to be contributing members of their supply chains. Not all suppliers need development assistance, but to reach the optimum collaborative relationship, frequently development assistance is required. Even suppliers recognized as the "best of the best" require investment on the part of the buying firm to realize the full benefit of the collaborative relationship. In major corporations such as General Electric, John Deere, Chrysler, and Honda of America, supply managers help their suppliers "improve quality, enhance delivery performance, and reduce costs."[17]

At Silicon Valley's Sun Microsystems, management treats development of suppliers as a two-way effort. In the late 1990s, Sun started having suppliers measure their performance using scorecards modified from the ones Sun used to evaluate their suppliers. The benefits to Sun were numerous. Suppliers felt as if Sun finally started to listen to their ideas instead of trying to push the Sun way of doing business on them. Suppliers could then contribute to developing Sun and improving the collaborative relationship to find "win-win" opportunities. The proactive efforts of Sun present the issue that the term "development" is actually a misnomer that implies a one-way interaction where the buying firm "develops" the supplying firm. The process of development should also include the development of the buyer in collaborative relationships.

In many instances, the buying firm may be unable to identify a world-class supplier that is willing (or able) to meet its needs. If the requirement is sufficiently important, the buying firm will select the most attractive supplier and then develop the supplier into one capable of meeting its present and future needs. Training in project management, teamwork, quality, production processes, and supply management may prove to be a worthy investment. Such training has been provided by several leading customer firms for well over a decade.

Managing Suppliers

The challenging issue of managing suppliers is dealt with in Chapter 2. At this point, however, it is essential to recognize that the supply manager has many responsibilities associated with the management of his or her suppliers. Satisfying these responsibilities should ensure that suppliers perform as required or that appropriate corrective action is taken to upgrade or eliminate them from the firm's supplier base.

In addition, supply management must, on a periodic basis, analyze its suppliers' abilities to meet the firm's long-term needs. Areas that deserve particular attention include the supplier's general growth plans, future design capability in relevant areas, the role of supply management in the supplier's strategic planning, potential for future production capacity, and financial ability to support such growth.

If present suppliers appear to be unlikely to meet future requirements, the firm has three options: (1) It may assist the appropriate supplier with financing and technological assistance, (2) it may develop new sources having the desired growth potential, or (3) it may have to develop the required capability internally.

Additional Strategic Issues

The buying firm must consider many factors in selecting sources of supply. This section presents several areas of concern. All of the issues described in this section have impacts beyond the supply management department. Accordingly, other departments must be involved in the decision making, and approval of the strategic outcomes should be sought from management above the functional areas impacted.

Early Supplier Involvement

As presented in Chapter 5, new product development and Chapter 6, specifications and standardization chapters, early supplier involvement (ESI) is an approach in supply management to bring the expertise and collaborative synergy of suppliers into the design process. Early involvement of the supplier seeks to find "win-win" opportunities in developing alternatives and improvements to materials, services, technology, specifications and tolerances, standards, order quantities and lead time, processes, packaging, transportation, redesigns, assembly changes, design cycle time, and inventory reductions. Today, early supplier involvement is an accepted way of life at many proactive firms and a requirement for "Proactive Supply Management" status. ESI helps in developing trust and communication between suppliers and the buying firm. ESI normally, but not always, results in the selection of a single source of supply. At most progressive companies, this selection process is the result of intensive competition between two or three carefully prequalified potential suppliers. The company selected becomes the single or primary source of supply for the life of the item using its material. The results of ESI translate into tangible cost savings. According to Dave Nelson, former Senior Vice President of Purchasing and Corporate Affairs for Honda of America, suppliers helped design the 1998 Accord and, as a result, lowered the cost of producing the car by over 20%.[18]

Supply Base Reduction

One of the interesting transitions taking place in supply management is the shift from enlarging the firm's supply base to downsizing the base. Reduction of the supply base is usually achieved through both reducing the variety of items procured and consolidating items previously procured from different suppliers with one supplier.

Supply base reduction success stories are abundant. For example, Xerox reduced its supply base by 92% in the early 1980s—from 5000 to 400 suppliers. Chrysler winnowed its supplier base from a mass of 2500 in the late 1980s to a lean, long-term nucleus of 300. During the 1990s, before its ill-fated acquisition by Daimler, suppliers loved working for Chrysler, and for obvious reasons: The company's production volume was growing rapidly. Chrysler included suppliers in development activities from day one and listened eagerly to their suggestions for design improvements and cost reductions. Chrysler had replaced its adversarial bidding system with one in which the company designated suppliers for a component and then used target pricing to determine with suppliers the component prices and how to achieve them. Most parts were sourced from one supplier for the life of the product.[19]

Applied Materials reduced its supply base from 1200 suppliers in the early 1990s to 400 by 2001. Applied Materials cites that the reduction has resulted in significant cost reductions in manufacturing and supply chain operations. Applied Materials develops and works with preferred suppliers that pass a series of qualification criteria. The reduced risk of being a preferred supplier empowers the company to invest in long-term strategies, such as early supplier involvement in design, materials research, process value analysis, and workforce education.

Similar supply base reductions have occurred at other major corporations. IBM uses 50 suppliers for 85% of its production requirements. Sun Microsystems uses 40 suppliers for 90% of its production material needs.[20] Such accomplishments generally require significant research and development expenditures or capital investment.

Two benefits of supply base reduction cited by John Deere are increased leverage with suppliers and better focus and supplier integration in product development. According to Deere, the increased leverage primarily results from the increased volume of business with the supplier.[21] However, as observers of Deere over the last several years, the authors of this book can safely state that the increased leverage is also due to the increased involvement with the suppliers, which builds goodwill and trust.

Single vs. Multiple Sourcing

Few strategic sourcing issues ignite more debate in corporate boardrooms around the world than the single vs. multiple source issue. Such decisions are larger than the supply management department, the commodity team, or the cross-functional team responsible for source selection. These decisions may affect the success—or even the survival—of the firm.

The major argument for placing all of a firm's business with one supplier is that in times of shortage, this supplier will give priority to the needs of a special customer. Additionally, single sources may be justified when:

▲ Lower total cost results from a much higher volume (economies of scale).
▲ Quality considerations dictate.[22]
▲ The buying firm obtains more influence—clout—with the supplier.
▲ Lower costs are incurred to source, process, expedite, and inspect.

▲ The quality, control, and coordination required with JIT manufacturing require a single source.

▲ Significantly lower freight costs may result.

▲ Special tooling is required, and the use of more than one supplier is impractical or excessively costly.

▲ Total system inventory will be reduced.

▲ An improved commitment on the supplier's part results.

▲ Improved interdependency and risk sharing result.

▲ More reliable, shorter lead times.[23]

▲ Time to market is critical.[24]

A common approach to multiple sourcing that can still yield many of the benefits of single sourcing is the "70–30" approach. Through the award of 70% of the volume to one supplier and 30% to a second supplier, economies of scale are obtained from the "big supplier" while the "little supplier" provides competition. Using the "70–30" strategy, when the 70% supplier "misbehaves," its volume is reduced and the smaller supplier is awarded an increase. (An interesting approach to discipline!)

Although the "70–30" strategy is reported to have started in Japan in the 1970s with JIT firms, the strategy is firmly established in many world-class companies. For example, visits in 2001 by one of the authors verified that Solectron, Applied Materials, and Cisco Systems all cognitively use the "70–30" approach in sourcing selected materials. Dual or multiple sourcing may be appropriate:

▲ To protect the buying firm during times of shortages, strikes, and other emergencies.

▲ To maintain competition and provide a back-up source. To meet local content requirements for international manufacturing locations.

▲ To meet customer's volume requirements.

▲ To avoid lethargy or complacency on the part of a single-source supplier.

▲ When the customer is a small player in the market for a specific item.

▲ When the technology path is uncertain.

▲ In areas where suppliers tend to leapfrog each other technologically.[25]

Note that the term "collaborative relationship" implies neither the presence nor the absence of a single-source relationship. That is, the buying firm may have one, two, or three "partners" for the same item, although the trend is toward single sourcing.

Share of Supplier's Capacity

Many highly regarded firms try not to exceed more than 15 to 25% of any one supplier's capacity. The percentage refers to the company's entire capacity and not one individual product or service. For example, Palm may contract for 15% of Flextronics production capacity, but purchase 100% of a specific product from Flextronics. Palm might reason that if its purchases represent too large a share of the supplier's business and they discontinue a product or purchase an item from another supplier, they could put Flextronics in a very difficult financial situation.

This issue became all too real in the early 2000s with the economic downturn. Many companies cancelled orders that had long supplier lead times, which resulted in suppliers being caught with, in some cases, hundreds of millions of dollars of work-in-process. Cisco Systems, for example, had outstanding orders that totaled around 2 billion! Unlike many companies during the early 2000s, Cisco paid for almost all of its orders and maintained its reputation as a good business partner. If Cisco had not paid its suppliers, the financial ruin that would have spread would have been even more staggering than it was. Still, Cisco's suppliers had to contend with few, if any, follow up orders because demand for Cisco's products had greatly diminished.

Local, National, and International Sourcing

Before the discovery process for building a supply base, the company must consider the issues of local vs. national vs. international sourcing. Local sourcing implies that the firm's headquarters and all facilities are located in the city or region (such as Northern California) where the materials or services will be used. Local sources are usually relatively small in contrast to national and international sources. National sourcing implies that the source is headquartered within the country and has facilities in multiple regions throughout the country. National sources are also larger companies as defined by the buying firm. For example, a national firm may be defined by a company as having a presence in at least two regions other than the one of the buying firm, at least 2000 employees, and annual revenues in excess of U.S. 100 million. The delineation between what constitutes a national source and an international source also should be defined by the buying firm. The common definition is that an international source is headquartered outside of the buying firm's country, but this does not define where the company has its operations. For example, Flextronics is headquartered in Singapore, but has virtually no manufacturing in Singapore.[26] The lines between local, national, and international sourcing have become blurred in the last 30 years.

Local Buying

Most supply managers prefer to patronize local sources whenever such action is prudent. A Stanford University research study found that approximately three-fourths of 152 supply managers surveyed indicated a preference to buy from local sources whenever possible. Many of them were willing to pay slightly higher prices to gain the advantage of better service and immediate availability of materials offered by some local suppliers, thereby lowering total cost.

JIT manufacturing requires dependable sources of defect-free materials that arrive within a very tight time frame. Suppliers to JIT customers meet their requirements in three ways: (1) They are located close to their customers,[27] (2) suppliers are implementing responsive manufacturing systems, and (3) they are taking aggressive action to control the transportation of their materials to their customers. In summary, local buying has the following advantages:

1. Closer cooperation between buying and selling firms is possible because of close geographical proximity. JIT deliveries are thus facilitated.
2. Delivery dates are more certain because transportation is only a minor factor in delivery.
3. Lower total costs can result from consolidated transportation and insurance charges. A local supplier, in effect, brings in many local buying firms' orders in the same shipment.
4. Shorter lead times frequently can permit reductions or the elimination of inventory. In effect, the seller produces JIT.
5. Rush orders are likely to be filled faster.
6. Disputes usually are more easily resolved.
7. Implied social responsibilities to the community are fulfilled.

Buying Nationally

National buying has the following advantages:

1. National sources, as a result of the economies of scale, in some situations can be more efficient than local suppliers and offer higher quality or better service at a lower price.
2. National companies often can provide superior technical assistance.
3. Large national companies have greater production capacity and therefore greater production flexibility to handle fluctuating demands.
4. Shortages are less likely with national companies because of their broader markets.

Buying Internationally

This important sourcing consideration is discussed in detail in the Chapter 14 on global supply management. It is mentioned here only to note that it is an important factor in supplier selection.

Manufacturer or Distributor

In deciding whether to buy from a manufacturer or distributor, a supply manager's considerations should focus largely on the distributor's capabilities and services, not on its location. In the steel industry, for example, distributors pay the same prices for steel as other buying firms. Distributors, however, buy in carload lots and sell in smaller quantities to users whose operations do not justify carload lot purchases. The distributors realize a profit because large lots sell at lower unit prices than do small lots. If a buying firm wishes to purchase steel directly from the mill and bypass the distributor, they are perfectly free to do so; however, when they do, they usually forgo certain special services that a competent distributor is equipped to offer. Distributors, for example, have cutting and shaping tools, and skilled personnel to operate them. They maintain large, diverse inventories. They also are able to perform numerous customer services.

When the materials ordered from a distributor are shipped directly to the user by the manufacturer (a *drop shipment*), an additional buying decision becomes necessary. In this situation, the distributor does not handle the materials physically; it acts only as a broker.[28] Under such circumstances, a supply manager is strongly motivated to buy directly from the manufacturer—if the manufacturer will sell to his or her firm.

Supply managers should be aware that distributors stock many manufacturers' products. Hence, ordering from a distributor can significantly reduce the total number of orders a supply manager must place to fill some of his or her materials requirements. If there were no distributors, orders for production as well as maintenance, repair, and operating (MRO) requirements would all have to be placed directly with many different manufacturers. This obviously would increase direct supply management costs. Furthermore, for every additional purchase order placed, an additional receiving, inspection, and accounts payable operation is created.

In the final analysis, the manufacturer-distributor decision centers on one critical fact: The functions of distribution cannot be eliminated. The supply manager needs most of these functions; therefore, the supply manager should pay for them once—but he or she should not pay for them twice. Either the distributor or the manufacturer must perform the essential distribution functions of carrying the inventory, giving technical advice, rendering service, extending credit, and so on. The supply manager must decide for each individual buying situation how to best purchase the functions needed. The supply manager must answer the question: Is it my company, the distributor, or the manufacturer that can perform the required distribution services satisfactorily at the lowest cost?

"Green" Supply Management

Environmentally sensitive supply management can make good business sense. Many of us have heard of the story of young Henry Ford. It seems that Mr. Ford was very explicit in the dimensions and quality of the lumber used in constructing the packing crates his suppliers used to ship parts to Ford. One day, one of the suppliers asked a Ford employee why a throwaway packing crate had to be made to such explicit specifications. The answer was, "Because we use the wood to build the floor boards of our Model T." Was Mr. Ford an environmentalist or a good businessman? Quite obviously, he was both!

Environmentally sensitive supply management has two components: (1) the purchase of materials and items that are recyclable and (2) the environmental and liability issues associated with the use and discharge of hazardous materials—anywhere in the supply chain.

Green supply management also addresses the goal of reducing the use of energy by purchasing energy efficient equipment and materials that require less energy to convert to use.[29] The emphasis on green operations must come from the goals of an organization as opposed to the goal of the purchasing department.

All environmentalists have at least one battle cry—reduce carbon emissions—and the U.K. seems to be the leader in establishing emission caps by establishing the Carbon Reduction Commitment (CRC) program, which complements the European Union (EU) emissions trading scheme.[30] These programs apply to the buyer's company products and to their suppliers. Suppliers can assist the buying company in early design involvement and in their own conservative programs. Wal-Mart has demanded its top 1000 Chinese suppliers improve their green footprint (remember the mercury- and lead-laced toys).[31] *Newsweek* has developed a scoring system that combines an environmental impact score, green policies score, and a reputation survey score to arrive at a total green score; Dell was the winner in 2010.[32] Reduction methods include everything from less energy (LED) to recycling waste.

Baxter International, a global healthcare company in Deerfield, Illinois, is a good example of green purchasing. In 2009, 35,000 suppliers in more than 100 countries accounted for an estimated 38% of Baxter's total carbon footprint.[33] This analysis stimulated development of its global supplier sustainability program that identifies 20 green criteria. This criteria's major segments are: environmental/sustainability programs: surveys of suppliers, ISO 1400 compliance, EMAS-EU, U.S. EPA smart way transport planner, etc; reductions in carbon footprint: goals to reduce greenhouse gas emissions and programs for renewable power; reductions in natural resource use: less energy, packaging waste, and water use; enhanced product stewardship: less use of hazardous substances, registration, evaluation and authorization of chemicals (REACH) compliant, less heavy materials in packaging, etc.[34]

One interesting result from this program involved changing the manufacturing method used to provide a rubber component. Baxter achieved reduced energy and raw material cost savings by changing the technique used to mold the product. This is a classic VE/VA technique that not only reduced carbons, but also saved money for both Baxter and the supplier. Baxter states that collaboration and incentives help "displace preconceived notions within the organization and among suppliers that green costs more."[35]

There are now several programs and sources to help you start an effective green purchasing program:[36]

1. Partnerships and Supply Chain for the EPA Climate Leaders Program, Environmental Protection Agency (EPA), Washington, D.C.
2. Carbon Disclosure Project's (CDP) Supply Chain Initiative, questionnaires to gather greenhouse gas (GHG) emissions.
3. EPA's Smart Way Program. Data on alternative fuels, federal & state laws, strategies to reduce GHG. See www.epa.gov/smartway.
4. Grants, see http://grants.gov.
5. National Clean Diesel Campaign. See www.epa.gov/cleandiesel/grantfund.htm.
6. Alternative Fuels and Advanced Data Center. See www.ardc.energy.gov/2pdc/laws/federal_summary.
7. EPA's Energy Star.
8. EPA's Green Suppliers Network.

Green purchasing should now be an important criteria to include in your sourcing efforts. It also reinforces our repeated recommendations regarding, trust, collaboration, training, and partnership efforts to have sustainable GHG programs with your suppliers. It may be obvious, but the buyer cannot simply order any suppliers to "just go do it"; green efforts must be joint programs. When developed in the correct way, these programs will be win-win for both buyers and sellers. Marketing is driving a lot of the green movement. For example, one major frozen food manufacturer proudly states on their packages "made from recycled material"; they obviously had to find a packaging supplier capable of providing this kind of packaging material. Even more dramatic is the program by Coke and Pepsi to develop plant-based packaging to replace the plastic bottles currently used in bottling their products. Neither firm could possibly make this conversion without full collaboration and joint packaging engineering efforts with their suppliers.[37]

Environmental and Liability Issues

Supply management, the firm's environmental engineer (or environmental consultant), and the firm's attorney should study the firm's value chain to identify the possible uses and disposal methods for environmentally hazardous substances and materials. It is entirely possible, for example, that a supplier who disposes of hazardous waste in an environmentally unsafe manner, while producing a product for the buying firm, may subject the buying firm to financial liability should the supplier have limited financial resources. Current statutes cover present and previous operators and owners.

Additionally, supply management has a responsibility to ensure that a supplier's salvage and disposal contractors meet OSHA standards both prior to award and during performance under the contracts. One way of dealing with this challenging issue is to require the supplier to post adequate performance and liability bonds.

Minority- and Women-Owned Business Enterprises

Many forces motivate a buying firm to develop and implement programs designed to ensure that minority- and women-owned business enterprises (MWBE) receive a share of the firm's business. These motivators include federal and state legislation, set-aside quotas in government appropriations, the actions of regulatory bodies such as the state public utilities commission, chambers of commerce, civil rights activists, and a firm's "corporate social consciousness." Perhaps one of the most significant motivators is the recognition by a firm's management that *its customer base* includes MWBEs and their employees. Companies that can demonstrate minority supplier content in their products are more likely to receive business from minority customers.[38] In addition to a sense of social responsibility, MWBEs should be focused on bottom-line profitability and good business sense.[39]

Ethical Considerations

Supply managers must be aware of potential conflicts of interest when selecting suppliers. A conflict of interest exists when supply managers must divide their loyalty between the firm that employs them and another firm. In supply management, this situation can occur when a supply manager is a substantial stockholder in a supplier's firm or when he or she makes purchases from close friends or relatives. The subject is introduced here solely to remind the reader that such conflicts always should be avoided in all source selection decisions.

Supply managers should keep themselves as free as possible from unethical influences in their choice of suppliers. It is very difficult to maintain complete objectivity in this matter, for it is only human to want to favor one's friends. On occasion, friends can make unusually good suppliers. They will normally respond to emergency needs more readily than suppliers without a strong tie of personal friendship. However, supply managers tend not to discipline friends who perform poorly to the same degree as they do other suppliers. A detailed discussion of ethical issues and social responsibilities is contained in Chapter 21.

Reciprocity

When supply managers give preference to suppliers that are also customers, they are engaging in a practice known as *reciprocity*.[40] The practice can be illegal, and the line between legal and illegal reciprocal practices frequently is very thin. It is entirely legal to buy from one's customers at fair market prices, without economic threat, and without the intent of restricting competition. A key criterion used by the courts in determining illegality is the degree to which reciprocal activity tends to restrict competition and trade. Hence, those who engage in reciprocal practices must do so with care and legal consultation.

Most supply managers disapprove of the practice of reciprocity, even when legal, because it restricts their ability to achieve competition among potential suppliers. However, the proponents of reciprocity contend that it is simply good business. They believe that if a supply manager buys from a friend, both the supply manager and the friend will profit in the long run. They maintain that service is better from suppliers who are also customers. They argue that reciprocity is a legitimate way to expand a company's markets. Consequently, some U.S. firms proclaim a reciprocity policy similar to this one: "When important factors such as quality, service, and price are equal, we prefer to buy from our customers."[41]

In the final analysis, reciprocity is neither a marketing problem nor a supply management problem; rather it is a management problem. If management believes that it can expand its markets permanently and add to the firm's profit *legally* by reciprocity, then this is the decision management should make. Conversely, if management believes buying without the constraints of reciprocity will increase profit, then that is the policy management should adopt. Although reciprocity can benefit a firm, no economist would argue that it benefits a nation's total economy.

Consortium Purchasing and Group Buying

Historically, group or cooperative buying as it was commonly called, was largely restricted to collective contracting or pooled requirements of non-profit entities such as schools, hospitals, and local government units. This procurement method was based on the obvious principle of leveraged volume discounts of standard "shelf items" and generic commodities. For example, all of the secondary schools in a local district could combine their requirements for desks, lockers, janitorial supplies, and computers and have a team negotiate a master contract with a supplier to achieve the lowest price for the combined quantity of new items and materials. The groups were very careful to obtain exemptions from federal and state anti-trust law, which forbids conspiracy to restrict competition, set prices, and other such restrants of trade as explained in Chapter 19. In recent years, many procurement personnel call group buying the collaborative model. The e-commerce boom of the 1990s triggered profit organizations such as manufacturers to form collective buying organizations for vertical buying within one industry. Often called exchanges or consortiums, they primarily used reverse auctions to drive the price to the lowest point for any particular "event," as they called their auction rounds. There were consortiums for utilities, aircraft parts, and many other industries.

The most dramatic example was Covisint, the rather strange name for the joint effort by Ford, GM, and Chrysler to standardize selected automotive parts and suppliers for all three and then hold reverse auctions and negotiate to achieve the lowest possible price. This enormous effort started with a rather large staff in Southfield, Michigan around 2002 and failed in late 2004. The major reasons for this failure and other similar vertical consortiums appear to be:

1. The difficulty of forcing long-time competitors to agree on the same specifications.
2. Reluctance and even refusal of many major suppliers to participate in a scheme to drive the price down in a return to adversarial purchasing.
3. The lack of enthusiasm and support by plant level supply-purchasing managers.
4. The added "middleman" expense of the consortium buying group.
5. The lingering (and very legitimate) fear of federal and state anti-trust action. Many corporate attorneys were horrified at the thought of such obvious conspiracy among competitors to fix prices.

Several purchasing executives have said that many of the vertical industry consortiums either have ceased operation or have evolved into data collection services tracking prices and sources for materials.[42]

However, the third-party horizontal group buying ventures, those that combine volume requirements of indirect material from firms in several different industries, appear to be successful. It is interesting to note these ventures do not use the term consortium. One of the best examples is Corporate United, Inc. in Cleveland, Ohio. Since 1997, Corporated United has managed indirect purchases for its member companies such as MRO supplies and services such as human resources, marketing, communication, IT, telecom, engineering, laboratory, and even consulting. Using several sourcing methods including personal negotiation, bidding, and, when appropriate, reverse auctions, Corporate United provides member companies with pre-negotiated master agreements and then manages these contracts throughout the contract life cycle.[43]

Disaster Plans

Your key supplliers should have written disruption contingency plans, especially if they are in high-risk areas for hurricans, flooding, earthquakes, and other "acts of God." Do they have state-of-the-art fire prevention plans? Have they backup suppliers? Do they have alternate logistics plans? There is more on this subject in the Endnotes of Chapter 22 under the topic of purchasing risk management and in the appendix for Chapter 22.

Summary

A supply manager's first responsibility in source selection is to develop and manage a viable supply base. Before development of the supply base, many strategic issues need to be addressed, such as supply base size, single vs. multiple sourcing, supplier involvement in design, negotiation vs. bidding, share of supplier's capacity, whether to internationally source, manufacturer or distributor, whether to pursue "green" supply management, and policies related to diversity, ethics, and reciprocity. These issues should be addressed in the strategic sourcing plan where relevant. In the plan, details about how suppliers will be discovered, evaluated, selected, developed, and managed should be committed to paper.

Clearly, the activity of developing a strategic sourcing plan impacts many functional areas as well as supply chain members. Therefore, the strategic sourcing plan should not be developed within a supply management vacuum. The plan should be developed in a collaborative environment that includes all relevant functional area representatives and supply chain members.

Selection of the right source is more important today than ever before, because more firms are entering into long-term collaborative relationships with a single source of supply. The benefits of such collaborative relationships are many, but the risks are great. Careful selection of suppliers and the professional management of the relationships are essential as supplier performance has a significant impact on customer satisfaction.

Appendix

Illustrative Plant Survey

After conducting preliminary surveys of potential suppliers of critical materials, equipment, or services, frequently it is desirable to conduct a plant visit to one or two of the most attractive candidates. The purpose of such visits is to gain firsthand knowledge of the supplier's facilities, personnel, and operations. Such a visit normally is conducted by a team from the buying firm. Each member will study his or her area of expertise at the potential supplier's operation.

The plant survey shown next is used by one high-tech manufacturer. The evaluation form calls for yes or no answers to specific questions and evaluation ratings for all questions, asking the team member to evaluate how well the supplier is doing in a given area. The evaluation ratings are described next.

Rating	Description
10	The provisions or conditions are extensive and function is excellent.
9	The provisions or conditions are moderately extensive and function is excellent.
8	The provisions or conditions are extensive and are functioning well.
6/7	The provisions or conditions are moderately extensive and are functioning well.
4/5	The provisions or conditions are limited in extent but are functioning well.
2/3	The provisions or conditions are moderately extensive but are functioning poorly.
1	The provisions or conditions are limited in extent and are functioning poorly.
0	The provisions or conditions are missing but needed.

Assume that the buying firm's quality manager is rating a potential supplier on the sections entitled quality management and quality information. When these phases of the study are complete, the quality manager is in a position to assign an average or overall rating for quality management on the first sheet of the survey.

Such an evaluation is used in at least two ways: (1) it now is possible to compare competitors' operations, with major emphasis on quality, and (2) actual or potential problem areas for an otherwise attractive supplier may be identified. The buying firm then may require that the area of concern be corrected or upgraded before award or as a condition of award.

Design Information

Control of design and manufacturing information is essential for the control of the product. It consists of making sure that operating personnel are furnished with complete technical instructions for the manufacture and inspection of the product. This information includes drawings, specifications, special purchase order requirements, engineering change information, inspection instructions, processing instructions, and other special information. A positive recall system is usually considered necessary to ensure against use of superseded or obsolete information.

1. (　)　　How well do procedures cover the release, change, and recall of design and manufacturing information, including correlation of customer specifications, and how well are procedures followed?

2. (　)　　How well do records reflect the incorporation of changes?

3. (　)　　How well does quality control verify that changes are incorporated at the effective points?

4. (   )   Is the design of experiments employed to ensure robust designs prior to the release of designs to manufacturing and supply management?

5. (   )   How well is the control of design and manufacturing information applied to the procurement activity?

6. (Y / N)   Is there a formal deviation procedure, and how well is it followed?

7. (Y / N)   Does your company have a written system for incorporating customer changes into shop drawings?

8. (Y / N)   Does your company have a reliability department?

9. (Y / N)   Are reliability data used in developing new designs?

10. (Y / N)   Is quality history fed back to engineering for improvements in current or future designs? Does quality management review new designs?

11. (Y / N)   Does your company have a sample or prototype department?

12. (Y / N)   Does QA review sample prototypes?

13. (Y / N)   Is this information used in developing shop inspection instructions?

14. (Y / N)   Are customer specifications interpreted into shop specifications?

15. (Y / N)   Do drawings and specifications accompany purchase orders to suppliers?

16. (Y / N)   Are these reviewed by quality management?

17. (Y / N)   Are characteristics classified on the engineering documents as to importance?

18. (Y / N)   Does QA review new drawings with the intent of designing gauging fixtures?

Procurement—Control of Purchased Material

It is essential for the assurance of quality that outside suppliers meet the standards for quality imposed on the firm's own operations department. Sources should be under continuous control or surveillance. Incoming material should be inspected to the extent necessary to assure that the requirements have been met.

19. (   )   How well are potential suppliers evaluated and monitored?

20. (   )   How well are quality requirements specified?

21. (   )   How well are inspection procedures specified, and how well are they followed?

22. (   )   How adequate are inspection facilities and equipment?

23. (   )   Have you certified (approved) key suppliers' design manufacturing and quality processes so that their shipments to you do not require inspection and testing?

24. (   )   How adequate are "certifications," which are used in lieu of inspection?

25. (   )   How well are certifications evaluated by independent checking?

26. (   )   How well are inspection results used for corrective action?

27. (Y / N)   Do you have an incoming inspection department? (If yes, list personnel) Inspectors _____ Supervisors _____ Quality Engineers _____
28. (Y / N)   Are purchase orders made available to incoming inspection?
29. (Y / N)   Is there a system for keeping shop drawings up-to-date?
30. (Y / N)   Are written inspection instructions available?
31. (Y / N)   Is sample inspection used?
32. (Y / N)   Is gauging equipment calibrated periodically?
33. (Y / N)   Is gauging equipment correlated with suppliers' equipment?
34. (Y / N)   Are suppliers' test records used for acceptance?
35. (Y / N)   Are commercial test records used for acceptance?
36. (Y / N)   Is material identified to physical and chemical test reports?
37. (Y / N)   Are records kept to show acceptance and rejection of incoming material?
38. (Y / N)   Does your company have a supplier rating system?
39. (Y / N)   Is it made available to the supply management department?
40. (Y / N)   Is the supplier notified of nonconforming material?
41. (Y / N)   Does your company have an approved supplier list?
42. (Y / N)   Does your company survey supplier facilities?
43. (Y / N)   Does the incoming inspection department have adequate storage space to hold material until it is inspected?
44. (Y / N)   Is nonconforming material identified as such?
45. (Y / N)   Is nonconforming material held in a specific area until disposition can be made? Who is responsible for making disposition of nonconforming material?

Material Control

Control of the identity and quality status of material in-stores and in-process is essential. It is not enough that the right materials be procured and verified; they must be identified and controlled in a manner that will assure they are also properly used. The entire quality program may be compromised if adequate controls are not maintained throughout procurement, storage, manufacturing, and inspection.

46. (   )   How adequate are procedures for storage, release, and movement of material, and how well are they followed?
47. (   )   How well are incoming materials quarantined while under test?
48. (   )   How well are materials in-stores identified and controlled?
49. (   )   How well are in-process materials identified and controlled?
50. (   )   How well are materials in inspection identified and controlled?
51. (   )   How adequate are storage areas and facilities?

52. (   )   How well is access to material controlled?
53. (   )   How well do procedures cover the prevention of corrosion, deterioration, or damage of material and finished goods?
54. (   )   How well are they followed?
55. (   )   How well are nonconforming items identified, isolated, and controlled?

Manufacturing Control

In-process inspection, utilizing the techniques of quality control, is one of the most satisfactory methods yet devised for attaining quality of product during manufacture. Because many quality characteristics cannot be evaluated in the end product, it is imperative that they be achieved and verified during the production process.

56. (   )   How well are process capabilities established and maintained?
57. (   )   How well is in-process inspection specified?
58. (   )   How effectively is it performed?
59. (   )   How adequate are inspection facilities and equipment?
60. (   )   How well are the results of in-process inspection used in the promotion of effective corrective action?
61. (   )   How adequate are equipment and facilities maintained?
62. (   )   How adequate are housekeeping procedures, and how well are they followed?
63. (   )   Does your company have a process inspection function? (If yes, list on a separate sheet inspectors, supervisory and quality engineering personnel.) To whom does process inspection report?
64. (Y / N) Are inspection stations located in the production area?
65. (Y / N) Are shop drawings and specifications available to inspection?
66. (Y / N) Is there a system for keeping the documents up-to-date?
67. (Y / N) Are written inspection instructions available?
68. (Y / N) Is there a system for reviewing and updating inspection instructions?
69. (Y / N) Is sample inspection used?
70. (Y / N) Do production workers inspect their own work?
71. (Y / N) Are inspection records kept on file?
72. (Y / N) Is inspection equipment calibrated periodically?
73. (Y / N) Is all material identified (route tags, etc.)?
74. (Y / N) Is defective material identified as such?
75. (Y / N) Is defective material segregated from acceptable material until disposition is made?
76. (Y / N) Are first production parts inspected before a job can be run?
77. (Y / N) Is corrective action taken to prevent the recurrence of defective material?

78. (Y / N) Who is responsible for making disposition of nonconforming material?
79. (Y / N) Does your company use x-bar and *R* charts?
80. (Y / N) Does your company use process capability studies?
81. (Y / N) Are standards calibrated by an outside source that certifies traceability to NBS?
82. (Y / N) Are standards calibrated directly by NBS?
83. (Y / N) Are packaged goods checked for proper packaging?

Quality Management

The key to the management of quality lies in philosophy, objectives, and organization structure. The philosophy forms the primary policy and should include the broad principles common to good-quality programs. The objectives should be clearly stated in specific terms and should provide operating policies which guide the activity of the quality program. The organizational structure should clearly define lines of authority and responsibility for quality from top management down to the operating levels.

84. (   ) Does the potential supplier embrace total quality management?
85. (   ) How adequate is the quality philosophy, and how well is it explained in operating policies and procedures?
86. (   ) How adequate is the technical competence in the quality discipline of those responsible for assuring quality?
87. (   ) How well does the organizational structure define quality responsibility and authority?
88. (   ) How well does the organizational structure provide access to top management?
89. (   ) How adequate is the documentation and dissemination of quality control procedures?
90. (   ) How adequate is the training program, including employee records?

Does the quality department have:

91. (Y / N) Written quality policy and procedures manual?
92. (Y / N) Written inspection instructions?
93. (Y / N) A quality engineering department?
94. (Y / N) Person or persons who perform vendor surveys?
95. (Y / N) Incoming inspection department?
96. (Y / N) In-process inspection department?
97. (Y / N) Final inspection department? To whom does the inspection department report? _____

98. (Y / N)   A quality audit function?
99. (Y / N)   A gauge control program?
100. (Y / N)   A gauge control laboratory?
101. (Y / N)   Other quality laboratories? (If yes, specify type) _____
102. (Y / N)   A quality cost program?
103. (Y / N)   A reliability department?
104. (Y / N)   Does the quality department use statistical tools (control charts, sampling plans, etc.)? Explain _____
105. (Y / N)   Is government source inspection available to your plant? Resident _____ Itinerant _____ No _____

Quality Information

Records of all inspections performed should be maintained, and the data should be analyzed periodically and used as a basis for action. Quality data should always be used, whether it is to improve the quality control operation by increasing or decreasing the amount of inspection, to improve the quality of product by the initiation of corrective action on processes or suppliers, to document certifications of product quality furnished to customers, or to report quality results and trends to management. Unused or unusual data are evidence of poor management.

106. ( )   How well are records of inspections maintained?
107. ( )   How adequate is the record and sample retention program?
108. ( )   How well is quality data used as a basis for action?
109. ( )   How well is quality data used in supporting certification of quality furnished to customers? How well is customer and field information used for corrective action?
110. ( )   How well is quality data reported to management?

Calibration—Inspection and Testing

Periodic inspection and calibration of certain tools, gauges, tests, and some items of process control equipment are necessary for the control and verification of product quality. Controlled standards periodically checked or referenced against national standards will assure the compatibility of vendor and vendee measurements. Inaccurate gauges and testers can compromise the entire quality control program and may result in either rejection of good material or acceptance of defective material.

111. ( )   How well do internal standards conform to national standards or customer standards?
112. ( )   How well are periodic inspections and calibrations specified?

113. ( )   How adequate are calibration facilities and equipment?

114. ( )   If external calibration sources are utilized, how adequate is the program and how well is it executed?

115. (Y / N)   Does your company have a gauge control function?

116. (Y / N)   Does your company have written instructions for operating inspection and test instruments?

117. (Y / N)   Are all inspection instruments calibrated at periodic intervals?

118. (Y / N)   Are records of calibration kept on file?

119. (Y / N)   Is there a system to recall inspection instruments when they are due for calibration?

120. (Y / N)   Are the inspection instruments used by production calibrated?

121. (Y / N)   If not, are these instruments removed from use until they can be repaired or recalibrated?

122. (Y / N)   Are shop masters calibrated at periodic intervals to secondary standards traceable to NBS?

Inspection of Completed Material

123. (Y / N)   Does your company have a final inspection function? If yes, list inspection, supervisory, and quality engineers on a separate sheet. To whom does the final inspection department report? _____

124. (Y / N)   Are shop drawings and specifications available for inspection?

125. (Y / N)   Is there a system for keeping the documents up-to-date?

126. (Y / N)   Are written inspection instructions available?

127. (Y / N)   Is there a system for reviewing and updating inspection instructions?

128. (Y / N)   Is a sample inspection used?

129. (Y / N)   Are inspection records kept on file?

130. (Y / N)   Are records of inspection results used for corrective action purposes?

131. (Y / N)   Is inspection equipment calibrated periodically?

132. (Y / N)   Is all material identified (route tags, etc.)?

133. (Y / N)   Is defective material identified as such?

134. (Y / N)   Is defective material segregated from acceptable material until disposition is made? Who is responsible for making disposition of nonconforming material? _____

135. (Y / N)   Is reworked material submitted for reinspection?

Final Acceptance

Final inspection, testing, and packing are critical operations necessary to assure the acceptability of material. The specifications must form the basis for these activities. To the extent that certifications or in-process inspections are used, in lieu of final inspection, records of those activities should be reviewed to verify conformance.

136. How well are specifications used in determining the acceptability of material?

137. ( ) How well are certifications and in-process inspection records used in the final acceptance decisions?

138. ( ) How adequate are inspection procedures? How well are they followed?

139. ( ) How adequate are inspection facilities and equipment?

140. ( ) How well are inspection results used for corrective action?

141. ( ) How adequate are packing and order-checking procedures?

142. ( ) How well are they followed?

Financial Statement Analysis[44]

The following information is useful during preliminary sourcing. These ratios and measures are useful for analyzing company-specific trends and for making comparisons among competing suppliers. Comparative data for specific industries may be obtained from Dun and Bradstreet or Robert Morris Associates.

Liquidity Measures

Liquidity refers to a company's ability to pay its bills when they are due and to provide for unanticipated cash requirements. In general, poor liquidity measures imply short-run credit problems. From a supply-management perspective, short-run credit problems could signal possible decreases in quality or difficulties in meeting scheduled deliveries. Three common liquidity measures are:

1. Working capital = current assets – current liabilities
   Working capital measures the amount of current assets that would remain if all current liabilities were paid.
2. Current ratio = current assets/current liabilities
   The current ratio is a standardized measure of liquidity. In general, the higher the ratio, the more protection a company has against liquidity problems. However, the ratio can be distorted by seasonal influences and abnormal payments on accounts payable made at the end of the period.
3. Quick ratio = quick assets/current liabilities
   The quick ratio is a standardized measure of liquidity in which only assets that can be converted to cash quickly (e.g., cash, accounts receivable, and marketable securities) are included in the calculation.

Funds Management Ratios

The financial position of a company depends on how it manages key assets such as accounts receivable, inventory, and fixed assets. As a business grows, the associated expansion of these items can lead to significant cash shortages—even for companies that maintain profitable operations. As implied previously, short-term cash problems may signal future decreases in quality or delays in scheduled deliveries. Six frequently used measures of funds management are:

1. Receivables to sales = accounts receivable/sales
   In the absence of detailed credit information, the receivables-to-sales ratio can be used to analyze trends in a company's credit policy.
2. Average collection period = (accounts receivable/sales) × 365
   The average collection period is used to assess the quality of a company's receivables. The average collection period may be assessed in relation to the company's own credit terms or to the typical credit terms of firms in its industry.
3. Average accounts payable period = (accounts payable/purchases) × 365
   The average accounts payable period is used to assess how well a firm manages its payables. If the average days payable is increasing, or large in relation to the credit terms offered by the company's suppliers, it may signal that trade credit is being used as a source of funds.
4. Inventory turnover = cost of goods sold/average inventory
   The inventory-turnover ratio indicates how fast inventory items move through a business.
5. Average days in inventory = 365/inventory turnover
   The average days in inventory is a simple conversion of the turnover ratio to a more intuitive measure of inventory management.
6. Fixed asset turnover = sales/average fixed assets
   The fixed asset turnover provides a crude measure of how well a firm's investment in plant and equipment is managed relative to the sales volume it supports. Unfortunately, interpreting the fixed asset turnover ratio is not always a straightforward proposition. For example, a decrease in the firm's turnover ratio could result from poor management of fixed assets *or* from an investment in new technology (e.g., computer integrated manufacturing).

Profitability Measures

Profitability refers to the ability of a firm to earn positive cash flows and to generate a satisfactory return on shareholders' investments. Profitability measures provide an indication of a firm's long-term viability. Profitability measures may be used to infer quality in the sense that, to generate a satisfactory return, a firm must ensure that it provides quality products from year to year. Profitability measures may also be used, to some extent, to infer a company's pricing policies. Thus, they may be useful for negotiating contract prices and, in particular, the profit portion of such prices.

1. Profit margin = net income/sales
   The profit margin percentage measures the amount of net income earned on a dollar of sales.
2. Gross profit margin = gross margin/sales
   The gross profit margin percentage measures the gross profit earned on each dollar of sales. Thus, this ratio may be used to infer the typical markup percentage used by a supplier.
3. Return on assets = net income/average total assets
   Return on investment measures how efficiently assets are used to produce income.
4. Return on equity = net income/average stockholders' equity
   Return on equity measures the percentage return on the stockholders' average investment.

Measures of Long-Term Financial Strength

The ability to deliver quality products over time is contingent on the long-term financial strength of the supplier. Difficulties meeting long-term obligations may also signal insolvency. The disruptions caused by insolvency can cause major delays in shipments, decreases in quality, or complete inability to perform.

1. Debt to equity = total liabilities/stockholders' equity
   Since debt requires periodic interest payments and eventual repayment, it is inherently more risky than equity. The debt-to-equity ratio measures the proportion of the company that is financed by creditors relative to the proportion financed by stockholders.
2. Times interest earned = operating profit before interest/interest on long-term debt
   The times-interest-earned ratio measures the extent to which a company's operating profits cover its interest payments. A low times-interest-earned ratio may signal difficulties in meeting long-term financial obligations.

Planning a Facility Visit

In planning facility or plant visits, only a few outstanding potential suppliers' plants should be chosen for observation due to the time and costs involved. In addition to observing production equipment and operations, there are other compelling reasons for plant visits. It is vital, for example, to determine a supplier's managerial capabilities and motivation to meet contractual obligations. The buying firm wants suppliers whose management is committed to excellence. To make such a determination properly requires an overall appraisal.

Among the factors to be addressed are:

▲ Attitude and stability of the top- and middle-management teams
▲ R&D capability
▲ Appropriateness of equipment
▲ Effectiveness of the production control, quality assurance, and cost control systems
▲ Competence of the technical and managerial staffs

Other important factors include:

▲ Morale of personnel at all levels
▲ Industrial relations
▲ Willingness of the potential supplier to work with the buying firm
▲ Quantity of back orders
▲ Effectiveness of supply management and materials management operations
▲ Past performance:
 ▼ Past major customers
 ▼ General reputation
 ▼ Letters of reference

The plant visit should be planned carefully to provide the required level of knowledge and insight into the potential supplier's operations, capacity, and orientation. The efficiency with which the plant visit is planned and conducted reflects on the buying organization. To provide the reader with a sense of the detail in which some plant visits are conducted, one firm's evaluation form is reproduced in Appendix A of this chapter. This evaluative instrument is used by a high-tech firm that utilizes a number of single-source collaborative relationship arrangements.

Although it varies with the firm's size and organizational structure, the *initial orientation meeting* typically is attended by the sourcing team members and their management counterparts from the potential supplier's organization. In smaller firms, the president often leads the supplier group. In this session, the sourcing team provides general information and explains the interests of its company: its kind of business, a brief history, kinds of products, the importance of the item to be purchased, and volume, quality, and delivery requirements.

The prospective supplier usually is requested to provide additional information on the company's history, current customers, sales volume, and financial stability. If classified or confidential data might be involved during design and production operations, the sourcing team must review the supplier's security control system.

In the quality area, the supply manager and his or her sourcing teammates should attempt to understand the *supplier's attitude* toward quality by asking questions such as:

How do you feel about zero defects and total quality management?
Are you ISO 9000 certified?
Do you employ the design of experiments during new product development?
Do you employ statistical process control?
Have you adopted a total quality commitment plan?
How do you measure customer satisfaction?
Show us how you've implemented these concepts. Do you have a quality manual?

Copies of the manual and the policy should be reviewed by the buying team's quality representative. The potential supplier also should be asked to describe *its own* supplier quality control program. The more technical aspects of supplier quality are addressed in Chapter 21. An increasing number of firms, including Motorola and many other leading-edge manufacturers, require that potential suppliers be registered under the appropriate ISO 9000 quality standard. It is important that the supply manager and members of the sourcing team understand that ISO 9000 is a series of process standards, and that an ISO certificate is not necessarily a guarantee of high product quality.[45]

If the sourcing team is satisfied with the results of the introductory meeting, a tour of the facilities typically is made. Prior to the tour, the sourcing team should get permission to talk freely with various individuals working in the operation, not just hand-picked managers.

The potential supplier's management often assumes that satisfactory operating controls are in place. In reality, however, experience frequently shows that some of these controls may not have been implemented, or that they may have been discontinued. When management describes these things, the sourcing team should respond, "That sounds excellent, we'd like to see them." The team should also check controls by asking such questions as, "How do you ensure that the most current drawing is in use?" "How do you segregate rejected materials?"[46]

As a cross-check on specific information obtained in the initial meeting, perceptive team members often ask shop and staff personnel similar types of questions. When the potential supplier's managers are not present, operating personnel should be asked about *their* understanding of the firm's quality systems, schedules, cost control efforts, and related requirements. Workers can also be asked about working conditions and turnover, about the firm's commitment to quality, about quality tools, and about training.

When observing plant equipment, the sourcing team should determine whether the equipment is modern, whether it is in good operating condition, whether tolerances can be held consistently, and what the output rates are. The sourcing team also should look for special modifications or adaptations of equipment; these things often provide clues to the ingenuity of operating management personnel.

The first impression of a supplier's operation is generally obtained through observing the housekeeping of the plant itself. Is it clean and well organized? Are the machines clean? Are the tools, equipment, and benches kept orderly and accessible? Good housekeeping tends to be an indication of efficiency. Many sourcing team members responsible for source selection believe it is reasonable to expect that a firm displaying pride in its facilities and equipment will also take pride in the workmanship that goes into its products.

During the visit, responsible sourcing team members should investigate production methods and efficiencies. Is a JIT system utilized to a significant extent? Is material moving freely from storage to production areas? Are there any production bottlenecks? Is the production scheduling and control function organized and functioning well? Is reserve production capacity available? Is it available on a regular or an overtime basis? Does the potential supplier have a competent maintenance crew? Finally, the sourcing team should determine whether inventory levels for both production materials and finished goods are adequate for the company's needs.

Employee attitudes are extremely important. In the long run, production results often depend more on people than on the physical plant. Do the employees seem to work harmoniously with one another and with their supervisors? Are they interested in quality and in improving the products they make? Is enthusiasm at a reasonable level? In short, do the people take pride in their jobs and in the firm—or do they view it as an eight-to-five clock-punching operation?

Endnotes

1. The benefits of collaborative relationships and alliances were presented in Chapter 4.
2. For more insight into this important strategic issue, see C. K. Prahalad and Gary Hamel, "The Core Competence of the Corporation," *Harvard Business Review* May–June (1990): 79–91.
3. Kompass Publications, Ltd., Windson Court, East Grimstead House, East Grimstead, Sussex, RH19-IXD, England.
4. Roberta J. Duffy, "The Future of Purchasing and Supply: Supply Chain Partner Selection and Contribution," *Purchasing Today*, November (1999).
5. M. Ensign, "Breaking Down Financial Barriers," *Purchasing Today*, July (2000).
6. L. Forker, W. Ruch, and J. Hershauer, "Examining Supplier Improvement Efforts from Both Sides," *Journal of Supply Chain Management* Summer (1999): 40–50.
7. Ensign, op. cit.

8. P. L. Carter, J. R. Carter, R. M. Monczka, T. H. Slaight, and A. J. Swan, "The Future of Purchasing and Supply: A Ten-Year Forecast," *Journal of Supply Chain Management* Winter (2000): 14–26.

9. Caron H. St. John and Kirk C. Heriot, "Small Suppliers and JIT Purchasing," *International Journal of Purchasing and Materials Management* Winter (1993): 12.

10. A follow-up study to the 10 year forecast warns that companies need to consider factors beyond operational excellence. They predict companies will turn to "value-based" sourcing to leverage a supplier's full capabilities for competitive advantage. P. L. Carter, J. R. Carter, and R. M. Monczka, *Key Supply Strategies For Tomorrow*, (Chicago, IL: ATKearney, 2007).

11. R. E. Spekman, J. Kamauff, J. Spear, "Towards More Effective Sourcing and Supplier Management," *European Journal of Purchasing and Supply Management* 5(1999): 105.

12. Carter et al., 2000, op. cit.

13. On occasion, a supply manager may intend to use the initial proposal solicitation process to identify firms with which he or she plans to conduct follow-up negotiations. In this case, professional ethics as well as good business judgment dictate that the initial solicitation state clearly that follow-up negotiations will be conducted.

14. Larry Weinsteinm, "Single-Source Successes and Snafus," *Purchasing Today* April (2000).

15. Srinivas Talluri and Gary L. Ragatz, "Multi-Attribute Reverse Auctions in B2B Exchanges: A Framework for Design and Implementation," *Journal of Supply Chain Management,* 40(1)(2004): 52.

16. Supply Chain Management Forum: Focus on Supply Management, The University of San Diego, 2001.

17. Janet Hartley and Gwen Jones, "Process Oriented Supplier Development: Building the Capability for Change," *International Journal of Purchasing and Materials Management* 33(3)(1997): 24.

18. R. E. Spekman, J. Kamauff, and J. Spear, "Towards More Effective Sourcing and Supplier Management," *European Journal of Purchasing and Supply Management* 5(1999): 105.

19. James P. Womack and Daniel T. Jones, "From Lean Production to the Lean Enterprise," *Harvard Business Review* March–April (1994): 97.

20. James Carbone, "Evaluation Programs Determine Top Suppliers," *Purchasing* November 18, 1999.

21. Bill Butterfield, "Supplier Development at John Deere," Presentation at the 16th Annual Supply Chain Forum, The University of San Diego, San Diego, CA, November 2001.

22. James Dairs, Manager of Transportation Programs at G.E.'s Plastics Group, Pittsfield, MA, quoted in "The Winning Edge" by Somerby Dowst, *Purchasing* March 12, 1987, 57.

23. Larry Weinstein, "Single-Source Successes and Snafus," *Purchasing Today*, interview with Ron Reese of Haliburton Energy Services, April 2000.

24. Bob Bretz, former director of Corporate Purchasing for Pitney Bowes and 1994 Shipman Medalist, indicates that "'single sourcing' is much simpler. There's less effort on the part of the seller and it's easier to resolve issues." Patrick Robert Bretz, quoted in Patrick Flanagen, "The Rules of Purchasing Are Changing," *Management Review* March (1994): 30.

25. Several large corporations use a dual or multiple approach to sourcing items with dynamic technology. See "Buyers Beef Up Supplier Management Skills," *Purchasing* October 21, 1993, 28.

26. The tax benefits of headquartering in Singapore make it a popular choice.

27. For example, during the 1980s, Chrysler built additional buildings on-site and leased them to its suppliers. Robert M. Faltra, "How Chrysler Buyers Make Quality a Standard Feature," *Electronics Purchasing* 101 (1986): 62A15.

28. Manufacturers' representatives, who usually deal only in technical items, also affect deliveries by drop shipments, and they act as brokers. Manufacturers' reps also aid a supply manager by being able to furnish numerous product lines from a single source.

29. Stephen C. Rogers, *The Supply-Based Advantage: How to Link Suppliers to Your Organization's Corporate Strategy* (New York: AMACOM, 2009): 51–52.

30. Lisa Cooling, "The Challenge to Cut Carbon Emissions," *Inside Supply Management* December (2008): 20–23. See www.ism.ws.

31. "10 Big Green Ideas," *Newsweek Green Ratings 2010* October 25, 2010, 49.

32. Ibid, 52.

33. Tina Bova, "Green Expectations," *Inside Supply Management* February (2011): 32–33.

34. Ibid, 33.

35. Ibid, 93.

36. Lisa Arnseth, "Greening the Logistics Network," *Inside Supply Management* March (2011): 20–23. See www.ism.ws.

37. Jeremiah McWillians, "Bottle Wars: Coke, Pepsi Vie Over Plant-Based Packaging," *Cleveland PD*, March 16, 2011, 4.

38. Ginger Conrad, John F. Robinson, and Forrest Walker, Jr., "Conquering the M/WBE Challenge," *Purchasing Today* April (2000).

39. Debbie Newman, Patricia Richards, and Linda Butler, "Shared Commitment to MWBE Development," *NAPM Conference Proceedings,* Tempe, AZ, 1994, 300.

40. Reciprocity becomes more insidious when it involves more than one tier of suppliers. For example, A is asked to buy from B, not because B is A's customer, but because B is C's customer, and C is A's customer, and B wants to sell to A. Obviously, it is possible for reciprocal relationships to extend to four or more tiers.

41. Some firms use vague phrases such as "Buy from customers when doing so will contribute to the greatest economic good of the firm."

42. Richard L. Pinkerton, "Creating Value with Consortia and Reverse Auctions: Fact or Fiction?" A presentation given at the Technology Procurement Conference, July 21–23, 2004, New York. Sponsored by ICN, Winter Park, FL.

43. Telephone conversation between Richard L. Pinkerton and David B. Clevenger, Vice President, Corporate United, September 2, 2008. Also see www.corporateunited.com.

44. "Financial Statement Analysis" was prepared by Donn W. Vickrey of the University of San Diego.

45. "Does the ISO 9000 Need Fixing?" Industry Forum Supplement to the June 1994 issue of *Management Review.*

46. Warren E. Norquist, Director of Worldwide Purchasing and Materials Management, Polaroid Corporation, personal interview, October 1987.

Suggested Reading

Blanco, Edgar E. "Stay Ahead of the GHG Curve." *Inside Supply Management* April (2011): 32–33.

Benton, W.C. *Purchasing and Supply Management.* New York: McGraw-Hill Higher Education, 2007.

Choi, T.Y., and J.L. Hartley. "An Exploration of Supplier Selection Practices across the Supply Chain." *Journal of Operations Management,* 14(4)(1996): 333–343.

Blanchard David. *Supply Chain Management Best Practices,* 2nd ed. Hoboken, NJ: John Wiley & Sons, Inc., 2010, 203–214. An excellent chapter on going green, risk management, 59–60, and RFID 189–201.

de Boer, L., E. Labro, and P. Morlacchi. "A Review of Methods Supporting Supplier Selection." *European Journal of Purchasing and Materials Management* 7(2001): 75–89.

Dunn, Steven C., and Richard R. Young. "Supplier Assistance Within Supplier Development Initiatives." *Journal of Supply Chain Management* 40(3)(2004): 19.

Ellram, Lisa M. "The Supplier Selection Decision In Strategic Partnerships." *International Journal of Purchasing and Materials Management* 26(4)(1990): 8–14.

Wright Jonathan, and Seb Hoyle. "Greening the Logistics Footprint: Standard Guidelines to Track Carbon Emissions Throughout the Supply Chain Can Help You Meet Consumer's Green Demands." *Inside Supply Management* September (2010): 32–33.

Hartley, Janet L., and Thomas Y. Choi. "Supplier Development: Customers as a Catalyst of Process Change." *Business Horizons* 39(4)(1996): 37.

Krause, Daniel R., and Thomas V. Scannell. "Supplier Development Practices: Product- and Service-Based Industry Comparisons." *Journal of Supply Chain Management* 38 (2)(2002): 13.

Heriot, Kirk C., and Subodh P. Kulkarni. "The Use of Intermediate Sourcinc Strategies." *Journal of Supply Chain Management* 37(1)(2001): 18–26.

Kannan, Vijay R., and Keah Choon Tan. "Supplier Selection and Assessment: Their Impact on Business Performance." *Journal of Supply Chain Management* 38(4)(2002): 11.

Krause, D.R., T.V. Scannell, and R.J. Calantone. "A Structural Analysis of the Effectiveness of Buying Firms' Strategies to Improve Supplier Performance." *Decision Sciences* 31(1) (2000): 33–56.

Making Green Work, from the Sustainable Supply Chain Resource Management luncheon, March 8, 2011, Palto Alto, CA. Sponsored by *The Harvard Business Review*.

Nelson, David, Patricia E. Moody, and Jonathan R. Stegner. *The Incredible Payback: Innovative Sourcing Solutions that Deliver Extraordinary Results*, New York: AMACOM, 2005.

Paquette, Larry. *The Sourcing Solution: A Step-by-Step Guide to Creating A Successful Purchasing Program*, New York: AMACOM, 2004.

Sarkis, Joseph, and Srinivas Talluri. "A Model for Strategic Supplier Selection." *Journal of Supply Chain Management* 38(1)(2002): 18.

Seungwook, Park, and Janet L. Hartley. "Exploring the Effect of Supplier Management on Performance in the Korean Automotive Supply Chain." *Journal of Supply Chain Management*, Spring (2002).

Vonderembse, M., and M. Tracey. "The Impact of Supplier Selection Criteria and Supplier Involvement on Manufacturing Performance." *Journal of Supply Chain Management* 35(3)(1999): 33.

CHAPTER 14

Global Supply Management

Global Management Perspective

The supply management profession operates in an environment characterized by at least four important driving forces: (1) an economic landscape increasingly driven by transnational issues and concerns, a truly mutually interdependent "global economy;" (2) a supplier selection and development process that seeks collaborative and collaborative-based relationships with the very best suppliers worldwide; (3) the formulation and positioning of supply management strategies which reflect forward-looking, long-term global supplier relationships, and (4) an orientation toward supply chain management (SCM), which extends supplier selection and relationship management issues well beyond the traditional business perspective of buyer-seller to those characterized by multi-tiered, highly interdependent supplier selection and relationships that span the globe.

Future of Global Supply Management

In an ambitious consideration of what the future holds for the supply management profession, the Center for Advanced Purchasing Studies (CAPS) concluded in several studies that:

> much has happened to change the world of business and supply management in particular—with much more expected to come.[1]

The CAPS studies also stated that

> global sourcing is increasing. In 2000, firms in this study sourced between 21 percent and 30 percent of their total annual spend on a worldwide basis. In 2005, total nondomestic spend increased to between 31 percent and

40 percent. It is projected that in 2010 the total dollar amount of purchased items obtained from nondomestic sources will be between 41 percent and 50 percent.[2]

The CAPS studies support the conclusion that the main reasons for global sourcing are driven by a desire to reduce costs. "On average, respondents achieved cost reductions of 19 percent and a total cost-of-ownership reduction of 12 percent."[3] Consequently, "China, India, Eastern Europe and Brazil will continue to gain importance as sources of supply over the next five years, while sourcing from U.S., Canadian and Western European markets will decline."[4]

Supply managers face increased challenges in this new environment. Global turbulence and the risks inherent in many new supply chain practices will present increased supply chain risks. Changes in market segments will drive suppliers to be more selective about which customers they choose to service. One study predicts supply managers "will face more aggressive and powerful suppliers in some markets."[5]

Procuring products and services of foreign origin can be extraordinarily challenging. On the one hand, virtually all of the practices and procedures described in this book are applicable. At the same time, many new issues must be addressed if a supply team is to ensure that its organization receives the right quality, in the right quantity, on time, with the right services, at the right cost. As with other areas in purchasing and supply chain management, the process of global sourcing is going through massive change.

Stages to Global Supply Management

At a number of leading firms, international sourcing is being replaced with a broader terminology called "global supply management."[6] Joseph Carter suggests three stages of worldwide sourcing as follows:[7]

▲ Stage One: International Purchasing—Organizations focus on leveraging volumes, minimizing prices, and managing inventory costs. These areas are characteristic of an organization first entering the global purchasing arena.

▲ Stage Two: Global Sourcing—Organizations focused on global opportunities will put more emphasis on supplier capability, supporting production strategies, and servicing customer markets. Most of those that have sourced offshore for some time are at this stage.

▲ Stage Three: Global Supply Management—Here, organizations optimize supply networks through effective logistics and capacity management. These organizations have effectively minimized risks in offshore sourcing, and have sourced worldwide for technology leadership.

Reasons for Global Sourcing

Global sourcing requires additional efforts when compared with domestic sourcing, but the efforts can yield large rewards. One of the complexities of buying goods and services of foreign origin is the wide variability among the producing countries in characteristics such as quality, service, and dependability. Quality, for example, may be very high in products from one country but inconsistent or unacceptably low in products from a neighboring land. This is particularly apparent today in the quality and safety issues domestic toy maker Mattel is experiencing with products currently manufactured in China.[8] With this caveat in mind, let us look at six common reasons for purchasing goods and services from global sources.

▲ *Superior quality*   A key reason for global supply management is to obtain the required level of quality. Although this factor is declining in significance, supply managers in a variety of industries still look to global sources to fulfill their most critical quality requirements. Please refer to Chapter 7 on quality for more details.

▲ *Better timeliness*   A second major reason for global sourcing is to improve the certainty of the supplier in meeting schedule requirements. Order lead-time lengths and variability in the lead-time estimations may be better than those available from domestic sources. As with quality, the timeliness capability of suppliers around the globe has steadily improved. Once initial difficulties of the new business relationship have been overcome, many international sources have proved to be remarkably dependable in meeting schedules.

▲ *Lower total cost*   While lower cost is often the major reason for importing, it must be remembered that international sourcing generates expenses beyond those normally encountered when sourcing domestically. For example, additional communications and transportation expenses, import duties, and greater costs when investigating the potential supplier's capabilities all add to the buying firm's total costs. The total cost for goods to arrive at the point of use must be calculated, not just the unit price. To illustrate the point, one major computer manufacturer uses a rule of thumb that a foreign material's price must be at least 20% lower than the comparable domestic price to compensate for these additional costs. *Nonetheless,* after all of the additional costs of "buying international" are considered, in the case of many materials it frequently is possible to significantly reduce the firm's *total cost* of the material through global sourcing.

▲ *More advanced technology*   No country holds a monopoly on advanced product and process technology. Developing countries are expanding their capabilities into design, finance, and distribution.[9] Global sources in some industries are more advanced technologically than their North American counterparts. Not to take advantage of such product or process technologies can result in a manufacturer's losing its competitive position to competitors that incorporate the more advanced technologies.

▲ *Broader supply base* Sourcing globally increases the number of possible suppliers from which the buying firm can select. Increased competition for the buying firm's business will then better enable the firm to develop reliable, low cost suppliers. Broadening the supply base does not mean increasing the number of suppliers. Broadening the supply base actually increases the opportunity to find better suppliers, thereby enabling the buying firm to decrease the number of contracted suppliers and pursue collaborative or alliance relationships when appropriate.

▲ *Expanded customer base* Sourcing globally can create opportunities to sell in countries where the buying firm's suppliers are based. Where trade restrictions are minimal, the interaction itself may yield some of the sales opportunities. However, in some countries, trade restrictions are in place requiring non-domestic suppliers to procure materials in the buying country as part of the sales transaction. These arrangements commonly are called barter, offsets, or countertrade. The tying of *sales into a country* with *the purchase of goods from that country* makes both marketing and supply management far more challenging than when pure monetary transactions are involved. For a firm to compete and make sales in many countries, increasingly it is necessary to enter into agreements to purchase items made in those countries. Countertrade is discussed in detail later in this chapter.

Potential Problems

▲ *Cultural issues* Cultural issues can be a problem in global sourcing due to the wide variety of approaches to conducting business in different regions of the world. A colleague recently stayed in a hotel in Hong Kong that did not have a fourth floor. The floor was there, but it was completely empty, with no walls! The number four has a variety of connotations depending upon whom you speak to in Asia. The explanation given by management at the hotel was that four means "death" and as such it would be bad luck to have a fourth floor. Cultural issues are very real and should not be ignored in global sourcing. Cultural issues are discussed later in this chapter.

▲ *Long lead times* Variable shipping schedules, unpredictable time requirements for customs activities, the need for greater coordination in global supply management, strikes by unions, storms at sea (which can cause both delays and damage), and increased security procedures usually result in longer lead times. Airfreight may be used to offset some of the problems of variable shipping schedules, but probably at an increase in cost.

▲ *Additional inventories* The quantity of additional inventory required to buffer the effects of such long lead times can be difficult to determine. Quite often, however, the additional inventories are not as large as one might expect. Nevertheless, inventory-carrying costs must be added to the purchase price, the freight, and the administrative costs to determine the true total cost of buying from global sources. Occasionally, when a domestic industry is producing at full capacity, it is possible to get both faster delivery and lower prices from global sources. Routinely, however, additional lead times, which traditionally exceed 30 days, must be considered in planning foreign purchases when surface transportation is involved. It should be noted that some supply managers do *not* add buffer stocks, relying on airfreight in case of emergencies.

▲ *Lower quality*   As previously mentioned, global suppliers frequently are utilized because many of them can provide a consistently high level of quality. But problems do exist. For example, the United States is the only major non-metric country in a metric world. This frequently leads to manufacturing *tolerance problems* for buyers of U.S. products and U.S. buyers of products from metric countries. Additionally, non-domestic suppliers tend to be less responsive to necessary design changes than do their domestic counterparts. In many cases, there is the risk that production outside of the domestic firm's control can result in "off-spec" incoming materials. Potential rework or scrap costs could add substantially to the total cost of doing business with global suppliers.

▲ *Social and labor problems*   In Europe and the United States, unions and some politicians are pushing for retaliatory measures against exporting countries where workers lack clout and labor laws are either weak or routinely flouted. Retailers such as Levis, Nordstrom, and Reebok discern a greater tendency by some customers to shun production from "sweatshops." Documentaries by the news media of working conditions in some international plants have made U.S. retailers sensitive to working conditions in those plants. It seems highly likely that manufacturers in other developed countries will soon have similar concerns with their global suppliers.

▲ *Higher costs of doing business*   The need for translators, communications problems, the distances involved in making site visits, and so on all add to the cost of doing business with global suppliers. Port-order services are more complicated because of currency fluctuations, methods of payment, customs issues, and the utilization of import brokers and international carriers. Inadequate local (international) logistical support functions such as communication systems (telephones, fax machines, Internet), transportation systems, financial institutions, and so forth can complicate communications and product distribution.

▲ *High opacity*   Investors, chief financial officers, bankers, equity analysts, and supply managers involved in global activities have long been aware that the risk of conducting business in different countries varies. Recently, a risk factor called the "Opacity Index" has been developed to address the risk costs associated with conducting business in a specific country.[10] The Global Opacity Index addresses the following areas:

(1) corruption in government bureaucracy; (2) laws governing contracts or property rights; (3) economic policies (fiscal, monetary, and tax-related); (4) accounting standards; and (5) business regulations. China, for example, has high opacity in comparison to the United States. The United States has less bureaucracy, fewer government imposed restrictions, less monetary transaction constraints, and very little corruption.

Questions Before Going Global

Several years ago, Raul Casillas, of the Alps Manufacturing organization, suggested that to help in determining if a part, product, or process is a candidate for global sourcing, ask the following questions:[11]

▲ Does it qualify as high-volume in your industry?
▲ Does it have a long life (two to three years)?
▲ Does it lend itself to repetitive manufacturing or assembly?
▲ Is demand for the product fairly stable?
▲ Are specifications and drawings clear and well defined?
▲ Is technology not available domestically at a competitive price and quality?

If the answer to all six questions is yes, then the supply manager may want to evaluate the support network within his or her firm, asking the following questions:

▲ Does sufficient engineering support exist to efficiently facilitate engineering change orders (ECOs) when they occur?
▲ Will the buyer be able to allow sufficient time to phase out existing "in the pipeline" inventory?
▲ Will the supply manager's firm take the responsibility for providing the necessary education and training for those that will have to interact with and support foreign suppliers?
▲ Is the firm prepared to make a financial commitment for trips to the supplier?
▲ Is management willing to change the approach, in some cases even the policy, of how business and related transactions are conducted?
▲ Is the buyer aware of the environment, including current and forecasted exchange rates, general impact of tariff schedules, available technologies, and products from other countries, as well as their political climates, and leading economic indicators both domestically and abroad?

If the answers to both sets of questions are all positive, global sourcing may be a realistic possibility. A significant number of negative responses indicates the potential for real problems if a global sourcing arrangement is developed. Before a positive decision is made, however, the buyer needs to explore several issues with top management. First, do the required procedural and policy changes mesh satisfactorily with the firm's existing mode of operation? More important, is the global sourcing concept, and its underlying rationale, compatible with the firm's long-term plans? It is important that the program contribute positively to achievement of the firm's long-range goals and that the commitment is made as something more than a short-term-strategy decision. In some cases, unions must be consulted.

Supply Channels

The next step after deciding to source globally is to decide what supply channels to use. The lowest price method for procuring goods globally usually is to procure them directly. Direct procurement requires the buying firm to deal with all of the issues associated with getting the goods to its facilities. Although direct procurement may result in a low price, total costs may be prohibitive. In addition, limited resources in supply management may make direct procurement infeasible. The simplest way to source globally is through the use of an intermediary. The value of using intermediaries dissipates over time as learning by the buying firm increases.

Global Trade Intermediaries

Selection of the appropriate intermediary is a function of availability and the services required. The use of such intermediaries typically adds a significant cost to the overall cost of the transaction, but in most cases their use avoids many unforeseen problems.[12] The supply manager who is venturing into global sourcing is well advised to solicit the advice of colleagues from the local supply management association.

Some typical intermediaries are:[13]

▲ *Import merchants* buy goods for their own account and sell the following through their own outlets. Since they assume all the risks of clearing goods through customs and performing all the intermediate activity, their customers are relieved of import problems and, in effect, can treat such transactions as domestic purchases.

▲ *Commission houses* usually act for exporters abroad, selling in the United States and receiving a commission from the foreign exporter. Such houses generally do not have goods billed to them, although they handle many of the shipping and customs details.

▲ *Agents* or *representatives* are firms or individuals representing sellers. Since the seller pays their commission, their primary interests are those of the exporter. They generally handle all shipping and customs clearance details, although they assume no financial responsibility of the principals.

▲ *Import brokers* act as "marriage brokers" between buyers and sellers from different nations. Their commissions are paid by sellers for locating buyers and by buyers for finding sources of supply, but they are *not* involved in shipment or clearance of an order through customs. They also may act as special purchasing agents for designated commodities on a commission basis. Like agents, import brokers do not assume any of the seller's fiscal responsibility.

▲ *Trading companies* are large companies that generally perform all the functions performed individually by the types of agencies previously listed. The worldwide operations and know-how of such firms offer significant advantages and convenience. Standard directories and trade publications list such firms, their capabilities, and areas of service.

▲ *Subsidiaries* are established by multi-national corporations in countries where a physical presence is needed to improve competitive capability or meet host government restrictions. For example, Hitachi, a Japanese company, created a Hitachi Americas subsidiary to serve North American markets. Subsidiaries can increase sales and lower costs through employing a workforce with unique training and education, and reduced transport distances and tariffs. Subsidiaries usually start with a large percentage of local language capable expatriate managers, which reduces over time as qualified managers from the host country are developed. Subsidiaries serve to buffer the supply manager from both language and time zone problems. They offer to set prices in local currency and deliver material to buyers with all duties paid. Unfortunately, they are often remote from manufacturing and marketing decision makers and can be blockers in the flow of technical information. They can also add 5 to 35% to the cost for their services.

International Procurement Offices

When an organization's purchases in a foreign country or region warrant, consideration should be given to establishing an international procurement office (IPO), also called an international purchasing office.[14] An IPO is an office in a foreign country that is owned and/or operated by the parent company in order to facilitate business interactions in the foreign country and surrounding region.

Supply management professionals at an IPO quickly become familiar with qualified sources, thereby expanding the buying firm's potential supplier base. IPO personnel can physically and personally evaluate suppliers, negotiate for price and other terms, and monitor quality and job progress through direct site visits. IPO personnel are in a position to develop and maintain better information on local conditions such as materials shortages, labor issues, and governmental actions than are domestically based supply managers. The IPO facilitates payments to the suppliers, provides on-site support at the supplier's site if problems arise, and provides logistical support.

Expatriates who have worked for the domestic manufacturer, usually in a technical role, normally staff IPOs; however, this generalization is changing as the percentage of locals staffing IPOs is increasing. IPOs normally are established as cost centers, charging a percentage markup (typically 2%) for their services. Competition from other channels (foreign trade intermediaries and direct relations) tends to keep IPOs efficient. The one weakness of IPOs that has been observed is their tendency to represent the local supplier's interests over those of the parent company.

Deere and Company undertook an aggressive globalization program under the direction of Dave Nelson, former Vice President of Worldwide Supply Management. Under Nelson's leadership, Deere's International Supply Management Services established IPOs around the world. According to a John Deere newsletter, the International Supply Management Services "group's mission evolved into leveraging opportunities around the world for strategic sourcing teams and for all of Deere's 75 factories."[15] Three of the group's key responsibilities were: (1) "maintaining cross-cultural relationships and training sourcing team members in global supply management," (2) "serving as the main link between supply management activities and the Deere & Company functions that global trade requires (customs, law, finance, and others that deal with such murky issues as quotas and duties, world economic forecasts, business development, risk management, currency and taxes)," and (3) establishing and facilitating International Purchasing Offices. According to Dave Nelson, "These offices will link local manufacturing to common enterprise processes, work to improve supplier capability by accessing and applying proven John Deere programs (such as Achieving Excellence and Supplier Development), and facilitate understanding of in-country trade and regulatory requirements, as well as cultures."[16]

Direct Suppliers

Dealing directly with the supplier usually will result in the lowest *purchase price* (including transportation and import duties). It eliminates the markups of global trade intermediaries. But it requires an investment in travel, communications, logistics, and interpretation of costs. Direct relations with the supplier should be undertaken only after carefully conducting a cost/benefit analysis. It is important to note that conditions in developing countries are often problematic. Buyers should anticipate problems. For example, China, India, South America, and Eastern Europe have relatively poor transportation infrastructure systems in comparison to North America, Western Europe, and the Pacific Rim.

Eliminating Intermediaries

After the buying firm has gained confidence in the quality of the imported materials, and volume increases, the firm typically attempts to discontinue the use of global trade intermediaries for major procurements. Its major motivation is to avoid the intermediary's markup. The supply manager should inform its supplier of this new policy and then visit each of the manufacturers, *without the intermediaries,* to negotiate new contracts. While cost and the desire for direct dealings on technical issues may motivate the buying firm to deal directly with the supplier, the final decision will be made at the supplier's headquarters. The supply manager should anticipate resistance by both the intermediaries and their manufacturers. But this resistance normally can be overcome. In some cases, new suppliers may have to be developed due to the tight ties the global trade intermediaries may have with the existing supplier. Before taking such action, the supply manager must ensure that his or her company is set up to handle items such as traffic, customs clearance, and international payments.

Direct procurement requires the involvement of the company in all aspects of the transaction; when properly conducted, it eliminates the added profit of the middleman. Outside agencies may be engaged to perform specialized services. For example, *customs brokers* can be used to handle entry requirements, *export brokers* to handle foreign clearances, and *freight forwarders* to arrange for transport. Such agents do not take title to the goods. Most direct purchasers whose scale of activities does not warrant such in-house capability use outside agencies.

Definitions of terms that the supply manager may encounter when dealing globally may be obtained by visiting the following Web sites:

> CISG—Table of Contracting States http://www.cisg.law.pace.edu/cisg/countries/cntries.html
> Dictionary of International Trade Terms http://www.itds.treas.gov/glossaryfrm.html
> International Chamber of Commerce http://www.iccwbo.org/
> International Trade Administration of the U.S. Department of Commerce
> OANDA—The Currency Site http://www.oanda.com/
> The International Monetary Fund (IMF) http://www.imf.org/
> The U.S. Central Intelligence Agency's World Factbook https://www.cia.gov/library/publications/the-world-factbook/geos/us.html
> The World Trade Organization (WTO) http://www.wto.org/index.htm
> Understanding Incoterms

Identifying Direct Suppliers

Global trade intermediaries also are an excellent source of information. Unfortunately, these organizations have a vested interest in maintaining their position in the supply channel.

The best way to prepare to bypass the intermediary is to develop direct contacts with key players at the division performing the design, manufacture, and marketing of the item or commodity class. The supply manager should provide performance feedback *directly* to the supplier. The supply manager should tell the intermediary that he or she wants to visit with the supplier's key personnel the next time they are in the country or the next time the key personnel are in the buyer's country. International purchasing authority Dick Locke recommends meeting the supplier's key personnel and presents other tips in his paper "Get the Purchasing Channel You Want," summarized in the following list.

▲ Use the meeting to provide performance feedback and to explain your company's purchasing goals and values. Take care not to appear to be an unreasonable company to work with, even if you must deliver a critical message. Work to make foreign visitors to your company feel as welcome as possible.

▲ As part of the strategy, consider the timing of your request. The ideal time is when you are considering a change in suppliers or are selecting a supplier for a new project. The possibility of a major increase in business will give you more leverage.

▲ If you are dealing with a new supplier, state your intention to deal directly right from the start. Once a subsidiary or representative has started to handle your business, they are difficult to dislodge. It is easier to change your mind and start dealing through reps than the other way around.

▲ Once your company has established a relationship with the business and technical staff of the supplier, make the request to deal more directly. You might be requesting to deal through an IPO or you might be asking to deal directly. This request should go to the supplier's sales management, and specifically to an individual whom you already know.

▲ Be prepared to give reasons. These might be that you need a lower cost and believe that both parties can benefit by removing intermediaries. Another reason might be that the representative or subsidiary does not add enough value to the transaction to justify the markups it must be charging.[17]

Potential direct global suppliers can be located through a wide variety of sources. The chapter on sourcing presents a detailed section on discovering sources of supply. In Chapter 13 on sourcing, the increasing use of the Internet in identifying sources is discussed. The use of the Internet is particularly advantageous in discovering global sources.

Qualifying Direct Suppliers

Prior to investing additional energy in dealing with a global supplier, two issues should be addressed: *country and regional stability* and *the potential supplier's financial condition*.

For approximately $750, Dun & Bradstreet will prepare a Country Analysis Report for its clients. The report includes some 70 pages of in-depth research, information on both the current and historical economy and government, import and export practices, trading partners, and monetary policies.

Most experts recommend a survey of a region as well as the company and country because such factors as political and monetary stability, currency transfer laws, and trade and product liability policies may be crucial to doing business there. According to Heidi Jacobs and Barbara Ettorre,

> . . . The client should also ask what is needed to engage in commerce in a particular country. Credit professionals cite such factors as: required documentation for transactions, the transportation and distribution infrastructure, religious customs, quality standards and existing regulations that may restrict sale of the client's product or service. Will there be overseas agents to facilitate a deal? How reliable and experienced are they? Many a deal has been derailed by such cross-border questions as whether the desired country prohibits sales of products whose *components* originated in a certain country.[18]

The supply manager or buying team is cautioned not to judge the creditworthiness of the potential supplier by the ability of its key personnel to speak fluent English. A careful financial analysis must be conducted. Jacobs and Ettorre list the following sources of information for such analyses: Dun & Bradstreet, Owens On Line, and Justitia International Inc.[19] International credit specialists representing U.S. firms also caution their clients to familiarize themselves with the Foreign Corrupt Practices Act, which bars U.S. companies from engaging in bribery and other practices when doing business overseas.[20]

Preparing for Direct Relations

Preparation for direct relations includes all the issues raised in the source selection chapter, as well as intercultural preparation, the hiring of a competent translator, and an exhaustive technical and commercial analysis.

Cultural Preparation

Virtually all supply relationships with global suppliers are the result of negotiations. The success of each of these negotiations is influenced, in part, by the negotiator's ability to understand the needs of and the ways of thinking and acting of representatives of global firms. What is considered ethical in one culture may not be ethical in another. The intention of filling commitments, the implications of gift giving, and even the legal systems differ widely.

In addition to the conventional preparation for any negotiation, it is essential to conduct an extensive study of the culture. It is important to emphasize that this study should focus on the culture, not the language. The ability to understand a supplier's cultural background is of great practical advantage for several reasons. Negotiators perform more effectively if they understand the cultural and business heritage of their counterparts and the effect of this heritage on their counterparts' negotiation strategies and tactics. Cultural awareness on the part of the buyer or buying team puts the supplier off his or her guard. Talk with others who have experienced living or working in the culture. Learn what the holidays are, what the units of measure are, what the currency exchange is, what topics are taboo, and so on.

Another aspect of cultural preparation becomes important in cases where there is a strong likelihood of continuing relations (i.e., one or more transactions that would require a year or more for completion). Under such circumstances, the supplier's representatives (accompanied by their spouses) frequently visit the domestic firm. The buying firm's hosts should go to considerable lengths to become acquainted with their counterparts (and their spouses) on a social basis. Americans, for example, should entertain the visitors in their homes (a rarity in Europe and the Far East). This will give the Americans and their spouses an opportunity to develop good relations with their counterparts. This bank of goodwill, while not a means of co-opting the foreign supplier, projects a desire and willingness to understand, which frequently proves to be invaluable during subsequent transactions.

One other aspect of cultural preparation needs to be emphasized: It takes much longer to negotiate with foreign suppliers than with a supplier from North America. This is especially true if the supplier has not had extensive exposure to the buying firm's business practices and specifications. The time required varies based on the mode of operation. In the case of European firms, it usually takes at least twice as much time as with United States and Canadian firms, and up to six times as long is often required for Far East firms. As a result, U.S. negotiators must be aware of the requirement for additional time and plan accordingly.

Cultural preparation is specific to the country in which a supply professional is planning to conduct business. As a result, a detailed discussion is beyond the scope of this book. Several excellent resources are provided in the Endnotes to aid the reader in his or her efforts.[21]

Interpreters

Language frequently poses a significant barrier to successful global business relations. Bilingual business discussions usually require a third-party interpreter even when both of the principal parties are fluent in one of the two languages. Differences in culture, language, dialects, or terminology may result in miscommunication and cause problems. Both parties may think they know what the other party has said, but true agreement and understanding often may be missing. Think, for instance, of the confusion the simple word "ton" can create. Is it a short ton (2000 lb), a long ton (2240 lb), or a metric ton (2204.62 lb)? The use of textbook English raises innumerable interesting problems. For example, in the Far East, the word "plant" is interpreted to mean only a living organism, not a physical facility. And "yes" only means "I understand" to many Japanese, not "I agree."

When there are language differences between cultural groups, many busy executives believe that a competent interpreter is all that is necessary to overcome these differences. While a good interpreter can speed negotiations, an ineffective interpreter, or one ineptly used, can convert even simple matters into interminable wrangles. Complex discussions may simply grind to a halt amid a haze of miscommunication. The inexperienced supply manager risks wasting inordinate amounts of time for very little gain while acquiring the necessary communication skills. According to Hal Porter, a specialist with interpreters, "One or two words with a double meaning can certainly change the entire content of a statement." Executives experienced in international trade usually have learned these lessons, if only by trial and error. The use of native born interpreters, while allowing communication to take place, does not obviate the need for an understanding of the supplier's culture. Even when one overcomes the natural barriers of language difference, it is still possible to fail to understand and be understood. [22]

Technical and Commercial Analysis

Technical and commercial analysis is discussed in greater detail in Chapter 13 on source selection given earlier in the book. *Before* dealing with identified global candidate suppliers, the supply management team should:

- ▲ Prepare and review specifications and drawings.
- ▲ Pack samples or photos of required materials if they would help in communicating requirements.
- ▲ Clearly prepare the quality requirements.
- ▲ Identify specific scheduling requirements.
- ▲ Determine (as a group) what percentage of the annual requirements for the item can be placed offshore.
- ▲ Determine requirements for special packaging.
- ▲ Identify likely lead times.
- ▲ Develop a clear idea of the price objective.
- ▲ Prepare a briefing on your (the buying) firm. Frequently, much effort will be expended selling the potential suppliers on doing business with the buying firm. The briefing should include:
  - ▼ Information on the relevant product line and related lines
  - ▼ Actual and forecasted sales volume

▼ Customers
▼ Market share
▼ Unclassified corporate strategy information
▼ Annual reports
▼ An indication of why the buying firm is soliciting the potential global supplier's interest (quality? price?)
▼ Business cards in English on one side and in the relevant language on the other side.
▼ The North American's title on his or her card should be "adapted" to the situation. Most non-North Americans are extremely rank conscious. Thus, a supply manager may be titled: Supply Manager, Director, Vice President, etc. for a specific situation. One such supply manager has seven different business cards with seven different titles!
▼ Green purchasing as noted in the Chapter 13.
▼ Disaster or risk management plans (see the appendix).

The Initial Meeting

Adequate preparation as detailed in the previous section will increase the probability of a smooth, efficient, and successful initial meeting. At the initial meeting it is good to conduct a facility tour or visit of the potential supplier's facilities and meet with critical personnel. Be certain the plant you visit is the one that will make the parts you are buying. Plant visits are discussed in detail in Chapter 13 on source selection. For large procurements with complex specifications, the buying firm's technical people clearly must be part of the visiting team. The potential supplier will be judging the buying firm just as much as the buyer will be judging the potential supplying firm. Experience has shown that the controller of the target supplier usually occupies a very influential position. To gain his or her support, the supply manager or buying firm's team should describe how and when the supplier's firm would get paid. Currency and payment issues are discussed in the next section.

Currency and Payment Issues

From the buying firm's point of view, the preferred method of payment is after receipt and inspection of the goods. However, it is customary in many countries for advance payments to be made prior to commencing work. Such a provision ties up the purchaser's capital. Letters of credit also are common in global commerce. Again, the purchaser's funds may be committed for a longer period of time than if a domestic source were involved. Not surprisingly, a cost is incurred in obtaining the letter of credit.

Exchange Rates

The absence of fixed exchange rates can be a problem; it creates the following situations next.

Case 1

A contract calls for *payment in a foreign currency*. The exchange rate moves against the U.S. dollar during performance of the contract. For example, assume that a contract was awarded to a supplier in Germany for 1 million euros. Assume further that the rate of exchange was U.S. $1 = Eur 0.689;[23] that is, one U.S. dollar purchased .69 euros at the time the contract was awarded. Ignoring all other costs, the dollar cost to the U.S. buyer would be

$$\text{Eur } 1,000,000$$

$$\text{Eur } .69/\$ = \$1,451,590$$

Assume that the U.S. dollar strengthens to the point that $1.00 buys Eur 1. The cost in dollars then becomes

$$\text{Eur } 1,000,000$$

$$\text{Eur } 1.0/\$ = \$1,000,000$$

This is a decrease of $451,590, or a 30% decrease in the cost of the item in U.S. dollars. Note that the German supplier is no better off because it receives only Eur 1 million, while the U.S. purchaser has benefited from the 30% decrease in the cost of the item in *U.S. dollars.*

Case 2

A contract calls for *payment in a foreign currency* (Eur), and the exchange rate improves for the U.S. dollar so that $1 now buys Eur 1. The cost of the item in U.S. dollars now is

$$\frac{\text{Eur } 1,000,000}{\text{Eur } 1/\$} = \$1,000,000$$

The U.S. buyer has reduced its costs from the initial likely amount of $1,451,590 to $1,000,000, a 30% saving.

The issue of currency risk is examined in greater detail in the appendix.

Payments

Payments to a global supplier are simplified when a trade intermediary or an IPO is involved. When payment is to be made directly by the buying firm to the supplier, a letter of credit frequently is used.

Letters of Credit

As part of the negotiations, many global suppliers will request that the buying firm obtain a letter of credit from its bank. A letter of credit is an instrument issued by a bank at the request of a buyer. It promises to pay a specified amount of money upon presentation of documents stipulated in the letter. The letter of credit is not a means of payment, but merely a promise to pay. Actual payment is accomplished through a draft, which is similar to a personal check. It is an order by one party to pay another party. Documents commonly stipulated in the letter of credit include the bill of lading, a consular invoice, and a description of goods. In effect, if the purchaser defaults, then the bank has to foot the bill. Thus, any risk of payment is transferred to the bank. Frequently, the global supplier will use the purchase order (contract) together with the letter of credit as security when obtaining a loan for working capital for the required labor and materials.

Letters of credit are classified three ways:

▲ *Irrevocable vs. revocable* An irrevocable letter of credit can be neither canceled nor modified without the consent of the beneficiary.
▲ *Confirmed vs. unconfirmed* A bank that confirms the letter of credit assumes the risk. The best method of payment for an exporter in most cases is a confirmed, irrevocable letter of credit. Some banks may not assume the risk, preferring to take an advisory role. Such banks and their correspondents believe that they are better able to judge the credibility of the issuing bank than the exporter.
▲ *Revolving vs. non-revolving* Non-revolving letters of credit are valid for one transaction only. When relationships are established, a revolving letter of credit may be issued.

Obtaining a letter of credit may take three to five business days. A detailed application must be completed. Since a letter of credit is an extension of credit from, the bank, it is processed much as a loan is processed. If no line of credit has been previously established with the bank, the applicant must prepay the specified amount. Typical charges involved include an application *fee* (0.008% on a $125 minimum) plus a negotiation charge (0.0025% on a $110 minimum). In case of cancellation, a charge of $100 is common.

Countertrade

The term "countertrade" refers to any transaction in which payment is made partially or fully with goods instead of money. Countertrade links two normally unrelated transactions: the sale of a product into a foreign country and the sale of goods out of that country. Foreign governments normally impose countertrade requirements in an effort to gain foreign exchange or foreign technology.[24] Countertrade has several distinct definitions:

▲ *Barter* This form of transaction preceded the use of money. Goods are exchanged for other goods with no money involved. This is the simplest form of countertrade. If goods are bartered to save on transportation costs, the arrangement is called a swap.

▲ *Offset* Under this form of transaction, some, all, or even more than 100% of the value of the sale is *offset* by the purchase (or facilitation of purchases by others) of items produced in the buying country. Offsets are categorized as direct and indirect. A *direct offset* involves close technological ties between the items sold and purchased. For example, when the government of Australia purchased helicopters made by Boeing, Boeing agreed to buy ailerons for the 727 from an Australian supplier. An *indirect offset* involves the purchase or facilitation of sales of commodities unrelated to the purchasing country. When the Swiss purchased F-5 aircraft, the manufacturer (Northrop) facilitated sales of Swiss elevators and other non-aircraft products in North America.

▲ *Counterpurchase* With this type of transaction, unrelated goods are exchanged. The U.S. manufacturer purchases goods in the foreign country from a supplier who is paid in local currency by the buyer of the manufacturer's goods. Counterpurchase normally involves two separate, but linked, contracts: one for purchase and one for counterpurchase.

▲ *Buy-back/compensation* Buy-back (or compensation) is an agreement by the seller of turnkey plants, machinery, or other capital equipment to accept as partial or full payment products produced in the plants or on the capital equipment.

Laura Forker, in her 1991 report on countertrade, identifies the following advantages and disadvantages:

Countertrade's Advantages

Companies involved in countertrade frequently have enjoyed a variety of marketing, financial, and manufacturing advantages that have resulted in increased sales, increased employment, and enhanced company competitiveness. By accepting goods or services as payment instead of cash, countertrade participants have been effective in: (1) avoiding exchange controls; (2) selling to countries with inconvertible currencies; (3) marketing products in less-developed, cash-strapped countries (with centrally planned economies) that could not make such purchases otherwise; and (4) reducing some of the risks associated with unstable currency values. In overcoming these financial obstacles, countertrading firms have been able to enter new or formerly closed markets, expand business contacts and sales volume, and dampen the impact of foreign protectionism on overseas business.

Countertrade has also engendered good will with foreign governments concerned about their trade balances and hard currency accounts. Finally, Western participants in countertrade have enjoyed fuller use of plant capacity, larger production runs, and reduced per-unit expenses due to the greater sales volume. Their expanded sales contacts abroad have sometimes led to new sources of attractive components and, at other times, to valuable outlets for the disposal of declining products. Countertrade has opened up many new opportunities for American firms willing to become involved in it.

Countertrade's Disadvantages

Experienced companies have encountered a number of problems unique to or exacerbated by countertrade. Countertrade negotiations tend to be lengthier and more complex than conventional sales negotiations and must be conducted at times with powerful government supply agencies that enjoy negotiating strength. Additional expenses in the form of brokerage fees, additional transaction costs, higher supply management involvement, and transactions in goods problems reduce the profitability of countertrade deals. For example, countertrade contracts that use goods as payback often result in difficulties with the quality, availability, and disposal of the goods. Countertrade also introduces pricing problems associated with the assignment of values to products or commodities received in exchange. Commodity prices can vary widely over the lengthy negotiation and delivery periods, and trading partners may differ as to the worth of particular products. All of these drawbacks result in higher risk and greater uncertainty about the profitability of a countertrade deal.

Offsets entail further concerns in the form of technology transfer requirements, local procurement conditions that favor local suppliers, and rigidities that offsets introduce into the buying process. The result for Western firms is often increased competition. Offset customers can become competitors later on. In addition, some offset requirements divert a Western firm's resources to less-than-optimal suppliers. These additional costs must be considered when a proposed deal is being evaluated.[25]

Supply Management's Role

Historically, the firm's marketing people who are intent on making a sale have coerced reluctant supply managers to engage in a countertrade transaction. One of the authors was involved in such transactions during the 1970s. Little thought was given to the domestic seller's countertrade obligations until the purchasing government brought economic and political pressure to bear. At this point, the domestic firm's supply managers frantically began to see what could be purchased in the foreign country. As a result, a very uncomfortable relationship developed between the customer country and the seller.

Both marketing and supply managers must recognize that they need to work as a team if countertrade is to operate to the firm's benefit. When countertrade is used to facilitate sales, supply management should be involved *up-front*. Supply managers should review the items their company requires. Similar requirements must be levied on the firm's suppliers so that they are in a position to assist the manufacturer in meeting its present or potential obligations.

Creative Countertrade

Elderkin and Norquist define traditional countertrade as focusing "on existing goods to be brought out of the host country and sold in existing world markets. Traditional countertrade must deal with the limitations of fitting what already exists into unresponsive markets."

"Creative countertrade," on the other hand, with its focus on creating future goods for *new market niches,* has greater flexibility and wider possibilities. Creative countertrade is broader than traditional countertrade. It includes not only traditional countertrade, but also international investment and joint venture activities. It carefully analyzes the needs of all the major parties, including the potential development of new global suppliers, and creatively applies existing business tools to answer these needs.

Traditional countertrade provides quick-fix solutions to ongoing trade problems. But it lacks the depth and longer time horizons of creative countertrade. It seems likely that progressive firms will embrace creative countertrade as a means of both increasing sales and developing new dependable sources of supply.[26]

Political and Economic Alliances

Global political and economic changes are constant issues for supply managers to identify and address. Countries in various regions of the world have restructured trade laws and developed compromise-based agreements in efforts to stabilize trade, open markets, and create a body for addressing trade issues. These new laws and agreements have had and will continue to have an impact on global supply management. Among the more prominent economic alliances are: The European Union (EU), North American Free Trade Agreement (NAFTA), MERCOSUR, South American Free Trade Area (SAFTA),

Association of Southeast Asian Nations (ASEAN), and Asia Pacific Economic Cooperation (APEC).[27] Most of the discussion in this section is on the European Union because it is the largest economic alliance in the world.

European Union[28]

The European Union (EU) is based on a treaty that calls for "common foreign security and, eventually, defense policies, and a common central bank and single currency."[29] The inspiration for the EU is thought by many to have come from the example of the United States and the need to prevent future European wars. Others argue that the EU had its roots in the 1957 Treaty of Rome, and was first envisioned as a "Common Market of Western Europe." Regardless of how the EU has come into being, the impact of the EU is profound on the field of supply management, creating opportunities as well as new challenges. William L. Richardson, former director of Commercial Services for British Steel, Inc., in London, made the following comments in 1993 that still hold true today:

> . . . For the American purchaser, the European Single Market offers considerable opportunities and it makes purchasing easier. ...First, it will strengthen or create new effective alternatives to existing large manufacturers, be they in the U.S.A., Japan, or elsewhere. Second, it makes purchasing easier by virtue of the creation of European standards where as many as 12 different national standards can exist. This is a huge aid to the cost of reducing and simplifying the quality and performance comparisons purchasers have to make when evaluating the advantages of different supply sources.
>
> If the U.S.A. fears the European Single Market, it is a misplaced fear. In a sense Europe has looked at the U.S.A., seen how America has created a giant manufacturing base and said to itself, what is it that prevents us Europeans from achieving similar growth and prosperity? The European answer is, first, to tear down its own internal barriers and then to open up its market to world trade fairly conducted within international law.[30]

Richard L. Pinkerton, in the conclusion to his article reporting on the history and evolution of the European Community (EC) and the implications for purchasing managers, writes that:

> Supply managers should be prepared to join their firm's EC strategy/tactics team. Each supply management professional must investigate the specific EC technical directives and the implications of the ISO 9000 standards as they apply to his

or her firm. Subsequently, the development of implementation plans should be undertaken as an integral part of the firm's overall EC strategy and plan.

In many respects, the standardization directives and programs of EC 92 will facilitate trade with Europe by reducing a number of different codes into a single code. Not only does this "harmonization" reduce the need for 12 different sets of paperwork, including border documents, and for a variety of rules and regulations, it should also reduce the costs of products bought and sold. Additionally, a more efficient and uniform European transportation system is expected to develop. Although some product variation will always be present as a result of differing styles, tastes, languages, and other cultural nuances among the member nations, it is very clear that Europe is moving toward essentially the same type of free market that currently exists in the United States. Sourcing should be accomplished more quickly, with fewer suppliers, as customers in all 12 countries utilize a common set of standards and procedures, coupled with the growth of mass distribution centers.[31]

The words of Richardson and Pinkerton have not yet been proven wrong, but the advancement of the EU's objectives has been slow, although it is steadily gaining momentum.

The Euro

The European Monetary Unit (EMU) called the euro was launched on January 4, 1999 with 11 EU member countries voting to join: Austria, Belgium, Finland, France, Germany, Ireland, Italy, Luxembourg, The Netherlands, Portugal, and Spain. The United Kingdom and Sweden did not join and the other members were not qualified. Slovenia and Greece have now joined and as of March 2011, the euro is used by 23 countries, 6 of which are not members of the European Union.

Initially, it was an electronic currency, which could be bought and sold on markets in which consumers could establish bank accounts and credit cards. In January 2001, the euro (in cash and coin form) began circulating alongside national currencies for up to two months until national notes and coins were withdrawn.

Economically, the euro is meant to complete the European single market, bolstering cross-border mergers, improving price transparency, and eliminating exchange-rate risk. Enthusiasts also hope it will be a rival to the hegemony of the U.S. dollar. Though its performance has been rocky, the potential for the euro to become the first international reserve currency in the near future is being debated among economists. Former Federal Reserve Chairman Alan Greenspan gave his opinion in September 2007 by stating that the euro could indeed replace the U.S. dollar as the world's primary reserve currency. He said that it is "absolutely conceivable that the euro will replace the dollar as reserve currency, or will be traded as an equally important reserve currency."[32] On February 10, 2011, the dollar was posted at $1.373 against the euro.[33]

However, the 2010 debt crisis in Greece and Ireland with serious financial problems in Portugal, Spain, and Italy have caused major concerns over the future of the euro. The problem stems from the failure of these countries to keep their yearly deficits below 3% of Gross Domestic Product (GDP) and their total debt below 60% of GDP.[34,35]

The euro has both a political and an economic rationale, but several key European countries do not support it. The United Kingdom and Sweden have not adopted the euro as their official currencies; however, many stores accept the euro throughout countries. In fact, there is some talk about eliminating the use of the euro.[36]

What possible consequences could the euro and its adoption pose for supply management professionals in non-EU states? Scholar Richard L. Pinkerton studied the potential risks as well as advantages for U.S. supply managers.[37] The potential advantages of the euro for U.S. supply management personnel were given by Pinkerton as follows:

▲ *Greatly reduced transaction costs* The U.S. firm is now dealing with one exchange rate vs. twelve. This is especially significant when U.S. firms in the 12 EMU countries buy from one another, either on an intra- or extra-firm basis.

▲ *Increased competition* (The Level Playing Field Concept) Should produce lower prices as firms are forced to be more productive because of price transparency.

▲ *Reduced exchange rate risk* U.S. firms will only have to hedge against one vs. twelve countries. This also reduces transaction costs as noted previously.

▲ *Increased trade and capital movement* The Euro will create a greatly increased capital bond and stock market and reduce the historical EU reliance on government and bank loans. Price stability and lower interest rates with controlled inflation should stimulate capital investment, and as a result, a sustainable economic growth rate in the 11 countries. However, increasing unemployment within the EU, especially France and Germany, is a major concern.

In looking back over the first 50 years of this historic union of European countries, the *Economist* provides an excellent evaluation of the euro in the year 2007 in its article "Europe's Mid-Life Crisis: A Special Report."

"Within the euro area a debate is in progress over whether the single currency itself encourages or discourages reforms. Most of its progenitors had hoped for the first. The euro has clearly boosted intra-EU trade by somewhere between 5 and 15% according to Organization for Economic Cooperation and Development. It has also been a spectacular success from a technical point of view, establishing itself not just as a viable currency but as the only plausible rival to the dollar. For example, it now accounts for 25% of global foreign currency reserves"[38]

While the European Union struggles with such current issues as agreement on a constitution, grass roots support including mistrust of the union, the poor performance of its economies in recent years, and the 2010 debt crisis in several EU countries, supply executives must continue to study and understand this political and economic union as it represents the world's largest market and it is here to stay.[39]

North American Free Trade Agreement

In June 1990, the presidents of the United States and Mexico endorsed the idea of a comprehensive U.S.-Mexico Free Trade Agreement in order to guarantee the positive effects of export growth and industrial competitiveness, which had already begun, would continue to expand. By 1991, Canada joined the talks, leading to the three-way negotiation known as the North American Free Trade Agreement, or NAFTA. This agreement was designed to create a free trade area (FTA) comprising the United States, Canada, and Mexico. Consistent with World Trade Organizations (WTO) rules, all tariffs will be eliminated within the FTA over a transition period. NAFTA involves an ambitious effort to eliminate barriers to agricultural, manufacturing, and service trade; to remove investment restrictions; and to protect intellectual property rights effectively. In addition, NAFTA marks the first time in the history of U.S. trade policy that environmental concerns have been directly addressed in a comprehensive trade agreement. By accelerating the integration of the three markets, NAFTA should enable North American businesses to produce goods that are more competitive compared with goods produced in Asia and in the EU and will allow North American consumers to benefit from a greater selection of higher-quality, lower-priced goods.[40]

Implications of NAFTA

Canada and the United States have long been sources of supply to each other. Modern-day Mexico has pockets of expertise that are world-class. Many U.S. buyers already avail themselves of Mexican sources of supply.

When a global analysis of potential suppliers reveals that it makes sense to develop a world-class supplier in Mexico, a joint venture with carefully developed plans, objectives, action plans, and milestones is the appropriate way of developing the supplier. (Obviously, these principles apply to the development of suppliers in many parts of the world.) This approach brings together the social, political, and economic strengths of the supplier with the knowledge, technology, systems, and commercial expertise of the global buyer.

MERCOSUR

MERCOSUR was founded in 1988 as a free-trade pact between Brazil and Argentina. The modest tariff reductions in its first years led to an 80% increase in trade between the two partners. In 1990, Paraguay and Bolivia joined MERCOSUR, and in 2005, Uruguay gained admission. The ambitious goal is to invite other Latin American nations into the pact, to form the Union of South American Nations. This union will unite two existing free-trade organizations, the MERCOSUR and the Andean Community.

Association of Southeast Asian Nations

The Association of Southeast Asian Nations (ASEAN) was formed in 1967 by Indonesia, Malaysia, the Philippines, Singapore, and Thailand to promote political and economic cooperation and regional stability. The ASEAN Declaration, signed in 1976 by ASEAN leaders in Bali and considered ASEAN's foundation document, formalized the principles of peace and cooperation to which ASEAN is dedicated. Brunei joined in 1984, shortly after its independence from the United Kingdom. In the 1990s, Vietnam, Laos, Burma, and Cambodia became members of ASEAN as well.

Also in 1976, ASEAN heads of state signed the Treaty of Amity and Cooperation in Southeast Asia (TAC). The stated goal of the treaty is to foster a peaceful, cohesive region and to promote regional economic cooperation. In July 1998, ASEAN Foreign Ministers signed the Second Protocol to the TAC, which permits accession by non-Southeast Asian countries. ASEAN then invited, and has since been urging, the Dialogue Partners to accede to the treaty.

ASEAN has established 10 "Dialogue Partner" relationships with other countries. The two sides meet at a Post-Ministerial Conference (PMC), which follows the annual ASEAN Ministerial Meeting (AMM). In 1994, ASEAN established the ASEAN Regional Forum, which focuses on regional security issues. This left the PMC to deal with international economic and political issues and transnational issues.

Asia-Pacific Economic Cooperation

Asia-Pacific Economic Cooperation (APEC) was established in 1989 in response to the growing interdependence among Asia-Pacific economies. Begun as an informal dialogue group, APEC has since become the primary regional vehicle for promoting open trade and practical economic cooperation. Its members define the geographic littoral of the Asia-Pacific Basin. Its goal is to advance Asia-Pacific economic dynamism and sense of community. Today, APEC's 21 member economies had a combined gross domestic product of over US$18 trillion in 1999 and 43.85% of global trade.[41] This makes APEC the world's largest free-trade area. Of the many issues before APEC's membership is a focus on streamlining intergovernmental procurement policies.

> APEC members are now working individually and collectively (through the Government Procurement Experts Group, established in 1995 to manage APEC's work in this area) to fulfill these and other commitments articulated in the OAA. Indeed, this initiative, which aims to enhance the transparency of members' existing government procurement systems, is one of the agreed collective actions included in the OAA meant to serve the above objectives. Another is the development, completed in 1999, of a set of non-binding principles (NBPs) on government procurement (comprising transparency; value for money, open and effective competition; fair dealing; accountability and due process; and non-discrimination).[42]

Summary

A firm's approach to global supply management normally progresses from a reactive mode to a proactive one. Under reactive global sourcing, the firm reacts to opportunities in the supply marketplace. If an internationally produced good or service is the most attractive buy, then it is purchased. As the firm embraces a proactive approach to procurement, it develops supply strategies and supply plans for its requirements. The development of these strategies and plans calls for the analysis of all possible sources of supply—both domestic and international.

Perhaps it is the level of difficulty and the degree to which global perspectives may conflict with one another that has scared off in-depth studies of global supply management. The shift from a tactical to a strategic business focus is no less profound than the shift in perception implicit in the terms "purchasing" or "procurement," when contrasted with "supply management." Nonetheless, it is this very complexity that requires our serious attention as we attempt to enact global supply management strategies successfully.

Far from merely an "inorganic" study of the various "tools" involved in the practice of global purchasing, we must seriously examine the professional competencies required in order to be effective in the global supply management environment. Supply management professionals must have the ability to: (1) develop a strategic point of view with regard to global supply management; (2) deal with change and chaotic, shifting environments effectively; (3) deal with diverse cultures effectively; (4) work with and within distributed organizational structures; (5) work with others in teams and act as team leader/project manager; and (6) learn to communicate effectively with those who may follow cultural beliefs and exhibit values very different than their own.

In recalling Socrates' entreaty that we are all "citizens of the world," the time has come to actively improve our understanding of world events as influenced by powerful political, economic, social, and cultural influences. Not only will we become better supply management professionals, but better human beings as well.

Appendix

Currency Risk

Locke and Anklesaria write:

> U.S. purchasing departments are at a disadvantage compared to their more sophisticated counterparts . . . in countries [which] deal in foreign currencies as a matter of course . . . U.S. buyers' unfamiliarity in dealing in foreign currencies leads to higher costs in two ways. First, they attempt to put all currency risk on the supplier, which causes the supplier to include charges for hedging, or to add an extra margin for contingencies into the price. Second, in an attempt to avoid dealing in foreign currencies, buyers use suppliers' U.S. subsidiaries and representatives, who will accept payment in dollars, but who also charge high markups.
>
> . . . Buyers and finance staffs of firms should understand when to buy in foreign currencies and when to buy in U.S. dollars. They should know the measures to take to avoid major increases in dollar cost and to be flexible enough to get decreases when possible. They should understand methods of reducing short-term risk through hedging. They should have analytical tools available to help them choose between various hedging strategies.[43]

The biggest advantage comes from the choice of the best pricing currency (the currency in which prices are set). The payment currency (you may actually pay an equivalent amount of a different currency) does not make a big difference in prices. To choose a pricing currency, you must answer two questions.

First, what are you buying? Product prices can be divided into cost driven and market driven categories. Cost driven prices are those where the supplier can set prices based on his or her costs. Market driven prices are those where prices are set on a world market, usually in U.S. dollars, and the supplier cannot sell at a higher price. Second, where is the product built? Some countries have currencies that are pegged to the U.S. dollar. Other currencies float freely. If a currency is truly pegged to the dollar, there should be no need for currency protection. Table 14.1 shows the possibilities.

Table 14.1 Best Buying Currency

Pricing Driver Currency	Type of Currency	
	Pegged Currency	**Floating Currency**
Cost-based products	Dollars or supplier's currency	Supplier's product
Dollar market-based product	Dollars	Dollars

Floating Currency, Cost-Driven Product

These products are typically custom or semi-custom ones. An example would be a printed circuit assembly from Japan or South Korea. By pricing in the supplier's currency, the supplier is relieved of the currency risk. This should enable a buyer to negotiate a lower initial price than if the supplier were to take on the risk. It is better to start with the lower price because one doesn't know if the dollar will strengthen or weaken. The buying firm can protect itself against dollar cost increases by low cost hedging; an escape clause is needed in the purchase agreement.

Floating Currency, Market-Driven Product

These products are typically commodities whose price is nearly the same anywhere in the world. Examples are gold, oil, and DRAMs. For this type of product, a buyer should not hedge. The buying firm is better off negotiating one worldwide price and maintaining the price the same around the world. This works best if the firm has a purchasing presence in various regions, so that hedging does not work as it does with cost-driven parts. If the dollar strengthens, the price in another currency goes up.

Pegged Currency, Cost-Driven Product

Countries with pegged currencies are generally smaller ones. They include Taiwan, Thailand, Hong Kong, and Korea. There is little need to hedge these currencies because they are unlikely to move against the dollar. In addition, the foreign exchange market is thin and not well developed. Instead of hedging, a buyer should have an escape clause in the contract because these currencies do make occasional controlled changes in value against the dollar.

Pegged Currency, Market-Driven Product

If the market is dollar-based, these products need not be hedged. Similar techniques to those used for market-driven products from floating currency countries are the best choice.

Hedging

Hedging protects the dollar value of a future foreign currency cash flow. The reason to hedge is to protect against major swings in the value of a purchase. A buyer can achieve this via forward or futures contracts or via currency options. The buyer would enter into contracts to sell dollars for foreign currency at the time the supplier is paid. It is easiest to think in terms of using the foreign currency that was purchased in the hedge to pay the supplier, but this is not what happens. There is a profit or loss on a hedge contract that takes place behind the scenes. This profit or loss is applied to a material price variance that results from exchange rate changes and offsets higher or lower part costs.

Forward contracts give a fixed cost for foreign currency and therefore for foreign currency purchasing. If the interest rates in the foreign country are higher than they are in the United States, the forward rate is at a discount to the spot rate, and this reduces the dollar cost still more.

Forward contracts also have the advantage of being suitable for internal transactions. If the buying company exports to the country in which it is buying and wants to sell in local currency, purchasing in local currency reduces the company's currency exposure. The purchasing flow of funds offsets the sales office flow of funds. If an internal forward agreement is made between the two departments, only the difference between the two flows needs to be hedged at banks.

Options allow a buyer to take advantage of an increase in the value of the U.S. dollar but protect against a decrease. Unfortunately, they are expensive. A six-month option on a volatile currency typically costs about 5% and most people choose not to buy them. An added difficulty is that option prices for the European style options that buyers need are not well listed in financial newspapers.

Risk of Buying in Dollars

Buying in dollars is not as safe of a solution to global buying that many want to believe. A dollar buyer may start off with a higher price than necessary. If the dollar weakens, the buyer is paying even more. A more sophisticated competitor would be paying less. A supplier's competitors will soon let buyers know that they are paying too much. Other channels of distribution could also open up. Finally, supplier promises of fixed dollar pricing are often broken when the value of the dollar declines.

Length of Hedging

Hedging for too long a period with forward contracts can lead to the same problems as buying in dollars. If the dollar increases in value, a buyer will be paying too much. Hedging for too long with options is expensive because the option premium increases with time. Three months of orders plus three months of lead time gives six months hedging, a typical period.

Risks in Hedging

Hedging does involve some risks, but they are limited and can be controlled with simple attention to the fundamentals. Risk arises from forecast inaccuracy, and can lead to unexpected price variations, either up or down. If a company over-forecasts purchases and hedges with forwards, there will be larger profit or loss on the hedge than the variance on part cost. With over-forecasts, there will be a loss on forward contracts if the dollar strengthens and a gain if the dollar weakens. The total unexpected gain or loss will be approximately the percentage of over-forecasts times the percentage that the dollar changed. For example, a 20% over-forecast and a 15% currency strengthening will result in a 3% (15% of 20%) extra cost of the parts.

With under-forecasts, some of the parts must be purchased at the spot rate without an offsetting hedge. If the dollar weakens, they will be more expensive and if it strengthens, they will be cheaper.

Choosing a Hedging Strategy

The biggest gains in currency management will come from choosing the right currency. A good negotiator should be able to get an initial price reduction of 5% or more against a volatile currency like the yen or the peso. The next most consequential decision is whether to hedge. Not hedging opens the buyer to dollar price swings that are often as much as 20% in six months. This uncertainty is unacceptable to most companies.

The third decision is to choose a hedging strategy![44]

In a recent article, Joseph Carter and his co-authors demonstrated the benefits of choosing a hedging strategy based on a Bayesian statistical analysis of probable outcomes. In this study, Carter shows that choosing a hedge strategy would have saved 3.6% compared with paying in the supplier's currency.[45]

Endnotes

1. Philip L. Carter, Joseph R. Carter, and Robert M. Monczka, "Key Supply Strategies For Tomorrow." Research study done by A.T. Kearney in conjunction with CAPS Research, Institute for Supply Management, Tempe, AZ, 2007, 1.
2. Robert M. Monczka, Robert J. Trent, and Kenneth J. Petersen, "Effective Global Sourcing and Supply for Superior Results," CAPS Research, Institute for Supply Management, Tempe, AZ, 2006, 7.
3. Ibid.
4. Carter et al., op. cit., 8.
5. Ibid.

6. Robert M. Monczka and Robert J. Trent, "Global Sourcing: A Development Approach," *International Journal of Purchasing and Materials Management* Spring (1991): 3.

7. Joseph R. Carter, "The Global Evolution," *Purchasing Today* July (1997): 33.

8. "Mattel Recalls 9 Million Toys from China," *The Plain Dealer* August 14, 2007.

9. Carter et al., op. cit.

10. The Opacity Index: Launching a New Measure of the Effects of Opacity on the Cost and Availability of Capital in Countries World-Wide (Executive Summary). London: Price Waterhouse Coopers, January 2001, 1–13.

11. Raul Casillas, "Foreign Sourcing: Is It for You?" *Pacific Purchaser* November-December (1988) 9.
    Also, for a more recent analysis, see Robi Bendorf, "The Global Sourcing Process—On the Road to World-Class," 86th Annual International Purchasing Conference Proceedings, Orlando, FL, May 2001.

12. Dick Locke, "Get the Purchasing Channel You Want," *Electronics Components* October (1993) U-12. Note: Dick Locke also wrote *Global Supply Management: A Guide to International Purchasing* (Chicago, IL: Irwin Professional Publishing, 1996) which is an excellent resource on global supply mangement.

13. N.A. DiOrio, "International Procurement," *Guide to Purchasing*, The National Association of Purchasing Management, Tempe, AZ, (1987), 7.

14. Prior to publishing this book, the authors debated on whether to use another term for IPO, because IPO is popularly known as "initial public offering." The decision to keep the acronym IPO is based on its entrenchment in supply management literature and the lack of a better term. We considered the obvious, GPO (global procurement office), but quickly realized that GPO already means group purchasing organization.

15. "Global Surge," *Supply Management Linkages*, a Newsletter from John Deere, Summer (2000).

16. Ibid.

17. Locke, op. cit., U-11.

18. Heidi Jacobs and Barbara Ettorre, "Evaluating Potential Foreign Partners." Cited by permission of the publisher. From *Management Review* October (1993) 60. American Management Association, New York. All rights reserved.

19. Ibid., 61.

20. Ibid., 60.

21. Four sound investments to help prepare for cultural issues are:
    Dick Locke, *Global Supply Management: A Guide to International Purchasing* (Chicago, IL: Irwin Professional Publishing, 1996).
    Fons Trompenaars and Charles Hampden-Turner, *Building Cross Cultural Competence: How to Create Wealth from Conflicting Values* (New York: McGraw-Hill, 2000).

Fons Trompenaars and Charles Hampden-Turner, *Riding the Waves of Culture: Understanding Diversity in Global Business*, 2nd ed. (London: McGraw-Hill, 1998). Edward and Mildred Hall's *Understanding Cultural Differences*, (Intercultural Press, Boston, MA, 1990).

22. Hal Porter, "Interpreters: What They'll Do for You," *Across the Board* October (1993) 14.

23. Rate reflects exchange rate as of November 4, 2007 quoted on ADVFN Currency Converter (www.advfn.com); 1 EUR = 1.4516 USD.

24. The interested reader is encouraged to read *Creative Countertrade: A Guide to Doing Business Worldwide* by Kenton W. Elderkin and Warren E. Norquist, (Cambridge, MA: Ballinger Publishing Co., 1987) and the more recent study by Laura Forker, "Countertrade: Purchasing's Perceptions and Involvement," Center for Advanced Purchasing Studies/National Association of Purchasing Management, Inc., Tempe, AZ, 1991. Single copies are available gratis by written request to the Center for Advanced Purchasing Studies, P.O. Box 22160, Tempe, AZ, 85285-2160.

25. Forker, op. cit., 11–12.

26. Elderkin and Norquist, op. cit., 122–123.

27. Current maps of the countries included in the alliances listed above are readily available at a variety of Web sites, such as cnn.com as of August 2001. We have chosen not to include maps due to frequent changes in the alliance countries.

28. The European Community, now called the European Union, consists of the following countries (as of September 2007): Austria, Belgium, Bulgaria, Cyprus, the Czech Republic, Denmark, Estonia, Finland, France, Germany, Greece, Hungary, Italy, Latvia, Lithuania, Luxembourg, Malta, the Netherlands, Poland, Portugal, Republic of Ireland, Romania, Slovakia, Slovenia, Spain, Sweden, and the United Kingdom.

29. Sally Jacobsen, "Europe Finally to Be United, but Federation Is a Loose One," *The Arizona Republic*, October 13, 1993, A10.

30. Quoted in "EC 92: It's Official" by Richard L. Pinkerton, *NAPM Insights* July (1993): 25–27.

31. Richard L. Pinkerton, "The European Community—'EC 92': Implications for Purchasing Managers," *International Journal of Purchasing and Materials Management* Spring (1993): 25. *Note:* EC was renamed the European Union in 1994.

32. *"Euro Could Replace Dollar as Top Currency—Greenspan*, http://www.reuters.com/article/bondsNews/idUSL1771147920070917. Retrieved on September 17, 2007.

33. ECB: Euro foreign exchange reference rates. http://www.ecb.eu/stats/exchange/eurofxref/html/index.en.html.

34. Brian M. Carney and Anne Jolis, "Toward a United States of Europe," *The Wall Street Journal*, December 18–19, 2010, A17.

35. Niall Ferguon, "Murder on the EU Express," *Newsweek* April 11, 2011, 8–9.

36. "Breaking Up the Euro Area," *The Economist* December 2, 2010.

37. Richard L. Pinkerton, "Implications of The Euro Dollar for U.S. Supply Management Personnel." Presented at the NAPM Global Supply Management Conference, November 8–9, 1999, Phoenix, AZ, and published in the *World Market Series, Business Briefing: Global Purchasing and Supply Chain Management,* October (1999): 46–52 World Markets Research Centre, London.

38. "Europe's Mid-Life Crisis: A Special Report (on the European Union)," *The Economist,* March 17–23, 2007, 8.

39. Ibid, 5–6.

40. *NAFTA, the Beginning of a New Era,* Business America (partial extract), August 24, 1992, National Trade Data Bank, March 27, 1994.

41. APEC Home page. www.apec.org. August 31, 2001.

42. APEC Web site. "Government Procurement in APEC." August 31, 2001.

43. Richard Locke, Jr., and Jimmy Anklesaria, "Selection of Currency and Hedging Strategy in Global Supply Management," *Proceedings,* International Conference of Purchasing and Materials Management, May 1994, Atlanta, GA, 294–299.

44. Locke, Jr., and Anklesaria, op. cit., 294–299.

45. Joseph R. Carter, Shawnee Vickery, and Michael P. D'Itri, "Currency Risk Management Strategies for Contracting with Japanese Suppliers," *International Journal of Purchasing and Materials Management* Summer (1993): 19–25.

Suggested Reading

Assaf, Michael, Cynthia Bonincontro, and Stephen Johnsen. *Global Sourcing and Purchasing Post 9/11.* Fort Lauderdale, FL: J. Ross Publishing, 2006.

Ball, Donald, Michael Geringer, Paul Frantz, Wendell McCullock, and Michael Minor. *International Business: The Challenge of Global Competition.* New York: McGraw-Hill/Irwin, 2005.

Berry, John. *Offshoring Opportunities: Strategies and Tactics for Global Competitiveness.* Hoboken, NJ: John Wiley & Sons, Inc., 2006.

CIA Fact Book. U.S. Government Printing-Office, Washington D.C. See Current Issue.

Cook, Thomas A. *Global Sourcing Logistics: How to Manage Risk and Gain Competitive Advantage in a Worldwide Marketplace,* New York: AMACOM, 2007.

The *Economist Annual* edition of the World Report. The Economist Magazine.

European Union Data. The Web site for the New York EU office is www.europa-eu-un.org and in Washington, DC, www.eurunion.org. The DC office includes a large EU bookstore and most publications are free.

Hickman, Thomas K., and William M. Hickman, Jr. *Global Purchasing; How to Buy Goods and Services in Foreign Markets.* Homewood, IL: Business One Irwin, 1992.

Hinkelman, Edward G. *Dictionary of International Trade*, 7th ed. World Trade Press, Petaluma, CA., 2006.

"Global Business Trends" in each issue of *Inside Supply Management*, see www.ism.ws.

Global Logistics and Supply Chain Strategies (trade magazine), Keller International Publishing.

"Global Sourcing to Grow—But Slowly." *Purchasing* 18(2001): 24–32.

http://www.supplychainbrain.com/content/index.php. Global logistics and supply chain strategies (excellent Web site!)

http://www.intracen.org/tradstat/welcome.htm. International Trade Centre, international trade statistics.

http://www.i-b-t.net/incoterms.html. INCOTERMS 2010 Rules Chart of Responsibility.

Locke, Dick. *Global Supply Management: A Guide to International Purchasing*. New York: McGraw-Hill Trade, 1996.

Monczka, Robert M., Robert J. Trent, and Kenneth J. Petersen, "Effective Global Sourcing and Supply for Superior Results." CAPS Research, Institute for Supply Management, 2006.

Overholt, William H. *The Rise of China; How Economic Reform is Creating a New Superpower*. New York: W.W. Norton & Company, 1993.

Pooler, Victor H., David J. Pooler, and Samuel D. Farney, *Global Purchasing and Supply Management: Fulfill the Vision*, 2nd ed., Boston, MA: Academic Publishers, 2004.

Ramberg, Jan. *Guide to INCOTERMS 2000 (INternational COmmercial TERMS)*. New York: International Chamber of Commerce, 1999.

Schaffer, Richard, Beverley Earle, and Filiberto Agusti, *International Business Law and Its Environment*, Mason, OH: SouthWestern Publishing, 6th ed., 2005.

Schuster, Camille, and Michael Copeland. *Global Business; Planning for Sales and Negotiations*. Orlando, FL: Harcourt Brace and Company, 1996.

Schary, Philip B., and Tage Skjott-Larsen. *Managing the Global Supply Chain*. 2nd ed. Herndon, VA: Sopenhagen Business School Press, 2001.

Walker, Danielle, Thomas Walker, and Joerg Schmitz. *Doing Business Internationally: The Guide to Cross-Cultural Success*, 2nd ed. New York: McGraw-Hill, 2003.

CHAPTER 15

Total Cost of Ownership

Total cost of ownership (TCO) includes all costs related to the procurement and use of a product including any related costs in disposing of the item after its usefulness. This concept can be applied to a company's costs singularly, or viewed more broadly to encompass costs throughout the supply chain. This will be discussed in detail later in this chapter.

Three Components of Total Cost

There are three components of cost that must be captured when developing a TCO model: acquisition costs, ownership costs, and post-ownership costs.

Acquisition Costs

Acquisition costs are the initial costs associated with the purchase of materials, products, and services. They are not long-term costs of ownership, but rather represent an immediate cash outflow. Scrutinizing purchase price, planning costs, and quality costs to determine the lowest total cost of ownership/usage may provide significant savings.

Purchase Price

The price paid for direct and indirect materials, a product, or a service is frequently the major component of the item's total cost. Acquisition costs may include freight and delivery, site preparation (capital purchases), installation, and testing. Supply management professionals can reduce acquisition costs by negotiating effectively, obtaining quantity discounts, standardizing specifications, and completing a value analysis. In addition, strategic cost analysis offers methods to analyze and understand suppliers' costs—allowing for more fruitful negotiations and enhancement of supplier relationships. The purchase of used materials and equipment of acceptable quality is another way to lower acquisition costs. A supply management professional must not compromise long-term ownership costs by focusing on purchase price alone.

Planning Costs

Costs incurred during the acquisition process include the costs of developing requirements and specifications, performing price and cost analysis, supplier selection/sourcing, contract determination, initial order processing, and monitoring. Increasing the spending in these areas at times can reduce future ownership/use costs. For example, during the development phase of a new product, time spent with engineering representatives and the supplier to replace custom parts with standardized ones will generally reduce the initial purchase price, as well as facilitate future repair, replacement, and inventory carrying costs.

Subscribing to e-procurement, B2B e-commerce, or electronic supply networks provides many businesses with a means to lower acquisition costs by reducing or eliminating overhead such as the time-consuming research and paperwork often associated with ordering the best product or service to satisfy specifications. The higher initial development and start-up costs are negated by the potential benefits of better communication, more information, reduced clerical overhead, and lower purchase costs.

Quality Costs

The higher initial cost of engineering-in quality during the design phase generally lowers future ownership and post-ownership costs for both the purchaser and the customer. Selecting and certifying a supplier to obtain the optimal level of quality and monitoring the results using, for example, design of experiments and statistical process control ensures the achievement of the desired quality. In addition, the *quality* of the relationship established during this process can have long-term benefits. Long-term strategic relationships improve communications and may facilitate product innovation and cost reduction, especially in a cross-functional environment. Costs associated with quality are discussed in detail in Chapter 7.

Taxes

According to Richard Janis, a partner with KPMG LLP, a firm that sources internationally must address the impact of taxes, both direct (e.g., duties, processing fees) and indirect (e.g., foreign fuel taxes, tolls, facility fees), on the cost of procured materials and products.

> Companies spend endless hours haggling over freight rates, the cost of warehousing services, and the purchase price of goods. But they typically pay little attention to the hidden expenses that can inflate a supply chain's costs. One of

the most pervasive of these hidden expenses is taxes. Supply chain managers need to sit down with their tax colleagues to minimize the global impact of all taxes on supply chain operations.[1]

Janis adds that when sourcing nationally, the firm must consider differences in state and local taxes. Experienced tax professionals must be included in the cross-functional team when taxes are a concern. Janis provides examples of solutions to reduce acquisition costs by minimizing taxes:

▲ *Customs duties and tariffs* focus on compliance to eliminate penalties, and on planning to ensure the proper tariff classifications with the lowest rates
▲ *Regional trade agreements* source or produce in free-trade areas that reduce or eliminate duties on all or part of a product
▲ *Income-base shifting* use transfer pricing to legally shift income from high tax areas to lower tax areas

The impact of taxes can be significant. The added cost of addressing domestic and international tax issues, up front, may have a significant effect in reducing the purchase price.

Financing Costs

Whether purchasing inventory and materials, opening new facilities, or investing in equipment, the acquisition team should consider the quantitative and qualitative costs of financing alternatives, which are considered ongoing acquisition costs. A business can finance an acquisition using surplus cash, debt financing (secured and unsecured term loans, mortgages, revolving lines of credit, capital leases, sale-leaseback arrangements, bonds, securitization of receivables, etc.), or equity financing (issue different classes of stock, form new partnerships and joint-ventures, etc.). Each form of financing has costs and benefits. The creditworthiness of the firm (cash flow, profitability, debt load, future sales) and the expected return on the investment are key variables in making this determination. The cost of money is normally not a supply management professional's concern, but must be considered by the firm.

Ownership Costs

Ownership costs are the costs, after the initial purchase, associated with the ongoing use of a purchased product or material. Ownership costs are both quantitative and qualitative. Examples of costs that are quantifiable include: energy usage, scheduled maintenance, repair, and financing (lease vs. buy). Qualitative costs, although difficult to quantify, remain important considerations when making purchases. Examples of qualitative *costs* include: ease of use (is it a time saver?), aesthetic (psychologically pleasing to the eye), and ergonomic (maximized productivity, reduced fatigue). The sum of both types of costs may exceed the initial purchase price and have a significant bearing on cash flow, profitability, and even employee morale and productivity. Understanding and minimizing these costs can have strategic significance. The supply management professional considers the following additional cost categories before making a significant purchase decision.

Downtime Costs

Making a purchase decision based solely on purchase price may have long-term implications depending on the reason for the lower price. A seller may discount a premium item to move excess inventory and to increase sales. It may also want to dump a troublesome product on an unsuspecting purchaser. A new entrant in the market may discount an unproven product in an effort to gain market share. Often the selling price is representative of the quality of the product—presumably, the higher the price, the higher the quality. Whatever the reason, the long-run costs associated with a purchase may include non-value-added downtime. Costs associated with downtime include, for example:

▲ Reduced production volume and idle resources in a manufacturing environment. Downtime is often caused by unreliable or inflexible equipment, or direct materials that are substandard or wrongly specified in the design stage. Downtime for an automobile production line can run $27,000 or more per minute.[2]

▲ Opportunity cost of lost sales due to lower production volume.

▲ Goodwill costs due to undelivered or late orders, resulting in unhappy customers.

Careful scrutiny of reliability and dependability problems can reduce the cost of downtime.

Risk Costs

Weighing the risk of an inventory stock-out in a retail or manufacturing business against the opportunity cost of maintaining excess inventory is an important issue. Keeping extra inventory *just in case* can be a stopgap decision or a needlessly costly move. In JIT literature, just-in-case inventory is treated as a form of waste that a company should endeavor to reduce or eliminate. Some costs of excess inventory include those associated with financing, reduced cash flow, lost interest on cash flow, obsolescence, theft, and additional floor space.

Consider risk costs when purchasing from new suppliers (issue: dependability; risk avoidance maneuver: multiple sourcing); using new materials, processes, and equipment in manufacturing (issues: reliability, flexibility, suitability; risk avoidance maneuver: parallel processing); hiring new employees (issue: adaptability; risk avoidance maneuver: additional training and backup personnel), or choosing legal representation (issue: expertise; risk avoidance maneuver: multiple representation).

Careful investigation and the development of appropriate sources of supply will reduce the inherent risk associated with the unknown, untried, and unproven. The appropriate place to begin is in the planning or acquisition stage where a risk assessment study should be conducted. Spending upfront to reduce risk is an investment in the long run efficiency and profitability of any firm.

Cycle Time Costs

Whether decreasing a new product's time-to-market or increasing the number of items produced in an hour (throughput), reducing cycle time can increase profitability and return on investment via lower total costs. An organization with vision will apply the principles discussed throughout this text to shorten the time to complete all relevant purchasing and production activities. Practices that a supply management professional can employ that may have significant impact include: implementing JIT materials management, forming strategic alliances with key suppliers, and establishing cross-functional alliances within the organization. The higher initial cost of establishing and implementing these goals will provide long-run savings in the cost of direct material, direct labor, and manufacturing overhead. In addition, qualitative *savings* may accrue in the form of a smoother running, more *user-friendly* organization.

Conversion Costs

Buying the wrong material whether in quality, form, or design can increase the cost of conversion (the application of direct labor and manufacturing overhead to direct materials to create a product or service). As discussed earlier, material not optimized for the production process can increase labor and overhead usage and thus, because throughput is decreased and the cost of maintaining the quality of the finished product is increased, the total cost of production. In addition, machine time, labor requirements, scrap, and rework may add to the unit cost. Spending too little time and money in the acquisition of materials may result in spending more time and money during production.

Other areas that affect conversion costs include production methods (assembly lines vs. cells, labor-intensive vs. automated production), employee training and working environment, and the methods of accounting for product costs, especially in the application of overhead costs to units of product. A well-informed and well-trained supply management professional may have the ability to influence decision-making in these areas when working in a cross-functional environment.

Non-Value-Added Costs

Non-value-added costs flourish in most businesses. It is estimated that some 40% of all costs add little or no value. Examples of non-value-added activities that add costs to a product or service include:

▲ Moving and stockpiling batches of direct materials and work-in-process inventory due to a poor factory layout, scheduling, and a variety of wastes that increase uncertainty in the outputs of the system.

▲ Maintaining cumbersome operating procedures that duplicate efforts and steps for no apparent reason.

▲ Routing daily service appointments in a random fashion rather than designing routes that minimize travel time.

Total quality management (TQM), continuous improvement, activity-based costing (ABC), and activity-based management (ABM) are incremental change approaches that help identify non-value-added activities. Process reengineering is a more radical approach to change that focuses on simplification and elimination of wasted effort.

Supply management professionals with a background in management, operations, manufacturing, finance, information technology, and logistics are qualified to make suggestions to suppliers (and suppliers' suppliers) that reduce non-value-added costs. A successful strategy, when possible and cost effective, is to visit a supplier's manufacturing site and observe how production takes place. Careful scrutiny may reveal a number of non-value-added costs that the supplier can reduce or eliminate, thus allowing for negotiations on lowering the supplier's price. Observing a service provider's processes, either on-site or at its place of operation, may reveal non-value-added costs that, when eliminated, will provide savings to both parties.

Supply Chain/Supply Network Costs

"If you process-map a supply chain and examine the material movement alone, such as the ins and outs of material flow from one organization to another, you will find many opportunities to eliminate waste."[3] Of course, waste adds unnecessary costs to purchased materials and services, as well as to logistics.

James E. Morehouse, a Vice President for A. T. Kearney in Chicago, asserts that extended enterprises are beginning to develop and will hasten improved efficiency and cost reductions along the supply chain. He believes that

> organizations will be outsourcing transportation, purchasing operations, manufacturing, warehousing, order entry, and customer service. As a result, organizations will be more integrated with their suppliers and customers in order to manage the

total supply chain from raw materials to the ultimate customer, the only source of revenue.[4]

A supply management professional should consider the following interrelated areas for developing better cost reduction strategies:

1. *Forecasting*   Improving customer demand forecasting and sharing the information downstream will allow more efficient scheduling and inventory management.
2. *Administration*   Implementing EDI within an organization and between members of the supply chain will facilitate communication, thus reducing purchasing time, paperwork, and errors.
3. *Transportation*   Streamlining material movement through the chain will reduce supply chain cycle time.
4. *Inventory*   Embracing a JIT-type philosophy will help reduce unnecessary stockpiling and movement of inventory; suppliers can share inventory type and level information.
5. *Manufacturing* Improving capital budgeting procedures, and designing and developing manufacturing processes that provide quality, efficiency, and reliability will lower costs and improve quality.
6. *Customer service*   Listening to the customer will help identify supply chain inefficiencies and blockages.
7. *Supplier selection/relationships*   Determining the appropriate source of supply and type of relationship (transactional, collaborative, or strategic alliance) with each supplier will minimize administrative overhead and focus on the lowest cost at the required quality.
8. *Global sourcing*   Expanding sourcing internationally will provide cost savings and quality improvements by focusing on an international supplier's comparative advantage and utilizing EDI and available low cost transportation.

Well-trained supply management professionals armed with this knowledge can bring fresh insight to the table when developing TCO models and negotiating throughout the supply chain.

Post-Ownership Costs

In the past, salvage value and disposal costs were the major inputs required when estimating post-ownership costs of capital purchases. These costs could be estimated as cash inflows (the sale of used plant and equipment), or outflows (such as demolition of an obsolete facility). For many purchases, there was an established market that provided data to help estimate reasonable future values, such as the Kelly blue book for used automobiles. An appraiser of industrial equipment could help estimate the future worth of the plant and equipment. Often companies made investments with *absolute certainty* of future appreciation, although estimating actual appreciation required more information than was available (e.g., property in a major metropolitan area). Today, supply management professionals must address these issues. In addition, three other factors with potential long-term impact must be addressed when performing a TCO analysis on equipment, a plant, direct materials, a product, or a service: long-term environmental impact, unanticipated warranty and product liabilities, and the negative marketing implications of low customer satisfaction.

Environmental Costs

Gasoline stations in California have faced the unplanned expenditure of replacing their underground gas storage tanks with more environmentally friendly models. They have also been required to sanitize the soil near the tank if leakage has occurred. This expenditure has cost many independent operators their businesses, devalued the property (if polluted and unsafe), and increased the margin required on each gallon of gas sold to help recover such expenditures. This type of post-ownership cost is becoming more common as environmental problems persist.

Warranty Costs

A poorly designed and produced product may have unanticipated warranty-related costs. Tire tread difficulties that occurred between 1978 and 1980 and again in 2000 resulted in such a problem for Firestone. In 2000, General Electric, in cooperation with the Consumer Product Safety Commission, recalled selected dishwashers manufactured between April 1983 and January 1989 to rewire a defective slide switch (as an option, GE offered a rebate toward a new unit). After-sale costs such as replacements, returns, or allowances can accrue to service providers (e.g., a carpet cleaning company whose poorly trained employee inadvertently uses a cleaning solution that fades the carpet fibers), and retail companies (e.g., a department store that misrepresents a products capabilities). A well-trained supply management professional participating in a cross-functional team in product or service design may point out potential warranty/recall costs early enough so that more emphasis is placed on designing and producing a defect-free and reliable product or service.

Product Liability Costs

Companies engaged in all types of business have faced unanticipated product liability costs due to poorly designed or produced products and services. For example, fuel tanks that explode on impact due to poor design; tire treads that separate due to poor design or manufacture, inferior materials, or improper inflation by end users; faulty ignition switches that cause cars to stall at inopportune moments; ground beef with the *E.coli* bacillus due to improper processing; lawyers and accountants who have not performed their services according to professional standards; and retail outlets that sell defective merchandise. This list is long and the remedies usually require expenditures, often not covered by insurance reimbursement.

Customer Dissatisfaction Costs

Some 75% of field failures in consumer goods can be attributed to defects in purchased materials. Field failures lead to customer dissatisfaction. When a customer is dissatisfied with a product, he or she frequently shares this dissatisfaction with many friends and acquaintances, some of whom may be potential customers. This flow of negative publicity frequently results in lost sales or "customer dissatisfaction" costs.

TCO, Net Present Value (NPV) Analysis, and Estimated Costs

When trying to consider the true TCO, one method of evaluating a potential capital investment is to combine the present values of the initial expenditure (initial cash payment), the future revenue streams (cash receipts, or the reduction or elimination of expenses), and future expenditure streams (cash payments, or the reduction or elimination of revenues). Analysts discount the positive and negative streams using an interest rate usually referred to as the opportunity cost of holding capital—the minimum required rate of return a business expects to receive on its investment. The opportunity cost of capital is linked to the riskiness of the investment and the firms' capital structure.[5]

TCO and NPV analysis are very similar in philosophy. Both attempt to estimate and analyze the acquisition cost, operating costs, and post-ownership costs in terms of value likely to be received by the company-in addition, NPV analysis attempts to analyze revenues and other cash inflows. NPV uses the present value of a sum of future cash flows discounted by a required rate of return—the larger the positive net present value, the more likely the investment will return more than required over its life. A net present value of zero (NPV = 0) is the point of indifference. An NPV greater than zero usually suggests that the investment should be accepted. A negative NPV indicates that the overall return will be less than the minimum rate of return required by the company for the investment. An example of an NPV analysis is presented in Table 15.1.

Table 15.1 provides an example that demonstrates, based on actual and *accurately* estimated cash inflows and outflows that this machine would be an unattractive investment—the required rate of return is 20% and this investment returns 15.66%. Alternatively, the NPV is negative so the investment opportunity should be rejected. Other potential investments may be more attractive.[6]

Table 15.1 Net Present Value Analysis—Copier

Required Rate of Return	20.00%							
Year	Now	1	2	3	4	5	6	Present Value
Cost of machine including installation and testing (actual)	(120,000)							(120,000)
Manufacturer required overhaul (estimated)				(9,000)				(5,208)
Cash inflows generated by using machine (estimated)		40,000	40,000	40,000	40,000	40,000	40,000	133,020
Cash outflows incurred by using machine (estimated)		(7,000)	(7,000)	(7,000)	(7,000)	(7,000)	(7,000)	(23,279)
Salvage value (estimated)							7,500	2,512
Net present value of potential investment								(12,955)
(Alternative Method)								
Total of annual streams (from above)	(120,000)	33,000	33,000	24,000	33,000	33,000	40,500	
Required rate of return		20%	20%	20%	20%	20%	20%	
Sum of present value of annual streams equals net present value of potential investment	(120,000)	27,500	22,917	13,889	15,914	13,262	13,563	(12,955)
Internal rate of return	(120,000)	33,000	33,000	24,000	33,000	33,000	40,500	15.66%

462

Qualitative considerations, which cannot be used in this calculation, may either support or not support this purchase. Obviously, poorly constructed estimates of future cash flows and discount rates may provide meaningless information.

TCO like NPV requires an analysis of the holding period of the asset. TCO focuses on estimating and analyzing the ownership and post-ownership costs. The following formula represents a simplified approach to TCO analysis:

$$TCO = A + P.V. \sum_{i=1}^{n} (T_i + O_i + M_i - S_n)$$

where TCO = total cost of ownership
$\qquad A$ = acquisition cost
$\qquad P.V.$ = present value
$\qquad T_i$ = training costs in year i
$\qquad O_i$ = operating cost in year i
$\qquad M_i$ = maintenance cost in year i
$\qquad S_n$ = salvage value in year n

Using the data presented in Table 15.1, the TCO calculation is shown in Table 15.2.

This type of analysis can be repeated on competing copiers—the copier with the lowest TCO should be the best choice, other considerations being equal.

This analysis focuses on costs or cash outflows and not cash inflows. Additionally, it considers the present value of those outflows. Incorporating NPV analysis, when applicable, into a TCO analysis will provide additional input that will allow the analyst to make a sound recommendation between alternatives.

Table 15.2 Sample TCO Calculation

	Acquisition Cost = $120,000	Present Value Formulas
Present value of cash outflows for years 1–6 =	23,279	$7,000[1/.20 - (1/.20(1 + .20)^6)]^*$
Present value of overhaul in year 3 =	5,208	$9,000/(1 + .20)^{3\dagger}$
Present value of salvage value in year 6 =	(2,512)	$7,500/(1 + .20)^6$
TCO =	$145,975	

$^*PV_{Annuity} = CF[1/r - 1/r(1 + r)^t]$
CF = periodic cash inflow or oulflow (must be the same each period)
r = discount rate per period (annual rate divided by the number of periods in one year)
t = total number of periods
$^{\dagger}PV = FV/(1 + r)^t$
FV = future value of single cash inflow or oulflow
r = discount rate per period (annual rate divided by the number of periods in one year)
t = total number of periods

A note about estimated costs: since TCO and NPV analyses require estimates of future costs of cash outflows, their reliability in providing useful information is only as good as the quality of the input data. A well-conceived analysis should rely on inputs provided by cross-functional representatives with specific knowledge of and interest in the subject of the analysis. The most interested team member should provide the estimate. For example, supply management provides data on purchase price; plant engineering provides an estimate of potential downtime costs; marketing provides an estimate of support costs; and manufacturing provides an estimate of productivity or efficiency costs. A good rule of thumb is to include relevant participants who want to be part of the process.

Another method of arriving at estimates is for the cross-functional team leader to propose a cost figure and submit it for discussion. Frequently, one or more team members will react by saying, "No, x is too much. It would be closer to y." Two other approaches of arriving at estimated costs include parametric (several variables that affect costs are addressed) and Delphi (the cross-functional team members reflect on and refine an initial guesstimate). Both approaches, when applied in a strict sense, tend to be needlessly costly.

Neither of these methods can quantify many intangible variables that may color a choice between alternatives, or influence you to proceed with an investment. For example, in choosing between a sports car and a sports utility vehicle, a cost analysis is often not enough. Many people use their vehicle to make a statement about them, so ego may hold more substance than cost in this instance. Similarly, a company may want to project an image that only an automobile with a higher TCO can provide. This image may or may not be quantifiable in terms of increased future sales or executive image. It can be difficult, if not impossible at times, to quantify cash inflows on purchases—in these instances, a strict NPV analysis may be pointless.

The Importance of Total Cost of Ownership in Supply Management

As Ray learns in the opening vignette, purchase price is only one component of the cost of purchasing material, a product, or a service. TCO should be a permanent concept in every supply management professional's mind whether in a service, retail, or manufacturing firm. Overemphasis on acquisition cost/purchase price frequently results in a failure to address other significant ownership and post-ownership costs. Total cost of ownership is a philosophy for really understanding all supply chain related costs of doing business with a particular supplier for a particular good or service.[7]

Some typical ownership costs include those associated with processing inventory (direct materials), repair, maintenance, warranty, training, operating, inventory carrying, contract administration, and downtime cost for operating equipment. Post-ownership costs may include those of disposal and environmental cleanup. The addition of risk and its associated costs adds yet another dimension. These costs and others must be estimated and included in a total cost analysis.

TCO is relevant not only for the firm that wants to reduce its cost of doing business, but also for the firm that aims to design products or services that provide the lowest total cost of ownership to end customers. For example, some automobile manufacturers have extended the tune-up interval on many models to 100,000 miles, thus reducing the vehicle operating cost for the car owner.

TCO analysis draws from a variety of academic disciplines such as finance (NPV analysis), accounting (product pricing and costing), operations management (reliability and quality), marketing (understanding and meeting customer wants), information technology (systems integration and e-commerce), and economics (minimum average total unit cost of production). The best and, perhaps, only way of addressing all relevant costs is to employ a cross-functional team representing the key stakeholders.

Supply management personnel can facilitate the TCO analysis by combining their broad-based analytical skills with those of other team members to ensure that all relevant costs are considered. A brief examination of each type of business together with supply management's vision and role in minimizing total costs is a good place to begin.

Service Providers

Services firms provide seemingly intangible products to satisfy human wants and needs. Service providers run the gamut from accounting, legal, and medical professions to federal, state, and local governments to window washers, gardeners, and taxi drivers. Service providers procure capital equipment, products, and services, as well as hire employees and provide employee benefits such as health and life insurance.

Like all businesses, service firms enhance profitability by increasing sales at a faster rate than costs, maintaining sales and reducing costs, or increasing sales and reducing costs, while maintaining the desired quality and timeliness. Understanding what drives the cost of overhead expenditures is crucial to any service business. Service revenue must cover the direct costs, material and labor, and overhead in order to generate a profit. Depreciation is a key element of overhead. We define depreciation as the systematic transfer of the cost of a capital expenditure (an asset on the balance sheet) to expense (the income statement). Another important cost element is the cost of maintaining capital and operating equipment. A TCO analysis of equipment purchases may help reduce the expenditures for maintenance and parts over the lives of the investments.

Another important consideration in service businesses, as well as in those of retail and manufacturing, is the total cost of maintaining the employee base. Paying the lowest wage does not necessarily result in the most cost-effective employee. The cost of getting a new and inexperienced employee "up to speed" can be high, and the learning curve long. Paying more for an experienced person with a short "ramp up" time may be the best long-term solution for a given position. A total cost/total benefit analysis of company-sponsored health insurance programs can reap rewards in terms of lower per person total costs, greater benefits for covered employees, and improved morale.

Retail

The considerations that apply to service businesses also apply to retailers. Retail businesses sell a product that often must be ordered, received, inventoried, sold, and perhaps, delivered to the customer. The choice of a system that facilitates the processes involved in inventory ordering and turnover will influence the total cost of inventory ownership. Many major retailers have empowered select suppliers to manage their product inventory for them,[8] thus reducing purchasing overhead and inventory carrying costs without necessarily increasing the product cost. Embracing the JIT philosophy is another way to improve QCT (quality, cost, and time) while reducing TCO. Lowering the cost of goods sold and the overhead costs associated with procurement, inventory carrying costs, and sales improves the bottom line. It is often easier to lower costs than to increase sales in a competitive business environment.

A retail business may own a product for a short time, but may be responsible for after-sale adjustments, warranty claims, and maintaining general customer satisfaction for an indefinite period. If a retailer selects an item or product for sale solely based on price,[9] thus ignoring reliability and product liability issues, customer satisfaction and future sales may suffer. Further, the retailer may incur an increased risk of financial or moral liability. Retailers must know their customer base and tailor the products they sell to satisfy that base. Retailers must also consider the long-term effects of every purchase made for resale.

Manufacturing

Manufacturing businesses are concerned with the same TCO issues as are service and retail firms. In addition, they procure direct materials (raw materials, products, sub-assemblies, etc.) and incur overhead in the production of their finished goods inventory and other activities required when conducting business. Managerial accountants place emphasis on the variance between what something "should cost" or is expected to cost and what it actually costs. Price variance analyses are often misleading. For example, a favorable price variance (the material cost less than anticipated), when compared to an unfavorable quantity variance (more material was used than anticipated) may indicate that although the material was less expensive, it was of a lower quality and, therefore, more was used. If you compare this to an unfavorable labor efficiency variance (more labor was used to work with the material than anticipated), it becomes clear that a lower acquisition price may translate into higher production costs. Considering all costs simultaneously, supply management professionals and other members of the product development team can better determine the right specifications for the material and ensure that suppliers meet these guidelines.

The accurate allocation of manufacturing overhead is a major factor in calculating the true unit cost of a product. Using the wrong cost driver (process or activity that creates the need for overhead) can make a product seem more or less expensive than it actually is. Activity-based costing (ABC), although initially somewhat complicated and expensive to implement, can return long-term benefits by providing more accurate unit cost information that serves as the basis of better decisions. Careful budgeting and procurement of overhead items from the purchase of capital equipment to that of lubricants used in production, and the implementation of systems to ensure the timely availability of accurate information are methods used in obtaining the lowest total cost of production.

Supply Chains/Supply Networks

A supply chain is a set of three or more entities (organizations or individuals) directly involved in the upstream and downstream flows of products, services, finances, and information from a source to a customer.[10] A supply network is a less linear, more flexible, virtual system linked by advanced communication systems and enhanced supplier relations. A supply management professional/organization can apply the philosophy and practice of TCO to the strategic optimization of costs within the chain or network. For example, an American company with the option of making a new product in Asia (potentially lower manufacturing costs) and shipping it to its customer base in the United States (higher transportation costs) or manufacturing it in the United States (potentially higher manufacturing costs) with minimal shipping (lower transportation costs) will have to determine the total cost of each alternative before making a decision. This TCO analysis should include[11] the study of:

- ▲ The manufacturability of the product (value engineering/value analysis).
- ▲ The manufacturing infrastructure requirements (the basic facilities, services, and installations needed for the optimal functioning of the manufacturing operation).
- ▲ The decision to outsource or self-manufacture.
- ▲ The abilities/location/responsiveness of potential tier two, three suppliers relative to the manufacturing operation.
- ▲ The structure of foreign and domestic tariffs/duties/taxes.
- ▲ The costs of transportation and the timeliness of delivery.

▲ Foreign business/labor/environmental regulations.

▲ Foreign political/economic stability.

▲ Foreign exchange risk.

▲ Language/communication requirements.

▲ Volatility of end-customer demand and the responsiveness of the network to changes in that demand.

▲ Inventory carrying costs (investment vs. service levels).

▲ Inventory risk (relocation, damage, obsolescence, shrinkage).

▲ Quality costs.

Although much of this analysis is ultimately quantifiable in dollars, some elements will require a qualitative evaluation offering less certainty. In this example, at first Asia may seem the logical choice, but distance from the customer base and other international issues may guide the decision to domestic production.

Summary

TCO is an analytical tool and a philosophy that supports management decision-making. A supply management professional can modify the TCO approach to support each major purchase decision, as well as integrate it into strategic cost analysis to support make or buy (outsourcing), pricing and costing, critical direct material purchases, and other decisions that require analysis of costs over time. TCO is also a powerful adjunct, for example, in evaluating employee benefit programs and aiding in analyses such as the total cost of implementing an integrated activity-based costing system in a manufacturing business. Estimates are the basis of most ownership and post-ownership costs. The care with which a supply management professional on a cross-functional team estimates these costs will determine the effectiveness of the resulting analysis.

As a philosophy, TCO can become an active part of everyday decision-making. For example, TCO can help a family determine the total costs of maintaining a pet or choosing kitchen appliances. If the expression, "There's no such thing as a free lunch" is true, then everything we do has a tangible or intangible cost that can be analyzed, if necessary.

Appendix: Supply Management in Action

Implementing and Using TCO at Scott Paper:
Prepared by Robert Porter Lynch[12]

Prior to being acquired by Kimberly Clark several years ago, Scott Paper was one of the world's largest producers of paper, with 20 plants in nearly 20 countries. The paper making process requires that wet pulp slurry be deposited uniformly on a continuously moving fabric belt. The fabric belt (known as "fabric" in the industry) is approximately 12 feet wide and the belt is approximately 60 feet in circumference. Fabrics are woven to enable the water in the pulp to be drawn out through the fabric, so that when the pulp leaves the end of the belt, it is in a semi-solid, rather gelatinous form. Once leaving the fabric belt, the pulp goes onto other machines, which further dry and then press the pulp into paper. Fabrics cost approximately $25,000 each.

For years, each plant's procurement team had negotiated with fabric suppliers. Knowing that the sales price was about $25,000 (a simple "component cost" to Scott), the procurement directors were always rewarded for driving costs down. Every buyer was trained in being a tough negotiator; they all knew that 5 to 10% should be driven out of the sales price. Each year the procurement group aimed to push prices down, thereby driving down the profit for the fabric manufacturers. At the end of the year, rewards were allocated to those buyers who got the most favorable pricing. Moreover, for years, Scott Paper's procurement department patted itself on the back for doing a wonderful job at keeping both the supplier's profits and Scott's prices low.

Scott Paper's profitability was among the lowest in the industry, making it ripe as a takeover target. Nevertheless, the purchasing managers were all confident that they were doing their part to get costs down to the lowest level possible.

In 1994, a new Vice President of Procurement, Ted Ramstad, arrived on the scene and began challenging the traditional thinking. In an effort to understand the real cost of the fabric, Ted began conducting a reevaluation of cost. Internal data were gathered:

▲ While replacing fabrics, the paper machine must be shut down, at a cost of nearly $100,000 per day to the paper company (because paper manufacturing requires a continuous process, and the machine is considered efficient only when it runs 24 hours a day).

▲ It takes approximately 8 hours to put fabric on a paper machine.

▲ Fabrics lasted an average of 40 days.

▲ Most fabrics broke on the machines.

▲ When a fabric broke, it normally had less than 10% wear.

▲ Seventeen companies supplied fabrics to Scott around the world. Each plant manager had a "favorite" supplier, but there was no compelling reason for using one supplier over another. Procurement assumed it could use the large number of suppliers in a competitive manner to keep the costs low.

▲ Cost of goods sold (COGS) for most suppliers was about 35% and R&D was 3 to 5%.

▲ Most plants had 4 to 6 fabrics in inventory.

While most of the buyers were unconcerned about this information, Ramstad and his team, applying TCO thinking, began probing and asking more questions:

▲ How can we lengthen the time a fabric lasts on a machine?

▲ How long should a fabric last?

▲ Are we getting the *best* fabrics from our suppliers, or just the *cheapest?*

▲ What suppliers are providing the research and development to give us better performance from our fabrics?

▲ Would fewer suppliers give us volume-purchasing power?

▲ Could we build win-win incentives to get more value from our suppliers and their fabrics?

▲ Where is there significant "non-value added" in the system?

▲ What benchmarks should we be using to be "best in class?"

▲ If the "absolute component cost" of a fabric is $25,000, what is the "total cost of ownership," and how does this compare as a "relative competitive advantage (or disadvantage)?"

Armed with a new focus and an energetic spirit, Ramstad's team began a world-wide search for answers. Fabric suppliers were interviewed, and information was gathered regarding competitors, indicating:

▲ The industry average fabric life span was 60 days.

▲ The industry benchmark fabric life span for one paper producer was 470 days.

▲ Only three suppliers were interested in helping Scott increase fabric longevity.

▲ None of the suppliers believed their fabrics were at fault for Scott's low life span; all blamed either the operators or the machinery manufacturers.

A Crucial Juncture

Now came the real test. In a "simple accounting, component-cost" world, fabrics clearly cost $25,000 apiece. However, Ramstad stuck his neck way out and maintained that this was only true in a narrow, "absolute" sense. In a broader, "relative advantage" perspective, the formulation of cost looked radically different. Here's what Ramstad's TCO calculations looked like:

Fabric cost: If the highest standard benchmark life is 470 days and Scott's standard is only 40 days, the relative cost of the fabrics Scott was purchasing is really 470/40, or 11.75 times that of the highest benchmarked competitor.

Therefore, 11.75 × 25,000 unit purchase cost = $293,750.

(To understand relative competitive advantage, think of relative motion. Consider the analogy of driving down a highway at 40 miles per hour in the right-hand lane. The average competitors are in the middle lane, passing you at 60 mph. But the best-in-class competitor flies by in the left lane at 470 mph. This is the relative competitive advantage view of costing.)

Down-time cost: Add an additional 8-hour portion of $100,000 per day to reflect the downtime for changing the fabric. (8/24 h × $100,000 = $33,000) Relative to the best-in-class competitor, Scott has to make 11.75 changes of the fabric to the best-in-class competitor's one change.

Additional relative cost to Scott is 11.75 × $33,000 = $387,750.

Burdened labor costs: It takes two men 8 hours to change a fabric. At a burdened labor rate of $45 per hour, the labor costs are 2 × $45 × 8 = $720. Relative to the best-in-class competitor, Scott has to make 11.75 changes to the best-in-class competitor's one change.

Additional relative cost to Scott is 11.75 × 720 = $8,460.

Incremental cost of purchasing the low-price fabric: Adding these figures, the results are overwhelming. Relative to the best in class, Scott's "relative disadvantaged cost" is $689,969![13] Very different from what was thought by procurement to be a $25,000 belt.

The procurement group had naively engaged in myopic thinking; they were playing the game "too small." Squeezing the supplier for a 5 to 10% discount made no sense when the stakes were really about how to gain an advantage of nearly $700,000. This is a "strategic systems" view of cost vs. a "component cost" view.

Ramstad did not stop there. He saw the relative disadvantage to be multiplied by the number of plants globally. Therefore, by multiplying the "relative single plant competitive disadvantaged cost" by the 20 plants throughout the world, there was nearly $14 million of advantage to be gained on this single line item alone.

(*Note:* The standard accounting systems at Scott could not measure this factor, and therefore it was "invisible" to the Chief Financial Officer, who steadfastly called this accounting hocus-pocus.)

Undaunted, Ramstad pressed on. He advocated that the problem was even worse because much of this inventory was actually scrapped due to product redesign before the inventory was utilized. What's more, he took the position that if Scott bothered to add the time-value of money for financing the inventory of belts (because of frequent breakage several extra belts had to be kept on hand), the extra inventory was tying up capital. Eventually Ramstad was able to eliminate $20 million in inventory.

And it didn't stop there. By selecting the best-in-class suppliers, thereby reducing the number of suppliers to two globally, and negotiating long-term contracts, Ramstad was able to convince suppliers that they no longer needed to make sales calls to Scott's procurement officers. Because sales costs were 35% of the component price, he persuaded the remaining suppliers to lower their prices 25%, increase their R&D budgets to provide better products, provide technical support, and work with the machinery companies to improve sensing and tuning devices. Because of the higher volumes for the remaining two suppliers, their actual profits were substantially higher under the new model than before. In addition, by not handling a continuous stream of bidding and purchasing, which previously accounted for 3 to 5% of the cost of ownership, Ramstad was able to reduce the procurement force as well.

Endnotes

1. Janis, Richard, "Texes: The Hidden Supply Chain Cost." *Supply Chain Management Review* Winter (2000): 72–77.
2. Personal interview with R. David Nelson, former Senior Vice President of Purchasing of Honda America Manufacturing, October 14, 1998.
3. Leroy Zimdars, C.P.M., former Director of Supply Chain Management for Harley-Davidson, quoted in John Yuva, "Reducing Costs through the Supply Chain," *Purchasing Today* June (2000).
4. Ibid.
5. In financial management literature, the opportunity cost of capital is referred to as the "weighted average cost of capital" or WACC.
6. This example does not consider the income tax on the revenue generated by the copier or the tax savings that accrue by depreciating the copier (known as the *depreciation tax shield*). Including the tax effects in this analysis will change the NPV results.
7. Lisa M. Ellram, Seminar on Strategic Cost Management for Purchasing Professionals, NAPM Convention, San Diego, CA, May 1999.
8. The acronym for suppliers managing inventory at the buyer's site is VMI, which stands for vendor managed inventory.
9. Usually, this reflects the reliability and quality of the product, that is, "you get what you pay for."
10. Mentzer John T., D. I. Flint, and G. T. M. Hult, "Fundamentals of Supply Chain Management: Twelve Drivers of Competitive Advantage," *Journal of Business Logistics,* 22(2)(2000): 4.
11. The bulleted list provides some general consideration and is by no means exhaustive.
12. Case study contributed by Robert Porter Lynch, President, The Warren Company (http://www.warrenco.com).
13. Fabric cost ($293,750) + downtime ($391,663) + labor ($8,460) = $693,873.

Suggested Reading

Barkman, Sandra, and Bryon S. Marks. "Getting What You Pay For?—Total Cost of Ownership Model." Presented at the 91st Annual International Supply Management Conference, May 2006, San Antonio, TX.

Cavinato, J. "A Total Cost/Value Model for Supply Chain Competitiveness." *Journal of Business Logistics* 13(2)(1992): 285–301.

Degraeve, Z., and F. Roodhooft. "Effectively Selecting Suppliers Using Total Cost of Ownership." *Journal of Supply Chain Management* 35(1)(1999b): 5–10.

Ellram, L.M. "Total Cost of Ownership: Elements and Implementation." *International Journal of Purchasing and Materials Management* 29(4)(1993): 3–10.

Ellram, L.M. "A Taxonomy of Total Cost of Ownership Models." *Journal of Business Logistics* 15(1)(1994): 171–191.

Ellram, L.M. "Activity-Based Costing and Total Cost of Ownership: A Critical Linkage." *Journal of Cost Management* 9(1)(1995): 22–30.

Ellram, L.M., and S.P. Siferd. "Purchasing: The Cornerstone of the Total Cost of Ownership Concept." *Journal of Business Logistics* 14(1)(1993): 163–185.

Ellram, L.M., and S.P. Siferd. "Total Cost of Ownership: A Key Concept in Strategic Cost Management Decisions." *Journal of Business Logistics* 19(1)(1998): 55–76.

Ellram, Lisa M. "Total Cost Modeling in Purchasing." *Center for Advanced Purchasing Studies* 1994 CAPS Focus Study.

Ferrin, Bruce G., and Richard E., Plank. "Total Cost of Ownership Models: An Exploratory Study." *Journal of Supply Chain Management* 38(3)(2002): 18.

Harding, Mary Lu. "Defining and Calculating TCO." *NAPM InfoEdge* May (2000).

Harding, Mary Lu. "Understanding Total Cost of Ownership." *NAPM Info Edge* 5(3) (2000).

Harding, Mary Lu. "Total Cost of Ownership—MRO." *Purchasing Today* 12(8)(2001): 15.

Milligan, B. "Tracking Total Cost of Ownership Proves Elusive." *Purchasing* 127(3)(1999): 22–23.

Roberts, Julie S. "The Supply Chain of Dollar and Cents." *Inside Supply Management* June (2002): 38, www.ism.ws/ResourceArticles/2002/060238.cfm.

CHAPTER 16

Price and Cost Analysis

Introduction

Obtaining materials at the right price can mean the difference between a firm's success and failure. Price, also referred to as acquisition cost, is frequently the largest component of total cost. Professional supply managers define the right price as a price that is fair and reasonable to both the buyer and the seller. Unfortunately, there is no magic formula for calculating precisely what constitutes a "fair and reasonable price." The right price for one supplier is not necessarily the right price for another supplier, either at the same time or at different points in time. To determine the right price for any specific purchase, a number of constantly changing variables and relationships must be evaluated. This chapter discusses the most important of these variables and their relationships.

General Economic Considerations

Conditions of Competition

Economists of the classical school speak of a competitive scale that includes three fundamental types of competition: pure, imperfect, and monopoly. At one end of the scale is *pure (or perfect) competition*. Under conditions of pure competition, the forces of supply and demand alone, not the individual actions of either buyers or sellers, determine prices. In this environment, producers are price takers—they have no control over the price they receive for their products.[1]

At the other end of the competitive scale is a monopoly. Under conditions of a *monopoly*, one seller controls the entire supply of a particular commodity, and thus is free to maximize its profit by regulating output and forcing a supply-demand relationship that is most favorable to the seller. In this environment, producers are price makers and exert varying degrees of control over the price they receive for their products.

The competitive area between the extremes of pure competition and monopoly is called *imperfect competition*. Imperfect competition takes two forms: (1) markets characterized by few sellers and (2) markets in which many sellers operate. When there are just a few sellers, an *oligopoly* is said to exist. The automobile, steel, and tobacco industries are examples of oligopolies. Generally, oligopolistic firms produce relatively few different products.

In contrast to oligopolies, the second form of imperfect competition exists where many sellers produce many products. This form of competition is referred to as *monopolistic competition*. Most of the products sold in this market are *differentiated* (distinguished by a specific difference), although some are not. Sellers, however, spend large sums on major promotional efforts to persuade buyers that their products are different. The majority of the products sold in the United States are traded in this market. These economic principles are portrayed in Figure 16.1.

In practice, the three categories of competition are not mutually exclusive; meaning they can overlap. When one considers both the buying and the selling sides of the total market, it is apparent that the number of market arrangements between individual buyers and sellers is very large.

It is frequently suggested that oligopolists conspire and act together as monopolists to thwart price competition. In fact, the U.S. Department of Justice does uncover a few conspiracies every year. The facts, however, indicate that price conspiracies among oligopolists are not the normal order of business. Any buyer who has purchased in oligopolistic markets knows that both price and service competition can be intense. Chevrolet strives intensely to outsell Ford, and vice versa. This is not to say that oligopolistic industries do not periodically exercise monopolistic tendencies to their own advantage. They do. For example, in times of recession, it requires only a basic knowledge of economics, not a conspiracy, for oligopolies to lower production rates and thus direct a balance of the forces of supply and demand in their favor. (One might reflect on how OPEC uses supply to affect oil prices!)

It should be noted that oligopolistic industries frequently hold firmly to their prices for long periods and appear noncompetitive. This appearance may be deceptive, however. Frequently, in order to gain a competitive advantage without notice, oligopolists shift their competitive efforts to other areas, such as service. Sellers may agree to perform such additional services as carrying customers' inventories, extending the payment time of their bills, or absorbing their freight charges. Such indirect price reductions often are not advertised.

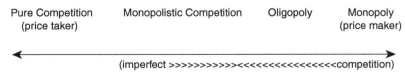

Figure 16.1 Categories of competition.

Consequently, the amount of service a firm is able to obtain usually correlates directly with the perception and skills of its supply management personnel. Foreign competitors also greatly influence the freedom of U.S. oligopolies to raise their prices above fair market prices. For example, the freedom of U.S. automobile, steel, and electronic companies to raise prices is noticeably restrained because of foreign imports. For a number of items, specialty suppliers also compete effectively with oligopolies. Consequently, the competent buyer who learns to operate successfully within the practices of oligopolistic industries can definitely influence the firm's total cost of materials.

It is important to understand that oligopoly is not characteristic of industry as a whole. Most firms and industries operate somewhere in the area of imperfect competition. Millions of people, working in thousands of factories, produce hundreds of thousands of products substantially without governmental or any other outside direction. The firms that make up this market exercise almost complete control over their prices, and price conspiracies in this market are extremely rare. In fact, aside from utilities, transportation, and some manufacturing industries, the concentration of oligopolistic power is rare.

Most prices are subject to adjustment. It is because most firms are free within broad limits to adjust their prices at will that competent buyers can obtain better prices in direct proportion to their ability to analyze costs, markets, and pricing methodologies. Prices can be negotiated very little with firms in the markets of pure competition or monopoly. They can be negotiated a great deal with firms operating in the markets of imperfect competition. The question then is: What proportion of the nation's total market falls within the area of imperfect competition? What percentage of a buyer's total purchases is subject to price flexibility?

Studies made at the Graduate School of Business at Stanford University show the nation's economy to be approximately 70% free and existing in the area of imperfect competition. The results of similar studies by other authorities support this conclusion. Buyers in most purchasing situations, therefore, have considerable latitude for negotiating both price and service with their suppliers. At the same time, however, every buyer should guard against the inducement of illegal price concessions. Buying personnel must understand the operation of federal and state restraint of trade laws.

Variable-Margin Pricing

Most industrial firms sell a line of products rather than just a single product. Very few firms attempt to earn the same profit margin on each product in the line. Most firms price their products to generate a satisfactory return on their whole line, not on each product in the line. Such a variable-margin pricing policy permits maximum competition on individual products. The profits from the most efficiently produced and "successfully priced" items are often used to offset the losses or the lower profit margins of the inefficiently produced items.

Recently, during a research project at a printed circuit manufacturer, one of the authors traced overhead for 188,000 boards consisting of approximately 4000 designs. The overhead generated by production was applied by the company to the boards on an equal basis; however, approximately 20% of the boards drove 80% of the overhead. As expected, the 20% were the low-volume boards produced in small lots. Since the company based its prices on the cost estimates, the high-volume boards were significantly overpriced.

An understanding of the theory of variable-margin pricing is essential if buyers are to obtain the right price. Whenever possible, sellers use average profit margins for pricing orders because it is usually advantageous to them. In some cases, this practice results in prices that sophisticated buyers realize are too high—particularly when low-cost, efficiently produced items are being purchased. Invariably when average margins are used, prices considerably above fair prices result for large, long-term purchases. When dealing with large, multiproduct firms that utilize this pricing approach, a buyer must also know which of the items purchased is a high-margin and which is a low-margin item. Such facts are learned by noting the differences in volumes, manufacturing skills, and costs of the various producers.

The following case illustrates the practical concepts of the preceding discussion. A large, high-technology research firm successfully negotiated a $2.8 million annual contract for medical and scientific supplies. At the outset, the seller proposed that the contract be priced at cost plus the firm's annual gross profit margin of 19%. After several hours of negotiation, the contract was priced at cost plus a 6% profit margin. Had the supply manager not understood the concept of variable-margin pricing, and not known which items the seller produced efficiently, this contract would have cost her company an additional $320,000.

In their search for optimum prices, competent buyers are aided by analyzing the pricing methods of both full-line and specialty suppliers. Regrettably, only buyers from a few progressive firms actually make such in-depth analyses of the entire product line of the industries in which they do business. Rather, most buyers focus their analyses on just one product at a time (the product presently being purchased). Buyers are rewarded by directing their efforts toward the development of savings produced by recurring long-term cost reductions, rather than focusing on savings from short-term cost reductions. In short, optimal pricing comes to buyers who understand the pricing processes for complete product lines, in all firms, in all industries from which they buy.

In the long run, a firm must recover all of its costs or go out of business. In the long run, for any given item, the price is roughly equal to the cost of the least efficient producer who is able to remain in business. In the short run, however, prices in the free, competitive segment of the economy (roughly 70% of the whole) are determined primarily by competition, that is, by supply and demand, and not by costs.

Product Differentiation

Many basic differences exist between the kinds of products marketed in the various segments of the economy. Some products in the competitive segment are *undifferentiated* (not distinguished by specific differences), while others are *differentiated*. In some cases, the products are intrinsically different (differentiated); in others, manufacturers are successful at making their similar products appear different from those of their competitors. Even in those cases where a product cannot be made different in substance, producers can still get premium prices if they can persuade customers *to believe* that their products are superior. Some producers spend huge sums of money on sales personnel and advertising to accomplish such a purpose. In the jargon of the economists, "They attempt to make the demand curve for the products of their firm somewhat inelastic." If their efforts are successful, they can charge higher prices for their products. On the other hand, if their efforts are defeated by counter efforts of competitors, as is frequently the case, price competition comparable to that in pure competition can result. Grocers, for example, are well acquainted with this economic fact.

For both differentiated and undifferentiated products, producers compete on quality and service as well as price. The consumer market is more susceptible to producers' advertising claims than are buyers in the industrial market; therefore, the major portion of advertising effort is directed toward the consumer market. Nonetheless, industrial supply management professionals must be aware of advertising and sales tactics and be very careful that they determine quality from an analysis of facts, not from unsupported claims.

Six Categories of Cost

The supply professional knows that price = cost + profit. He or she also must understand variable, fixed, semi-variable, total, direct, and indirect costs, and how these costs influence prices.

Variable Manufacturing Costs

These are items of cost that *vary directly and proportionally with the production quantity* of a particular product. Variable manufacturing costs include direct labor (unless fixed by contract), direct materials (includes raw materials, sub-assemblies, etc.), and variable manufacturing overhead (e.g., plant utilities if they vary with machine use/output). For example, if a specific cutting tool costs $10 and lasts for 100 cuttings, each cut represents a variable cost of 10 cents. If three cuts were required in machining a specific item, then the variable cost for cutting would be 30 cents. Variable costs may decrease due to economies of scale (e.g., purchase discounts on direct materials), and increase due to diseconomies of scale (e.g., too many workers in a confined workspace). Variable costs not only exist in the manufacturing environment, but also in the selling, general, and administrative areas (non-manufacturing costs). In summary, variable costs are fixed per unit, but vary in total as the activity level changes.

Fixed Manufacturing Costs

Fixed costs *do not vary with volume*, but change over time. Fixed costs are costs sellers must pay simply because they are in business. They are a function of time and are not influenced by the volume of production.[2] Fixed costs generally represent either money the seller has already spent for buildings and equipment (e.g., depreciation) or money the seller will have to spend for unavoidable expenses (e.g., rent and insurance) regardless of the plant's volume of production. For example, if the lathe that held the cutting tool in the preceding example depreciates at a rate of $250 a month, this is a fixed cost. The seller has this $250 expense every month, whether any turnings are made during that period. Fixed selling and general and administrative costs may include advertising and research and development—these are classified as non-manufacturing costs. Fixed costs may be increased or decreased from one time period to another, regardless of production volume. Fixed costs are fixed in total, but vary per unit as the activity level changes (fixed costs per unit decrease as more units are produced).

Semi-Variable or Mixed Manufacturing Costs

Generally, it is not possible to classify all production costs as being either completely fixed or completely variable. Many others, termed semi-variable or mixed costs, fall somewhere between these extremes. Costs such as maintenance, utilities, and postage are partly variable and partly fixed. Each is like a fixed cost because its total cannot be tied directly to a particular unit of production. Yet, it is possible to sort out specific elements in each of these costs that are fixed as soon as the plant begins to operate. When the fixed portion is removed, the remaining elements frequently vary closely in proportion to the production volume. For example, if a plant is producing an average of 5000 items a month, it might have an average light bill of $700 a month. Should the number of units produced be increased to 8000, the light bill might increase by $100 to $800. The $100 increase is not proportional to the production increase because a certain segment of the light bill is fixed whether any production occurs. Above this fixed segment, however, light costs may vary in a fairly consistent relationship with production volume. Mixed selling and general and administrative costs may include selling salary (fixed) and commission (variable), telephone service (fixed) and metered local and long distance calls (variable), etc.

Total Production Costs

The sum of the variable, fixed, and semi-variable costs comprise the total costs. As the volume of production increases, total costs increase. However, the cost to produce *each unit* of product decreases. This is because the fixed costs do not increase; rather, they are simply spread over a larger number of units of product. Suppose, for example, that a single-product firm has the following cost structure:

Variable manufacturing costs, per unit	$ 2.25
Fixed manufacturing costs, per month	$1,200.00
Mixed manufacturing costs:	
Variable portion, per unit	$ 0.30
Fixed portion, per month	$ 450.00
Variable selling, general and administrative costs, per unit	$ 0.35
Fixed selling, general and administrative costs, per month	$ 700.00

Under these circumstances, Table 16.1 shows how unit costs change as volume changes; this example assumes a $7 per unit selling price. A contribution format income statement used by managerial and cost accountants may help to illustrate these costs. Notice how manufacturing cost per unit and total cost per unit decrease as production increases to plant capacity. Also, notice how the income generated per unit increases as production increases. To understand the intricacies of the cost-volume-profit relationship fully it is essential to understand variable, fixed, and semi-variable costs.

Because it is difficult to allocate costs specifically as fixed, variable, and semi-variable, accountants generally classify costs in two categories—direct costs and indirect costs. These are discussed briefly next.

Direct Costs

These costs are specifically traceable to or caused by a specific project or production operation. Two major direct costs are direct labor and direct materials. Although most direct costs are variable, conceptually, direct costs should not be confused with variable costs; the two terms are rooted in different concepts. Direct costs relate to *traceability* of costs to specific operations, while variable costs relate to the *behavior* of costs as volume fluctuates. The salary of a production supervisor, for example, can be directly traceable to a product even though he or she is paid a fixed salary regardless of the volume produced. Returning to the illustration of the cutting tool, if a firm pays a worker 15 cents for making the three cuts required for each item, direct labor costs are 15 cents. If the value of the piece of metal being cut is 85 cents, direct costs for the item are $1.00.

Table 16.1 Cost, Volume, Profit Relationships

Number of Units Produced	Per Unit	500	1,000	1,500	2,000	2,500
Revenue	7.00	$3,500.00	$7,000.00	$10,500.00	$14,000.00	$17,500.00
Variable Manufacturing Costs	2.25	1,125.00	2,250.00	3,375.00	4,500.00	5,625.00
Variable Portion of Mixed Manufacturing Costs	0.30	150.00	300.00	450.00	600.00	750.00
Variable Selling, General, and Administrative Costs	0.35	175.00	350.00	525.00	700.00	875.00
Total Variable Cost	2.90	1,450.00	2,900.00	4,350.00	5,800.00	7,250.00
Contribution Margin (the remainder, after deducting variable costs from revenue, to cover fixed costs and return a profit or minimize a loss)	4.10	2,050.00	4,100.00	6,150.00	8,200.00	10,250.00
Fixed Manufacturing Costs		1,200.00	1,200.00	1,200.00	1,200.00	1,200.00
Fixed Portion of Mixed Manufacturing Costs		450.00	450.00	450.00	450.00	450.00
Fixed Selling, General, and Administrative Costs		700.00	700.00	700.00	700.00	700.00
Total Fixed Cost		2,350.00	2,350.00	2,350.00	2,350.00	2,350.00
Operating Income		−$300.00	$1,750.00	$3,800.00	$5,850.00	$7,900.00
Total Cost Per Unit		7.60	5.25	4.47	4.08	3.84
Manufacturing Cost Per Unit		5.85	4.20	3.65	3.38	3.21
Income Generated Per Unit		−0.60	1.75	2.53	2.93	3.16

Assumptions:
- Selling price remains constant
- All Manufacturing and Nonmanufacturing Costs are Included
- Variable Costs do not change with volume
- Fixed Costs do not change within the relevant range of production, 0–2,500 units
- All units produced are sold
- Taxes are not considered

Indirect Costs (Overhead)

Indirect costs are associated with or caused by two or more operating activities "jointly," but are not traced to each of them individually. The nature of an indirect cost is such that it is either not possible or practical to measure directly how much of the cost is attributable to a single operating activity.[3] Indirect costs can be fixed or variable, depending on their behavior (property taxes are fixed, but the portion of energy consumption that varies with the level of production is variable). Therefore, it is important that the reader not confuse indirect costs with fixed costs.

Regulation by Competition

From a buyer's point of view, competition is the mainspring of good pricing. As previously discussed, most producers do not have the same real costs of production. Even when their costs are the same, their competitive positions can be quite different. Hence, their prices can also be quite different. Consider the following example. Assume that a supply manager is ready to purchase 10,000 specially designed cutting tools. He sends the specifications to five companies for quotations. All five respond. For the sake of simplicity, assume that direct costs in these five companies are identical. Further, assume that each company uses the same price-estimating formula; overhead is figured as 150% of direct labor, and profit is calculated as 12% of total cost. Each company could then lay out its figures as follows:

Cost of materials	$12,000	
Cost of direct labor	3,000	
Cost of overhead	4,500	(150% of direct labor)
Total cost	$19,500	
Profit	2,340	(12% of total cost)
Price	$21,840	

To simplify the example, assume all overhead is classified as fixed; that is, it remains constant over a given range of production.

Even with all the controlling figures fixed, the companies more than likely would not quote the same price because *the cost of production and profit are only two of the factors a seller considers in determining an asking price.* In the final analysis, it is the factors stemming from competition, that determine the exact price each firm will quote. That is, when faced with the realities of competition, the price any specific firm will quote will be governed largely by *its need for business* and by *what it thinks its competitors will quote,* not by costs or profits.

Who is responsible for final determination of the price to be quoted? Generally, it is a marketing executive; in some cases, it is the president of the company. Pricing is one of the most important management decisions a firm must make. As an objective, a firm tends to seek the highest price that is compatible with its long-range goals. What is the possible price range for the order in the preceding example? The out-of-pocket (variable) costs for this order are $12,000 for materials and $3,000 for direct labor, a total of $15,000. This is the lowest price any company should accept under any circumstances. The highest price is $21,840, based on the assumption that a profit in excess of 12% is not in the long-range interest of the firm. (Such a profit may attract additional competition to enter this market, which in turn would erode the profitability of the market.)

What could cause one of the firms to consider a price of $17,000? Keen competition among suppliers could. On the other hand, keen competition among buyers could drive the price higher. This is why competition, as a leveler, is such a dominant factor in pricing. If the firm had been unable to obtain a satisfactory volume of other business, it would gladly take this order for a price of $17,000. As a result of the order, the $15,000 out-of-pocket costs would be covered, the experienced work force could be kept working, and a $2,000 contribution could be made to overhead. Remember that the fixed overhead would continue whether or not the firm received this order.

In the long run, a firm must recover all costs or go out of business, for in the long run, plant and machinery must be maintained, modernized, and replaced. *In the short run*, however, it is generally better for a firm to recover variable costs and some portion of overhead rather than undergo a significant decline in business. This would not be true if such additional business would affect the pricing of other orders the firm has already filled or is going to fill.

Business in good times is not ordinarily done at out-of-pocket (or variable cost) prices. A more common situation would be for each of the five firms to bid prices above the total cost figure of $19,500. How much above this figure each would bid would depend on the specific economic circumstances and expectations applicable to each firm. Those firms hungry for business would bid just slightly above the total cost figure of $19,500. Those with large backlogs and growing lists of steady customers (and therefore not in need of new business in the short run) would bid a larger profit margin (perhaps 14%). Sellers can be expected to evaluate competitive situations differently, depending on how much they want or need the business. Therefore, even with the simplifying assumption of identical costs, it is reasonable to expect bids in this situation to range from approximately $19,700 (1% profit) to $21,840 (12% profit). Prices close to out-of-pocket costs could be offered if the seller were attempting to obtain a desirable, prestigious account, if the supplier desired to gain experience in a situation wherein additional large orders are expected to follow, or if the supplier desired to keep its workforce employed.

Varying Profit Margins

A seller must recover *all* costs from his or her total sales to make a profit. However, *each* product in the line does not have to make a profit, and not all accounts have to yield the same profit margin. Bearing these thoughts in mind, the principal cost/competition implications of pricing can be summarized as follows: *Sound pricing policy dictates that sellers, in accordance with their interpretation of the prevailing competitive forces, quote prices that are high enough to include all variable costs and make the maximum possible contribution toward fixed costs and profit.*

Similarly, sound pricing policy dictates that, for any given purchase, supply professionals should use their knowledge of products, markets, costs, and competitive conditions to estimate the price range at which sellers can reasonably be expected to do business. Finally, with this information, a knowledge of the value of the buyer's ongoing business to a seller, and an appreciation of the value of this specific order, the supply professional applies all relevant purchasing principles and techniques to purchase at prices as close as possible to the bottom of the estimated price range.

Price Analysis

Some form of price analysis is required for every purchase. The method and scope of analysis required are dictated by the dollar amount and circumstances attending each specific purchase. Price analysis is defined as the examination of a seller's price proposal (bid) in comparison with reasonable price benchmarks, without examination and evaluation of the separate elements of the cost and profit making up the price.

A supply professional has six tools that can be used to conduct a price analysis: (1) analysis of competitive price proposals; (2) comparison with regulated, catalog, or market prices; (3) use of Web-based e-procurement; (4) comparison with historical prices; (5) supply and demand factors, and (6) use of independent cost estimates.

Competitive Price Proposals

Chapter 13 describes the conditions that should be satisfied before using competitive bidding as a means of selecting the source of supply. When this approach is employed, and the following additional conditions are satisfied, the resulting low bid normally provides a fair and reasonable price:

▲ At least two qualified sources have responded to the solicitation.
▲ The proposals are responsive to the buying firm's requirements.
▲ The supplier competed independently for the award.
▲ The supplier submitting the lowest offer does not have an unfair advantage over its competitors.
▲ The lowest evaluated price is reasonable.

The buyer cannot apply this approach to pricing in a mechanical manner. He or she clearly must use common sense and ensure that the price is reasonable when compared with past prices, with independent estimates, or with realistic rules of thumb.

Regulated, Catalog, and Market Prices

Prices Set by Law or Regulation

When the price is set by law or regulation, the supplier must identify the regulating authority and specify the regulated prices. With regulated prices, some governmental body (federal, state, or local) has determined that prices of certain goods and services should be controlled directly. Normally, approval of price changes requires formal review, hearings, and an affirmative vote of the regulatory authority. No supplier may charge more or less than the approved price.

Catalog Price

An established catalog price is a price that is included in a catalog, a price list, or some other form that is regularly maintained by the supplier. The price sources must be dated and readily available for inspection by potential customers. The buyer should request a recent sales summary demonstrating that significant quantities are sold to a significant number of customers at the indicated price before accepting a catalog price.

Market Price

A market price results from the interaction of many buyers and sellers who are willing to trade at a given (market) price. The forces of supply and demand establish the price. A market price is generally for an item or a service that is generic in nature and not particularly unique to the seller. Eggs and lumber, for example, are priced based on the market. Normally, the daily market price is published in local newspapers or trade publications that are independent of the supplier.

Internet/e-Commerce II

Advanced communications using the Internet allow supply management personnel to view up-to-date pricing, as well as catalogs, specification sheets, video presentations, and other information the seller has on a material, product, or service. Since the Internet does not have geographical constraints, the information is available worldwide. Of particular interest here are buying exchanges, reverse auctions, and the search capability of the Internet.

Buying exchanges (often referred to as B2B e-commerce) offer purchasing firms a list of pre-approved sellers offering identical or similar products or services, usually within a specific category, from which to choose. Normally, prices or discounts from list prices are provided.

Reverse auctions identify materials, equipment, or services required, and request carefully prequalified suppliers to submit bids. Potential suppliers are able to see prices submitted by their competitors and revise their bids until the preestablished closing time for the auction. Caution must be used when employing reverse auctions because they ignore the relationship dimension of the transaction.

Tailored global searches allow expanded Internet search capabilities. Supply management professionals can investigate products or services by simultaneously scanning all *relevant* public and private Web sites worldwide. Obtaining the right price, quality, and delivery is becoming easier and faster as the Internet expands and more procurement-specific portals are developed.

Historical Prices

Price analysis can be performed by comparing a proposed price with historical quotes or prices for the same or similar item. It is essential to determine that the base price is fair and reasonable (as determined through price analysis) and is still a valid standard against which to measure the offered price. The fact that a historical price exists does not automatically make it a valid basis for comparison. Several issues must be considered:

- ▲ How have conditions changed?[4]
- ▲ Were there one-time engineering, setup, or tooling charges in the original price?
- ▲ What should be the effect of inflation or deflation on the price?
- ▲ Will the new procurement create a situation in which the supplier should enjoy the benefits of learning? (The concept of learning curve analysis is discussed at the end of this chapter.)

Supply and Demand Factors

The Institute for Supply Management (ISM) Report on Business is a monthly report on new orders, production, employment, supplier deliveries, and inventory conditions for commodities and non-manufacturing services. The ISM purchasing manufacturing index (PMI) is highly regarded as is their non-manufacturing (service) index (NMI). See the current monthly issue of *Inside Supply Management* magazine published by ISM (www.ism.ws).

Independent Cost Estimates

When other techniques of price analysis cannot be utilized, the supply manager may use an independent cost estimate as the basis for comparison. He or she must determine that the estimate is fair and reasonable.[5] If price analysis is impractical or if it does not allow the buyer to reach a conclusion that the price is fair and reasonable, then cost analysis should be employed.

Cost Analysis

Cost analysis should be employed when price analysis is impractical or does not allow the supply management professional to reach the conclusion that a price is fair and reasonable. Cost analysis is generally most useful when purchasing nonstandard items and services. This chapter now focuses on the application of cost analysis to the acquisition cost of materials, products, and services.

Cost Analysis Defined

We have seen that price analysis is a process of comparisons. *Cost analysis* is a review and an evaluation of actual or anticipated costs. This analysis involves the application of experience, knowledge, and judgment to data in an attempt to project reasonable estimated contract costs. Estimated costs serve as the basis for buyer-seller negotiations to arrive at mutually agreeable contract prices.

The purpose of cost analysis is to arrive at a price that is fair and reasonable to both the buying and selling firms. Estimates can be made with the help of one's engineering department or by analyzing the estimates submitted by the seller. To analyze a supplier's costs, a supply manager must understand the nature of each of the various costs a supplier incurs. The supply manager must compare the labor hours, material costs, and overhead costs of all competing suppliers as listed on their cost-breakdown sheets. Most important, he or she must determine the reasons for any differences, focusing on three principle elements of cost: direct, indirect (overhead), and profit.

A supply manager should always be conscious of the fact that costs vary widely among manufacturing firms. Some firms are high-cost producers; others are low-cost producers. Many factors affect the costs of specific firms, as well as the cost of individual products within any given firm. Some of the most important elements affecting costs are:

▲ Capabilities of management
▲ Efficiency of labor
▲ Amount and quality of subcontracting
▲ Plant capacity and the continuity of output

Each of these factors can change with respect to either product or time. For this reason, a specific firm can be a high-cost producer for one item and a low-cost producer for another. Similarly, the firm can be a low-cost producer one year and a high-cost producer another year. These circumstances make it extremely important for a supply manager to obtain competition among potential suppliers, when appropriate. Competition can be a supply manager's key to locating the desired low-cost producer.

Capabilities of Management

The skill with which management plans, organizes, staffs, coordinates, and controls all the personnel, capital, and equipment at its disposal determines the efficiency of the firm. Managements utilize the resources available to them with substantially different degrees of efficiency. This is one basic reason why finding the correct supplier (and price) is so rewarding for astute supply management professionals.

Efficiency of Labor

Anyone who has visited a number of different suppliers surely has noticed the differences in attitudes and skills that exist between various labor forces. Some are cooperative, take great pride in their work, have high morale, and produce efficiently, while others do not. The skill with which management exercises its responsibilities contributes greatly to these differences between efficient and inefficient labor forces. Supply managers are well rewarded for pinpointing suppliers with efficient labor forces.

Amount and Quality of Subcontracting

When a contract has been awarded to a supplier (the *prime contractor*), the supplier frequently subcontracts some of the work required to complete the job. The supplier's subcontracting decisions are important to the buying firm because they may involve a large percentage of prime contract money. The first decision a prime contractor must make regarding subcontracts is which items should be made and which specific items should be bought. Should the prime contractor decide to buy some of those items that could be made more efficiently, and vice versa, the buyer suffers financially. Even if the prime contractor makes the correct "make" decision, it still is responsible for selecting those subcontractors that are needed for the "buy" items. Subcontractor prices and performance directly influence the prices the buying firm pays the prime contractor. Hence, the prime contractor's skills in both making and administering its subcontracts are of great importance to the supply manager. For this reason, supply managers must periodically review their major suppliers to ensure that they have effective supply management and subcontracting capabilities of their own.

Plant Capacity

A plant's overhead costs are directly influenced by its size. A plant can get too large for efficient production and, as a result, lose its competitive ability. On the other hand, plants with large capital investments, or those manufacturing products on a mass production

Table 16.2 How Production Volume Affects Fixed Costs, Variable Costs, and Profit

Production Quantity	Selling Price	Sales Revenue	Fixed Costs	Variable Costs	Total Cost	Total Profit	Profit per Unit of Added Production
0	$20	$0	$4,000	$0	$4,000	–$4,000	$15
100	20	2,000	4,000	500	4,500	–2,500	15
200	20	4,000	4,000	1,000	5,000	–1,000	15
300	20	6,000	4,000	1,500	5,500	+ 500	15
400	20	8,000	4,000	2,000	6,000	+ 2,000	15
500	20	10,000	4,000	2,500	6,500	+ 3,500	15
600	20	12,000	4,000	3,000	7,000	+ 5,000	14
700*	20	14,000	4,000	3,600	7,600	+ 6,400	12
800*	20	16,000	4,000	4,400	8,400	+ 7,600	10
900*	20	18,000	4,000	5,400	9,400	+ 8,600	

*Plant begins to strain capacity and administrative capabilities. The result is less efficient operation; that is, overtime is required, less experienced workers are utilized, scheduling and handling of materials become less efficient, and variable costs per unit rise. Consequently, although profit continues to increase beyond a production quantity of 600, it increases at a decreasing rate.

basis, can be too small to attain the most efficient production levels. The supply professional must be alert to detect firms whose operations are adversely affected by size.

Plant output is clearly one of the controlling elements in the cost/profit picture. Table 16.2 illustrates this concept numerically. Note how volume affects profit when variable costs change because of inefficient use of facilities beyond optimum plant capacity. Note also that while total profit continues to increase as production output increases, beyond a certain output profit increases at a decreasing rate. This relationship is an important one for supply professionals to keep in mind.

Sources of Cost Data

There are three primary sources of cost data: (1) from potential suppliers as a precondition of submitting proposals and bids, (2) from suppliers with whom the firm has developed preferred or strategic supplier relationships/alliances, and (3) cost models.

Potential Suppliers

When a supply manager anticipates that cost analysis will be required, he or she should include a request for a cost breakdown with each request for quotation. *This is the proper time to make such a request, not after negotiations have started.* This is critical! Suppliers cannot complain that making this breakdown is an extra burden at this time because they must perform such an analysis to prepare their bids. A simple procedure used by a number of progressive firms for obtaining cost breakdowns is to include the following statement with their request for quotations: "We will not consider any quotation not accompanied by a cost breakdown." Not all suppliers readily provide cost-of-production information; however, the number refusing to do so for nonstandard items is declining. An example of a typical cost breakdown request form is shown in Figure 16.2.

Supply Partners

As firms develop open relationships built on trust and collaboration, the purchasing firm shares information on forecasts, schedules, the way purchased items integrate into its product or process, etc. The supply partner shares information on its design, production, and quality processes and on its *design and production costs.*

Cost Models

On some occasions, it may not be possible to obtain cost data from the supplier. In other cases, the cost data obtained may appear unrealistic or support prices that are unacceptable. Under these conditions, it may be necessary for the purchasing firm to develop its own cost models to estimate what the supplier's costs *should be.* (The U.S. Navy and Air Force call such models "Should-Cost Models." Both services have found that this approach to pricing is extraordinarily powerful.) The development of such models requires the application of both accounting and industrial engineering skills, and is beyond the scope of the presentation in this chapter. It is sufficient at this point simply to say that this approach, although not extensively used by small- and medium-sized firms, is commonly used by leading-edge firms that have the technical resources available.[6]

Direct Costs

Except in industries with heavy fixed capital investments, direct costs are normally the major portion of product or service costs and are the most easily traceable. As such, they generally serve as the basis on which sellers allocate their overhead costs. The astute supply manager, therefore, must carefully investigate a seller's direct costs. *A tiny reduction here (because they are relatively large) is worth more (price-wise) to the buying firm than a major reduction in the percentage of profit* (which is relatively small).

COST ANALYSIS	CHECK APPROPRIATE BOX ESTIMATED COST ☐ HISTORICAL COST ☐ PERIOD COVERED:		
NAME OF SUPPLIER	INQUIRY OR PURCHASE REQUISITION NO.		
ADDRESS (Street, City, State)	QUANTITY	AT $ EACH	AMOUNT $
ARTICLE			
TERMS AND DISCOUNT	NET TOTAL OF QUOTATION $		
ANALYSIS OF COST AS OF _____ , 20 ___ INDICATE WHETHER: COST PER ITEM ☐ OR TOTAL COST ☐			

ITEM	AMOUNT	PERCENT OF COST
1. DIRECT MATERIAL		
2. LESS SCRAP OR SALVAGE		
3. NET DIRECT MATERIAL		
4. PURCHASED PARTS - FROM SUBCONTRACTORS		
5. DIRECT PRODUCTIVE LABOR HOURS AT $		
6. DIRECT FACTORY CHARGES:		
(A) TOOLS AND DIES		
1. DIRECT WAGES HOURS AT $		
2. TOOLING BURDEN		
3. MATERIALS		
(B) SPECIAL MACHINERY		
(C) MISCELLANEOUS		
7. INDIRECT FACTORY EXPENSES (Burden), ON BASIS OF See Note[a]		
8. ENGINEERING AND DEVELOPMENT EXPENSES • DIRECT:		
(a) SALARIES AND WAGES HOURS AT $		
(b) BURDEN		
(c) OTHER		
TOTAL MANUFACTURING COST		
9. GENERAL AND ADMINISTRATIVE EXPENSE: PERCENT OF See Note[b]		
10. SELLING EXPENSE See Note[c]		
11. CONTINGENCIES See Note[d]		
12. OTHER EXPENSES See Note[e]		
13.		
14.		
15.		
16.		
17. TOTAL COST		
18. SELLING PRICE		

19. (a) Are the wage rates used in estimating the direct labor of the unit cost break-down
    the same as those now prevailing?
   (b) If "No", explain difference and indicate approximate amount thereof.

20. (a) What operating rate has been used in calculating the above estimate?
    Hours of operation per week?
   (b) At what rate is your plant now operating?
    Hours of operation per week?

_____   _____
      (Supplier)   (Signature and title)

_____
      (Date)

[a] State basis of allocation.
[b] State nature of expenses included and basis of allocation.
[c] State nature of expenses included and amount of advertising, if any, separately, and basis of allocation.
[d] Explain in detail.
[e] State nature of expenses, basis of allocation, and why related to the cost of this item.

Figure 16.2 A typical request for a cost breakdown.

Table 16.3 Direct Costs and Prices

Cost Elements	Situation 1	Situation 2
Material	$8.00	$8.00
Direct labor	8.00	6.00
Fixed overhead at 150 percent of direct labor	12.00	9.00
Manufacturing cost	$28.00	$23.00
General and administrative overhead at 10 percent of manufacturing cost	2.80	2.30
Total cost	30.80	25.30
Profit at 10 percent of total cost	3.08	2.53
Price	$33.88	$27.83

Referring to Table 16.3, a 25% reduction in the $8 direct labor cost of situation 1 to the $6 direct labor cost of situation 2 results in a $6.05 ($33.88 − $27.83) reduction in price. A 25% reduction in profit would result in only a 77-cent reduction in price (0.25 × $3.08 = $0.77).

Direct Labor

During the development and production phase of a new item, a supplier typically experiences a heavy design and production engineering effort. These efforts will peak and then decrease. As they do, tooling and setup efforts increase, peak, and decline. Machining, assembly, and test effort then become the predominant users of labor. The supply professional should be cognizant of these factors and should analyze a supplier's estimate to ensure that it is based on proper planning, applying reasonable expectations of efficiency in this regard. When analyzing direct manufacturing labor estimates, the supply manager should pay particular attention to the following:

▲ Allowances for rework
▲ Geographic variations
▲ Variations in skills

Allowances for Rework

The supply manager should carefully review a bidder's estimate of rework costs. Modern production techniques now make it possible to drastically reduce scrap rates. Effective purchasing by a supplier's organization can reduce the defect rates on incoming materials as much as 95%. The combined effect of reduced incoming quality problems and improved production and quality systems should reduce a supplier's requirement for rework markedly.

Geographic Variations

Wage rates vary significantly from one country to another, as well as within a country's borders. A buyer must ensure that the wage rates proposed are, in fact, the wage rates applicable in the areas where the work is to be performed. The Bureau of Labor Statistics provides current wage rates in the United States for a variety of trades in different locations.

Variations in Skills

The supply manager, with assistance from the firm's industrial engineering or production departments, should review the types of labor skills proposed, to ensure that they are relevant and necessary for accomplishment of the required tasks.

Direct Materials

Direct materials are consumed or converted during the production process. Sheet metal, fasteners, electrical relays, and radios for automobiles are all examples of direct materials. In most cases, such materials are normally purchased from a wide variety of suppliers. In some cases, the materials may have been produced or partially processed in other plants or divisions of the supplier's operation. The resulting costs should be scrutinized carefully for internal transfer charges and markups.

Further analysis of proposed materials costs frequently reveals a difference between the buyer's cost estimate for a given bill of materials and the supplier's estimate for the same bill. In such cases, the supply manager should request supporting data from the supplier. In some cases, the labor component of the proposed materials costs should reflect a learning effect as more units are produced (this topic is discussed next). In any case, careful analysis and discussion should help identify the source of the variance.

Tooling Costs

Most procurement authorities advocate that the buying firm pay for and take title to special tooling. Such an approach allows the buying firm maximum control. Analysis of production costs is easier, and the tooling can be moved if circumstances dictate.

There should be an inverse relationship between the investment in tooling and the number of hours required to produce a unit of output. The supply manager should ensure that the supplier plans to use sufficient tooling to minimize labor hours, but at the same time avoids investments that are not recovered through labor savings.

Learning Curves

In many circumstances, labor and supervision become more efficient as more units are produced. The *learning curve* (sometimes called the *improvement curve)* is defined as an empirical relationship between the number of units produced and the number of labor hours required to produce them. Production managers can use this relationship in scheduling production and in determining labor requirements for a particular product over a given period of time. Supply managers can use the relationship to analyze the effects of production and management "learning" on a supplier's unit cost of production.

Traditionally, the learning curve has been used primarily for purchases of complex equipment in the aircraft, electronics, and other highly technical industries. Recently, its use has spread to other industries. The learning curve is useful in both price and cost analysis. It is probably most useful in negotiations, as a starting point for pricing a new item. In addition to providing "buyer's insurance" against overcharging, the learning curve is also used effectively by government and commercial supply professionals in developing (1) target costs for new products, (2) make-or-buy information, (3) delivery schedules, and (4) progress payment schedules for suppliers.

Cumulative Curve and the Unit Curve

In practice, two basic forms of the learning curve exist. The first curve, "the cumulative average cost curve," is commonly used in price and cost analysis. This curve plots cumulative units produced against the average direct labor cost or *average labor hours required per unit for all units produced.* The second, "the unit or marginal cost curve," is also used in labor and cost-estimating work. The unit curve plots cumulative units produced against the *actual labor hours required to produce each unit.* Figure 16.3 illustrates and compares the two types of curves.

Selection of the learning curve technique to use tends to be based on an organization's past experience. Ideally, whether one should use a cumulative or unit curve is a function of the production process itself. Some operations conform to a cumulative curve; others conform to a unit curve. The only way to know which to use is to record the actual production data and then determine which type of curve fits the data best. The relationship is strictly an *empirical* one.

Target Cost Estimation

If a new product is custom-made to unique specifications, what should be paid for the 50th item? The 500th item? Obviously, costs should decline—but by how much? Analysis of the learning curve provides an answer. Cost reductions and estimated prices can be obtained merely by reading figures from a graph.

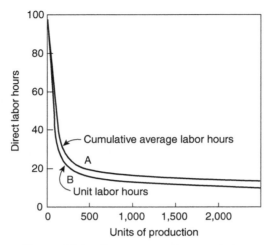

Figure 16.3 Comparison of a cumulative average learning curve (A) and a unit learning curve (B), plotted on an arithmetic grid.

The learning curve is a quantitative model of the commonsense observation that the unit cost of a new product decreases as more units of the product are made because of the learning process. The manufacturer, through the repetitive production process, learns how to make the product at a lower cost. For example, the more times an individual repeats a complicated operation, the more efficient he or she becomes, in both speed and skill. This, in turn, means progressively lower unit labor costs. Familiarity with an operation also results in fewer rejects and reworks, better scheduling, possible improvements in tooling, fewer engineering changes, and more efficient management systems.

Suppose a supply manager knows that it took a supplier 100 hours of labor to turn out the first unit of a new product, as indicated in Figures 16.4 and 16-5. The supplier reports that the second unit took 80 hours to make, so the average labor requirement for the two items is 180 ÷ 2 = 90 hours per unit. The production report for the first four units is summarized in Table 16.4.

Observe that the labor requirement dropped to 74 hours for the third unit and to 70 hours for the fourth unit. Column 4 shows that the average number of labor hours required for the first four units was 81 hours per unit. Investigation of the learning rate shows the following relationships:

▲ As production doubled from one to two units, *the average labor hours required per unit* dropped from 100 to 90, a reduction of 10%.
▲ As production doubled from two to four units, *the average labor hours required per unit* dropped from 90 to 81, a reduction of 10%.

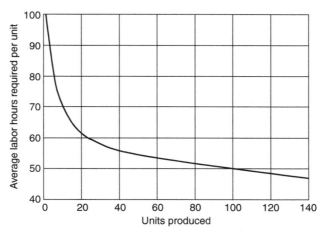

Figure 16.4 A 90 percent cumulative average learning curve, plotted on an arithmetic grid.

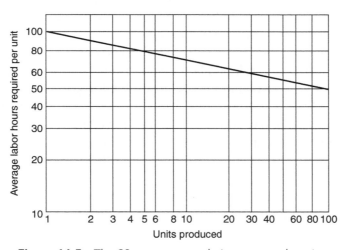

Figure 16.5 The 90 percent cumulative average learning curve of Figure 16.4, plotted on a log-log grid.

Table 16.4 Ninety Percent Cumulative Learning Curve Data

Unit Produced	Labor Hours Required	Cumulative Labor Hours Required	Average Labor Hours Required per Unit
1st	100	100	100.0
2nd	80	180	90.0
3rd	74	254	84.7
4th	70	324	81.0

Figure 16.4 indicates that the same learning rate continues as production of the new item increases. Each time production doubles, the average labor requirement for all units declines by 10%. Thus, the product is said to have a 90% learning rate, or a 90% learning curve. Note that this is based on the *cumulative average* learning curve phenomenon. The basic point revealed by the learning curve is that a *specific and constant percentage reduction in the average direct labor hours required per unit results each time the number of units produced is doubled.* It is an established fact that specific learning rates occur with reasonable regularity for similar groups of products in many different industries.

Studies made in the aircraft, electronics, and small electromechanical subassembly fields indicate that learning rates of 75 to 95% are typical. However, learning curves can vary anywhere within the practical limits of 50 to 100%. As more units are produced, the effect of a constant learning rate on unit costs gradually diminishes. After several thousand units, the absolute reduction in cost from learning becomes negligible. Note in Figure 16.4 how the curve flattens out as the number of units produced increases. This is why learning curve analysis is of greatest value for new products.[7]

Most analysts prefer to plot the data for learning curves on log-log graph paper, as in Figure 16-5. The logarithmic scales on both the horizontal and vertical axes convert the curve of Figure 16.4 into a straight line (because a log-log grid plots a constant rate of change as a straight line). The straight line is easier to read, and it simplifies forecasting because a constant learning rate always appears as a straight line on log-log coordinates. To verify the fact that both graphs represent the same thing, look at the number of hours needed to produce 100 units in Figures 16.4 and 16.5; both figures indicate about 50 hours per unit.

In addition to determining the direct labor component of price, the labor hour data also have the following purchasing applications. See Appendix A for additional applications of Learning Curves.

Direct Materials

Experience indicates the organizations "learn" how to purchase materials more efficiently as quantities increase and time progresses. It is the authors' experience that a 95% rate of learning applies to the purchase of the same material over time.

Estimating Delivery Times

Since the learning curve can be used to forecast labor time required, it is possible to estimate how many units a supplier can produce over a specified time with a given labor force. This information can be extremely helpful to a buyer in scheduling deliveries, in planning his or her firm's production, and in identifying suppliers who obviously cannot meet desired delivery schedules.

Supplier Progress Payments

Since the learning curve reflects changing labor costs, it provides a basis for figuring a supplier's financial commitment on any given number of units. This information is important because suppliers often operate in the red during the initial part of a production run, until learning can reduce costs below the average price. Supply managers can minimize supplier hardship by using the learning curve to break down an order into two or more production lots—each with successively lower average prices—and then set up progress payments based on the supplier's costs.

Indirect Costs

Indirect costs represent 30 to 40% of many suppliers' total costs of production. Five of the most common indirect cost pools are engineering overhead, materials overhead, manufacturing overhead, general and administrative expense, and selling expense.

Engineering Overhead

This is the cost of directing and supporting the engineering department and its direct labor staff. Costs include supervisory and support labor, fringe benefits, indirect supplies, and fixed charges such as depreciation.

Material Overhead

This category of overhead usually includes the indirect costs of purchasing, transporting incoming materials, receiving, inspection, handling, and the storage of materials.

Manufacturing Overhead

This category includes all production costs except direct materials, direct labor, and similar costs that can be assigned directly to the production of an item.

Manufacturing overhead includes:

- The cost of supervision, inspection (some firms charge quality assurance or inspection as a direct cost), maintenance, custodial, and related personnel costs
- Fringe benefits such as social security and unemployment taxes, allowances for vacation pay, and group insurance
- Indirect supplies such as lubricating oils, grinding wheels, and janitorial supplies
- Fixed charges, including depreciation, rent, insurance, and property taxes
- Utilities

General and Administrative

General and administrative (G&A) expenses include the company's general and executive offices, staff services, and miscellaneous activities.

Selling

Selling expenses include sales salaries, bonuses, and commissions, and the normal costs of running the department.

Recovering Indirect Costs

The supplier allocates overhead costs to specific operations. Normally, this allocation is based on a product's age in terms of its life cycle. For instance, mature products generally incur lower G&A expenses than new products do, which require more development and marketing effort. This allocation results in the use of an overhead rate for each indirect cost pool. The rate is determined by management personnel, who select an appropriate base (or cost driver that causes the incurrence of the overhead costs) and then develop the ratio of the indirect cost pool dollars to that base. For example, the following allocation formula might be used for manufacturing overhead:

$$\frac{\text{Manufacturing overhead pool dollars}}{\text{Manufacturing direct labor hours}} = \frac{\$5,000,000}{\$1,000,000} = \frac{\$5 \text{ per direct}}{\text{manufacturing labor hours}}$$

Overhead rates are generally established annually, typically before the start of the accounting period.

When determining the reasonableness of overhead rates, a buyer should not look only at the rate. He or she must consider the reasonableness of the indirect costs in the overhead pool and the appropriateness of the overhead allocation base. Since rates typically are established annually, the supply manager should also ensure that the allocation rate used by a supplier is applied consistently.

Activity-Based Costing

Activity costing can be traced back to the late 1700s. During the intervening 200 years, management focused on labor costs because they represented over 50% of total costs. The allocation of overhead costs based on the number of hours required to produce a product was relatively realistic and certainly easy.

But as direct labor costs have shrunk to 10% or so of total costs, they have become a less logical and realistic basis on which to allocate indirect costs. During the 1980s, a band of accounting academics, led by Professors Robert Kaplan and Robin Cooper of the Harvard Business School and Claremont McKenna College, respectively, developed what has become known as activity-based costing (ABC)—a tool that more accurately identifies and allocates indirect costs to the products they support. In the 1990s, ABC evolved into activity-based management (ABM). ABM essentially integrated ABC, continuous improvement, and business process analysis.

ABC or ABM can be used to identify opportunities to reduce the supplier's indirect costs. ABC goes beyond identifying and allocating these indirect costs to products by identifying the drivers of these costs.[8] Some examples of cost drivers are the number of orders, length of setups, specifications, engineering changes, and liaison trips required. This identification allows management to identify and implement cost savings opportunities. Quite obviously, if the supplier's management does not implement the required changes, an alert supply manager can "encourage" such action.

When analyzing a supplier's cost breakdown, it is imperative that the supply management professional understand how the supplier estimates and applies overhead to the product being purchased. It is also important to be aware of how well the supplier understands his or her own overhead cost structure. (A supply professional should develop the expertise to understand and motivate a selling firm with poor cost control to improve its system of collecting and applying costs to products.) As overhead becomes a greater proportion of product cost and as supply managers continue to seek ways to reduce acquisition cost, a small error in estimating and applying overhead can significantly affect the final cost. A note regarding accuracy: greater accuracy may not only allow for a lower purchase price, but may also lead to a higher price because true costs are now known.

Target Costing

A number of years ago the late management guru, Peter F. Drucker, wrote of five deadly business sins—avoidable mistakes that will (and in many cases have) harm(ed) a business. Drucker's third deadly sin is cost-driven pricing. He argued, "The only thing that works is price-driven costing. . . . The only sound way to price is to start out with what the market is willing to pay . . . and design to that price specification."[9] Dr. Drucker's comments are as applicable to procurement today as they were then.

Some 40 years ago, the Ford Motor Company employed price-based costing in the development of its highly successful Mustang. The car was designed to retail at $1,995. This pricing objective drove design engineering to focus on cost, as well as performance and aesthetics. In turn, this drove engineers and purchasing and supply management to identify target prices for items to be purchased from suppliers. Members of these two functions then worked with their potential suppliers to develop processes and procedures to produce the required materials and components at these target prices. Curiously enough, American management largely reverted to cost-based pricing during the intervening 40 years, while its Japanese competition adopted price-based costing. Dr. Drucker points out that "cost-based pricing is the reason there is no American consumer-electronics industry anymore.[10]

It is heartening to see a growing number of organizations replace their "adversarial bidding system with one in which the company designates suppliers for a component and then uses target pricing . . . to determine with suppliers the component prices and how to achieve them."[11]

Profit

There are no precise formulas that can be used to help form a positive judgment concerning the right price, of which profit is one component (price = cost + profit). There are, however, certain basic concepts of pricing on which scholars and practitioners do agree. One objective of sound purchasing is to achieve good supplier relations. This objective implies that the price must be high enough to keep the supplier in business. The price must also include a profit sufficiently high to encourage the supplier to accept the business in the first place and, second, to motivate the firm to deliver the materials or services on time. What profit does it take to get these two desired results? On what basis should it be calculated?

If profit were calculated on a percentage-of-cost basis, the high-cost, inefficient producer would receive the higher profit (in absolute terms). To make matters even worse, under the cost concept of pricing, producers who succeeded in lowering their costs by attaining greater efficiency would be "rewarded" by a reduction in total profit. For example, if an efficient producer has costs of $1,000 and a fair profit is agreed to be 10% of cost, its profit would be $100. If an inefficient producer has costs of $1,500, its profit on the same basis would be $150. If by better techniques the efficient producer should lower its costs to $800, the reward would be a $20 loss in profit—from $100 to $80. Obviously, the concept of determining a fair profit as a fixed percentage of cost is unrealistic.

Another concept on which profit might be determined is the relationship of capital investment required to produce the profit. Profit might be calculated as a percentage of capital investment. However, under this system, it would still be possible for the inefficient producer to receive the greater reward. For example, suppose firm A makes a capital investment of $2 million to produce product X. Firm B, on the other hand, invests only $1 million in its plant to produce product X successfully. From the buyer's point of view, there is no reason why firm A, simply because of its greater investment, should receive a higher profit on product X than firm B. Firm B, in fact, is utilizing its investment more efficiently. Thus, profit calculated as a fixed percentage of a firm's capital investment is not a satisfactory method for a buyer to use in determining a fair profit.

In a competitive economy, the major incentive for more efficient production is greater profit and repeat orders. A fair profit in our society cannot be determined as a fixed percentage figure. Rather, it is a flexible figure that should be higher for the more efficient producer than it is for the less efficient one. Low-cost producers can price lower than their competitors, while simultaneously enjoying a higher profit. Consequently, one of a buyer's greatest challenges is constantly to seek out the efficient, low-cost producer.

Considerations other than production efficiency can also rightly influence the relative size of a firm's profit. Six of the most common considerations are discussed briefly next.

1. Profit is the basic reward for risk taking as well as the reward for efficiency; therefore, higher profits justifiably accompany extraordinary risks, whatever form they take. For example, great financial risk usually accompanies the production of new products. For this reason, a higher profit for new products is often necessary to induce a seller to take the risk of producing them.

2. A higher dollar profit per unit of product purchased on small special orders is generally justified over that allowed on larger orders. The justification stems from the fact that the producer incurs a fixed amount of setup and administrative expense, regardless of the size of the order. Consequently, the cost of production for each unit is greater on small orders than on large orders. Since producers incur this cost at the request of the buyer, they usually demand a proportionately higher absolute profit before accepting an order that forces them to use their facilities in a less efficient manner than they might otherwise do.

3. Rapid technological advancement creates a continuing nationwide shortage of technical talent. The cost in dollars and time of training highly technical personnel frequently makes it necessary to pay a higher profit on jobs requiring highly skilled people.

4. In the space age, technical reliability can be a factor of overriding importance. A higher profit is generally conceded as justified for a firm that repeatedly turns out superbly reliable technical products than for one producing less reliable products. Good quality control, efficiency in controlling costs, on-time delivery, and technical assistance that has resulted in better production or design simplification all merit profit consideration.

5. On occasion, because of various temporary unfavorable supply-demand factors (e.g., excessive inventories, a shortage of capital, a cancellation of large orders), a firm may be forced to sell its products at a loss in order to recover a portion of its invested capital quickly or to keep its production facilities in operation.

6. A firm that manufactures a product according to the design and specifications of another firm is not entitled to the same percentage of profit as a firm that incurs the risk of manufacturing to its own design. In the first instance, the manufacturing firm is assured of a sale without marketing expense or risk of any kind, provided only that it

fulfills the terms of the contract. In the second instance, the manufacturing firm is without assurance that its product can be sold profitably, if at all, in a competitive market.

In summary, there is no single answer to the question: What is a fair profit? In a capitalistic society, profit generally is implied to mean the reward over costs that a firm receives for *the measure of efficiency it attains* and the *degree of risk it assumes.* From a purchasing viewpoint, profit provides two basic incentives. First, it induces the seller to take the order. Second, it induces the seller to perform as efficiently as possible, to deliver on time, and to provide all reasonable services associated with the order. Except in those temporary cases where a firm is willing to sell at a loss, the profit is too low if it does not create these two incentives for the seller.

Resisting Arbitrary Price Increases

It is not uncommon for a supplier to use a legitimate increase in the material cost to capitalize on taking the increase on the last total price paid. The fallacy of this reasoning is that while one component cost of the suppliers product may have gone up, the increase should not be taken over the total cost of the product or service as other costs may have remained stable or even gone down.

For example, if the cost of steel in a manager's desk has risen 10%, that increase should not be applied to the direct labor, overhead, or profit components of the price. This opportunistic pricing strategy will result in an exponential price increase over time that outweighs the justifiable increase of the steel. See Table 16.5 for the Desk example.

Thus, the legitimate price increase would be 3% vs. 10% resulting from applying the increase across the board of all cost components! This is another reason you need to know the cost components of what you are buying, even if you have to estimate it!

Table 16.5   Sample of an Arbitrary Price Increase-Desk

Cost Components	Last Price Paid	Price + 10% Overall	Price + 10% on Steel Only
Steel	$100	$110	$110
Laminate Top	50	55	50
Hardware	20	22	20
Direct Labor	30	33	30
Overhead	60	66	60
Profit	52	57.20	52
Total	$312	$343.20	$322
% Increase		10%	3%

There are several other recommendations regarding how a buyer handles attempted, mid-contract price increases:

1. Challenge every attempt to raise prices by asking for complete documentation as to the reasons for the increase.
2. This is another reason you should have performed a target should-cost-analysis as previously discussed.
3. As demonstrated in the desk analysis, buyers must know the cost structure of the major products and services they purchase.
4. Remember that labor and overhead costs do not go up in the short-run. There are obvious labor contracts at the supplier's plant and they usually have long-term contracts for materials and services. Therefore, we must have evidence of unexpected and unusual cost increases at the supplier's plant.
5. Most sales representatives have no idea of the cost structures for the product and services that they sell. While there are some exceptions, the corporate management of the selling firm often do not want their sales representatives to know the cost breakdown including the profit for fear that they might reveal too much.
6. Never, never offer or accept formula pricing which includes "price escalation clauses" as it implies the price can only go up. The language should read "price change warning clauses" which means the price could go up or down. There is a naïve assumption that prices can only go up when we know about productivity increases, learning curve cost reduction, and changes in supply and demand. The various indices that we have described can actually reveal the need for a price decrease, but you will never get it if you don't ask for it. Warren Norquist, former Director of Purchasing at Polaroid, referred to this as "opportunistic pricing."
7. Remember, not all components of the producer price index (ppi) or other similar indices are relevant to the particular product you are buying. Never allow an index to dictate potential changes in pricing unless you know the index components are relevant.
8. If you must sign clauses allowing "surcharges," just make sure the surcharges are documented. For example, if the sellers are asking for fuel surcharges, what percentage of the total cost of the product or service is attributed to fuel? This is a similar case as our desk sample. Grant the fuel surcharge only on the fuel and only for the specific cost at the suppliers' plant. In other words, don't use a national average.

Summary

Obtaining the *right price* is one of supply management's most important responsibilities. When focusing on price, insight into the current economic environment and knowledge of the cost elements that underlie a selling price will support the most favorable procurement.

In addition, the supply management professional should be aware of the various means available to locate potential suppliers, make price comparisons, and utilize competitive bidding and negotiation.

When price analysis is not possible, cost analysis becomes the basis of obtaining a *fair and reasonable* price. Armed with an understanding of cost principles, the buyer now is in a position to conduct an analysis of a potential supplier's proposal. Many costs are known and understood, but all companies have hidden costs that often reside in overhead. A supply professional needs an understanding of costs, cost systems, and overhead composition and allocation. Other key elements of cost analysis are labor efficiency, subcontracting, plant capacity, experience, cost modeling, and profit and they should be permanent concepts in the minds of all supply management professionals.

On a final note, the purchase or acquisition price should be evaluated in the context of the total cost of ownership, and do not forget to ask for the appropriate discounts as discussed in Appendix B.

Appendix A: Application of Learning Curves

Before applying a learning curve to a particular item, a supply manager must be certain that learning does in fact occur at a reasonably constant rate. Many production operations do not possess such properties. Gross errors can be made if a learning curve is misapplied; therefore, buyers must be alert to the following problems.

Nonuniform Learning Rates

Learning curve analysis is predicated on the assumption that the process in question exhibits learning at a reasonably constant rate. Direct labor data from such a process should plot in a straight line on a log-log grid. If a straight line cannot be fitted to the data reasonably well, the learning rate is not uniform and the technique should not be used.

Low-Labor-Content Items

Continued learning occurs principally in the production of products entailing a high percentage of labor. The learning opportunity is particularly high in complex assembly work. On the other hand, if most work on a new item involves machine time, where output tends to be determined by machine capacity, there is little opportunity for continued learning.

Small Payoffs

Obtaining historical cost data to construct a learning curve entails much time and effort, particularly when a supplier uses a standard cost accounting system. Therefore, learning curve analysis is worthwhile only if the amount of money that can be saved is substantial.

Incorrect Learning Rates

Learning varies from industry to industry, plant to plant, product to product, and part to part. Applying one rate just because someone in the industry has used it can be misleading. Intelligent use of learning curves demands that learning rates be determined as accurately as possible from comparable past experience.

Established Items

If a supplier has previously made the item for someone else, a supply manager should not use the learning curve even if the product is nonstandard and new to the buying firm. Since most of the learning has already been done on previous work, any additional cost reduction may well be negligible.

Misleading Data

Not all cost savings stem from learning. The economies of large-scale production spread fixed costs over a larger number of output units, thus reducing the unit cost of the item. However, this phenomenon has nothing to do with the learning curve.

An Example of Learning Curve Application: The Cumulative Average Curve

The following simplified example shows a basic application of the cumulative average learning curve concept in labor cost analysis and contract pricing.

The ABC Corporation has purchased 50 pieces of a specially designed electronic component at $2,000 per unit. Of the $2,000 selling price, $1,000 represents direct labor. An audit of product costs for the first 50 units established that the operation is subject to an 80% cumulative average learning curve. What should ABC pay for the purchase of 350 more units?

Solution

1. Using log-log paper, plot 50 units (on the horizontal axis) against $1,000 direct labor cost on the vertical axis (see Figure 16.6).
2. Double the number of units to 100 on the horizontal axis and plot against a labor cost of $800 (80% as high as the original $1,000 cost).
3. Draw a straight line through the two cost points. The line represents an 80% learning curve, constructed on the basis of labor cost data for the first 50 units of production.
4. Locate 400 units on the horizontal axis (the total expected production of 50 original units plus 350 new ones). Read from the curve the labor cost of approximately $510. This is the *average* expected labor cost per unit for the total production of 400 units.

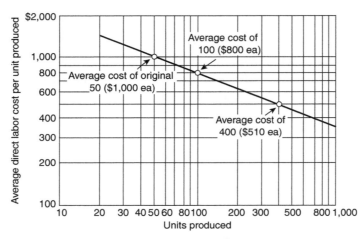

Figure 16.6 Estimating labor cost for the new contract.

5. To find the labor cost for 400 units, multiply 400 by $510, the direct labor cost per unit. The total is $204,000.

6. Subtract the labor paid in the original order to determine the labor cost of the new order of 350 units. Hence, subtract $50,000 (50 × $1,000) from $204,000. The answer is $154,000, the labor cost which should be paid for the new order of 350 units: $154,000 ÷ 350 units = $440 per unit labor cost, as compared with the original $1,000 per unit.

7. Now determine the cost for materials, overhead, and profit on the 350 units. Add this figure to the labor cost determined in step 6 to obtain the total price ABC should pay for the additional 350 units.

An Example of Learning Curve Application: The Unit Curve

The preceding application dealt with the use of a *cumulative average* learning curve. To illustrate the application of a *unit* learning curve, consider the following hypothetical situation.

Assume that a manufacturer receives an order to produce 515 units of a new product. After the necessary production and tooling design work is completed, the manufacturer begins production. Prior experience with moderately similar products leads the manufacturing manager to believe that a *unit* learning curve phenomenon probably will be experienced as production operations proceed.

To investigate this possibility, for the first 32 units produced, he or she records the production data shown in Table 16.6, columns 1 and 2. Then he or she makes the calculations shown in column 3. As production doubled (1 to 2, 2 to 4, 4 to 8, etc.), in each case it is clear that a significant learning effect was experienced. (The second unit took 85% of the time required by the first unit, the fourth unit required 84.3% of the time required by

Table 16.6 The Manufacturer's Production Data

Column 1 Unit Produced	Column 2 Labor Hours Required to Produce the Corresponding Unit in Column 1	Column 3 Labor Hours Required as a Percentage of Those Required for the Preceding Unit
1	60	—
2	51	85.0%
4	43	84.3
8	37	86.0
16	31	83.8
32	26	83.9

the second unit, and so on, as indicated in column 3). Although the rate varies slightly, the manager concludes that a unit learning curve of approximately 85% is a good indicator of the manner in which the process will behave during the production of the remainder of the order. (That is, that the sixty fourth unit will require 85% of the time required by the thirty second unit, that the hundred twenty eighth unit will require 85% of the time required by the sixty fourth unit, and so on.)

Consequently, the supply professional constructs the curve on a log-log grid—and reads directly from the graph that the five hundred twelfth unit will require approximately 16.6 direct labor hours for its production. Similar determinations for all units produced permit him to calculate the total number of labor hours for the complete job. Using this information, he can schedule production efficiently, as well as estimate the total labor cost the job will incur.

Appendix B: Discounts

Discounts are frequently considered a routine, prosaic part of pricing. Perceptive supply professionals recognize that this is not always the case. As will be illustrated in the discussion to follow, discounts can sometimes succeed as a technique for reducing prices after all other techniques have failed. The four most commonly used kinds of discounts are *trade discounts, quantity discounts, seasonal discounts,* and *cash discounts.*

Trade Discounts

Trade discounts are reductions from list price given to various classes of buyers and distributors to compensate them for performing certain marketing functions for the original seller (usually the manufacturer) of the product. Trade discounts are frequently structured as a sequence of individual discounts (e.g., 25, 10, and 5%), and in such cases, they are called series discounts. Those who perform only a part of the distribution functions get only one or two of the discounts in the series. If the retail price of an item with such discounts is $100, the full discounted price is calculated as follows: 25% of $100 = $25; 10% of ($100 − $25) = $7.50; 5% of ($100 − $25 − $7.50) = $3.38. The manufacturer's selling price, then, is $100 − $25 − $7.50 − $3.38 = $64.12.

An industrial supply professional who purchases through distributors must, as a result of the very nature of series trade discounts, be certain that the buy is from the right distributor (i.e., the distributor obtaining the most discounts). The general guideline is to get as close to the manufacturer as is practical. For example, a large buyer normally should not purchase paper requirements from a janitorial supply house, which usually does not obtain all discounts in the series for paper. If an account is sufficiently large, the buying firm should purchase from a paper distributor that normally does obtain all discounts in the series. Such a supplier can, at the same profit margin as the janitorial supply house, offer buyers lower prices. Buyers with very large accounts should purchase directly from paper manufacturers.

Quantity Discounts

These price reductions are given to a buyer for purchasing increasingly larger quantities of materials. They normally are offered under one of three purchasing arrangements:

1. For purchasing a specific quantity of items at one time
2. For purchasing a specified dollar total of any number of different items at one time
3. For purchasing a specified dollar total of any number of items over an agreed-upon time period

The third type of quantity discount noted previously is called a cumulative discount. The period of accumulation can be a month, a quarter, or, more commonly, a year. For large-dollar-value, repetitive purchases, buyers should always seek this type of discount. Also, because unplanned increases in business occur with regular frequency, supply professionals should include in all quantity discount contracts a provision that if the total purchases made under the contract exceed the estimated quantities, then an additional discount will be allowed for all such excesses.

The quantity discount concept originally stemmed from the unit cost reductions inherent in large-volume manufacturing operations. In a traditional mass or batch production operation, a large production run of a single product spreads the fixed costs over the number of items produced and results in a lower production cost per item. With the continuing improvement of flexible manufacturing systems, however, the dilution of fixed setup costs becomes rather small compared with other product-specific cost elements.

Consequently, if a quantity discount is based solely on the distribution of setup and order processing costs over the volume of production, there is clearly a declining incentive for such a supplier to offer this form of quantity discount.

Seasonal Discounts

Based on the seasonal nature of some products (primarily consumer products), producers commonly offer discounts for purchases made in the off-season. For example, room air conditioners usually can be purchased at a discount during the fall or winter seasons.

Cash Discounts

In many industries, sellers traditionally offer price reductions for the prompt payment of bills. When such discounts are given, they are offered as a percentage of the net invoice price. When suppliers extend credit, they cannot avoid certain attendant costs, including the cost of tied-up capital, the cost of operating a credit department, and the cost of some bad-debt losses. Most sellers can reduce these costs by dealing on a short-term payment basis. Therefore, they are willing to pass on part of the savings to buyers in the form of a cash discount.

Supply professionals should be aware of the importance of negotiating the highest possible cash discount. The most commonly used discount in practice is 2% 10 days, net 30 days. In industries where prompt payment is particularly important, cash discounts as high as 8% have been allowed. A cash discount of 2/10, net 30 means that a discount of 2% can be taken if the invoice is paid within 10 days, while the full amount should be remitted if payment is made between 10 and 30 days after receipt of the invoice.

A 2% discount, viewed casually, does not appear to represent much money. In one sense, however, it is the equivalent of a 36.5% annual interest rate. Because the bill must be paid in 30 days and the discount can be taken up to the tenth day, a buyer not taking the discount is paying 2% of the dollar amount of the invoice to use the cash involved for 20 days. In a 365-day year, there are 18.25 twenty-day periods (365/20 = 18.25). A 2/10 discount translates into an *annual* discount rate of 36.5% (2% times 18.25).[12] If a firm does not have sufficient cash on hand to take cash discounts, the possibility of borrowing the needed money should be investigated. Under normal conditions, paying 10 to 15% for capital that returns 36.5% is good business. Capable buyers understand the time value of money. In some situations, generous cash discounts can be obtained either for prepayment or for 48-hour payment.

Various other types of cash discounts are in use. One common type is the *end-of-month (EOM)* dating system. This system of cash discounting permits the buying firm to take a designated percentage discount if payment is made within a specified number of days after the end of the month in which the order is shipped. If materials are shipped on October 16 under 2/10 EOM terms, a 2% discount can be taken at any time until November 10.

Lower prices, in the form of higher cash discounts, are an ever-present source of price reduction that supply professionals should always explore. Frequently, sellers who will not consider reducing the prices of their products will consider allowing higher cash discounts. Such action accomplishes the identical result for the purchasing firm. For example, a major petroleum company recently was able to gain a 6% price reduction on the purchase of a complex testing machine—a machine the manufacturer had never before sold below its listed $92,000 selling price. The $5,520 price reduction was achieved by the supply manager's offering to pay one-half of the purchase price one week in advance of the machine's delivery to his company's testing laboratory.

Endnotes

1. Pure competition exists only under the following circumstances: The market contains a large number of buyers and sellers of approximately equal market power. The products traded are homogeneous (a buyer would not desire one particular seller's product over any other's). The buyers and sellers always have full knowledge of the market. The buyers always act rationally, and sellers are free to enter and leave the market at will.

2. Fixed costs are usually "fixed" over the relevant range of production. The relevant range extends from the minimum to the maximum capacity/output of a given manufacturing facility. To increase production beyond maximum capacity requires the purchase of additional plant capacity and results in increased fixed costs.

3. Many costs treated as indirect by organizations are traceable and could become direct costs through improvements in the measurement system. The authors contend that the reasons for not tracing costs directly have diminished with information technology advances.

4. The Bureau of Labor Statistics in Washington, D.C., provides thousands of different price indexes every month. Available are indexes by stage of processing, industry, and individual commodity grades. A commonly used series for these purposes is the producer price index (PPI). In addition, import and export price indexes broken down in the major subcategories are available. These indexes allow the price analyst to adjust historical prices by appropriate changes over time.

5. The development and use of independent cost estimates is described in detail in D. N. Burt, W. Norquist, and J. Anklesaria, *Zero Base PricingTM: Achieving World Class Competitiveness through Reduced All-in-Cost* (Chicago, IL: Probus Publishing, 1990) Chap. 4.

6. The interested reader is referred to Burt et al., op. cit., Chap. 8.
7. Different types of labor generate different percentages of learning. Assembly-type labor generates the most rapid improvement and fabrication-type labor the least. Fabrication labor has a lower learning rate because the speed of jobs dependent on this type of labor is governed more by the capability of the equipment than the skill of the operator. The operator's learning in this case is confined to setup and maintenance times. In some situations, therefore, when a precise analysis is desired, a learning curve should be developed for each category of labor. Also, it should be noted that different firms within the same industry experience different rates of learning.
8. The OSD Comptroller iCenter is a comprehensive online site related to the U.S. Department of Defense budget process, financial management, and best practices. Visit this site for educational information on the use of ABCosting, 2002.
9. Peter F. Drucker, "The Five Deadly Business Sins," *Wall Street Journal*, Oct. 21, 1993, 14.
10. Ibid.
11. James P. Womack and Daniel T. Jones, "From Lean Production to the Lean Enterprise," *Harvard Business Review* March–April (1994): 97.
12. A complete analysis of this situation must include the opportunity cost of early payment. If a firm's internal cost of capital is 15% per year, the net saving generated by the 2/10, net 30 discount is 36.5%–15%, or 21.5% on an annualized basis.

Suggested Reading

Brimson, James A. *Activity Accounting: An Activity Based Approach.* New York: John Wiley & Sons, 1991.

Burt, David N., and Richard L. Pinkerton. "Special Secondary Source Techniques for Estimating Cost Components." In: *A Purchasing Manager's Guide to Strategic Proactive Purchasing.* New York: AMACOM, 1996, 282–289.

Ellram, L.M. "Total Cost Modeling in Purchasing." The Center for Advanced Purchasing Studies (CAPS), Tempe, AZ, 1994.

Ellram, Lisa M. "Implementation of Target Costing in the United States: Theory versus Practice." *Journal of Supply Chain Management* 42(1)(2006): 13–26.

Graw, Leroy H. *Cost/Price Analysis: Tools to Improve Profit Margins.* Van Nostrand Reinhold Company, New York, NY, 1994.

Lopoukhine, Serge R., Kathy English, and Patricia A. Cox, "Cost/Price (Plus) Analysis = Opportunity." *Purchasing Today* 8(9)(1997): 43.

Newman, Richard G. "Monitoring Price Increases with Economic Data: A Practical Approach." *International Journal of Purchasing and Materials Management* Fall (1997): 35–40.

Newman, Richard G. *Supplier Price Analysis: A Guide for Purchasing Accounting and Financial Analysis.* Quorum Books, Westport, CT, 1992.

Newman, Richard G., and John M. McKeller. "Target Pricing—A Challenge for Purchasing." *International Journal of Purchasing and Materials Management* Summer (1995): 12–20.

Smith, Michael E., Lee Buddress, and Alan Raedels. "The Strategic Use of Supplier Price and Cost Analysis." Abstract from the 91st Annual International Supply Management Conference, Minneapolis, MN, May 2006.

Stueland, Valerie J. "The Lowest Price is Not Always the Best." *Inside Supply Management* February (2006): 12–14.

Whittington, Elaine. "Anatomy of a Price." Abstract from the 86th Annual International Supply Management Conference, Orlando, FL, 2001.

Yuva, John. "Achieving Sustainable Cost Savings." *Inside Supply Management* 15(2)(2004): 22.

CHAPTER 17

Methods of Compensation

Introduction to Compensation Arrangements

A wide selection of contract compensation arrangements is necessary to provide the flexibility needed for the procurement of a large variety of materials and services. The compensation arrangement determines (1) the degree and timing of the cost responsibility assumed by the supplier, (2) the amount of profit or fee available to the supplier, and (3) the motivational implications of the fee portion of the compensation arrangements. The following examples are introduced in an effort to portray visually the seller's and, in turn, the buyer's problem of dealing with uncertainty.

Example 1: Low Level of Uncertainty   In this example, the seller's likely cost for a project is $1,000,000. The seller is confident that the lowest possible cost will be $950,000 and the highest cost $1,050,000. This information is portrayed in Figure 17.1. One can see that the seller is virtually certain that costs will be within the range of $950,000 to $1,050,000, with the most likely outcome near $1,000,000.

If the seller adds 10% for profit to the most likely cost outcome, it may be willing to agree to a firm fixed price of $1,100,000. Note that the supplier's actual profit will be in the range of $150,000 (if actual costs are $950,000) to $50,000 (if actual costs are $1,050,000). The most likely profit is $100,000 [$1,100,000 (the price) − $1,000,000 (the most likely cost outcome)]. In this example, the use of a firm fixed price contract (as discussed next) seems appropriate.

Example 2: High Level of Uncertainty   In this example, assume that the range of likely cost outcomes is much wider—say, $500,000 to $1,500,000. (Such an extreme range of cost outcomes is highly unlikely but is used here to introduce the concept that the type of compensation agreement should be appropriate for the amount of uncertainty present.) Again, the most likely cost outcome is $1,000,000. This example is portrayed in Figure 17.2.

Most sellers are very risk averse; that is, they are unwilling to accept large amounts of uncertainty unless they are able to transfer the uncertainty to the buyer in the form of

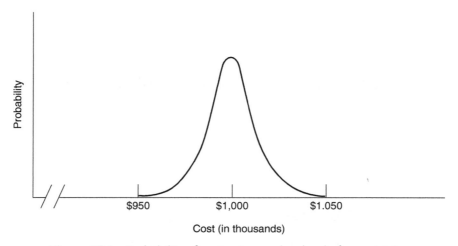

Figure 17.1   Probability of cost outcome: low level of uncertainty.

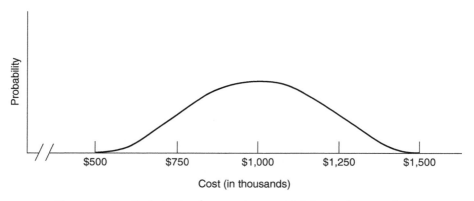

Figure 17.2   Probability of cost outcome: high level of uncertainty.

higher prices. In this case, the seller studies the distribution of likely cost outcomes and concludes that 9 times out of 10, the actual cost will be $1,400,000 or less. Accordingly, if the buyer is unwise enough to insist on using a firm fixed method of compensation, the seller may demand a firm fixed price of $1,540,000 [$1,400,000 + $140,000 (10% profit on this cost)].

Table 17.1 portrays the supplier's profit under various cost outcomes for this second example. It is fairly obvious that if the buyer could assume the risk inherent in this procurement, a lower price would be possible—except in the extreme case in which actual costs exceeded approximately $1,540,000.

Table 17.1 Supplier's Profit Using FFP Under Various Cost Outcomes

Actual Cost	Actual Profit	Fixed Price
$ 500,000	$1,040,000	$1,540,000
1,000,000	540,000	$1,540,000
1,400,000	140,000	$1,540,000
1,500,000	40,000	$1,540,000

Let us assume that the buyer and supplier agreed to a cost plus fixed fee contract with a target cost of $1,000,000 and a fixed fee of $50,000. Note that the fee is relatively low: 5% of target cost. This is because the supplier will not incur any risk. Table 17.2 portrays the price to be paid by the buyer under different cost outcomes.

The sophisticated reader may say: "Interesting. But a fee or profit of $50,000 on a cost of $500,000 (see line 1, Table 17.2) is 10%. That's too high, considering that there's no cost risk! Why not apply a fixed percent fee to costs?" Such an approach is shown in Table 17.3. One does not have to be a rocket scientist to see that such an approach motivates the supplier to increase costs because the higher the cost, the higher the fee!

Table 17.2 Buyer's Profit Using CFFP Under Various Actual Cost Outcomes

Actual Cost	Fixed Fee or Profit	Price Paid
$ 500,000	$50,000	$ 550,000
1,000,000	50,000	$1,050,000
1,400,000	50,000	$1,450,000
1,500,000	50,000	$1,550,000

Table 17.3 Fee and Price Outcomes at Various Cost Outcomes with Profit as a Fixed Percentage of Cost

Actual Cost	Fee (10% of Actual Cost)	Price Paid
$ 500,000	$ 50,000	$ 550,000
1,000,000	100,000	$1,100,000
1,400,000	140,000	$1.540,000
1,500,000	150,000	$1,650,000

Observation

Without going into further detail at this point, it should be apparent that selection of the right type of compensation arrangement can save money. Insightful readers may be asking themselves: "What would happen if, instead of paying a fixed fee or a fixed percent of cost, we were able to incentivize the supplier to control costs?"

The supply manager has a range of compensation arrangements designed to meet the needs of a particular procurement. At one end of this range is the firm fixed price contract where the supplier assumes all cost responsibility and where, therefore, profit and loss potentials are high. At the other end of this range is the cost plus fixed fee contract where the supplier has no cost risk and where the fee (profit) is fixed, usually at a relatively low level. In between these two extremes are numerous incentive arrangements that reflect a sharing of the cost responsibility.

Contract Cost Risk Appraisal

The degree of cost responsibility a supplier reasonably can be expected to assume is determined primarily by the cost risk involved. It is to the supply manager's advantage to estimate this risk prior to negotiations. Since the majority of contracts are "forward priced," that is, priced prior to completion of the work, some cost risk is involved in each of them. The degree of cost risk involved will depend on how accurately the cost of the contract can be estimated prior to performance. The accuracy of the cost estimate and the degree of cost risk usually are a function of both technical and contract schedule risk.

A supply manager should insist on a fixed price contract unless (1) the risks will result in a contract price containing large reserves for contingencies that may not occur, (2) the risks result in reliable suppliers refusing to agree to a fixed price contract because a significant loss might be incurred, or (3) the use of a fixed price contract could result in the supplier "cutting corners" in order to avoid taking a loss.

Technical Risk

Technical risk is that risk associated with the nature of the item being purchased. Appraisal of technical risk includes analysis of the type and complexity of the item or service being purchased, stability of design specifications or statement of work, availability of historical pricing data, and prior production experience. Analysis of technical risk in a complex system may include appraisals by a team with members from the user group, the engineering staff, and the purchasing and supply management group. Think, for example, of the technical risk involved in the Apollo mission to put a man on the moon: leaving the earth's gravity, sustaining life in gravity-less environment, landing on the unknown surface structure of the moon's crust, and reentering the earth's stratosphere without burning up.

Technical risk is reduced as the job requirements, production methods, and pricing data become better defined and the design specifications or statement of work become more stable. Research and development contracts, in particular, have a rather high technical risk associated with them. This is due to the ill-defined requirements that arise from the necessity to deal beyond, or at least very near, the limits of the current technology.

Contract Schedule Risk

In addition to technical risk, schedule risk must be assessed in determining the supplier's cost risk. Preferred procurement practice calls for forward pricing of contract efforts. This practice attempts to anticipate material and labor cost increases during performance of the contract. These estimates, along with possible schedule slippage, are always subject to error.

General Types of Contract Compensation Arrangements

Compensation arrangements can be classified into three broad categories: (1) fixed price contracts, (2) incentive contracts, and (3) cost reimbursement contracts.

Fixed Price Contracts

Under a fixed price arrangement, the supplier is obligated to deliver the product called for by the contract for a fixed price. If, prior to completion of the product, the supplier finds that the effort is more difficult or costly than anticipated, the supplier is still obligated to deliver the product. Further, the supplier will receive no more than the previously agreed-on amount. The amount of profit the supplier receives will depend on the actual cost outcome. There is no maximum or minimum profit limitation in fixed price contracts. A fixed price arrangement is normally used in situations where specifications are well defined and cost risk is relatively low.

Incentive Contracts

Incentive contracts are employed in an effort to motivate the supplier to improve cost and possibly other stated requirements such as schedule performance. In an incentive contract, the cost responsibility is shared by the buyer and the seller. This sharing addresses two issues: (1) the desire to motivate the supplier to control cost and

(2) an awareness that if the supplier assumes all or most of the risk when significant uncertainty is present, a contingency allowance will be required, thereby inflating the contract price.

Incentive contracts are of two types: (1) fixed price incentive and (2) cost plus incentive fee. With a fixed price incentive contract, the ceiling price is agreed to (or fixed) during negotiations. Under the cost plus incentive fee arrangement, the supplier is reimbursed for all allowable costs incurred, up to any prescribed ceiling. Obviously, the supplier's cost accounting system must meet commonly accepted standards and be open to the customer for review when employing incentives based on costs incurred

Cost-Reimbursement Contracts

Under a cost-reimbursement arrangement, the buyer's obligation is to reimburse the supplier for all allowable, reasonable, and allocatable costs incurred, and to pay a fixed fee. Again, the supplier's cost accounting practices must meet commonly accepted standards and be open to the customer. Most cost arrangements include a cost limitation clause that sets an administrative limitation on the reimbursement of costs. Generally, under a cost-reimbursement arrangement, the supplier is obligated only to provide its "best effort." Usually, neither performance nor delivery is guaranteed. Cost-reimbursement arrangements are normally used when:

▲ Procurement of research and development involves high technical risk
▲ Some doubt exists that the project can be successfully completed
▲ Product specifications are incomplete
▲ High-dollar, highly uncertain procurements such as software development are involved

Specific Types of Compensation Arrangements

There are a number of specific types of compensation arrangements under each of the previous categories.

1. Fixed price compensation arrangements:
   ▼ Firm fixed price
   ▼ Fixed price with economic price adjustment
   ▼ Fixed price redetermination
2. Incentive arrangements:
   ▼ Fixed price incentive
   ▼ Cost plus incentive fee

3. Cost-reimbursement arrangements:
- ▼ Cost reimbursement
- ▼ Cost sharing
- ▼ Time and materials
- ▼ Cost plus fixed fee
- ▼ Cost plus award fee

The applicability, elements, structure, and final price computation for the various compensation arrangements are discussed in the following paragraphs.

Firm Fixed Price Contracts

The most preferred contract type, if appropriate for the procurement, is the firm fixed price contract. A firm fixed price (FFP) contract is an agreement to pay a specified price when the items (services) specified by the contract have been delivered (completed) and accepted. The contracting parties establish a firm price through either competitive bidding or negotiation. Since there is no adjustment in contract price after the work is completed and actual costs are known, the cost risk to the supplier can be high.

An FFP contract is appropriate in competitive bidding where the specifications are definite, there is little schedule risk, and competition has established the existence of a fair and reasonable price. An FFP contract also can be appropriate for negotiated procurements if a review reveals adequate specifications and if price and cost analysis establish the reasonableness of the price.

As previously stated, under an FFP contract there is no price adjustment due to the supplier's cost experience. Because the supplier has all cost responsibility, the actual outcome will show up in the form of profit or losses. Therefore, the supplier has maximum incentive to control costs under an FFP contract. If the supplier incurs expenses beyond the buyer's obligation, the seller must find the required funds elsewhere. Conversely, if the supplier reduces costs, all savings contribute to the supplier's profit. This dollar-for-dollar relationship between expenditures and profit is the greatest motivator of efficiency available. An FFP contract has only one contract compensation arrangement element: total price. Although negotiations may involve the discussion of costs and profit, the contractual document reflects only total price. This structure can be seen in Figure 17.3, which depicts an FFP contract for $20,000.

In this example, cost is shown as the independent variable (x-axis), and profit, since it is a function of cost, as the dependent variable (y-axis). The graph depicts the one-to-one relationship between costs and profit by showing that as costs increase by $1, profit decreases by $1.

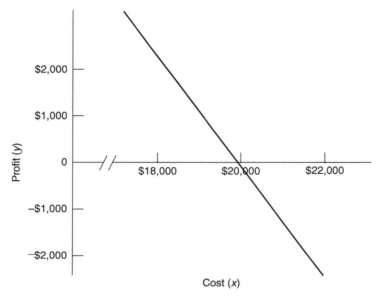

Figure 17.3   Firm fixed price contract—$20,000.

Computing the final price in an FFP-type contract is a simple matter. If a $20,000 FFP contract is awarded, on contract completion the supplier will receive $20,000 whether costs were $15,000 or $25,000 or any other amount. It should be noted that a fixed price does not always stay fixed. A supplier who is losing money *may* request and get some relief, if any of the following apply:

1. The customer in some way has contributed to the loss.
2. The customer badly needs the items and other suppliers are not willing to provide them at the established price.
3. The supplier has unique facilities and time is too short to do anything but to get the product at an increased cost from the initial supplier.
4. The customer's representatives do not employ sound supply management practices.

As discussed previously, early supplier involvement (ESI) with a decision to rely on one supplier during and after development results in many benefits based on the early matching of process and product. But ESI may result in cost overruns and higher costs if the supplier cannot perform at the fixed price due to unforeseen (usually technical) reasons. It should be recognized that when a supplier fails to perform under a fixed price contract and the buying firm is forced to turn the business into a cost plus type of contract, the supplier has damaged its chance for future business with the customer and other potential customers. The supply manager, on the other hand, can use the prospect of continued future business to keep the price well below what the supplier's leverage of the moment might suggest.

Variations of the FFP contract have been developed to meet special circumstances. One such variation is the FFP level-of-effort contract. This arrangement calls for a set number of labor hours to be expended over a period of time. The contract is considered complete when the hours are expended, although normally a report of findings is also required. The FFP level-of-effort contract is appropriately used when the specification is general in nature and when no specific end item (other than a report) is required. This arrangement is most frequently used for research and development efforts under $100,000 and for "get our foot in the door" consulting contracts.

Fixed Price with Economic Price Adjustment Contracts

Fixed price with economic price adjustment (FPEPA) contracts (sometimes called formula pricing contracts) are used to adjust for economic uncertainties, such as unstable labor or market conditions that would prevent the establishment of an FFP contract without a large contingency for possible cost increases or decreases in the unit cost of labor or materials. An FPEPA contract is simply an FFP contract that includes economic price adjustment clauses. Such provisions are common when purchasing items that contain precious metals and for construction services.

Economic price adjustments (EPA) or escalator/de-escalator clauses provide for both price increases and decreases to protect the buyer and supplier from the effects of economic changes. If such clauses were not used, suppliers would include contingency allowances in their bids or proposals to eliminate or reduce the risk of loss. With a fixed contingency allowance in the contract price, the supplier is hurt if the changes exceed its estimate, and the buyer will overpay if the input unit cost increases do not materialize.

An EPA clause may be used for fixed price–type arrangements resulting from both competitively bid and negotiated contracts. Price adjustments normally should be restricted to contingencies beyond the control of the supplier. Under an FPEPA contract, specific contingencies are left open subject to an EPA clause, and the final contract price is adjusted, depending on what happens to these contingencies. Where cost pass-through or escalator clauses cover specific materials or labor, the buyer should be sure that the price increase does not occur until the higher-cost material is used or until the labor contract increase takes effect.

The use of EPA clauses varies with the probability of significant price fluctuations. Their use also increases when purchasing strategy favors early supplier involvement, longer-term contracts, fewer supplies, and more single-source suppliers. An EPA clause should recognize the possibility of both inflation and deflation in determining price adjustments. Further, labor and material costs subject to economic adjustment must reflect the effects of learning on both labor and material costs. It takes considerable purchasing skill to use EPA clauses well. Decisions must be made on what items to include and which price/cost index or benchmark is best for each item.

The cost elements to adjust are high-value raw materials, specific high-value components, and direct labor. The professional buyer generally should oppose including costs within the supplier's control such as development, depreciation, fixed expenses, other overhead items, and profit in the base subject to escalation. In selecting indexes for price adjustment clauses, the following rules are suggested:

▲ Select from the appropriate Bureau of Labor Statistics category.
▲ Avoid broad indexes; use the lowest-level classification that includes the item.
▲ Develop a weighted index for materials in a product.
▲ Select labor rate indexes by type and location.
▲ Define energy indexes by fuel type and location.
▲ Analyze the past history of each proposed index vs. the actual price change of the item being indexed.

Using a broad index can produce strange results. One marketing executive used the producer price index (PPI) to adjust the purchase price of electronic apparatus, not recognizing that the PPI consists of about 40% food and fuel components with only 3% electronics input.

The details of the EPA clause must be thought through with various scenarios in mind. When will adjustments be made? Under what conditions can the contract be renegotiated? How will it be audited? By whom?

Fixed Price Redetermination Contracts

These contracts provide for an FFP for an initial contract period with a redetermination (upward or downward) provision at a stated time during contract performance [FPR (prospective)] or after contract completion [FPR (retroactive)]. The FPR (prospective) is usually used only in those circumstances calling for quantity production or services where a fair and reasonable price can be negotiated for initial periods but not for subsequent periods. The FPR (retroactive) is used in those circumstances where, at the time of negotiation, a fair and reasonable price cannot be established and the amount involved is so small and the performance period so short that use of any other contract type would be impractical.

The data shown in Table 17.3 are also applicable to an FPR contract. As was observed, the supplier is motivated to increase costs because, the higher the cost, the higher the fee!

Incentive Arrangements

FFP and cost plus fixed fee (CPFF) contracts are extremes of the range of contract compensation arrangements because, in either case, all of the cost responsibility falls on only one party. In between these two extremes are a number of contract arrangements where the cost responsibility is shared between the customer and the supplier. These are called incentive-type contracts.

Incentives are applied to contracts in an attempt to motivate the supplier to improve performance in cost, schedule, or other stated parameters. By far the most frequent application of incentives is in the area of cost control. However, this is not the only type of incentive. The specific type of incentive applied depends on the desired outcome. For example, if the primary interest is in developing a high-performance read head, it would be logical to reward the supplier for development and production of a read head that exceeds the minimum specifications. If the same read head were needed to meet a crash development effort, schedule may be the appropriate basis of an incentive. For the same read head, funds may be a real constraint due to budgetary limitations, and a production unit cost incentive would be appropriate. If a combination of performance and cost objectives were of concern, a multiple-incentive contract could be developed.

In this book, the discussion of incentive arrangements is limited to cost incentives. The focus will be on the two most frequently applied cost incentive compensation arrangements: the fixed price incentive (FPI) contract and the cost plus incentive fee (CPIF) contract. A general discussion of how a simplified incentive contract is structured precedes analysis of the specific elements and structure of these two compensation arrangements. The elements of a simplified incentive contract include (1) the target cost, (2) the target profit, and (3) the sharing arrangement.

Target Cost

The target cost for an incentive contract is that cost outcome which both the buyer and the supplier feel is the most likely outcome for the effort involved. The target cost should be based on costs that would result under "normal business conditions." Although the target cost is thought to be the most likely, it is recognized that the probability of the supplier's final costs being very close to the target cost is low. After all, if there were a high probability that the target cost would be close to the final cost, an FFP contract would be appropriate. The target should be that cost point where both parties agree that there is an equal chance of cost going above or below the target.

Target Profit

In addition to a target cost, a target profit is developed. The target profit in an incentive contract is a profit amount that is considered fair and reasonable, based on all relevant facts.

Allocating Costs Above or Below Target

Since an incentive contract recognizes that the target most likely will *not* be met, a method of allocating cost increases above or decreases below target is necessary. The method is a sharing arrangement that reflects the sharing of the cost responsibility between the buyer and the supplier. This arrangement should reflect the cost risk involved as evidenced by the magnitude of potential increases and decreases for the specific effort. In addition, the sharing arrangement must address two questions: "What percentage of the savings below target will be required to motivate the supplier to perform as efficiently as possible?" and "What percentage of the cost overrun—cost above target—charged to the supplier (in the form of lower profit) will cause the supplier to perform as efficiently as possible?"

How is the magnitude of a potential cost increase or decrease established? It is developed through an assessment of possible cost outcomes, based on varying circumstances a supplier might face during contract performance. In addition to developing a target cost and profit outcome, the parties establish cost outcomes and associated profits for a "best case" and a "worst case" situation. The best-case cost outcome is referred to as the most optimistic cost (MOC) point, and its related profit is referred to as the most optimistic profit (MOPr) point. The worst-case cost outcome is referred to as the most pessimistic cost (MPC) point, and its profit is referred to as the most pessimistic profit (MPPr) point.

The difference between the target point and the most optimistic point provides the supply manager with the magnitude of a potential cost decrease. The difference between the target point and the most pessimistic point provides the supply manager with the magnitude of a potential cost increase. One normally would not expect these magnitudes to be equal because the potential for things to go wrong is usually higher than the potential for things to go better than expected. Another way of looking at the magnitude of potential cost increase is that it provides an estimate of the cost risk a supplier faces if the target cost is not met. This cost risk and the supplier's assumption of this risk are reflected in the sharing arrangement.

Fixed Price Incentive Fee

Table 17.4 shows how a fixed price incentive fee contract is structured. The cost and profit outcomes shown in Table 17.4 were agreed on by the buying and selling firms. These data are portrayed in Figure 17.4.

Computing of the final payment under an incentive arrangement is more complex than under either an FFP or CPFF contract. Under an incentive arrangement, the supplier's profit will be adjusted to reflect performance in the cost area. If the supplier has incurred costs above target, the profit will be decreased by the supplier's share of the cost

Table 17.4 Fixed Price Incentive Fee

	Estimated Dollar (in thousands)	Price (in thousands)
Target cost (TC)	$1,000	
Target profit (TPr)	80	$1,080
Most optimistic cost (MOC)	800	
Most optimistic profit (MOPr)	120	920
Most pessimistic cost (MPC)	1,200	
Most pessimistic profit (MPPr)	0	1,200
Ceiling price	1,200	1,200

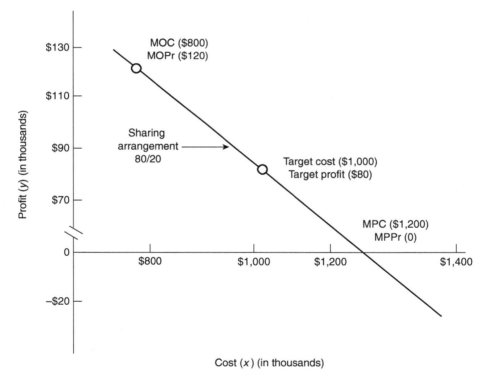

Figure 17.4 Fixed price incentive fee.

above target cost up to the ceiling price. Conversely, if the supplier's costs are below target, its profit is increased. The final price outcome would be the supplier's cost plus profit. The supplier's profit equals the sum of the target profit plus or minus the supplier's share of the cost savings or cost increase.

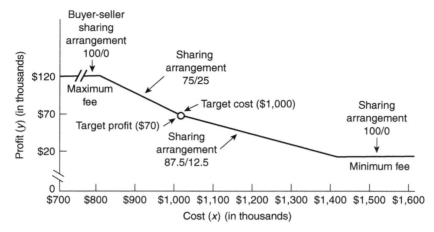

Figure 17.5 CPIF arrangement.

Cost Plus Incentive Fee Arrangements

CPIF contracts combine the incentive arrangement and the CPFF arrangement. Under a CPIF arrangement, an incentive applies over part of the range of cost outcomes. The fee structure resembles a CPFF contract at both the low-cost and high-cost ends of the range, as shown in Figure 17-5. This information is also depicted in Table 17.5. Thus, if cost were $800,000 or less, the fee would be $120,000. If cost were $1,400,000 or more, the fee would be $20,000.

A CPIF arrangement is used in those circumstances where the cost risk warrants a cost-type arrangement but where an incentive can be established to provide the supplier with positive motivation to manage costs. CPIF arrangements are most suitable for advanced

Table 17.5 Cost Plus Incentive Fees

Target cost	$1,000,000
Target profit	70,000
Optimistic cost	800,000
Optimistic and maximum profit	120,000
Pessimistic cost	1,400,000
Pessimistic and minimum profit	20,000
Sharing below target (customer/supplier)	75/25
Sharing above target (customer/supplier)	87.5/12.5

development efforts and for initial production runs. In these circumstances, risk may be too high to warrant use of a fixed price arrangement or an FPI arrangement, but not high enough to require a CPFF arrangement to get a reliable supplier.

The CPIF contract is structured in a manner very similar to the FPI compensation arrangement. Cost and fee outcomes are established for target, most optimistic, and most pessimistic points. These cost and fee outcomes are used to establish the sharing arrangement for cost decrease and increase situations. The difference between the structure of the FPI and CPIF arrangements is that under the CPIF arrangement, the contract converts to a CPFF contract at both the most optimistic and the most pessimistic fee points.

The computation of the final price to be paid to the supplier on contract completion follows the same steps as in the FPI arrangement. However, in a CPIF contract a comparison is made between computed fees and minimum and maximum fees prior to the calculation of the final price. For example, using the CPIF contract structured in Figure 17.5, the supply manager would compute the final price, based on a final cost of $700,000, as follows:

Target cost	$1,000,000
Target profit	70,000
Maximum fee	120,000
Minimum fee	20,000

1. Cost savings = target cost − final cost

$300,000 = $1,000,000 − $700,000

2. Supplier's share of cost savings = cost savings × supplier share

$75,000 = $300,000 × 0.25

3. Computed fee = savings fee + target fee

$145,000 = $75,000 + $70,000

Since there is a maximum limitation on fee, a comparison is made between the computed fee and the maximum fee. In this case, the supplier receives only the maximum fee, $120,000.

4. Final price = final cost + maximum fee

$820,000 = $700,000 + $120,000

The CPIF contract is an incentive arrangement that converts to a CPFF contract at both the maximum and minimum fee points. This type of contract provides the supplier some incentive to control cost outcomes in the area over which the sharing arrangements apply, called the range of incentive effectiveness.

Cost Plus Fixed Fee Arrangements

Under a CPFF contract, the buyer agrees to reimburse the supplier for all allowable, reasonable, and allocable costs that may be incurred during the performance of the contract. Moreover, the buyer agrees to pay the supplier a fixed number of dollars above the cost as the fee for doing the work. The fee changes only when the scope of work changes. Under the CPFF, the supplier has no incentive to reduce or control costs.

The contractual elements of this arrangement include an estimated cost and a fixed fee. The estimated cost represents the best estimate of the customer and the supplier for the work involved. The fixed fee is the amount of fee the supplier will receive regardless of cost outcome. Because the supplier has no cost risk under a CPFF contract, the profit potential is relatively low. Normally there is a limit on the customer's total liability.

The supply professional should remember that the final cost should be audited. Supply management departments spend hours negotiating the right to inspect the actual invoices for material and the hours worked. Many, however, do not conduct an appropriate audit. It is a good use of time to look at the details even though one might not expect to find any inappropriate charges. The knowledge gained will often prove helpful in future negotiations.

Computing the final payment due the supplier under a CPFF contract is simply a matter of adding the incurred costs (assuming that an audit has found them to be reasonable, allowable, and allocable) to the fixed fee. In the case of a CPFF contract with an estimated cost of $1,000,000 and a fixed fee of $50,000, some possible final contract price outcomes are shown in Table 17.6.

The supply manager must remember that in a cost-type contract, the limit is on the fee, not on total customer obligation. Obviously, the CPFF-type contract should be used only when one cannot get a more favorable arrangement or when the presence of great uncertainty and risk would result in inclusion of a large contingency in an FFPcontract. The CPFF contract also is appropriate in circumstances where the technical and schedule risks are so high that the cost risk is too large for the supplier to assume. This type of contract is designed chiefly for use in research or exploratory development when the

Table 17.6 Cost Plus Fixed Fee: Sample Price Outcomes

	Possible Outcomes (in thousands)			
Final cost	$800	$900	$1,000	$1,200
Fixed fee	50	50	50	50
Price to be paid to supplier	$850	$950	$1,050	$1,250

uncertainty of performance is so great that a firm price or an incentive arrangement cannot be set up at any time during the life of the contract. Costs normally are audited by the buyer before final payment.

Cost Plus Award Fee (CPAF)

The cost plus award fee (CPAF) was pioneered by NASA when the agency was purchasing highly complex hardware and professional services in support of the space program. The award fee is very applicable to the procurement of software developed for the buying company, and for janitorial, landscaping, and similar services where the ability to reward the supplier for non-quantitative aspects of its performance on a subjective basis makes good business sense. The award fee is a pool of money established by the buyer to reward the supplier on a periodic basis for the *application of effort in meeting the buyer's stated needs*. The key difference between the award fee and other fees is that the supplier's receipt of the fee is based on the buying firm's *subjective evaluation* of how well the supplier applies its *efforts* in meeting the buyer's needs. The subjective aspect provides flexibility to contracting situations in an uncertain environment. When properly used, the award fee benefits both the buyer and the supplier. Superior performance receives superior rewards in the form of a superior fee. The award fee also introduces an element of flexibility because the buyer can change the areas receiving supplier attention by providing advance guidance for any performance period. The award fee gives the buying firm's management a flexible tool with which to influence performance.

Cost without Fee

Nonprofit institutions, such as universities, frequently do research work for both government and industry, without the objective of making a profit. Such research is done under cost-type contracts without a fee. Because universities do much of the nation's pure research, as distinguished from applied research done by industry, a growing number of contracts of this type are being used. Naturally, the universities recover all overhead costs, which generally include facilities costs and remuneration for personnel who work on the contracts. In recent years, high-technology firms have increased their use of this contract type.

Cost Sharing

In some situations, a firm doing research under a cost type of contract stands to benefit if the product developed can be used in its own product line. Under such circumstances, the buyer and the seller agree on what they consider to be a fair basis to share the costs (most often it is 50-50). The electronics industry has found this type of contract especially useful.

Time and Materials

In certain types of contracts, such as those calling for repairs to machinery, the precise work to be done cannot be predicted in advance. For instance, it cannot be known exactly what must be done to a large malfunctioning pump aboard a ship until it is opened and examined. Perhaps only a new gasket is required to put it in good working order. On the other hand, its impeller could require a major job of balancing and realignment. The time and materials contract is one method of pricing this type of work. Under this type of contract, the parties agree on a fixed rate per labor hour that includes overhead and profit, with materials supplied at cost.

Suppose a mechanic working on the ship's pump is paid $30 per hour. Assume also that overhead is calculated as 100% of the labor cost and profit is set at 10% of total cost. A billing rate for this mechanic for one hour would be calculated as shown in Table 17.7.

If it took the mechanic two days (16 hours) to repair the pump, using $320 worth of material, the job price would be $1,376 (16 × $66 + $320).[1] Profit should be paid only on labor and overhead costs. Note that the supplier is motivated to increase the number of hours for two reasons: (1) The $30 of overhead commonly is a fixed cost. Thus, every additional hour contributes $30 to an overhead which probably has been amortized. In effect, this $30 becomes a gift of $30 additional profit. (2) The stated profit of $6 increases with each additional half hour.

A variation of the time and materials type of contract is called a labor-hour contract. In this type of contract, materials are not supplied by the seller; however, other costs are agreed to as in time and materials contracts.

Table 17.7   Time and Materials: Billing Rate

Direct labor cost, per hour	$30
Overhead at 100% of labor	30
Total cost	$60
Profit at 10% of total cost	6
Billing rate, per hour	$66

Letter Contracts and Letters of Intent

Letter contracts are used in those rare situations in which it is imperative that work start on a complex project immediately. Letter contracts are *preliminary contractual authorizations* under which the seller can commence work immediately. The seller can prepare drawings, obtain required materials, and start actual production. Under letter contracts, the seller is guaranteed reimbursement for costs up to a specified amount. Letter contracts should be converted to definitive contracts at the earliest possible date.[2]

Considerations When Selecting the Method of Compensation

With such a wide variety of compensation methods available, the supply professional must exercise considerable care in selecting the best one for a particular use. If a bid or a quoted price is reasonable, this will help the supply manager decide to use an FFP contract. On the other hand, if the fairness of the price is in doubt, a fixed price contract could entail excessive expense to the buying firm and would be a poor choice. If price uncertainty stems from unstable labor or market conditions, fixed price with escalation may be a solution. If the uncertainty is due to a potential improvement in production effort, an incentive contract may be the best answer. Plainly, the many factors that affect procurement costs can themselves guide the supply professional in his or her selection of the best type of contract for a given purchasing situation.

The specific nature of the materials, equipment, or services to be purchased also can frequently point up advantages of one contract type over another. The more complex or developmental the purchased item, the greater the risks and difficulties in using a fixed price contract. Any uncertainty in design affects a seller's ability to estimate costs, as does a lack of cost experience with a new item. The details of any given purchase will themselves indicate the magnitude of the price uncertainties involved. A full understanding of these uncertainties permits supply managers to allocate the risks more equitably between their firm and the supplier's firm through the proper choice of compensation method. Timing of the procurement quite frequently is a controlling factor in selection of the compensation method. Allowing potential suppliers only a short time to prepare their bids can reduce the reliability of the cost estimates and increase prices. A short delivery period usually rules out the effective use of incentive contracts. On the other hand, a long contract period allows time to generate and apply cost-reducing efficiencies, an ideal situation for an incentive contract. The facts of each procurement must be considered individually in determining which compensation method the supply manager should use.

Business practices in specific industries frequently can provide additional clues as to the best choice of compensation method. The construction industry, for example, traditionally accepts a wider range of competitive fixed price jobs than does the aerospace industry. The lumber industry accepts prices established by open auctions. Architects and engineers frequently will not enter into price competition with one another for architectural or engineering services; neither will some management consulting firms compete with one another on price. Business factors such as these help determine the best compensation method in a great many purchasing situations.

The scope and intensity of competition definitely can influence the method of compensation to use. If competition is intense and the prices bid are close, then the buyer can justifiably feel that the prices are fair and reasonable and use an FFP contract. On the other hand, if competition is not adequate and the supply manager has doubts concerning the reasonableness of prices, an incentive or cost contract may be appropriate.

In short, *the supply professional's basic preference for an FFP contract is just the starting point for an analysis of alternative compensation choices.* As the supply manager considers all available compensation methods, he or she must weigh the preference for fixed prices against the risks involved, the time available, the degree of competition involved, experience with the industry involved, the apparent soundness of the offered price, the technical and developmental state of the item being purchased, and all the other technical and economic information that affects the purchase transaction. Determination of the best compensation agreement for a given situation requires a careful analysis of all the factors relevant to that situation.

Summary

A supply manager can, in most instances, enter into an FFP contract even when significant cost risk is present. But if risk is high, either of two equally unsatisfactory scenarios frequently results: (1) The contract price will include a large contingency or (2) the supplier could incur a loss. The possibility of a loss may result in (1) reduced quality in an effort to minimize the loss, (2) a request to renegotiate, (3) refusal to complete the work, (4) insolvency, resulting in the loss of a good supplier, or (5) a "grin and bear it" approach.

The selection of the contract compensation arrangement to be used for a specific contract is an important determination. The selection must be based on the uncertainties and risks involved and the circumstances surrounding the procurement. The compensation agreement selected must result in a reasonable allocation of the cost risk and should provide adequate motivation to the supplier to assure effective performance. In addition, the compensation arrangement selected must be compatible with the supplier's accounting system. Sound application of these compensation methods will significantly reduce expenditures when cost risk is present.

Endnotes

1. The alert reader will observe that this is really a cost plus percentage of cost contract. If the mechanic is a good worker, he might complete the job in 10 hours ($60 profit to his employer). If he is a poor worker, he could take 20 hours, and his employer would receive $120 in profit. Obviously, supply managers must exercise close control over this type of contract to be sure that inefficient or wasteful methods are not used.

2. For more on this topic, see William A. Hancock, "Using Letters of Intent to Provide a Framework for Relationships with Suppliers," *The Purchaser's Legal Adviser* (Chesterland, OH: Business Laws, 1994) (800-759-0920). Also see Judi Coover, "Letter of Intent," *Inside Supply Management* November (2006): 32–33.

Suggested Reading

Michels, William L. and George E. Cantrell. "An Ongoing Plan for Volatile-Priced Commodities." *Inside Supply Management* August (2004): 22.

Ore, Norbert J. "Making the Most of the ROB." *Inside Supply Management* May (2002): 36.

Whittington, Elaine M. "Types of Contracts." Abstract from a presentation at the 79th Annual ISM International Conference Proceedings, Atlanta, GA, 1994.

Web Sites

www.bls.gov/cpi/—Consumer Price Index

www.bls.gov/ppi/—Producer Price Index

http://www.bls.gov/ppi/ppifaq.htm—Information on how to use the producer price index for contract escalation

CHAPTER 18

Negotiations

Introduction

Negotiation is one of the most important as well as one of the most interesting and challenging aspects of supply management. In industry, and at most levels of government, the term "negotiation" frequently causes misunderstandings. In industry, negotiation is sometimes confused with "haggling" and "price chiseling." In government, negotiation is frequently visualized as a nefarious means of avoiding competitive bidding and of awarding large contracts surreptitiously to favored suppliers.

Webster's dictionary defines negotiation broadly as "conferring, discussing, or bargaining to reach agreement in business transactions."[1] Herb Cohen describes negotiation as a pervasive process in which people ultimately attempt to reach a joint decision on matters of common concern in situations in which there is initial disagreement. Thus, a negotiation always requires both shared interests and issues of conflict. Obviously, without commonality there is no reason to achieve resolution.[2] To be fully effective in purchasing, negotiation must be utilized in its broadest context—as a part of a decision-making process. In this context, negotiation is a process of planning, reviewing, and analyzing used by a buyer and a seller to reach acceptable agreements or compromises. These agreements and compromises include all aspects of the business agreement, not just price.

Negotiations differ from a ball game or a war. In those activities, only one side can win; the other side must lose. In most successful business negotiations, both sides win something. Popular usage calls this approach "win-win negotiation." The "winnings," however, are seldom equally divided; invariably, one side wins more than the other. This is as it should be in business—superior business skills merit superior rewards.

Increasingly, negotiations are conducted by cross-functional teams. These teams must be well coordinated in order to function as an integrated entity. Accordingly, we will use the term *negotiator* to refer to an individual supply management professional and also to a cross-functional negotiating team.

Truly professional negotiators are constantly on the search for opportunities to "enlarge the pie." Instead of focusing on how to "share a fixed pie," they focus first on enlarging the pie.

Objectives of Negotiation

Several objectives are common to all procurement/sales negotiations:

- ▲ To obtain the quality specified
- ▲ To obtain a fair and reasonable price
- ▲ To get the supplier to perform the contract on time

In addition, the following objectives frequently must be met:

- ▲ To exert some control over the manner in which the contract is performed
- ▲ To persuade the supplier to give maximum cooperation to the purchasing company
- ▲ To develop a sound and continuing relationship with competent suppliers
- ▲ To create a long-term relationship with a highly qualified supplier

Quality

In most cases, the negotiator's objectives require obtaining the quality specified by design engineering or by the user group. In some cases, however, quality itself may be a variable. For example, assume that the cost per unit for a critical item of material, with a guarantee of no more than 1000 defective parts per million incoming parts, is $5. Further, assume that the cost per unit with a guarantee of no more than 100 defective parts per million is $10. Does the higher unit price result in higher or lower *total* costs? Quite obviously, a highly advanced and accurate management information system must be available to provide the necessary data to allow the negotiator to make an optimal decision.

Fair and Reasonable Price

In many cases, the establishment of a fair and reasonable price for the desired level of quality becomes the principal focus of the negotiation process. This aspect of negotiation ranges in complexity from the use of price analysis to the more complex analysis of the potential supplier's cost elements. While the negotiator may focus on obtaining a fair price, this must be done within the context of obtaining the lowest *total* cost, as previously discussed.

On-Time Performance

Inability to meet delivery schedules for the quality and quantity specified is the single greatest supplier failure encountered in supply management operations. This results primarily from (1) failure of requisitioners to submit their purchase requests early enough to allow for necessary supply and manufacturing lead times and (2) failure of supply management personnel or the negotiating team to plan the delivery phase of negotiations properly. Because unrealistic delivery schedules reduce competition, increase prices, and jeopardize quality, it is important that supply management personnel or the negotiating team obtain realistic delivery schedules that suppliers can meet without endangering the other requirements of the purchase.

Control

Deficiencies in supplier performance can seriously affect, and in some cases completely disrupt, the operations of the buying firm. For this reason, on important contracts negotiators should obtain controls that will assure compliance with the quality, quantity, delivery, and service terms of the contract. Traditionally, controls have been found to be useful in areas such as labor-hours of effort, levels of scientific talent, special test equipment requirements, the amounts and types of work to be subcontracted, and progress reports.

Cooperation

Cooperation is best obtained by rewarding those suppliers who perform well with future orders. In addition to subsequent orders, however, good suppliers also expect courtesy, pleasant working relations, timely payment, and cooperation from their customers—cooperation begets cooperation.

Supplier Relationship Management

Negotiators should recognize that current actions usually constitute only a part of a continuing relationship. Conditions that permit the negotiator to take unfair advantage of sellers invariably, with time, change to conditions that allow sellers to "hold up" the purchasing firm. For this reason, a negotiator must realize that any advantage *not honestly won* will, in all likelihood, be recovered by the supplier at a later date—probably with interest. Thus, as a matter of self-interest, negotiators must maintain a proper balance between their concern for a supplier's immediate performance on the one hand and their interest in the supplier's long-run performance on the other. This balancing is an important aspect of supplier relationship management.

In summary, the objectives of negotiation require investigation, with the supplier, of every area of negotiable concern—considering both short-term and, normally, long-term performance. Negotiation is used not only to reach an agreement on price but is a process employed throughout the procurement cycle. The negotiator's major analytical tools for negotiating prices were discussed in Chapter 16—such things as price and cost analysis, and learning curves. Additional negotiating tools, as well as the development of strategy and tactics for negotiation, are discussed throughout this chapter.

When to Negotiate

Negotiation is the appropriate method for determining the reasonableness of a price when competitive bidding or reverse auctions are impractical. Some of the most common circumstances dictating the use of negotiation are noted as follows:

▲ When any of the five prerequisite criteria for competitive bidding are absent. The criteria are discussed in Chapter 13.

▲ When many variable factors bear not only on price but also on quality and service. Many high-dollar-value industrial and governmental contracts fall into this category.

▲ When early supplier involvement (as described in Chapter 5) is employed.

▲ When the business risks and costs involved cannot be accurately predetermined. When supply management seeks competitive bids under such circumstances, excessively high prices inevitably result. For self-protection, most suppliers factor every conceivable contingency into their bids. In practice, many of these contingencies do not occur. Hence, the customer firm unnecessarily pays for something not received.

▲ When a customer firm is contracting for a portion of the seller's production capacity, rather than for a product the seller has designed and manufactured. In such cases, the customer firm has designed the product to be manufactured and, as an entrepreneur, assumes all risks concerning the product's specifications and salability. In buying production capacity, the objective is not only to attain production capability but also to acquire such control over it as may be needed to improve the product and the production process. This type of control can be achieved only by negotiation and the voluntary cooperation of the supplier.

▲ When tooling and setup costs represent a large percentage of the supplier's total costs. For many contracts, the supplier must either make or buy many costly jigs, dies, fixtures, molds, special test equipment, gauges, and so on. Because of their special nature, these jigs and fixtures are primarily limited in use to the customer firm's contract. The division of special tooling costs between such firms and a customer is subject to negotiation. This negotiation includes a thorough analysis of future buyer or seller use of the tools, the length and dollar amount of the contract, the type of compensation appropriate, and so on.

▲ When a long period of time is required to produce the items purchased. Under these circumstances, suitable economic price adjustment clauses must be negotiated. Also, opportunities for various improvements may develop; for example, new manufacturing methods, new packaging possibilities, substitute materials, new plant layouts, and new tools. Negotiation permits an examination and evaluation of all these potential improvements. Competitive bidding does not. What supplier, for example, would modify its plant layout to achieve increased efficiency to produce the buying firm's unique material without assurance of sufficient long-term business to cover the cost involved and assurance of a reasonable profit for the effort?

▲ When production is interrupted frequently because of numerous change orders. This is a common situation in fields of fast-changing technology. Contracts in these fields must provide for frequent change orders; otherwise the product being purchased could become obsolete before completion of production. The ways in which expensive changes in drawings, designs, and specifications are to be handled and paid for are subjects for mutual agreement arrived at through negotiation.

▲ When a thorough analysis is required to solve a difficult make-or-buy decision. Precisely what a seller is going to make and what it is going to subcontract should be decided by negotiation. When free to make its own decision, the seller often makes the easiest decision in terms of production scheduling. This may well be the most costly decision for the customer in terms of price.

▲ When the products of a specific supplier are desired to the exclusion of others. This can be either a single or a sole sourcing situation. In this case, competition is minimal or totally lacking. Terms and prices, therefore, must be negotiated to minimize unreasonable dictation of such terms by the seller.

In all of these situations, negotiation is essential; and, in each case, quality and service are as important as price.

Supply Management's Role in Negotiation

Depending on the type of purchase, a supply management professional plays one of two distinct roles in negotiation. In the first role, he or she is the company's sole negotiator. In the second, the supply professional leads a cross-functional team of specialists that collectively negotiates on behalf of its company.

The Supply Management Professional Acting Alone

For low-dollar-value, non-critical items, the supply management professional normally acts alone. Typically, for this type of purchase, a negotiation conference is held in the supply management professional's office with the supplier's sales manager (the seller's sole negotiator). These two persons alone negotiate all the important terms and conditions of the contract.

A supply manager's "solo negotiation" is not limited to periodic formal negotiating sessions. Rather, such negotiations continue on a daily basis with both current suppliers and visiting salespersons wishing to become suppliers. Consider several typical examples. A supplier calls on the telephone and informs the supply manager that prices are to be raised 20% within 60 days. The supply manager responds with the thought that production in her company's plant is slack and that a price rise as high as 20% could well trigger a "make" decision, in lieu of what is now a "buy" decision. The supply manager is negotiating!

A seller's value analyst discovers a substantially less expensive method of manufacturing one of the purchasing firm's products. However, there is one drawback: an expensive new machine is required for the job. The supplier's sales representative informs the supply manager of the discovery. The supply manager and an engineer study the concept and determine it is a good one. The two thank the sales representative for introducing the new idea. At the same time, the supply manager explains that his or her company is financially unable, at this particular time, to invest in the required new machine. Further, the supply manager conjectures that if the seller's company were to purchase the machine, then the supply manager could get his company to reconsider the rejected long-term contract the seller proposed last year. Informal negotiations are being conducted.

A salesperson calls, and the supply manager says, "I have been thinking about your contract with us. Under the contract, our purchases now total roughly $60,000 per year, primarily for valves. But, your company also manufactures a number of fittings that we use. If these fittings were combined with the purchase of the valves, what benefits would your company be able to grant us?" Another informal negotiation is under way.

The preceding year a supply manager purchased $300,000 worth of liquid oxygen in individual cylinders from a single supplier. Because of its high dollar value, the supply manager began to analyze oxygen usage requirements thoroughly. In this analysis, an interesting fact was discovered. If a bulk storage tank was installed at the purchasing firm's plant (at a cost of $160,000) and the supplier's personnel delivered the required liquid oxygen to the tanks, $70,000 could be saved per year. When the supplier's salesperson called, she was informed of the supply manager's study and was given the supply manager's worksheets for her review and study. Negotiations were under way, and the total cost of liquid oxygen would soon be reduced.

The Supply Management Professional as the Negotiating Team Leader

The complexity of a purchasing contract frequently correlates directly with the complexity of the item being purchased. For high-value, technically oriented contracts (such as those developed for the purchase of high-technology products, capital equipment, and research and development projects), and for the development of long-term relationships where the supplier's production flows into the buying firm's operation, the supply manager typically is no longer qualified to act as a sole negotiator. His or her role, therefore, shifts from that of sole negotiator to that of negotiating team leader. A typical team consists of two to eight members, depending on the complexity and importance of the purchase to be made. Team members are selected for their expertise in the technical or business fields needed to optimize their team's negotiating strength. Customarily, members are from fields such as design engineering, manufacturing engineering, cost analysis, estimating, finance, production, traffic, supply management, and legal affairs. Frequently, such team members have limited knowledge and skill in *professional* negotiations. Perhaps the largest challenge faced by the negotiation team leader is the belief by many of his or her teammates that they "know" how to negotiate. The truth of the matter is that most of us have had experience "haggling," but not professional negotiation. Thus, the team leader must coach his or her teammates in the use of professional practices.

In the team approach to negotiation, the supply manager frequently serves as the leader of the team (and is called the negotiator). In this capacity, he or she functions as the coordinator of a heterogeneous group of specialists from several functional areas who can be expected to view similar matters differently. As leader of the team, the negotiator must meld the team members into an integrated whole. The team leader must draw on the specialized knowledge of each team member and combine this expertise with his or her own. To accomplish this, it is very important that an overall strategy be developed by the team and that each team member is assigned a specific role. Additionally, mock negotiations should be included as one of the final steps in team preparation. Mock negotiations normally constitute the best possible insurance to preclude the team's committing the most costly error of negotiation—that of the members speaking out of turn and thus revealing their firm's position to the seller's team. In this way, the negotiating team develops a sound, unified approach to uncover, analyze, and resolve (from a company-wide point of view) all the important issues applicable to the contract under negotiation.

The Negotiation Process

In the broadest sense, negotiation begins with the origin of a firm's requirements for specific materials or services. The ultimate in purchasing value is possible only if design, production or operations, supply management, and marketing are able to reconcile their differing views with respect to specifications or the Statement of Work (SOW). The more open the specifications, the greater the leverage of the negotiating team. Negotiators must always think in terms of total cost and total value, not in terms of price alone.

The prelude to a negotiation normally begins with supply management's request for proposals from potential suppliers. The formal negotiation process consists of three major phases: (1) preparation, (2) face-to-face discussions, which result in agreement on all items and conditions of a contract or a decision not to enter into an agreement with the potential supplier, and (3) the debriefing wherein the negotiating team members review both the preparation and face-to-face discussions for lessons learned.

Preparation

Ninety percent or more of the time involved in a successful negotiation is invested in preparation for the actual face-to-face discussions. The negotiator must (1) possess or gain a technical understanding of the item or service to be purchased, (2) analyze the relative bargaining positions of both parties, (3) have conducted a price or cost analysis (as appropriate), (4) know the seller, (5) be aware of cultural nuances, and (6) be thoroughly prepared.

Know the Item or Service

The negotiator does not need to understand all the technical ramifications of the item being purchased. But it is essential that he or she has a general understanding of what is being purchased, the production or services process involved, and any other issues that will affect quality, timeliness of performance, and cost of production. The negotiator should understand the item's intended use, any limitations, and the existence of potential substitutes. The buyer should be aware of any prospective engineering problems that may arise. The negotiator should be aware of the item's procurement history and likely future requirements. Ideally, the negotiator will be familiar with any phraseology or customs relevant to the industry. A similar level of knowledge is appropriate when purchasing equipment and services.

The Seller's Bargaining Strength

The seller's bargaining strength usually depends on three basic factors: (1) how badly the seller wants the contract, (2) how certain he or she feels of getting it, and (3) how much time is available to reach agreement on suitable terms.

The negotiator should encounter no difficulty in determining how urgently a seller wants a contract. The frequency with which the salesperson calls and general market conditions are positive indicators of seller interest. The seller's annual profit and loss statements, as well as miscellaneous reports concerning backlog, volume of operations, and trends, are valuable sources of information about individual sellers. Publications such as the Department of Commerce's "Economic Indicators," *The Federal Reserve Bulletin*, industrial trade papers, the Institute for Supply Management™ (formerly NAPM) *Report on Business*, and local newspapers provide a wealth of basic information about potential suppliers and their industries in general.

The less a seller needs or wants a contract, the more powerful its bargaining position becomes. The presence of an industry boom, for example, places it in a strong position. On the other hand, when a seller finds itself in a general recession or in an industry plagued with excess capacity, its bargaining position is decidedly weakened.

If a seller learns that its prices are lower than the competition's or learns from engineering, production, or services personnel that it is a preferred or sole source of supply, it naturally concludes that its chances of getting the contract are next to certain. In these circumstances, a supplier may become very difficult to deal with during negotiations. In extreme situations, it may be unwilling to make any concessions whatsoever. When this happens, the negotiator sometimes has only one alternative—to accept the supplier's terms.

When trapped by such circumstances, a negotiator can threaten a delay to search for other sources. Such threats are likely to be ineffective, unless the seller knows that alternative sources are actually available and interested in the business. An alternative source of possible power that may be effective when patents are not involved is the threat to manufacture the needed item in the buyer's plant. When made realistically—when the supplier believes the buyer has the technical capability, the determination, and the capacity to make the product or service—such a threat frequently gains concessions.

A buying firm's negotiating position is always strengthened when the company has a clear policy that permits only members of its supply management department to discuss pricing, timing, and other commercial terms with sellers. Most pre-negotiation information leaks that give sellers a feeling of confidence about getting a contract occur in the technical departments of a firm. Such leaks can be extremely costly, and because they are often undetected by general management, they can be a continuing source of unnecessarily high costs.

Short lead times drastically reduce one's negotiating strength. Conversely, they significantly increase the seller's bargaining strength. Once a supplier knows that a negotiator has a tight deadline, it becomes easy for the supplier to drag its feet and then negotiate favorable terms at the last minute when the negotiator is under severe pressure to consummate the contract.

The Buyer's Bargaining Strength

The buyer's bargaining strength usually depends on four basic factors: (1) the extent of competition present among potential suppliers, (2) the adequacy of cost or price analysis, (3) the logic and reasoning behind the challenged cost, and (4) the thoroughness with which the buyer and all other members of the buying team have prepared for the negotiation.

Intense supplier competition always strengthens a buyer's position. Competition is always keenest when a number of competent sellers eagerly want an order. General economic conditions can bear heavily on the extent to which a firm really wants to compete. A firm's shop load, its inventory position, and its back-order position, or demand for its services are typical factors that also bear heavily on the ever-changing competitive climate.

When necessary, a supply manager can increase competition by developing new suppliers; making items in-house rather than buying them; buying suppliers' companies; providing tools, money, and management to competent but financially weak suppliers; and, above all, hiring highly skilled supply managers.

The Adequacy of Cost or Price Analysis

A comprehensive knowledge of cost analysis and price analysis is one of the basic responsibilities of all supply managers involved in negotiations. When an initial contract is awarded for a portion of a supplier's production capacity rather than for a finished product, cost analysis becomes vital. In this situation, the negotiators are not prepared to explore with the supplier the reasonableness of its proposals until after a comprehensive analysis of all applicable costs has been completed. Cost analysis in such purchases, in a very real sense, is a substitute for direct competition. Price analysis is usually sufficient to assure that prices are reasonable for contracts for common commercial items. In the aggregate, the greater the amount of available cost, price, and financial data, the greater the chances for a successful negotiation.

Caution must be exercised not to over-focus on price. If a collaborative, long-term relationship is desired, price must be placed in the proper context: it is only one aspect of the negotiation!

Know the Seller

Professional negotiators should endeavor to know and understand both the prospective supplier firm and its representatives. World-class supply managers prepare for critical negotiations by reviewing financial data and articles dealing with prospective suppliers. These supply managers and all negotiating team members know how the supplier's business is faring, key personnel changes, and so on. One of the keys to successful negotiations is to put oneself in the other person's shoes. Understand his or her wants and needs. This level of preparation pays dividends when conducting face-to-face negotiations.

Cultural Nuances

A clear understanding of the effects and nuances of cultural similarities and differences is an essential skill for the contemporary business manager. Effective global executives are those with the ability to: develop and use global strategic skills; manage change and transition; manage cultural diversity and function, within flexible organizational structures; work with others and in teams; communicate; and learn and transfer knowledge in an organization.[3]

The Thoroughness of Preparation

Knowledge is power. The more knowledge the negotiator acquires about the theory and practice of negotiation, the seller's negotiating position, and the product or service being purchased, the stronger his or her own negotiating stance will be. A negotiator without a thorough knowledge of the product being purchased is greatly handicapped. A negotiator is similarly handicapped if he or she has not studied and analyzed every detail of the supplier's proposal. Whenever feasible, before requesting proposals, the negotiator should develop an estimate of the price and value levels for the items or services being purchased. Knowledge of current economic conditions in the market for the product or service in question is also an essential element of preparation.

Prior to the face-to-face negotiating session, all members of the negotiating team must evaluate all relevant data and carefully assess their own and their supplier's strengths and weaknesses. From this assessment, they develop not only a basic strategy of operation, but also specific negotiating tactics. Alert suppliers readily recognize negotiators who are not prepared. They gladly accept the real and psychological bargaining advantage that comes to them from lack of preparation by members of the buying firm's negotiating team.

Establishing Objectives

The outcome of contract negotiations hinges on relative buyer-seller power, information, negotiating skills, and how both perceive the logic of the impending negotiations. Each of these controlling factors can be influenced by adroit advanced planning. This is why proper planning and preparation is, without question, the most important step in successful negotiations.

As part of the preparation process, the negotiating team should establish objectives. Negotiation objectives must be specific. General objectives such as "lower than previous prices," "good delivery," or "satisfactory technical assistance" are inadequate. For each term and condition to be negotiated, the negotiating team should develop three specific positions: (1) an objective position (or target), (2) a minimum position, and (3) a maximum position. Using the cost objective as an example, the minimum position is developed on the premise that every required seller action will turn out satisfactorily and with minimum cost. The maximum position is developed on the premise that a large number of required seller actions will turn out unsatisfactorily and with maximum cost. The objective position is the best estimate of what they expect the seller's actual costs plus a fair profit should be.

In developing concrete objectives, desired or required dates are established for delivery schedules, desired or required numerical ranges for quality acceptance, and dollar levels for applicable elements of cost. The major elements of cost that traditionally are negotiated—and for which objective, maximum, and minimum positions are developed—include quantity of labor, wage rates, quantity of materials, prices of materials, factory overhead, engineering expense, tooling expense, general and administrative expense, and profit. In addition to determining a position for each major element of cost, the supply manager and the negotiating team members must estimate the objective, maximum, and minimum positions of the seller. Determining the seller's maximum position is easy; it is the offer made in the seller's proposal.

In addition to costs and prices, delivery schedules and acceptable quality levels, specific objectives should be established for all items to be discussed during the negotiation, including:

- All technical aspects of the purchase
- Types of materials and substitutes
- Buyer-furnished material and equipment
- Mode of transportation
- Warranty terms and conditions
- Payment terms (including discount provisions)
- Liability for claims and damage
- F.O.B. point
- General terms and conditions
- Details on how a service is to be performed

Other objectives may include:

- Progress reports
- Production control plans
- Escalation/de-escalation provisions
- Incentive arrangements
- Patents and infringement protection
- Packaging
- Title to special tools and equipment
- Disposition of damaged goods and off-spec (non-conforming) materials

Simulations and role playing negotiation exercises conducted at the University of San Diego during a 10-year period demonstrate that negotiators who establish a demanding objective, *which is within the realm of reasonableness,* normally achieve a more favorable outcome than do those who enter the negotiation with a less demanding objective.

Identify the Desired Type of Relationship

We find it extremely desirable to determine the type of relationship that we hope to establish and maintain during and after the face-to-face discussions. The three primary approaches are transactional, collaborative, and alliance. These relationships were described in detail in Chapter 4. As we shall see shortly, the type of relationship desired affects the tactics employed.

Five Powerful Preparation Activities

The BATNA

Perhaps the most important aspect of preparation is the development of the firm's BATNA, the acronym for the "best alternative to a negotiated agreement," a term coined years ago by Roger Fisher and Bill Ury in their book *Getting to Yes*. The firm's BATNA describes what it would do if the negotiation were unsuccessful. The BATNA may be an alternative supplier (at a most likely price), a decision to "make," or incorporation of a substitute material (at a most likely total cost of ownership).

Insight into or an accurate estimate of the supplier's BATNA on each issue is of equal importance. A Fortune 50 invests considerable resources (including the use of consulting services) developing its suppliers' BATNAs on key procurements.

The Agenda

Successful negotiators spend considerable time and effort developing their agendas prior to face-to-face discussions. They place major emphasis on the sequence in which they plan to address issues during all phases of the face-to-face discussions. Experience indicates that easier to agree on issues should be addressed early. Seasoned professionals sprinkle "throw aways" throughout their agendas. But they know that they never "give" something for "nothing"; that is, they expect something in return!

All negotiations center on specific issues. One of the difficult tasks of negotiation is to define fully the important issues that are to be included on the agenda and then to be sure that the discussion is confined to these issues. Most authorities believe that the issues should be discussed in the order of their probable ease of solution. With this priority system, an atmosphere of cooperation and momentum can develop that may facilitate solving the more difficult issues.

"Murder Boards" and Mock Negotiations

Experienced negotiators frequently finalize their preparation through the use of "murder boards" and mock negotiations. A murder board consists of senior supply management, finance, manufacturing, quality, engineering, operations, and general management personnel. The negotiating team presents its agenda, objectives, and tactics for the forthcoming negotiations. Members of the murder board dissect the negotiating plan in an effort to identify avoidable problems.

Mock negotiations allow the members of a negotiating team to prepare for the negotiation through a simulation of what is likely to occur during the face-to-face discussions. Other members of the organization (preferably from general management) play the roles of the supplier's negotiating team members during the simulated negotiation. Suppliers generally do a good job of preparing for critical negotiations through the use of murder boards or mock negotiations. It is essential that the buying team be equally prepared.

Murder boards and mock negotiations enhance the negotiating team's level of preparation. Further, the processes result in general management's being aware of the negotiating team's agenda, objectives, and tactics. Should the subsequent negotiations deadlock, management is in a position to step in and revitalize critical negotiations.

Crib Sheets

Experience gained through 10 years of directing negotiation courses demonstrates that the development and use of a "crib sheet" is an extraordinarily powerful preparation tool. The process of developing the crib sheets reinforces and upgrades the preparation process. Of equal importance, the negotiating team members function much more professionally when they have a single document to which they can refer. (Obviously, this sheet must be protected from "the other side"!)

Draft Agreements

The development of one or more "ideal" agreements also is a powerful preparation tool. The process of developing such agreements helps hone the negotiating team. In addition, possessing such a documents has been shown to expedite closure of the face-to-face process.

Face-to-Face Discussions

Establishing trust is a key ingredient in effective negotiations. Leigh Thompson lists several activities that can facilitate the development of a reasonable level of trust: agree on a common goal or shared vision, expand the pie, use fairness criteria that everyone can buy into, capitalize on network connections, find a shared problem or shared enemy (perhaps another supply chain), focus on the future (instead of the past), and use shared procedures.[4]

Fact Finding

During the initial phase of a meeting with a potential supplier, professional negotiators limit discussions to fact finding. Any inconsistencies between the supplier's proposal and the negotiator's information are investigated. Fact finding should continue until the negotiator has a complete understanding of the supplier's proposal. Questions of a how, what, when, who, and why nature are used by the negotiator. Experience has shown that when the negotiator limits this initial phase to fact finding, a satisfactory agreement normally results with a minimum of hassle and disagreement. During the fact-finding process, the negotiator should gain a better understanding of both the supplier's interests and the supplier's strengths and weaknesses.

The buying and selling representatives should disclose their interests, not their objectives. Altogether too much time is wasted on haggling over positions. Professional negotiators quickly learn their opposite's interests. It is much easier to satisfy interests than it is to move one's opposites off their positions. On completion of the fact-finding process, the negotiator should call for a recess or caucus.

Recess

During the recess, the negotiating team should reassess its relative strengths and weaknesses, as well as those of the supplier. It may also want to review and refine its cost estimate and any other estimates or assumptions. The team then should review and revise its objectives and their acceptable ranges. Next, the team should reorganize the agenda it desires to pursue when the two teams return to the negotiating table.

Narrowing the Differences

When the formal negotiations reconvene, the negotiator defines each issue, states the facts (and any underlying assumptions), and attempts to convince the supplier's representative that the negotiator's position is reasonable. If agreement cannot be reached on an issue, the negotiator moves on to the next issue. Frequently, discussions on a subsequent issue will unblock an earlier deadlock.

During this phase of the negotiating process, problem solving and compromise are used to find creative solutions wherein both parties win. For example, the buying team's manufacturing engineers may identify a more cost-effective process than the supplier had planned to use. Small acceptable changes in packaging, schedule, or tolerances; offers by the customer to furnish a material or an item of equipment; and payment terms (including possible advance payments) can unblock negotiations to the benefit of both parties.

The package approach lends itself to situations wherein several issues are on the negotiating table. Let us assume there are five issues: A, B, C, D, and E. The two parties address issue A and reach a tentative agreement. They then address B. No agreement. One party then says, "Let's put B on hold." On to C—agreement. On to D—no agreement. And no agreement on E. Then, one party constructs and proposes a package (addressing all five points) that it feels is fair and appropriate for consideration. Further, the proposing party suggests that the package approach be looked at as a full bucket of water. If something flows into the bucket (e.g., a higher price, longer lead time, etc.), then something flows out (improved warranty, better service, etc.). Experience has shown that the combination of the package approach and the full bucket of water is a most attractive way of avoiding deadlocks.

In most instances, it is possible to reach a satisfactory agreement through the use of these procedures. If a satisfactory agreement cannot be reached, the negotiating team has the choice of adjourning (an attractive alternative for the buyer if another supplier is waiting in the wings) or moving on to hard bargaining.

Hard Bargaining

Hard bargaining, the last resort, involves the use of take-it-or-leave-it tactics. Its use is limited to one-time or adversarial situations where long-term collaborative relationships are not an objective. The negotiating team should carefully and professionally review and revise its objectives and, if absolutely necessary, give the supplier the option of accepting or rejecting its final proposal. Possession of a BATNA protects the buying firm from entering into an unwise agreement. The experienced negotiator does not bluff unless willing to have the bluff called. Unless a one-time purchase of an item already produced (e.g., an automobile on a dealer's lot) is involved, the wise negotiator avoids having the seller feel that it has been abused or treated unfairly. Such feelings set the stage for future confrontations, arguments, unsatisfactory performance, and possible claims.

Techniques

Negotiation techniques (tactics) are the negotiator's working tools. The negotiator uses them to achieve his or her goals. In the hands of a skillful negotiator, these tools are powerful weapons. In the hands of a novice, they can be dangerous booby traps. Competent negotiators, therefore, spend a great deal of time studying and perfecting the use of these techniques. There are so many negotiating techniques that all cannot be discussed here. Those selected for discussion represent some of the techniques that have proved to be most important and most effective for the authors, their colleagues, their students, and working professionals entrusted to their guidance.

The objective of negotiation is agreement. Even though agreement is the fundamental goal of negotiation, sometimes negotiations end without agreement. In the short run, not reaching an agreement is better than reaching an unsatisfactory agreement. Generally speaking, however, experienced negotiators seldom let negotiations break down completely. They do not intentionally maneuver or let their opponents maneuver themselves into take-it-or-leave-it or walkout situations unless they are involved in a one-time or adversarial relationship.

Negotiating techniques may be divided into three categories: (1) those that are universally applicable, (2) those that are applicable to transactional (and, frequently, adversarial) dealings, and (3) those that are applicable to collaborative and alliance relationships.

Universally Applicable Techniques

These techniques are applicable to all negotiations, whether transactional or collaborative ones.

Getting to Know You

Not only is this the title of an old hit tune, it is a powerful and effective negotiation technique! The negotiator is not dealing with abstract representatives, but rather with human beings. If possible, he or she should get to know the individuals representing the seller before the face-to-face phase of the negotiation begins. Americans tend to be too anxious to rush into negotiations without getting to know and understand the other side's representatives. We have much to learn from members of other cultures who spend a good amount of time becoming acquainted with those with whom they are to negotiate before entering into the formal face-to-face phase of negotiations. These negotiators find ways to meet the seller's representatives informally. If possible, they arrive early before the face-to-face negotiation is scheduled to begin and stay late after it ends.

Use Diversions

On the human side of negotiations, the negotiator who knows the seller personally, or has carefully studied his or her personal behavior patterns (as should be the case), has an advantage. When tempers start to get out of hand, as they occasionally do, the experienced negotiator quickly diverts attention away from the issue at hand. At such times a joke, an anecdote, or a coffee (or tea) break can be an effective means of easing tensions. This type of diversion is usually more easily accomplished when the participants know which situations are most irritating to their opposites.

Use Questions Effectively

The wise use of questions is one of the most important techniques available in negotiation. By properly timing and phrasing questions, the negotiator can control the progress and direction of the negotiation. A perceptive question can forcefully, yet tactfully, attack the supplier's position. Similarly, the negotiator can effectively defend his or her own position by asking the seller to evaluate certain carefully chosen data the negotiator has developed.

The technique of answering questions properly sometimes is as important as the technique of asking them properly. The successful negotiator knows when to answer, when not to answer, when to answer clearly, and when to answer vaguely. Not all questions require an answer. Many questions are asked for which the seller knows there is no answer; therefore, a reply is not really expected.

The correct answer to questions in negotiation is not governed by the same criteria governing the correct answer to questions in most other situations. For negotiation questions, the correct answer is the answer that furthers either the negotiator's short-term tactics or long-range strategy. Labor leaders and politicians are experts at asking and answering questions. Their questions and their answers are made to correlate with their strategic plans (strike platforms, party platforms, and so on). To an uninformed observer, it often appears that the answers given by politicians and labor leaders do not relate to the questions that they are asked. These observations are only partially correct. When answering questions, politicians and labor leaders tell their listeners what they want them to know about their platforms, whether or not the response fully answers the questions asked.

Successful negotiators realize that negotiation sessions are not like the classroom, where precise answers earn high marks. In negotiation, the purpose of questions and answers is not to illustrate to the seller how smart the negotiator is. Rather, it is to ferret out the seller's objectives and to learn as much as possible about how the seller's representatives intend to maneuver to achieve them. For this purpose, precise answers are sometimes the wrong answers. The correct degree of precision is dictated by the particular circumstances of each negotiation.

Avoid naive questions such as "Do you have good service?" The supplier will obviously say "yes." Good questions on service capability include: How many service technicians do you have? What is the response time? What is their technical training? Where are they located? What kind of problems can they fix/solve? Will you allow a contract clause that guarantees a certain response time? What is the proper procedure to get a service call? Finally, what are the labor charges and price of the most commonly replaced parts and are they readily available?

Use Positive Statements

As with sophisticated questions, perceptively used positive statements can favorably influence the course of negotiations. For example, assume a negotiator knows that certain questions will evoke an emotional reaction from the seller. The questions are asked, and an opportunity is created for the proper use of a positive statement. A competent negotiator would say something like this: "I see your point, and I understand how you feel about this matter. Your point is well taken." Contrast the effect of this type of positive response with that of an emotional, negative response in which the negotiator tells the seller that he or she is "dead wrong." When a negotiator tells a seller that the seller's viewpoint is understood and considered reasonable, even though the negotiator does not agree with it, the seller is more likely to consider the negotiator's viewpoint objectively.

Machiavelli, in *The Prince*, gave the world some unusually sage advice concerning the use and misuse of positive statements:

> I hold it to be proof of great prudence for men to abstain from threats and insulting words toward anyone, for neither . . . diminishes the strength of the enemy; but the one makes him more cautious, and the other increases his hatred of you, and makes him more persevering in his efforts to injure you.[5]

Be a Good Listener

Generally speaking, salespeople thoroughly enjoy talking. Consequently, negotiators should let them talk. While talking, they very often talk themselves into concessions that a negotiator could never gain through negotiation. Listening, per se, recognizes a basic need of a seller. Additionally, listening carefully to a seller's choice of words, phrases, and tone of voice, while at the same time observing his or her gestures and other uses of body language, can be rewarding. By observing such actions, a negotiator can gain many clues regarding a seller's negotiating position.

Be Considerate of Sellers

A small number of negotiation experts contend that negotiations are best won by negotiators who are as brutal and as arbitrary as possible. This is definitely a minority opinion. Unquestionably, there are some purchasing situations in which a merciless frontal assault can be a proper and successful negotiating technique. However, for the vast majority of firms—those that seek profitable, continuing relationships with the seller—a more considerate and reasoned technique is recommended. Professionals lose no negotiating advantages whatsoever by being fully considerate of sellers personally, by letting them save face, and by reasonably satisfying their emotional needs.

Transactional Techniques

Much negotiating literature is based on traditional—even adversarial (win-lose)—approaches. Two effective books addressing this approach are Gerald Nierenberg's *The Complete Negotiator*[6] and Herb Cohen's *You Can Negotiate Anything*.[7] Two of the traditional tactics that deserve special emphasis are: (1) keep the initiative and (2) never giving anything away.

Keep the Initiative

The negotiator should strive never to lose the initiative automatically obtained when the supplier's proposal is received and reviewed. There is a good deal of truth in the old saying that a good offense is the best defense. The negotiator should constantly "carry the game" to the supplier, keep the supplier on the defensive by confronting its representatives with point after point, making the supplier continually justify its position. For example, if the supplier states the cost of materials in dollars, the negotiator should ask the seller's representative to justify the figures with a bill of materials, appropriate scrap rates, and a full explanation of the manufacturing processes to be used. The more the negotiator bores in and the more pressure he or she maintains, the better will be his or her bargaining position. If the supplier's position seems sound, the negotiator can offer a counterproposal. In either case, the negotiator starts with the initiative, and should work hard to retain it.

Never Give Anything Away

As a matter of strategy, a successful negotiator periodically lets the seller maneuver him or her into accepting one of the seller's proposals. This does not mean that the negotiator gives something away. He or she never "gives anything away." Professional negotiators always expect to get a concession in exchange. On the other hand, the negotiator does not feel obligated to match every concession made by the seller. Consequently, in the exchange process, the successful negotiator makes fewer concessions than his or her less successful adversary. Through a continuation of this exchange process, a position close to the objectives of both parties is usually reached. Mutual concessions benefit both parties, and a contract so negotiated is *mutually* advantageous; but it is not *equally* advantageous.

Frame the Question

Negotiating authority Herb Cohen points out that too many of us allow "the other side" to frame the issue. In effect, we are doing business on their terms. For example, the seller says, "Do you want alternative A or B?" The professional negotiator responds, "Those certainly may be viable options, but let's develop some others."[8] As previously mentioned, avoid naïve questions.

The Dynamics of a Transactional Negotiation

Typically, the two parties' positions appear as shown in Figure 18.1. The seller's positions are generally all higher than the corresponding positions of the buying firm. The closer the two objectives are initially, the easier the negotiations. As negotiations proceed, the seller tends to make concessions from its maximum position toward its objective. Simultaneously, the buying firm's negotiators reduce their demands, moving from the minimum position toward their objective.[9] Usually, little difficulty arises during this preliminary skirmish. This is not to say that this part of the negotiation process is easy or that it does not take time. Normally, vigorous testing is required to convince each party that the other is actually at his or her objective. Each party attempts to convince the other that the objective has been reached before this has, in fact, occurred. As each party approaches its objective, negotiation becomes difficult. The distance between the buying firm's objective and the supplier's objective can well be called the "essence" or "heart" of the negotiation (see Figure 18.1). Any concession made by either party from its objective position will appear unreasonable based on the previous analysis of the facts. Changes in position, therefore, must now be the result of either logical persuasion and negotiating skills (entailing further investigation, analysis, and reassessment of the facts) or the pressure of brute economic strength.

The skillful negotiator stands out in the area of objective persuasion. He or she makes progress by uncovering new facts and additional areas of negotiation that permit the supplier to reduce its demands. For example, an analysis of the supplier's manufacturing or services operations might reveal that if lead time were increased by only one week, the job could be done with fewer machines or human resources. This change could substantially reduce the supplier's setup and scheduling costs, thus permitting a price reduction. Additional lead time might be made available by a slight modification in the buyer's production schedule. The cost of making this change could well be much less than the seller's savings from the longer production runs. Thus, both parties would profit from the

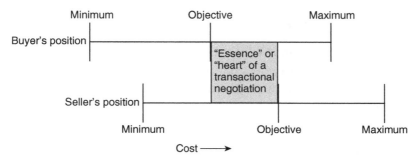

Figure 18.1 Dynamics of transactional negotiation.

change. It is this type of situation that competent negotiators constantly seek to discover and exploit in their attempt to close the gap between the seller's objective and their own. Such situations have the highly desirable effect of benefiting each party at no expense to the other party.

In some sole-source negotiations, the seller's objective is to maximize its position at the expense of the buying firm. In these situations, a continuing relationship is of secondary interest to the seller; therefore, it uses its bargaining strength to maximize price, rather than to achieve a mutually advantageous contract that will lead to continued business. The negotiator who senses such a situation should start negotiations by attacking the reasonableness of the seller's cost breakdown, using his or her own prepared cost estimates as the basis for such challenges. In the absence of competition, this is a negotiator's most logical and effective plan of action. If the supplier refuses to divulge its cost data, the negotiator has only three available courses of action. He or she can appeal to the seller's sense of reason, pointing out the potential negative long-run implications of such actions. A second approach is to fight force with force by threatening to use substitutes, or to redesign and manufacture the product or service. A third alternative is to further develop and refine the firm's cost estimating models and utilize them more forcefully in pursuing the original course of negotiating action.

When faced with this type of problem, the negotiator must attempt to bring the supplier's price as close to the objective as possible. In the short run, the negotiator usually pays the seller's price. In the long run, the negotiator works toward the development of competing sources, substitute products, and compromises with the supplier.

If a seller's negotiation objective is to resolve issues as quickly as possible by employing logical analysis rather than economic bargaining power, it is sometimes reasonable for the negotiator to start negotiations by proposing his or her actual objective as the counteroffer to the seller's proposal. In fact, in industrial situations where continuing relationships are the rule, each successive negotiation brings the objectives of both parties ever closer together. Under these conditions, representatives of the buying and selling firms need develop only their objective positions; there is no need for maximum and minimum positions.

Collaborative and Alliance Negotiating Techniques

Getting to Yes by Roger Fisher and William Ury[10] is the most widely read book written on collaborative negotiations. The authors introduce what they call "the principled negotiation method of focusing on basic interests, mutually satisfying options, and fair standards, resulting in a wise agreement."

Fisher and Ury's method calls for the use of four powerful techniques:

▲ Separate the people (negotiators) from the problem (quality, price, etc.).
▲ Focus on interests, not positions.

▲ Invent options for mutual gain.
▲ Insist on using objective criteria.

Experience demonstrates that applying these techniques to collaborative (or win-win) negotiations will result in wise agreements and the basis of success for long-term relationships. Many of these relationships blossom into preferred partnerships and even into strategic supply alliances.

Separate the People from the Problem

Successful negotiators divide the negotiation into two components: the people issues and the technical issues such as quality, time, price, etc. People issues require negotiators to understand where the other party is coming from. In effect, the negotiators must walk in the shoes of the seller's representatives. Professional negotiators work at ensuring that they understand the other party, frequently rephrasing what they have heard to ensure understanding. In addition, they frequently ask the seller's representatives to describe in their own words what the negotiator has said.

Emotions frequently get in the way of successful negotiations. Both parties to a negotiation have a right to get upset or angry and to express such emotions. Wise negotiators allow their opposites to let steam off without taking offense or allowing the negotiation to become disrupted. The perception or belief that the "other side" is stubborn or irrational is likely to manifest itself with exactly such behavior. Stop and walk in their shoes! The most constructive negotiations occur when the representatives of the buying and selling organizations work together in a search for a fair agreement in which both sides are better off than if there were no agreement.

Focus on Interests

During the fact-finding phase of face-to-face negotiations, the professional negotiator learns the seller's interests while disclosing his or her own interests (but not objectives!). During the third phase (narrowing the differences), both parties work at reconciling and satisfying interests, not positions. Since both buyer and seller normally have multiple interests, it is wise to identify all of them and then work at developing a solution (agreement) that satisfies most or all of these interests. This approach calls on creativity and frequently results in increasing the size of the pie—the package of benefits to be shared by the two parties.

James K. Sebenius of the Harvard Business School writes:

> …interest-driven bargainers see the process primarily as a reconciliation of underlying interest: you have one set of interest, I have another and through joint problem solving we should be better able to meet both sets of interest and thus create new value.[11]

Invent Options for Mutual Gain

Fisher and Ury's third principle flows from the use of creativity to develop many options. When both parties become involved in creativity or brainstorming, they generate creative solutions wherein both parties benefit. Two rules of thumb are (1) develop many options and (2) remain in option generation past the point of comfort. Many of the most creative ideas require time and even discomfort to develop. Only after a list including one or more truly creative ideas has been developed should the negotiators attempt to select from the list.

Use Objective Criteria

When a long-term relationship is an objective of both parties, the use of objective criteria will avoid much positional negotiation—and the possibility of disrupting or destroying the relationship. For example, if price is the issue under discussion, then possible objective criteria could include (1) the supplier's agreed-to allowable costs plus a reasonable profit, (2) development of a cost model to be used as the basis of the price, (3) a market-based pricing methodology, or (4) target (design-to-cost) pricing. (Remember, these issues were all discussed in Chapter 16 on cost analysis.) Having identified four possible objective criteria, the issue now becomes a discussion of which criteria, or combination of them, should be applied.

Benefits Are Not Divided Equally

In some supply management circles, a common misunderstanding exists that successful negotiation means an equal distribution of the benefits. While both buyer and seller should benefit from a well-negotiated contract, the benefits are seldom divided 50-50. Based on years of observation, 60 to 70% of the benefits of a typical negotiated contract go to the more skillful negotiator, leaving 30 to 40% for the less skilled negotiator. But notice, both parties are better off than were there no agreement. (This is the meaning of a "win-win" outcome.)

The Debriefing: An Incredible Learning Opportunity

Thompson points out that "most people have little opportunity to learn how to negotiate effectively. The problem is not lack of experience but a shortage of accurate and timely feedback."[12]

Experience gained in academia demonstrates that a 15-minute debriefing conducted by the negotiating team provides an incredible opportunity for learning and improvement in future negotiations. On completion of each negotiation, the negotiating teams are required to conduct a debriefing of their preparation process. They must identify both what was done well during their preparation process and lessons learned (e.g., what could/ should have been done more professionally.)

The team members then are required to analyze their "face-to-face" activities. They identify (reinforce) what went well. Then they identify weaknesses in their processes. (Feedback is at both an individual and team level.) Both undergraduates and graduates alike maintain that the debriefing is a marvelous tool, which they plan to implement in the "real world."

Documentation

Personnel turnover and the frailties of the human memory make accurate documentation of the negotiation essential. The documentation must permit a rapid reconstruction of all significant considerations and agreements.

Documentation begins in the supply management office with the receipt of a purchase requisition and continues with the selection of potential suppliers and their proposals. Documentation of the actual negotiation must be adequate to allow someone other than the buyer to understand what was agreed to, how, and why. Burt, Norquist, and Anklesaria suggest the following format for the documentation of negotiations:[13]

Subject   This is a memorandum designed for readers with many different orientations. This section, together with the introductory summary, should give the reader a complete overview of the negotiation, including information such as the supplier's name and location, the contract number, and a brief description of what is being purchased.

Introductory summary   The introductory summary describes the type of contract and the type of negotiation action involved, together with comparative figures from the supplier's proposal, the buyer's negotiation objective, and the negotiated results.

Particulars   The purpose of this section is to cover the details about what is being bought and who is involved in the procurement. This should be done without duplicating information that was included in the subject section.

Procurement situation   The purpose of this section is to discuss factors in the procurement situation that affect the reasonableness of the final price.

Negotiation summary   This section shows the supplier's contract pricing proposal, the buying firm's negotiation objective, and the negotiation results, tabulated in parallel form and broken down by major elements of cost and profit. Whether these are shown as summary figures for total contract value, summary for the total price of the major item, units price for the major items, or some other form of presentation depends on how negotiations were conducted. The general rule is to portray the negotiation as it actually took place.

Online Negotiation[14]

Studies have compared the establishment and maintenance of business relationships based on forms of communication. Those relationships founded solely on written communication tend to founder. Those that include written and telephone communication are sustainable, but the best relationships are those that go beyond other forms of communication to involve face-to-face meetings.

In face-to-face communication, one party can easily discern from facial expressions and shifts in body positioning the reaction of the other party to a stated position, or even a misinterpretation. Immediate steps can be taken to remedy any misunderstandings or to provide further explanation or support for a negotiation position. When communication takes place in written form, these remedies may not be available.

A carelessly drafted e-mail may damage a buyer-supplier relationship without the sender even being aware of the impact on the receiver. One such situation involved a buyer who needed a prompt response to an e-mail message, so he drafted the entire message in capital letters to signify urgency. The foreign supplier interpreted the message as insulting and an attempt at intimidation, the equivalent of shouting. Needless to say, the transaction did not proceed smoothly. The list of examples is nearly endless. All of us have received e-mail messages containing spelling and grammatical errors. The unfavorable impression elicited by such messages extends not only to the sender, but to the organization as well, and may be virtually irreparable.

Examples abound of the different interpretations that a receiver may generate of communications delivered face-to-face, and the same communications delivered in written form. To some extent, this can be attributed to the fact that not all of the content is contained in just the words; it is also in *how* we say those words. Our tone of voice may be hard or flat, soft, cheerful, loud, or strident and insistent. In response to a statement, we might say, "Oh, right," but while the words are the same in each case, if our tone signals enthusiastic agreement the message received is considerably different from the message delivered with a tone of sarcasm.

A partial remedy exists in resorting to oral communication, but we have a device for accomplishing this in the telephone, and do not need to resort to the Internet (except, perhaps, as a means to avoid long-distance telephone charges). Even this solution is only partial, and its inadequacy is well displayed in the frequency of business travel.

Beyond the presentation of our message, we also tend to convey substantially different content between what we write and what we might say. For example, think about the difference between writing a letter to a friend and telling the person the same information over the phone. In written communication, there is no sense of timing, as in delivering the punch line to a joke. It is easy to sense the urgency in someone's voice, but more difficult to convey that gravity in writing. How do all of the jargon and slang words we routinely use in conversation look in print? Run-on sentences are hardly noticeable when spoken, but are painfully obvious in writing.

Further, even if we advance to video conferencing over the Internet, many of these concerns persist. Perhaps timing and presentation will begin to approximate that of face-to-face delivery when the technology has advanced enough. However, as pointed out by Lanier,[15] we have not yet reached that point. Further, even with the advancement of technology to provide similar capacity, it is likely that the use of the technology will continue to change how we communicate, only some of which may be coming into investigative focus at this juncture.

Advantages to Online Negotiation

Experience indicates that electronic communication can help one focus on the issues separately from personalities. There is some evidence that groups using electronic communication tools may be more productive at brainstorming. Online communication can free the buyer and supplier from location dependency, and perhaps even the requirement to find a common time for conducting the negotiations.

An Example of a Successful Application of Negotiating Online

A recent *Wall Street Journal* article on thriving B2B software companies cites the following successful application.

"A South San Francisco–based company began marketing software this year to help automate dealings among close business partners. One of its biggest installations is Provision X, a Chicago-based meat-trading exchange unveiled this spring by five U.S. poultry, pork and beef suppliers.

Before the exchange, a buyer for a supermarket chain might call up sales managers at one or two meat suppliers and request price quotes. Those sales managers, in turn, might call up pricing managers inside their own companies who are plugged into a range of factors that shape the quantity, quality and price of their products. Armed with data from the pricing experts, the sales staff goes back and negotiates with the buyers.

Provision X transfers those processes online, creating an automated way of negotiating. Instead of trading phone and fax messages, buyers and sellers log on to create and solicit orders and price quotes, look up price lists, check sales performance against purchase contracts and quickly generate reports summarizing the results of recent activities. Throughout these activities the relationships and hierarchies are maintained, right down to specifying the names of people who have rights to see certain data or make certain transactions.

"We're not trying to take away the people-centric focus of this business," says Kevin Nemetz, Provision X's chief executive officer. "We are just trying to take those relationships online."[16]

Drawbacks to Online Negotiation

It is far easier to say no in writing than it is in person. The psychological separation that goes with the lack of personal contact makes the termination of the relationship much easier. On one hand, this is a distinct advantage because it curtails problems that may be associated with negotiating with people that we know and like, including reluctance to disagree for fear of damaging the relationship.

Online negotiators are likely to feel a need to be more persuasive or more convincing. However, we must take care to avoid excessive stridency in our persuasive efforts, or our labors can easily backfire. Because of a lack of other cues, our use of language takes on particular importance, and we carefully consider the impact of each word. How might connotations or interpretations differ? Often, the same word can be taken in a number of different ways, and the limitations of online communication as currently realized make it difficult to evaluate the perception of the receiver.

Buddress, Raedels, and Smith[14] propose the following hypotheses:

▲ The more important the issue, the more likely it is that it will be negotiated face-to-face.
▲ The more politically sensitive the issue, the more likely it is that it will be negotiated face-to-face.
▲ If either negotiator will be personally affected by the outcome, that person may want to conduct the negotiation in person.
▲ If the topic involves issues of firm sensitivity, such as trade secrets or core competencies, the negotiation is more likely to take place face-to-face.
▲ Buyer-supplier relationships will be perceived as more distant, the more online communication and negotiation are used.
▲ Less formal planning will occur prior to online negotiations than for those conducted face-to-face.

Negotiating for Price

Historically, price is the most difficult of all contract terms to be negotiated. Based on its high relative importance and its complexity, negotiation for price will be used as an example to illustrate what is involved in negotiating many other terms of the contract. If the reader understands what is involved in negotiating price, he or she can easily visualize what is involved in negotiating other issues. When negotiating price, the negotiator must concurrently consider the method of compensation to be used. Compensation method and the negotiation of price are directly related; hence, they must be considered together.

To assure buying at favorable prices, negotiators strive to develop the greatest practical amount of competition or enter into fact-finding discussions with representatives of preferred suppliers or "partners" about their costs and cost drivers. Therefore, whenever it is possible, the initial step for a negotiator seeking successful negotiations based on competition is to get an adequate number of proposals from among those potential suppliers who are genuinely interested in competing for the contract. When cost negotiations are likely, requests for proposals usually ask for not only the total price but also a complete breakdown of all supporting costs.

For every negotiated purchase, either price analysis or cost analysis, or both, is required. Which analysis is best to use and the extent of the analysis required are determined by the facts bearing on each specific purchase being negotiated. Generally speaking, price analysis is used for lower-dollar-value contracts and cost analysis for higher-dollar-value contracts. A discussion of the applicable uses of both price analysis negotiation and cost analysis negotiation follows.

Price Analysis Negotiation

Price analysis negotiation (often referred to simply as "price negotiation") is the most commonly used approach when negotiating only for price. Some proponents of cost negotiation disparage price negotiation, referring to it as "unsophisticated" and "emotional." In the many cases where price negotiations are undertaken in an unprofessional manner, such criticism is fully justified. Banging on the table and shouting "I want lower prices" or "I can get it cheaper from another supplier" is certainly not professional price negotiation.

On the other hand, in many specific cases where pricing data are developed and utilized with professional skill, price negotiation can be just as advantageous as cost negotiation, or more so. Compared with cost negotiation, price negotiation has three distinct advantages: (1) negotiation time is shorter, (2) support of technical specialists is seldom needed, and (3) pricing data are relatively easy to acquire.

The traditional sources from which supply management professionals get pricing data are federal government publications, purchasing trade publications, newspapers, and business journals. Competing suppliers are excellent sources of pricing data. They can provide the buyer with price lists, catalogs, numerous special pricing data, and formal price quotations. From these competing suppliers' data, the supply management professional can readily determine two very important facts: the nature of the market (competitive or noncompetitive) and the extent of supplier interest in this particular purchase. Historical pricing data and engineering estimates also provide a sound basis for price analysis.

Price Comparison

The negotiator's first step in price analysis is to determine the extent of market competitiveness and supplier interest. The second step is to examine in detail the absolute and relative differences existing among the various prices quoted by the competing suppliers. From this examination, a buyer detects that differences in prices among suppliers exist but does not learn the causes of these differences. The search for causes begins in the supply management department's supplier information file.

The price proposals of the competing suppliers are compared with past prices of similar purchases from the supplier information file. The causes of all significant variations are pinpointed and analyzed. Adjustments are made for changes in factors such as specifications, quantities ordered, times of deliveries, variations that have taken place in the general levels of business activity and prices, and differences that may have resulted from learning experience. After these adjustments are made, the negotiator (sometimes with the help of an engineering estimator or a price analyst) determines whether the prices offered are reasonable. From this determination, the negotiator decides on the target objective to use for his or her negotiating position.

Trend Comparisons

Historical prices paid for purchases of similar quantities can be analyzed to disclose helpful price trend information. For example, if prices have been increasing, it is reasonable to expect that the seller will attempt to maintain a similar pattern of increase. Hence, by carefully analyzing the reasons for all price increases, the negotiator can structure a bargaining position on the basis of any invalidities uncovered.

Similarly, the negotiator can analyze decreasing prices to determine whether the price decrease is too little or too much. If the negotiator determines that the decrease is too large, he or she must determine whether the trend is creating, or is likely to create, quality or service problems in contract performance. If the decrease is too little, then the negotiator must determine whether the benefits of improved production processes are being proportionally reflected in lower prices.

Even a level price trend offers opportunities for price analysis. For example, the negotiator may ask whether level prices are justified, considering the many manufacturing improvements have been made. Did the supplier charge too much initially? Has the supplier's competitive position in the industry changed? If the negotiator's analysis indicates that costs have fallen because of reductions in the supplier's cost for materials or because of improvements in the production processes, his or her negotiating position is clear. The professional negotiator obtains reductions reflecting these changes. (It must be noted that under collaborative or alliance relationships, these savings should be shared.)

Cost Analysis Negotiation

As previously stated, price analysis negotiation is more commonly used than cost analysis negotiation. In cost negotiations, each applicable cost element is negotiated individually, that is, design engineering cost, tooling cost, direct materials cost, labor hours, labor rates, subcontracting, overhead cost, other direct costs, profit, and so on. Cost analysis negotiation (commonly referred to as "cost negotiation") is steadily growing in use. It has been used successfully for decades by many large firms such as General Electric, Ford, and Honda, and in recent years it has been employed increasingly by small- and medium-size firms.

When sophisticated collaborative or alliance supply relationships are utilized by a firm, careful detailed analysis of the supplier's costs (both present and projected) replaces the role of competition in the marketplace. Both the buying and selling firms' representatives must see themselves as members of a supply chain competing with other supply chains for the customer's purchasing dollar. Thus, discussions about costs, cost allocations, cost drivers, cost reductions, possible cost avoidance, and profits must be seen in context: if our supply chain becomes noncompetitive, then we will fail to attract the customer's purchase dollars and we both lose! Discussions can and often do get heated—conflict can be healthy—but the discussions should be conducted in the context of what is in the joint best interests of the two parties. In reality, they should see themselves as members of the same supply chain team.

Characteristics of a Successful Negotiator

The characteristics of successful negotiators should now be clear. These people are skillful individuals, with broad business experience. They possess a good working knowledge of all the primary functions of business, and they know how to use the tools of management—accounting, human relations, economics, business law, and quantitative analysis. They are knowledgeable about the techniques of negotiation and the products and services their firms buy. They are able to lead meetings and conferences and to integrate specialists into smoothly functioning teams. In addition to being well educated and experienced, successful negotiators excel in good judgment. It is good judgment that causes them to attach the correct degree of importance to each of the factors bearing on the major issues. Combining their skills, knowledge, and judgment, they develop superior plans. Additionally, they consider problems from the viewpoint of the firm as a whole, not from the viewpoint of a functional manager. The successful negotiator is pragmatic in use of negotiating techniques. Always searching for a collaborative or "win-win" experience, the pragmatic negotiator can adapt when the other party uses hard ball or win-lose techniques. The successful negotiator has high self-esteem and is always most interested in professionalism and the best interest of the enterprise. The successful negotiator is ethical and honest and not influenced by friendship or gratuities.

Successful negotiators share four common attributes:

1. All realize that specialized training and practice are required for an individual to become an effective negotiator. Although some people have stronger verbal aptitude than others, no one is born with negotiating knowledge and skills.
2. All habitually enter into negotiations with more demanding negotiating objectives than their counterparts, and generally they achieve them.
3. All are pragmatic and flexible in their capability to deal with different negotiation techniques from "hardball" to "collaborative."
4. All are included, or are destined to become included, among an organization's most highly valued professionals.

Summary

Negotiation is free enterprise at its very best! When traditional negotiations (win-lose) are appropriate, negotiation matches the skills of determined buyers against those of equally determined sellers. Both explore ways to achieve objectives that tend to maximize the self-interest of their organizations. In short, in such circumstances negotiation is a powerful supply management tool that competent professionals use to achieve maximum value at minimum cost. By rewarding efficiency and penalizing inefficiency, the negotiation process not only benefits the negotiating firms, but also benefits the nation's economy as a whole.

The increasingly common collaborative approach to negotiations that is required with collaborative and alliance supply relationships substitutes a win-win approach for the more traditional, transactional one. With this approach, both parties are better off entering into the negotiated deal than were they not to reach agreement. This approach substitutes the expertise of the buying and selling firms' representatives for the forces of marketplace competition. Thus, costs must be driven to their lowest possible levels (without adversely affecting quality or service) in order to ensure the survival and success of the buyer and seller's supply chain in the marketplace.

Endnotes

1. Leigh Thompson in the second edition of her insightful book, *The Mind and Heart of the Negotiator*, provides the following definition: "an interpersonal decision-making process by which two or more people agree how to allocate scarce resources." Leah Thompson, *The Mind and Heart of the Negotiator* (Upper Saddle River, NJ: Prentice Hall, 2001), 2.

2. Herb Cohen, presentation at the Learning Annex, San Diego, CA, May 22, 2001.

3. Darlington, Gerry, "Culture: A Theoretical Review," in *Managing Across Cultures: Issues and Perspectives*, ed. Pat Joynt, and Malcom Warner. (London, England: International Thompson Business Press, 1996). Cited in a paper by James D. Reeds, "Understanding Cultural Diversity: the Influence of National Culture in Global Purchasing," 1998 International Conference of the Australian Institute of Purchasing and Materials Management, Paramatta, New South Wales, Australia, October 19, 1998.

4. Thompson, op. cit., 136.

5. Niccolo Machiavelli, *The Prince*, Great Books of the Western World, *Encyclopedia Britannica*, vol. 23, Chicago, IL, www.Britannica.com. 1982.

6. Gerald Nierenberg, *The Complete Negotiator*, (New York: Nierenberg & Zeif Publishers, 1986.)

7. Herb Cohen, *You Can Negotiate Anything*, 2nd ed. (New York: Bantam Books, 1982).

8. Cohen, op. cit.

9. If the negotiator believes there is a possibility of actually achieving the minimum position, he or she should open with a position below this point—*provided such a position can be logically supported.*

10. R. Fisher and W. Ury, *Getting to Yes*, first published by Houghton Mifflin Company in 1981 and now published by Penguin Books, New York.

11. James K. Sebenius, "Six Habits of Merely Effective Negotiators," *Harvard Business Review*, April (2001): 87–95.

12. Thompson, op. cit., 5.

13. David N. Burt, Warren E. Norquist, and J. Anklesaria, Zero Base Pricing™: *Achieving World Class Competitiveness through Reduced All-in-Cost*, (Chicago: Probus Publishing, 1990) chap. 13.

14. Appreciation is expressed to Professors Lee Buddress and Alan Raedels of Portland State University and Professor Michael Smith of Western Carolina University for much of the material included in this section.

15. J. Lanier, "Virtually There: Three-Dimensional Tele-Immersion May Eventually Bring the World to Your Desk," Scientific American (2001): 14.

16. Don Clark, "Perception, Reality," *Wall Street Journal*, May 21, 2001, R16.

Suggested Reading

Brett, Jeanne M. *Negotiating Globally: How to Negotiate Deals, Resolve Disputes, and Make Decisions Across Cultural Boundaries.* San Francisco, CA: John Wiley & Sons, 2007.

Dawson, Roger. *Secrets of Power Negotiating.* Franklin Lakes, NJ: Book-Mart Press, The Career Press Inc., 2001.

Fisher, Roger, Bruce M. Patton, and William L. Ury. *Getting to Yes: Negotiating Agreement Without Giving In.* New York: Houghton Mifflin Company, 1991.

Karrass, Chester L. *In Business as in Life—You Don't Get What You Deserve, You Get What You Negotiate.* New York: Harper Business Publishing Inc., 1996.

Karrass, Chester L. *The Negotiating Game, and Give and Take.* Crowell Publishing, 1990. For information on their renowned seminars, visit www.bothwin.com or call 1-323-866-3800, Beverly Hills, CA.

Ghauri, Pervez N. and Jean-Claude Usunier. *International Business Negotiations.* Emerald Group Publishing, 2003.

Lewicki, Roy J., Bruce Barry, David M. Saunders, and John W. Minton. *Essentials of Negotiation,* 4th ed. New York: McGraw-Hill Company, 2007.

Lewicki, Roy J. and Alexander Hiam. *Mastering Business Negotiation: A Working Guide to Making Deals and Resolving Conflict.* San Francisco, CA: Jossey Bass, A Wiley Imprint, 2006.

Lewicki, Roy J. David M. Saunders, and Bruce Barry. *Negotiation,* 5th ed. New York: McGraw-Hill/Irwin, 2006.

Nussle, Ron, and Jim Morgan. *Integrated Cost Reduction.* New York: Reed Press, 2004.

Stark, Peter B., and Jane S. Flaherty. *The Only Negotiating Guide You'll Ever Need: 101 Ways to Win Every Time in Any Situation.* New York: Random House, 2003.

Ury, William. *Getting Past No.* New York: Bantam Books, 1993.

Watkins, Michael. *Breakthrough Business Negotiation: A Toolbox for Managers.* San Francisco, CA: Jossey-Bass, 2002.

CHAPTER 19

Contract Formation
and Legal Issues

Litigation Prevention

It may be persuasively argued that the best way to deal with legal disputes is to make sure one does one's best to avoid them in the first place! The maxim that an ounce of prevention is worth a pound of cure certainly comes to mind in this context. In professional life, most supply management professionals seldom—if they're fortunate—become involved in litigation. Yet their daily activities are subject to two major areas of the law—the law of agency and the law of contracts.

A supply manager or a buyer acts as an agent for his or her firm. Legally, this relationship is defined and governed by the law of agency. When a firm buys materials and services from other firms, each purchase involves the formation of a purchase contract. In fact, the courts assume buyer agents are familiar with the applicable laws pursuant to their particular buying situation, industry, trade custom, and practice. Should a serious disagreement arise between the purchaser and the supplier, the conflict becomes subject to the disputes resolution process, which ranges from negotiation, to mediation, to litigation—whether in front of an arbitration panel or a court. In any event, the dispute would be resolved by applying the law of contracts.

A supply manager's basic responsibility is to conduct the firm's procurement business as efficiently and expeditiously as possible. Buying policies and practices therefore are predicated primarily on business requirements and business judgment, rather than on legal considerations. From a business standpoint, contractual disputes can normally be resolved much more effectively and with less cost by negotiation. A lawsuit almost always alienates a good supplier. Additionally, the outcome of any court case is usually uncertain. Litigation is also costly, even in the event of a favorable decision. The total cost of legal fees and expenses, executive time diverted to the dispute, and disrupted business operations is seldom recovered from damage awards. For these reasons, in resolving disputes, most business firms utilize litigation only as a last resort.

The fact that a supply management executive tries to avoid litigation, however, does not mean that he or she can overlook the legal dimensions of his or her role. On the contrary, a basic knowledge of relevant legal principles is essential to success. Unless a supply manager understands the legal implications of his or her job—and actions—legal entanglements are almost certain to crop up from time to time.

The purpose of this chapter is to review briefly some of the principal legal concepts as they relate to a supply management professional's responsibilities. The chapter does not attempt to provide a complete discussion of these concepts. Most supply professionals should acquire some depth in the field through selected studies in commercial law.

Dispute Resolution

Serious contract disputes are a rarity in the lives of most supply managers. But in a complex business operation, they do arise from time to time. When a dispute does arise, after doing the appropriate "homework," the first step in the resolution process is to discuss the problem with the supplier. When attempting to resolve a dispute, it pays to keep in mind six considerations: (1) time, (2) money, (3) complexity/formality of method of dispute resolution, (4) stress, (5) visibility, and (6) damage to the relationship.

Negotiation

Most disputes are best resolved through negotiation and compromise. In most cases, the executives involved want to avoid further confrontation simply because it may be too time-consuming, too costly, too messy and complicated, too stressful, too embarrassing, and too damaging to the parties' relationship to do otherwise. In the event a satisfactory solution cannot be determined by the two parties, limited alternatives remain. They can mediate or litigate.

Mediation

If negotiation fails, the parties can consider mediation, which involves introducing a third-party into the discussion. The mediator's role is to listen, sympathize, empathize, coax, cajole, and persuade. Depending on the level of trust and credibility the mediator has with the parties, he or she may even propose, suggest, or encourage possible solutions. For that reason, the more trusted and respected a mediator is, the more likely he or she will be able to help resolve the dispute. One thing the mediator may not do, however, is decide anything.

Litigation

Litigation may be brought before an arbitrator or before a court. There are plusses and minuses to each approach, and it is fair to say that one size does not fit all.

Arbitration

Arbitration may take many forms, but its basic feature is that the outcome of a dispute is no longer in the hands of the parties themselves, but rather in the hands of a third-party. Unlike negotiation or mediation, where the parties attain resolution by agreeing on the outcome, arbitration vests the decision-making authority with the arbitral panel, which typically consists of one or three arbitrators. Under such circumstances, a professional or agreed upon arbitrator will hear testimony and study evidence from both sides, then make a decision based on the facts and the law. If the contract contains a mandatory arbitration provision, then the parties are *obligated* to arbitrate all disputes within the scope of that provision. However, like with mediation, the parties can agree, even after entering into a contract without an arbitration clause, to arbitrate any dispute that may arise.

Courts

Litigation may also be brought in an appropriate state or federal court, depending on the location of the parties and the amount in dispute. Lawsuits may be heard by a judge or by a jury. In either case, relevant court rules of procedure will prescribe how the litigation will proceed, and the relevant rules of evidence will prescribe what evidence may and may not be presented to the Trier of fact.

So, how does one best avoid ending up in litigation? As suggested at the beginning of this chapter, the best approach is avoidance and prevention of destructive legal disputes. And just how does one best avoid and prevent destructive legal disputes? This is where a fundamental understanding of certain basic legal principles can prove invaluable. One benefits most from knowing the law so as to avoid and prevent the destructive legal disputes that may ensue from ignorance of the law. After all, how many times has one heard the comment "ignorance of the law is no defense"?

Development of Commercial Law

Historically, each state developed its own body of statutes and common law to deal with the problems prevalent in its particular spheres of activity. Individual state development ultimately led to the creation to a series of commercial laws that varied widely from state to state—a situation that obviously produced difficulties for businesses involved in interstate commerce.

In an attempt to promote uniformity among the laws applicable to business transactions, the American Bar Association created a committee known as the National Conference of Commissioners on Uniform State Laws (NCCUSL) that worked together with the American Law Institute (ALI) on the development of a model act for a uniform set of laws governing aspects of commercial transaction.

The resulting code, entitled the Uniform Commercial Code (UCC), was published in 1952; revised versions of the code followed, with the most recent being published in 2001. Although the UCC deals with a wide range of commercial transactions, Article 2 deals specifically with the sale and purchase of goods. The UCC does not apply to the purchase of services, but a judge may elect to use it as a guide, and when purchases include both goods and services, the UCC may apply. The language in the UCC does not prevent the code from being used to rule on services even though the UCC was written for the sale of goods. However, buyers must remember that ordinarily, the UCC or the particular state version of it does not govern services. This means a state's common law of contracting will apply and each state will have variations regarding warranties, damages, and other performance issues. Therefore, the statement of work (S.O.W.) must detail all the provisions normally covered in the UCC, plus unique conditions such as behavior rules, security, substance abuse, personal conduct, and attire.[1] In addition, the supply manager must reserve the right to approve all subcontractors. Issues of dispute resolution such as arbitration, mediation, and cancellation rights all must be spelled out in the S.O.W. In some cases, the contract will include both goods and services such as an installation, a situation often called a "hybrid contract"; equipment contracts are often hybrid. In case of conflict, most courts will use the "predominant purpose" guideline to determine which laws prevail.

Each state could determine how much or little of the UCC it wished to adopt. Article 2 has been adopted by the District of Columbia and all the states with the exception of Louisiana. The UCC has effectively eliminated a majority of the important differences between the commercial laws of the various states and also has provided new statutory provisions to fill many of the gaps in the prior laws. It should be noted, however, that the code is silent on some matters covered by earlier laws. Consequently, unless superseded by provisions of the UCC, laws dealing with matters such as principal and agent, fraud, mistakes, coercion, and misrepresentation continue to be in effect.

Electronic Contract Considerations: Cyber Law

The UCC was drafted before widespread use of software and the advent of e-commerce. As a result, Article 2 has been applied in many states by analogy to govern disputes concerning software licensing and e-commerce.

Many have argued that e-commerce has spawned the need for a new or expanded set of laws to cover cyber-transactions. Accordingly, in 1995, the NCCUSL and the ALI set about drafting a proposed Article 2B to the UCC. The purpose of 2B was to govern licenses, including computer software. The completed draft, however, resulted in a split between the NCCUSL and the ALI. While the NCCUSL was satisfied with the draft, the ALI regarded many of the provisions as being too favorable to software licensors, at the expense of consumers.

After that split of opinion, the NCCUSL unilaterally decided to publish its draft separate and apart from the UCC. That model code became known as the Uniform Computer Information Transactions Act (UCITA) and, beginning in 1999, the NCCUSL encouraged states to enact it. Soon afterward, however, a majority of state attorney generals asked the NCCUSL to revise the UCITA because they believed that it undermined consumer protection by classifying consumer software as non-goods. Only two states have adopted UCITA: Maryland and Virginia. Another major problem caused by e-commerce (Internet and e-mail transactions) is the issue of electronic records and signatures as they pertain to the Statute of Frauds requirements for written records. Consequently in 1999, the NCCUSL published its proposed Uniform Electronic Transactions Act (UETA) that almost all the states have adopted.

If the contracting parties have agreed to honor electronic signatures, section 7 of UETA recognizes this deviation from the Statute of Frauds requirements for hard copy personal signatures, that is, "a record or signature may not be denied legal effect or enforceability solely because it is in electronic form."[2] Section 14 of UETA also recognizes contracts "formed by machines functioning as electronic agents for parties to a transaction."[3] It is important to remember that there are certain legal instruments such as wills that are exempted from UETA. In addition, many business lawyers and supply professionals believe that major contracts should have the traditional formal signing ceremony to reinforce the significance of the agreement.

In 2000, the U.S. Congress enacted The Electronic Signatures In Global and National Commerce (E-Sign) ACT that incorporates much of the UETA language. E-Sign also allows states who have adopted UETA to preempt or substitute the act for the relevant UETA provisions.[4] It appears that E-Sign is also based on the rational developed in the United Nations Model Law on Electronic Signatures of The United Nations Commission on International Trade Law as amended and distributed, December 12, 2001.[5]

Finally, most firms use some form of cryptographic (code) signature methods such as public key infrastructure (PKI) to secure, protect, and authenticate online signatures. One additional caveat: although the content is basically the same, states can and have made structural changes to the UCC applicable only to their state. The problem with this is that it undermines the uniform aspect of the UCC making it problematic to transact interstate commerce. Obviously, the whole issue of electronic contracting is still in a state of flux.

Attempts to Revise the UCC

Early in this decade, the NCCUSL and the ALI proposed revising significantly the provisions of Article 2 to make it current with contemporary practices. The proposed changes would radically alter a number of provisions, including (1) the minimum amount required for an agreement to be in writing (from $500 to $5,000); (2) revising the "battle of the forms" so that additional or conflicting terms in an acceptance or acknowledgment do not become part of the contract unless the parties otherwise have agreed to such terms; (3) delivery terms (e.g., F.O.B.) are deleted in favor of a reference to the International Chamber of Commerce's Incoterms; (4) revising the provision allowing for a seller to cure a nonconforming delivery; and (5) revising the statute of limitations from four years to five years in certain circumstances. To date, no state legislature has adopted any of the proposed revisions. This probably has resulted from the view that the current version, although flawed in some ways, is more or less consistent across the United States and that enactment of the proposed revisions once again would throw sand in the gears of interstate commerce.[6]

Topics treated throughout the rest of this chapter therefore reflect the provisions of the UCC where applicable, as well as the provisions of earlier laws that have not been displaced by the code.

Basic Legal Considerations

Status of an Agent

In the legal sense, an agent is a person who, by express or implied agreement, is authorized to act for someone else in business dealings with a third party. Regardless of the job title, this is precisely what supply managers and buyers do. A "supply manager" is not a legal party to his or her business transactions, but rather acts as the representative of the company. In this capacity, the law requires the agent to be loyal to the employer (the principal) and use reasonable care, skill, and judgment in performance of his or her duties. The agent owes the principal a fiduciary duty; therefore, the employer may hold a purchasing agent or supply manager personally liable for any secret advantages gained for him or herself.

The authority under which a buyer, purchasing agent, or supply manager functions is granted by the employer. Since the law requires him or her to operate within the bounds of this authority, it behooves such individuals to know as precisely as possible the types of transactions in which he or she can and cannot legally represent the firm. In practice, the degree of authority delegated to buyers, purchasing agents, and supply managers varies significantly among companies. Hence, it is difficult for sales representatives to know the exact limits of a particular buyer's authority. Consequently, under the law, such individuals operate under three types of authority—actual or express authority, apparent authority, and implied authority. A third party dealing with an agent is entitled to rely on the buyer's actual or express authority. Issues arise, however, when the buyer does not have actual or express authority but holds himself or herself out as having the requisite authority to make an agreement (known as apparent authority), or when the third party reasonably believes that the buyer has authority comparable with that of similar agents in similar companies (known as implied authority).

If the third party reasonably relied on the buyer's apparent or implied authority, then the buyer's principal will likely be held to the terms of the contract entered into by the buyer—even if the buyer lacked the requisite authority. However, the buyer's principal, in turn, can bring suit against the agent for acting beyond the limit of his or her actual authority.

Just as supply professionals occupy the legal status of buying agents for their firms, sales representatives similarly hold the status of selling agents for their firms. Buyers, purchasing agents, and supply managers, however, usually are classified as general agents, while salespeople typically are classified as special agents, having somewhat more restricted authority. Consequently, in most cases a salesperson does not have the authority to bind a company to a sales contract or to a warranty. The courts usually hold that, unless otherwise stated, as special agents sales representatives have authority only to solicit orders. It is important that buyers recognize this fact. On important jobs, to ensure that a legally binding contract does in fact exist, a buyer should require acceptance of the order by an authorized company officer, normally one of the supplier's sales managers who customarily serve as the company's general agent for this purpose.

It is critical to spell out in the written purchasing policy and procedures manual, and in the welcome pamphlet for suppliers, who has what kind of purchasing authority; in particular, internal personnel requesting material (the users) that they cannot commit the company in order to avoid backdoor selling and buying, which is a problem in many organizations.

The Purchase Contract

Although a legalistic approach to purchasing is in most cases unnecessary, every buyer, purchasing agent, and supply manager nevertheless must protect his or her company against potential legal problems. The buyer's major responsibility in this regard is to ensure that each purchase contract is satisfactorily drawn and legally binding on both parties. To be valid and enforceable, a contract must contain four basic elements: (1) agreement ("meeting of the minds") resulting from an offer and an acceptance; (2) consideration, or mutual obligation; (3) competent parties; and (4) a lawful purpose.

Offer and Acceptance

When a buyer, purchasing agent, or supply manager sends a purchase order to a supplier, this act usually constitutes a legal offer to buy in accordance with the terms stated in the order. Agreement does not exist, however, until the supplier accepts the offer; when this occurs, the law deems that a "meeting of the minds" exists regarding the proposed contract. In the event that a buyer requests a quotation or a bid from a supplier, the supplier's quotation usually constitutes an offer. Agreement then exists when the buyer accepts the quotations (often by subsequently sending a purchase order to the supplier).

Under the Uniform Sales Act, the law required acceptance of an offer in terms that were identical with the terms of the offer—the mirror image concept. The UCC, however, eliminates this stringent requirement. The code states that "conduct by both parties which recognizes the existence of a contract is sufficient to establish a contract or sale although the writings of the parties do not otherwise establish a contract." The code also recognizes suppliers' standard confirmation forms and acknowledgment forms as a valid acceptance, even if the terms stated thereon are different from the terms of the offer.

In the case of a contract for the sale of goods, when the terms of an acceptance differ from the terms of the offer—the so-called battle of the forms—the terms of the acceptance will automatically be incorporated in the contract unless one of three conditions exists: (1) They materially alter the intent of the offer, (2) the offeror objects in writing, or (3) the offer explicitly states that no different terms will be accepted. What happens when an offer and an acceptance contain conflicting terms, and yet none of the preceding conditions exist? All terms except the conflicting terms become part of the contract, and the conflicting terms are simply omitted from the contract. In that case, the UCC's "gap-filler" provisions (covering price, time and place of delivery, and time of payment) may apply to replace any conflicting provisions that have been knocked-out. In the case of a contract not covered by the UCC, the so-called "mirror image" rule may apply, in which case any contrary or additional terms in an acceptance will not operate as an acceptance but will be a counter-offer that the buyer may accept expressly or by performance.

The UCC contains another important provision relating to the acceptance of an offer to buy. The UCC recognizes as valid the communication of an acceptance in "any manner and by any medium reasonable to the circumstances." Consequently, when a supplier receives an order for the purchase of material for immediate delivery, it can accept the offer either by prompt acknowledgment of the order or by prompt shipment of the material. The code thus permits prompt supplier performance of such proposed contracts to constitute acceptance of the offer. The contract becomes effective when the supplier ships the material.

A longstanding principle of commercial law stated that an offer could be revoked by the offeror at any time before it had been accepted, regardless of the time period stipulated in the offer. The UCC has changed this principle with respect to the purchase or sale of goods. The code states that an offeror may offer to buy or sell material in a signed writing and expressly state therein that the offer remains valid, either for a stipulated time period or if no time period is stipulated, the offer remains valid for a "reasonable" period of time, not to exceed three months. In either case, the offeror may not revoke the offer during the validity period.

This provision of the code has significant implications for industrial purchasers and their potential suppliers. Purchasers can use suppliers' firm quotations in making precise manufacturing cost calculations and rely on the fact that the quotations cannot be revoked before a certain date. Without the code, no such assurance existed. On the other hand, this provision limits a buyer's ability to cancel a firm order, without legal obligation, prior to acceptance. To maintain firm control, it is now doubly important that the buyer state in the order the length of time for which the offer is valid (or the date by which acceptance of the order is required.)

Consideration

In addition to a meeting of minds, a valid contract must also contain the element of obligation. Most purchase contracts are bilateral; that is, both parties agree to do something they would not otherwise be required to do. The buyer promises to buy from the supplier certain material at a stated price; the supplier promises to deliver the material in accordance with stated contract conditions. The important point is the mutuality of obligation. The contract must be drawn so that each party (or promisor) is bound. If both are not bound, in the eyes of the law neither is bound. Hence, no contract exists.

A buyer is confronted with the practical significance of the "mutual obligation" concept when he or she formulates the terms of purchase. The statements regarding material quantity, price, delivery, and so on must be specific enough to bind both the buyer's firm and the supplier to definable levels of performance. In writing a blanket purchase order for pipe fittings, for example, it is not sufficient to state the quantity as "all company X desires." Such a statement is too indefinite to bind company X to any specific purchase. However, if the requirement were stated as "the quantity company X uses during the month of March," most courts would consider this sufficient to define X's purchase obligation. It is also prudent to qualify such a statement by indicating approximate minimum and maximum levels of consumption.

Similar situations arise in specifying prices and delivery dates. Some companies, for example, occasionally issue unpriced purchase orders. Aside from the questionable wisdom of such a business practice, a legal question concerning the definiteness of the offer also exists. From a legal standpoint, the question that must be answered is: Under existing conditions, can the price be determined precisely enough to define the obligations of both parties? The UCC provides more latitude in answering this question than did the

Uniform Sales Act. The code specifically says that a buyer and a supplier can make a binding contract without agreeing on an exact price until a later date. If at the time of shipment a price cannot be agreed on, the code includes provisions by which a fair price shall be determined. On such orders, however, a buyer should protect his or her firm by noting a precise price range or by stating how the price is to be determined.

The issue of predatory purchasing is an interesting situation where predatory bidding can harm competition. The recent Ross-Simmons Weyerhaeuser court case is a new case law on the subject.

> The key message of Weyerhaeuser for supply management professionals is that aggressive bidding for scarce supplies is lawful in the vast majority of cases. Only when a buyer intentionally bids up the cost of supplies to a point that leads to below-cost pricing in the output market, and it is probable that the loss can be recovered after competitors are eliminated, is an anti-trust claim against the buyer likely to succeed.[7]

Competent Parties

A valid contract must be made by persons who have full contractual capacity. A contract made by a minor or by an insane or intoxicated person is usually entirely void or voidable at the option of the incompetent party.

Legality of Purpose

A contract whose purpose is illegal is automatically illegal and void. A contract whose primary purpose is legal, but one of whose ancillary terms is illegal, may be either void or valid, depending on the seriousness of the illegality and the extent to which the illegal part can be separated from the legal part of the contract. The latter situation may occasionally have relevance for buyers. Such would be the case, for example, if a material were purchased at a price that violated restraint of trade or price discrimination laws.

The Written and the Spoken Word

Buyers should be aware of several basic concepts concerning the construction of a contract. Contrary to common belief, a contract is not a physical thing. A contract is actually a relationship that exists between the parties making the contract. When a contract is reduced to writing, the written document is not in fact the contract; it is simply evidence of the contract. Hence, a contract may be supported by either written or oral evidence. In most cases, courts hold an oral contract to be just as binding as a written one, although it may be substantially more difficult to prove the facts on which an oral contract is based. However, the law currently requires some types of agreements to be in writing. In the case of sales transactions between qualified "merchants,"[8] for example, the UCC specifically states that when a selling price of $500 or more is involved, the contract must be reduced to writing to be enforceable.[9]

In the event an oral contract between a supplier and buyer is later confirmed in writing, the written confirmation is binding on both parties if no objection is raised within 10 days. Hence, it is important to note that when a contract is reduced to writing, the written evidence supersedes all prior oral evidence. The courts generally hold that a contract expressed in writing embodies all preceding oral discussion pertinent to the agreement. Generally speaking, this means that a buyer cannot legally rely on a supplier's oral statements concerning a material's performance or warranty unless the statements have been included in the written agreement. Consequently, from a legal standpoint, a buyer should carefully consider the content of his or her oral negotiations with a supplier and ensure that all relevant data to be included in the contract have been reduced to writing. The buyer should also be aware that courts have ruled that written or typed statements take precedence over printed statements on the contract form, should conflicting statements appear in the document.

Letters of Intent

Letters of Intent (LOI) and Memorandums of Understanding (MOU) are a form of pre-contracting as if to say "we do not have a formal agreement but here is what we have agreed upon so far." Many parties to a contract use the LOI as a planning document to order material with long lead times, special tooling, or unique design work. Think of the LOI as a preliminary agreement with general open issues to be resolved. Both parties must be extremely careful as to the wording in LOIs and MOUs lest the language unintentionally create a legally binding contract. Understandably, the legal profession does not favor these instruments with the logical argument that you either have a contract or you do not.[10] However there are situations where their use is appropriate.

Special Legal Considerations

Inspection Rights

If a purchaser has not previously inspected the material purchased to ensure that it conforms to the terms of the contract, the law gives him or her a reasonable period of time to inspect the material after it is received. If the purchaser raises no objection to the material within a reasonable period of time, he or she is deemed to have accepted it. In court decisions on this matter, it has been largely industry practice that sets the standard for "reasonable" time.

Rights of Rejection

A purchaser has the right to reject material that does not conform to the terms of the contract. If an overshipment is received, the purchaser can either reject the complete shipment or reject the quantity in excess of the contract amount. When a buyer does not wish to accept defectively delivered material, he or she is required only to notify the supplier of this fact, describing specifically the nature of the defect or default. The buyer is not legally bound to return the rejected material. However, the buying firm is obligated to protect and care for the material in a reasonable manner. If the buyer neither returns the material nor notifies the supplier of rejection within a reasonable period of time, however, the buying firm is then obligated to pay for the material.

Title and Risk of Loss

From a legal point of view, the question of which party has title to purchased materials is normally answered by defining the F.O.B. point of purchase. In the case of an F.O.B. origin shipment, the buying firm becomes the owner when the material is loaded into the carrier's vehicle. When material is shipped F.O.B. destination, the supplier owns the material until it is off-loaded at the buyer's receiving dock.

Title is to be distinguished from the risk of loss of the goods. Put differently, risk of loss concerns which party assumes liability for goods damaged or destroyed before acceptance. It is a common misconception that risk of loss passes when title passes. The UCC, however, draws a distinction, the lack of knowledge of which is a trap for the unwary.

Risk of loss for *conforming* goods passes at the FOB point unless the contract specifies otherwise. Risk of loss for *nonconforming* goods, however, does not pass until the seller delivers conforming goods or the buyer accepts the nonconforming tender. A savvy buyer will include in the contract language to the effect that risk of loss does not pass until it accepts the goods, including conforming goods.

Warranties

The UCC identifies two specific types of warranties:

1. Implied warranty
2. Express warranty

An implied warranty is one that is read into the contract as a matter of law, unless the parties agree to exclude one or more of the implied warranties. The warranties implied by the UCC are:

Implied Warranty of Good Title: When a supplier agrees to sell a particular item, the firm implies that it (or its principal) has title to the item and hence has legal authority to sell it.

Implied Warranty of Non-infringement: A merchant seller warrants that the goods do not infringe a third party's intellectual property rights.

Implied Warranty of Merchantability: The supplier also implies that the item is free from defects in material and workmanship—that it is at least of "fair average quality." This means that the item meets the standards of the trade and that its quality is appropriate for ordinary use.

Implied Warranty of Fitness for a Particular Purpose: Another implied warranty a buyer may receive under certain circumstances is an implied warranty of fitness for a particular purpose. If a buyer communicates to a supplier what requirements the purchased material must satisfy, and subsequently relies on the skill or judgment of the supplier in selecting a specific material for the job, the material usually carries an implied warranty of fitness for the stated need. This assumes that the supplier is fully aware of the buyer's need and knows that the buyer is relying on guidance from the supplier's personnel. Hence, it should be amply clear why buyers, purchasing agents, and supply managers must insist that purchase orders and related material specifications be written clearly and completely.

An express warranty is a promise or representation that the goods, services, or subject matter of the contract will have certain characteristics or qualities. It is not necessary that an express warranty be labeled as such. If a supplier accepts a purchase order without qualification, descriptions of the material included on the order form—model number, size, capacity, chemical composition, technical specifications, and so on—become an express warranty. The supplier warrants that the material delivered will conform to these descriptions. Additionally, suppliers frequently make express warranties for their products in sales and technical literature. Such warranties typically refer to the material's performance characteristics, physical composition, appearance, and so on. If a buyer has no way of determining the facts of the matter and consequently relies on such warranties, the supplier is normally held liable for them. The buyer should also recognize that an express warranty nullifies an implied warranty to the extent that it conflicts with the implied warranty (with the exception of an implied warranty of fitness for a particular purpose).

Numerous factors influence the extent to which a buyer can rely on an implied warranty in a specific situation. The knowledge and conduct of buyer and seller, as well as the specific conditions surrounding a transaction, are taken into consideration by the court in resolving a dispute over warranty. Generally speaking, if a buyer acts in good faith and has no knowledge of conditions contrary to an implied warranty, the law holds a supplier liable for such implied warranties, unless otherwise stated in the contract.

Recent legislation[11] has tended to increase warranty protection for buyers by strengthening and expanding the liability of manufacturers and sellers with respect to warranty performance. A buyer should recognize, however, that the UCC permits a seller to exclude or modify the implied warranties for a product.

A supplier can disclaim or exclude the implied warranties in one of four ways: First, in the case of the implied warranty of merchantability, the seller must use the word "merchantability" in the disclaimer, and the disclaimer must be conspicuous. No magic words are required to disclaim implied warranty of fitness for particular purpose or other implied warranties. Second, the seller may disclaim implied warranties by use of a statement such as "This item is offered for sale 'as is'," or "There are no warranties which extend beyond the description on the face hereof." Interestingly, in several states such warranty disclaimers have been declared to be contrary to public policy—and, hence, invalid. Third, the buyer cannot rely on the implied warranties if it has inspected the goods or has refused to inspect the goods, if the inspection would have revealed a condition breaching an implied warranty. Fourth, the implied warranty may be excluded by industry custom or practice, or the parties' course of dealing or performance. As a general rule, however, a prudent buyer adopts a caveat emptor attitude in verifying the warranty protection he or she actually has in any given purchase.

Evergreen Contracts

Many firms issue blanket orders in which the exact quantities are not fixed or there are no termination dates. There are many different terms for such contracts such as corporate-wide agreements, requirements contracts, and evergreen contracts.[12] Such contracts have become more common as electronic transactions have become more popular using computer reorder systems. One problem with these contracts is how and when can they be amended as there are no fixed terms? In July 2007, the Ninth District Court of Appeals ruled "one party to a contract is not required to continuously monitor the terms posted on the other party's web site to make sure the other party hasn't surreptitiously changed the terms of the contract." This recent ruling reaffirms a rather long-standing legal principle that a party to a contract must be given direct notice of any proposed changes and be given a reasonable amount of time to respond.[13]

Order Cancellation and Breach of Contract

If a supplier fails to deliver an order by the delivery date agreed on in the contract, or if it fails to perform in accordance with contract provisions, legally the supplier has breached the contract. Not all breaches entitle a party to declare the other in default. Rather, only if the effect of the breach is to deny the non-breaching party the "benefit of the bargain" can the non-breaching party terminate the contract for breach and, if appropriate, seek damages.

Damages are measured based on which party breached and when the breach occurred. If the seller has breached prior to acceptance, the buyer is entitled to either (1) purchase the goods from a third party and charge the breaching seller with the difference between the contract price and the price paid to the replacement contractor; or (2) if the buyer does not replace the goods the seller had contracted to supply, the buyer is entitled to the difference between the contract price and the market price of conforming goods as of the date of the breach. The former remedy is commonly known as "cover." If the buyer has accepted the goods, the buyer is entitled to recover the difference between actual value of the goods accepted and their value if they had been delivered in conformance with the contract.

If the buyer has breached prior to acceptance, such as by repudiating the contract or wrongfully refusing to accept conforming goods, the seller has an action for difference between contract and market price if the seller has not resold goods. If the seller has resold the goods, the seller would be entitled to the difference between the contract price and price at which the seller resold the goods. If the buyer has breached after acceptance, the seller is entitled to be paid the contract price.

Liquidated Damages Provision

If it is evident at the time a major contract is drawn that breach of the contract would severely injure one or both parties and damages would be difficult to determine, it is wise to include a liquidated damages provision in the contract. Such provisions stipulate in advance the procedures to be used in determining costs and damages. In some cases, specific damage payments are stated. For example, if the contract is for the purchase of power-generating equipment to be used on a large construction project, the date of delivery may be critical for the purchaser. Perhaps installation of the equipment must precede other important phases of the construction work. If the project is delayed by late delivery of the generating equipment, the purchaser might incur heavy financial losses. Sound practice on such a contract is normally to include a liquidated damages clause that requires the supplier to pay the purchaser damages of a set amount per day for late delivery. It is essential, however, that the damage figure specified be a reasonable estimate of the probable loss to the buyer, and not be calculated simply to impose a penalty on the supplier. Courts generally refuse to enforce a penalty provision, even if the parties have agreed to it in their contract.

JIT Contracts

Because JIT purchasing and manufacturing operations are somewhat unique, they occasionally generate unexpected legal difficulties. The most common problems are reviewed briefly in the following paragraphs.

The major factor a buyer should keep in mind is that in most cases a JIT purchasing agreement requires different levels of supplier performance than the supplier typically has been used to. Consequently, it is important that communications be clear and complete. This includes oral discussions prior to the purchase, as well as the final written documents. Requirements for quality, delivery scheduling, inventory levels, and any other key factors should be spelled out in unequivocal terms, so there is little opportunity for misunderstanding that might lead to litigation.

Consider the following illustration. Assume that a buyer and a supplier have been doing business satisfactorily for several years. During the past year, approximately one-third of the shipments from the supplier arrived a week or so late, but the buyer accepted them without serious complaints. In the eyes of the law, these acceptances by the buyer may have set a precedent which waives the buyer's rights to timely delivery on future contracts—not good for a new JIT contract. What must the buyer do to regain his or her rights? The two legal requirements are:

1. Give explicit written notice to the supplier.
2. Provide the supplier a reasonable period of time to gear up to meet the new delivery requirements.

With respect to the timing of design or configuration changes, the contract should always specify the minimal lead time, in terms of days or weeks of material usage that the supplier will accept prior to supplying the modified material. Both parties should know what the supplier's planned inventory levels are so the firm is not likely to be left with an unusable stock built to the buyer's specifications. By the same token, the two must also agree on the minimum practical lead-time requirements for a delivery lot size increase or an accelerated delivery schedule.

The point is that JIT systems must be able to respond quickly to demand changes because of their tight scheduling and low inventory characteristics on the buyer's side. These requirements for flexibility must be built into the purchase contract to the extent possible. Although meshing the buyer's needs with the supplier's capabilities may be difficult, it is these issues that should be discussed ahead of time, agreed on, and stated in the contract.

Inspection and acceptance is another area that can pose problems. Many JIT shipments are delivered directly to the point of use, without first going through incoming inspection. In many cases, detection of nonconforming items does not occur until some time later, after the item has entered the production process. If no provision for this situation is made in the contract, legally the material may be considered to have been accepted when delivered. Consequently, this modified operating procedure should be detailed clearly in the contract. A satisfactory time frame for acceptance and the responsibility for subsequent rework costs should be stipulated.

In structuring JIT purchase orders and contracts, common sense tells the buyer to be conservative and to include ample detail about these unique issues in the contractual documents.

Honest Mistakes

When an honest mistake is made in drawing up a purchase contract, the conditions surrounding each specific case weigh heavily in determining whether the contract is enforceable as written. As a general rule, a mistake made by only one party does not render a contract unenforceable, unless the other party is aware or should have been aware of the mistake at the time of contract formation. The mistake must concern a basic assumption on which the contract was made, and must materially affect the agreed exchange of performance. To affect a contract, a mistake usually must be made by both parties as to some basic assumption of the contract or its subject matter. Even then, not every mutual mistake invalidates the contract.

Assume, for example, that a supplier intends to submit a quotation with a price of $260. Through an error, the price is typed on the quotation as $250 and is so transmitted to the buyer. In such cases, courts have held that if the buyer accepts the offer, without knowledge of the error, a valid contract exists. The magnitude of the error is deemed insufficient to affect agreement materially. On the other hand, if the $260 price were incorrectly typed as $26, the court would probably hold that a competent buyer should have recognized the error and sought clarification before accepting it.

Contrast the foregoing example with the following scenario in which the contract would not be enforceable as written because of the failure of a basic assumption: If the buyer and a supplier agree on the sale of specific machinery and unknown to either, the machinery has been destroyed or for some other reason is not available for sale, a mutual mistake exists and the contract would not be enforced.

Generally speaking, a buyer should not assume that a mistake, however innocent, will release his or her firm from a contractual obligation. In the majority of cases, it will not do so. A prudent buyer employs all reasonable means to minimize the possibility of committing contractual mistakes.

Patent Infringement

The law does not give a patent holder the exclusive right to manufacture, sell, and use the patented device for a specified number of years. Rather, the law gives a patent holder the right to *preclude* others from manufacturing, using, or selling the patented device for the term of the patent. A supplier who engages in any of these activities during the term of the patent without permission from the patent holder may be liable to the patent holder for patent infringement.

Buyers frequently have no way of knowing whether their suppliers are selling infringing materials with or without authorization from the patent holder. If a purchaser unknowingly buys an item from a supplier who has infringed the patent holder's rights, the purchaser may also be liable for infringement. To protect against such unintentional violations, most companies include protective clauses in their purchase contracts in which the supplier warrants that its materials do not infringe a third party's intellectual property rights[14] and that the supplier will indemnify the purchaser for all expenses and damages resulting from patent infringement. Clauses of this type do not prevent the patent holder from suing the customer. If properly stated, however, they can require the supplier to defend the customer in such legal proceedings and can give the customer legal recourse to recover any resulting losses from the supplier.

Restraint of Trade Laws

The Robinson-Patman Act, a 1936 amendment to the Clayton Act, is designed to prevent price discrimination that reduces competition in interstate commerce. Generally speaking, the Act prevents a supplier from offering the same quantity of a specific material to similarly situated *distributors* at different prices, unless (1) one distributor is offered a lower price because his or her purchases entail lower manufacturing or distribution costs for the supplier, or (2) one distributor is offered a lower price in order to meet the legitimate bid of a competing supplier.[15]

As Pinkerton and Kemp state, while the law was designed primarily to protect small retail food businesses from the large chain stores such as A&P, "the act could apply to typical industrial purchasing situations. There have been very few cases involving industrial buyers purchasing materials, equipment, and components for their own use".[16] However, there have been several major cases including Jacobs Manufacturing Company (manufacturer of industrial chucks used in portable machine tools), and the famous Minneapolis-Honeywell Regulator Company case of 1948.[17] While litigation is rather dormant at the present time, one must remember the Robison-Patman Act is "still on the books" and future federal administrations may aggressively pursue its enforcement.[18] In addition, some suppliers will use the Robinson-Patman Act as a scare tactic to not grant legitimate volume discounts; very legal under the various defenses as listed in this section and in the Pinkerton-Kemp article.

International Considerations

When a supply manager sources outside the United States, the chances are very good that a different set of laws will govern the related purchasing transactions. If stipulated and agreed on in the contract, the governing law could be U.S. law, or it could be the law of the supplier's country. Whether or not U.S. law is selected as the governing law, the United Nations' Convention on Contracts for the International Sale of Goods (CISG)—an international analog to the UCC, but with several provisions that materially differ from the UCC—could be read into the contract unless expressly waived.

In any case, in international or global procurement, it is particularly important to stipulate in the purchase order or contract which body of law is acceptable to both the buyer and the seller—and subsequently will govern the transaction. Likewise, it is important also to stipulate a mutually acceptable "choice of forum"—that is, the location in which disputes will be heard, in the event a legal dispute arises.

Many firms today select arbitration for dispute resolution under cross-border contracts. There are several factors that commend this approach over litigating a dispute in a foreign court. First, the parties can stipulate, among other things, who will supervise the arbitration (e.g., the London Court of International Arbitration, the International Chamber of Commerce, among many), the venue for the arbitration, the language in which the arbitration will be conducted, and the qualifications of the arbitrators. Second, countries that have signed the Convention on the Recognition and Enforcement of Foreign Arbitral Awards (known as the "New York Convention") agree to enforce arbitral awards wherever obtained.[19] Third, the parties can avoid the situation where one party is favored over the other by virtue of having the dispute litigated in its hometown.

The following sections provide an overview of two key topics for international purchasers—the CISG and the Foreign Corrupt Practices Act.

Contracts for the International Sale of Goods

During the early 1980s, the United Nations facilitated the development of a uniform body of law to govern contracts for the international sales of commercial goods. As noted previously, the title given to this body of law is the United Nations' Convention on Contracts for the International Sale of Goods, commonly known as the CISG. The CISG's objective is much like the objective of the Uniform Commercial Code, projected to the international level. The CISG does not apply to the purchase of services or to personal purchases of consumer goods.

Generally speaking, the CISG and the UCC have more similarities than differences. However, there are five significant differences that purchasing and supply professionals should know about:

▲ *Acceptance of an offer* The CISG requires that an offer be accepted in identical terms—the mirror image concept. If an acceptance contains terms that conflict with those in the offer, no contract exists and the acceptance is treated as a counter-offer.

▲ *Revocation of an offer* The CISG permits an offer to be revoked any time before an acceptance is received. One exception is

> if it was reasonable for the offeree to rely on the offer as being irrevocable and the offeree acted in reliance on the offer

then the offer cannot be revoked. This revocation provision is less stringent than its counterpart in the UCC.

▲ *Formation of a contract* Under the CISG, a contract is created at the time the acceptance is received by the offeror. Under the UCC, the contract is created when the acceptance is mailed or transmitted to the offeror.

▲ *Oral contracts* The CISG recognizes oral contracts as being valid and enforceable. In contrast with the UCC, contracts exceeding $500 in value do not require written evidence.

CISG use by American purchasers is placed in focus by the following statement of an internationally known legal authority:

> The similarities between the CISG and the UCC are sufficient enough so business executives do not have to make an issue out of which set of rules applies. On the other hand, one should always be aware that there are these two sets of rules, and in specific cases one may be preferable to the other. In any event, the CISG would appear to be preferable (from the U.S. standpoint) to agreeing to (the use of) another country's law.[20]

Foreign Corrupt Practices Act[21]

In the early 1970s, Congress and the American public learned about a number of questionable payments made by U.S. multinational corporations to foreign government officials to gain an advantage in bidding for business contracts awarded by those governments. As a result of the strong negative public reaction to these shady business dealings, appropriate federal agencies investigated the international activities of U.S. firms that appeared to involve the possibility of commercial bribery. The investigation identified several hundred major firms that had been involved in such questionable dealings with potential international customers.

As a result of these findings, in 1977 Congress passed the Foreign Corrupt Practices Act (FCPA) as an amendment to the Securities Exchange Act of 1934. The objective of the new act was to curtail U.S. corporate involvement in foreign commercial bribery activities—and more generally to enhance the image of the United States throughout the world.

The FCPA contains three major sections focusing on (1) antibribery issues, (2) record-keeping requirements, and (3) penalty provisions. The antibribery section makes it a crime for a U.S. firm to offer or to make payments or gifts of substantial value to foreign officials. The intent is to prohibit payments in any form to influence a major decision of a foreign government official.

Somewhat to the contrary, however, is the fact that the act does allow some forms of bribery—those that are considered to be minor and inconsequential in influencing important government decisions. It is permissible to make payments (known as "grease money") to operating officials with ministerial or clerical duties. Although the FCPA is vague with respect to the details of application, the Omnibus Trade Act of 1988 specifies what types of payments are acceptable and who may receive them. Such payments are termed "transaction bribes" and are intended to accelerate the performance of a routine function, such as loading and unloading cargo, processing goods through customs promptly, moving goods across country, processing papers, and so on. It is expected that these types of payments may speed up governmental actions by lower-level officials that, in time, would have occurred anyway.

All other types of bribes are considered illegal. As the law now stands, the Omnibus Trade Act holds a firm criminally liable if evidence indicates that one of its representatives had actual knowledge that an illegal payment was made to a foreign government official to secure a favorable decision on a major issue.

Clearly, the FCPA and the related Omnibus Trade Act were designed primarily to curb unacceptable international sales practices. At the same time, however, they apply to international procurement practices. Purchasing professionals engaged in international buying should understand the provisions of these acts, and they must know the difference between acceptable transaction bribes and bribes whose intent and motivation are illegal.

Summary

The purpose of this chapter is to alert buyers and supply management professionals to the most basic legal considerations that relate to the purchasing and supply function. Yet, there is danger in doing this. No author can briefly accomplish this objective without simplifying the issues. Such simplification may at times leave the reader with an incomplete understanding, which lulls him or her into a false feeling of security.

Even though adoption of the UCC by all but one of the states creates greater uniformity among state commercial laws than ever before, it is unreasonable to assume that interpretations of the laws by the various states will not vary. Only time can reveal how significant such variations will be. Moreover, the interpretation of circumstances surrounding each specific case weighs heavily in the analysis of that particular case. These factors virtually defy a definite and unqualified analysis of a legal controversy by anyone who is not a highly skilled professional in the legal field. Heinritz, Farrell, Guinipero, and Kolchin state the matter cogently in saying that "the person who tries to be his or her own lawyer has a fool for a client."[22] Perhaps the most important function of this chapter is to underscore the fact that supply professionals should seek sound legal counsel whenever potential legal problems arise.

Just as a lawyer is expected to exhibit skill in extricating his or her client from legal entanglements, so a purchasing and supply executive is expected to exhibit skill in avoiding legal controversies. A supply manager must understand basic legal concepts well enough to detect potential problems before they become realities. At the same time, the most powerful tool he or she can utilize to avoid legal problems is skill in selecting sound, cooperating, and reliable suppliers. Vigilance in this area of responsibility minimizes the need for legal assistance.

Endnotes

1. G. Ernest, and J. D. Gabbard, "Contracting for Services: What Are the Differences?" *Inside Supply Management* July (2005): 14–15.

2. Richard A. Mann and Barry S. Roberts, *Smith & Roberson's Business Law,* 13th ed. (Mason, OH: West Legal Studies in Business, an imprint of Thomson/South-Western, Thomson Higher Education, 2006) 254, 983.

3. Ibid., 983.

4. Ibid., 983 and see http://www.cybersign.com.

5. http://www.uncitral.org/uncitrazh/publications.html

6. See, generally, Dr. John Murray, Jr., "What's New in UCC Article 2," http://www/purchasing.com/article/CA337305.html.

7. Anthony A. Dean, "Predatory Purchasing: Part 2," *Inside Supply Management* May (2007): 48–49.

8. Section 2-104 of the UCC defines a "merchant" as one who deals in goods; one who holds himself out as having particular skill in the subject matter; or one who uses a person who holds himself out as having such knowledge or skill. Hence, it is generally held under this broad definition that almost every person in business, including a purchasing officer, is a "merchant." Even a person not in business may be classified as a "merchant" if he or she employs a purchasing officer or broker.

9. Prior to enactment of the UCC, the Statute of Frauds required contracts relating to personal property, for which neither delivery nor payment had been made, to be in writing if the value of the sale exceeded a specified amount; this specified amount varied widely among states.

10. Judi Coover, "Letter of Intent," *Inside Supply Management* November (2006): 32–33.

11. The Consumer Product Safety Act and the Federal Warranties Act, as well as the UCC.

12. Jane K. Winn, "Evergreen Customer Contracts," *Inside Supply Management* February (2008): 38–39.

13. Ibid.

14. Note that in a contract for the sale of goods it is not necessary to have an express warranty of non-infringement. As discussed on p. 12–14, under the UCC the supplier gives an implied warranty of non-infringement unless it is disclaimed.

15. A third condition also permits the offering of a discriminatory price—namely, one in which the marketability of goods is affected. Seasonal goods or those approaching obsolescence or deterioration fall into this category.

16. Richard L. Pinkerton and Deborah J. Kemp, "The Industrial Buyer and the Robinson-Patman Act," *International Journal of Purchasing and Materials Management* Winter (1996): 29–36. See also W.C. Benton, Jr., *Purchasing and Supply Chain Management*, 2nd ed. (Burr Ridge, IL: McGraw-Hill/Irwin 2010): 300–307. Deborah Kemp and Richard L. Pinkerton, "Does the Robinson-Patman Act Really Affect Purchasing," *NAPM Insights* October (1994): 16 (see www.ism.ws).

17. Ibid.

18. Ibid.

19. As of the date of publication, 142 countries have signed on to the New York Convention.

20. W.A. Hancock, "The UN Convention on the International Sale of Goods," *Executive Legal Summary* May (1993): 100–104.

21. This discussion is based on material presented by Glenn A. Pitman and James P. Sanford, "The Foreign Corrupt Practices Act Revisited: Attempting to Regulate Ethical Bribes in Global Business," *International Journal of Purchasing and Materials Management* Summer (1994): 15–20.

22. Stuart Heinritz, Paul Farrell, Larry Guinipero, and Michael Kolchin, *Purchasing Principles and Applications* (Englewood Cliffs, NJ: Prentice-Hall, 1991) 241.

Suggested Reading

Bouchoux, Deborah. *Protecting Your Company's Intellectual Property: A Practical Guide to Trademarks, Copyrights, Patents & Trade Secrets.* New York: AMACOM, 2002.

Focus on the Legal Aspects of Supply Management: A Collection of ISM Legal Articles. Tempe, AZ: ISM, 2003, www.ism.ws.

Gabriel, Henry D, and Linda J. Rusch. ABCs of the UCC. Article 2: Sales. American Bar Association, Chicago, IL: 1997.

ISM Technotes: Contract Terms and Conditions for Purchase Orders and Other Contractual Agreements. Tempe, AZ: ISM, 2001. www.ism.ws.

Mann, Richard A. and Barry S. Roberts. *Business Law and the Regulation of Business* 9th ed. Cincinnati, OH: Cengage Learning, 2007.

Pohlig, Helen M. *Legal Aspects of Supply Management*, 3rd ed. Tempe, AZ: ISM, 2008.

Potential changes to the UCC. http://www.law.upenn.edu/bll/ulc/ulc.htm

Uniform Commercial Code, The American Law Institute, 4025 Chestnut Street, Philadelphia, PA 19104; and *the National Conference of Commissioners on Uniform State Laws*, 211 E. Ontario St. Suite 1300, Chicago, IL 60611

200 – Official Text West Group

Box 61799, 620 Operman Drive, St. Paul MN 55164-0779

Uniform Commercial Code & Commercial Law. http://www.megalaw.com/
http://www.law.cornell.edu/uniform/ucc.html
http://lawcrawler.findlaw.com/

CHAPTER 20

Contract and Relationship Management

Need for Better Contract Management

Historically, the post-award phase of supply management has been a weak one at many organizations. Prior to the mid-1980s, large inventories were available to accommodate quality problems and late deliveries. Multiple sources of supply often allowed the buyer to live with little supplier management. But this mentality cost the buying firm dearly. Shorter production runs under a multiple-sourcing policy frequently resulted in higher prices, lower quality, and the receipt of items that were not completely identical. Late deliveries resulted in production disruptions, higher production costs, and broken delivery commitments to the firm's customers.

Today, most firms have reduced inventories as embodied in the just-in-time (JIT) philosophy. JIT creates even more need for professional post-award management. Under JIT, large inventories are no longer available to cushion the results of weak management at this stage in the process. JIT requires buyers and suppliers to work together to reduce the need for inventories. Tight schedule integration between supplier and customer must be maintained. Processes must be balanced. Waste must be reduced or eliminated. Communication must be in real-time. These requirements combine to make buyer-supplier collaboration essential. Such collaboration requires supply managers to take a proactive approach to managing both the contract and the supplier relationship. Professional supply relations management is a vital ingredient in several other settings: defense subcontracts, construction contracts, and the purchase of essential services.

The foremost prerequisite to successful contract and relationship management is a sound understanding by both parties of all aspects of the program. Early supplier involvement, as described throughout this text, greatly facilitates the development of this understanding.

This chapter discusses the many activities a supply manager must perform to ensure that the quality specified in the contract is received on time and that relations with key suppliers are managed carefully in an effort to ensure satisfaction with the supplier's performance now and in the future.

Preaward Conference: The Stage Has Been Set

When the dollar magnitude, complexity, or criticality of the work to be performed dictates, professional supply managers hold a conference with the prospective supplier immediately prior to award of the contract. The supply management team—consisting of the supply manager, subcontract administrator or expediter, design engineer, manufacturing engineer, internal customer, quality engineer, and inspector, as appropriate—should meet with the supplier's team to discuss the supplier's plans for satisfying the customer's needs.

The issues, presented in technical terms, will have been addressed in the request for proposal and the proposed contract. It is important that supply professionals are aware of the transformation of responsibility taking place within the supplying organization. The supplier's employees responsible for consummating the sale normally are not responsible for performance under the contract. Thus, a new team consisting of operations, demand planning and management, quality, and the supplier's supply management assumes responsibility for performance under the terms of the contract. The preaward conference is the vehicle the supply management professional and his or her team use to ensure that the contract provisions are fully understood and implemented. The following items, as appropriate, should be addressed:

- All terms and conditions.
- Delivery or operations schedule.
- Staffing and supervision.
- Site conditions, work rules, safety (if appropriate).
- Invoicing procedures and documentation (for incentive and cost contracts).
- Materials purchase procedures (for incentive, cost, and time and materials contracts).
- Background checks and security clearances.
- Insurance certificates.
- Permits.
- Possible conflicts with other work.
- Submission of time sheets (for incentive, cost, and T&M contracts).
- Buyer responsibilities. Buyer-supplied items such as customer furnished purchased materials, tools, equipment, facilities, and so on. Timeliness of buyer reviews and

approvals for studies, reports, plans and specifications, and so on must be established and accepted by both parties.

▲ Collaboration milestones. Potential points of collaboration that improve designs, processes, communications, quality, and delivery should be identified as early as possible, rather than after award of the contract.

▲ Key contact people to communicate all of the above with progress reports.

Major one-time projects, such as large construction or site development jobs, have their own, fairly unique, reporting requirements. These are described in Appendix A of this chapter.

Monitoring and Controlling Project Progress

Suppliers are responsible for the timely and satisfactory performance of their contracts. Unfortunately, the supply manager cannot rely entirely on the supplier to ensure that work is progressing as scheduled and that delivery will be as specified. Poor performance or late deliveries disrupt production operations and result in lost sales. Accordingly, supply management must monitor supplier progress to ensure that desired material is delivered on time. The method of monitoring depends on the lead time, complexity, and urgency of the order or contract. The level of monitoring depends upon the criticality of the supplied material or service and the demonstrated capability of the supplier in meeting the buying firm's requirements.

At the time a purchase order or contract is awarded, the supply manager should decide whether routine or special attention is appropriate. On many orders for non-critical items, simply monitoring the receipt of receiving and inspection reports may be adequate. On others, telephone confirmation that delivery will be as specified may be sufficient. But on orders for items critical to the scheduling of operations, more detailed procedures are in order.

When evaluating a supplier's progress, the supply manager is interested in *actual* progress toward completing the work. Data about progress may be obtained from a variety of sources: progress conferences over the phone or face-to-face, faxes, emails, field visits to the supplier's facility, and periodic operations progress reports by the supplier.

Many organizations assign a project manager to provide overall planning, supervision, and tracking of large and complicated projects.

Operations Progress Reports

In some instances, the supplier is required by the terms of the contract to submit a phased production schedule for review and approval. A phased operations schedule shows the time required to perform the production cycle—planning, designing, purchasing, tooling, plant rearrangements, component manufacture, subassembly, final assembly, testing, and shipping.

In many cases, the supply manager may include a requirement for production progress information in the request for proposal (RFP) and in the resulting contract. The ensuing reports frequently show the supplier's actual and forecasted deliveries, as compared with the contract schedule; delay factors, if any; and the status of incomplete preproduction work such as design and engineering, tooling, construction of prototypes, and so on. The reports also should contain narrative sections in which the supplier explains any difficulties, and action proposed or taken to overcome the difficulties. In designing the system, the supply manager should ask himself or herself, "What is really essential information?" in an effort to prevent the system from becoming a burden instead of a tool for good management.

Operations progress reports do not alleviate the requirement to conduct visits to the supplier's work site on crucial contracts. The right to conduct such visits must be established in the RFP and the resulting contract. On critical contracts, where the cost of such visits is justified, it may be desirable to establish a resident facility monitor to ensure the quality and timeliness of the work being performed at the supplier's facility.

When it is determined that an active system of monitoring the supplier's progress is appropriate, the first step in ensuring timely delivery is to evaluate the supplier's proposed delivery schedule for attainability. In their planning and control activities, most suppliers utilize a variety of graphic methods for portraying the proposed schedule, and then for monitoring progress against it. These are useful management tools that also can be reviewed and evaluated by the buying team. These visual presentations are forceful and usually can be updated economically. Two progress planning and control techniques commonly used for important projects and jobs are now discussed briefly.

Gantt Charts

Gantt charts are the simplest of the charting techniques for planning and controlling major projects and the materials deliveries that flow from them. Gantt charting requires that (1) first, a project must be broken into its elements; (2) next, the time required to complete each element must be estimated and plotted on a time scale; (3) the elements then must be listed vertically in time sequence, determining which elements must be performed sequentially and which concurrently; and (4) finally, actual progress is charted against the plan on the time scale.

In addition to the detailed charts maintained by the supplier, the supply manager also can construct a master chart to use in controlling the job. In this case, the supplier is asked

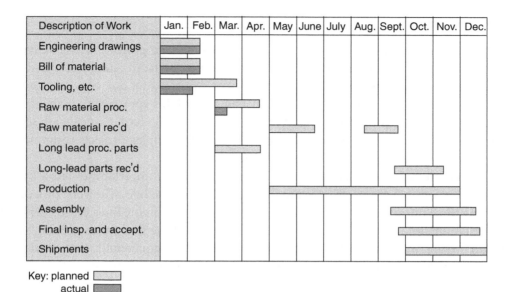

Figure 20.1 Production schedule and progress chart (Gantt chart).

to submit weekly or monthly progress data that are posted to the buying firm's master chart. Figure 20.1 illustrates a typical Gantt chart of this type.

The Gantt chart, then, portrays the plan, schedule, and progress together in one easy-to-use chart. It shows the status of project elements, or activities, and identifies which are behind or ahead of schedule. Unfortunately, Gantt charts fail to provide the full impact of an activity's being behind or ahead of schedule, and they do not provide sufficient detail to detect some schedule slippages in a timely manner.

As long as the project is not too large or complex, Gantt charts work fine. As the number of *interdependent* activities increases, however, a Gantt chart fails to tell the supply manager one important fact he or she needs to know to manage the project efficiently. If slowdowns occur as work progresses on the various parts of the project, is the timing of some activities more critical than others? The answer is yes—and the Gantt chart does not tell which ones! For every project, at least one group of interrelated activities makes up what is called the *critical path*. If any of these activities falls behind schedule, the total project will not be completed on time. The supply manager must monitor these potential bottleneck activities very carefully. To overcome this Gantt chart deficiency, a computer-based technique called critical path scheduling has been developed to provide the required management capability.

CPM and PERT

Critical path scheduling is a tool that can be used to manage project buying activities, construction projects, and research and development projects, to name just a few. The technique can be used for these and other complex projects of a one-time nature. The project's magnitude must justify the relatively high cost of this approach to management compared with more conventional methods such as the Gantt chart. Critical path scheduling is useful for planning, monitoring, and controlling complex projects composed of a large number of *interrelated* and *interdependent* activities.

The critical path approach quantifies information about uncertainties faced by the activities responsible for meeting a predetermined time schedule. The very process of analyzing these uncertainties focuses the manager's attention on the most critical series of activities in the total project from a timing perspective—those that constitute the "critical path." The critical path activities must be accomplished sequentially, thus representing the chain of activities that require the most time from start to finish of the project. When these are pinpointed, the manager can develop an appropriate control strategy to optimize the operating results.

When used for controlling contract performance, the posting of progress data permits the supply manager to compare actual accomplishments with those planned. He or she can then determine the likely implications of slippages with respect to critical path time requirements. The supply manager then works with the supplier to correct or minimize the resulting problems.

A variety of specific techniques have been derived from the basic critical path scheduling concept. The best known of these are *critical path method* (CPM) and *program evaluation and review technique* (PERT). CPM was originally developed in 1955 by the Du Pont and Remington Rand companies for use in coping with complex plant maintenance problems. PERT emerged in 1958 through the joint efforts of the United States Navy, the Booz, Allen & Hamilton consulting firm, and the Lockheed Missile and Space Division in connection with the Polaris weapons program. With the passage of time, PERT and CPM have become very similar in concept. Currently, they differ only with respect to various details of application.

In practice, the application of CPM/PERT generally is accomplished with a computer program that uses network diagrams to show time and dependency relationships between the activities that make up the total project. The purpose of the technique is to keep all the "parts" arriving on schedule so that the total project can be completed as planned. Using CPM/PERT data outputs, the supply manager can evaluate possible trade-offs if supplier resources were reallocated in alternative ways to improve the chances of meeting the time schedule. This technique quickly can determine the results of alternative courses of action, thus making the best choice of the alternatives available.

Most importantly, the CPM/PERT technique is both a planning and a control tool. The fact that the individual activities of a project are structured into a network requiring time and sequence determination is critically important; it is the basis for all subsequent monitoring and control activity. Use of the technique forces the supply manager to conduct this step-by-step planning in advance, and to periodically reexamine the logic of the decisions as the dynamics of the operation unfold. The mechanics of critical path scheduling are briefly described with an example in Appendix B at the end of this chapter.

Closed-Loop MRP Systems

Many manufacturers use a closed-loop MRP system to schedule and control production, inventory levels, and deliveries from outside suppliers. Supplier-furnished data can be used by the individuals controlling the firm's incoming materials schedules as they monitor their system reports. Daily or weekly status reports flow from the supplier to the scheduler, who then inputs the data for the next MRP run. See Chapter 10 for a full discussion of MRP systems.

A growing number of firms link their suppliers' computer-based systems into their own computer system so that real-time data are available to both parties. If supplier-furnished data indicate that the supplier's delivery dates will result in a disruption of the buying firm's schedule, the appropriate scheduler or supply manager can take action to modify the supplier's schedule or to adjust the buying firm's own production schedule, as appropriate.

Monitoring and Controlling Total Supplier Performance

The use of supplier performance evaluation systems is on the rise. A majority of major manufacturing firms, as well as increasing the number of services, either have established formal supplier-evaluation programs or are in the process of doing so.

Many progressive buying organizations monitor their critical suppliers' performance at both a contract and an aggregate level. This information is used to control a supplier's contract performance, and it is also used during source selection for follow-up procurements to ensure that only satisfactory performers are considered.

Supplier Performance Evaluation

After a major supplier has been selected and the buyer-supplier relationship has begun to develop, it is important to monitor and assess the supplier's overall performance. The purpose is to enhance the relationship and thereby control performance.

Many evaluation teams use a 3- to 6-month moving average for the aggregate evaluation of a supplier's performance. For example, with a 6-month window, a supplier's rating in June is an average of all the ratings accumulated between January and June. The moving average allows suppliers to start over at some point. The supplier's misdeeds don't haunt them forever. They are motivated to improve. The length of the window is important and should be case-specific. A shorter window may be ineffective because it lets suppliers off the hook too easily. A longer window may be punitive and self-defeating.

Three types of evaluation plans are common: the *categorical plan,* the *weighted point plan,* and the *cost ratio plan.* Each of these plans is reviewed briefly in the following pages.

Categorical Plan

Under this plan, personnel from various departments of the buying firm maintain informal evaluation records. The individuals involved traditionally include personnel from supply management, engineering, quality, accounting, and receiving. Each evaluator prepares a list of performance factors that are important to him or her for each major supplier. Each major supplier is evaluated against each evaluator's list of factors at a monthly or bimonthly meeting. After the factors are weighted for relative importance, each supplier is then assigned an overall group evaluation, usually expressed in simple categorical terms, such as "preferred," "neutral," or "unsatisfactory." This simple qualitative plan is easy to administer and has been reported by many firms to be very effective.

The Weighted Point Plan

Under this plan, the performance factors to be evaluated (often various aspects of quality, service, price, technology, and management skills) are given "weights." For example, in one circumstance, quality might be weighted 25%, service 25%, and price 50%. In another, quality could be raised to 50%, and price reduced to 25%. The weights selected in any specific situation represent supply manager or supply team judgments concerning the relative importance of the respective factors.

After performance factors have been selected and weighted, a specific procedure is then developed to measure actual supplier performance on each factor. Supplier performance on each factor must be expressed in quantitative terms. In order to determine a supplier's overall rating, each factor weight is multiplied by the supplier's corresponding performance number; the results (for each factor) are then totaled to get the supplier's final rating for the time period in question.

The following hypothetical case illustrates the procedure. Assume that a supply management department has decided to weight and measure the three basic performance factors as follows:

Weight	Factors	Measurement Formula
50%	Quality performance = 100% − percentage of rejects	
25%	Service performance = 100% − 7% for each failure	
25%	Price performance = $\dfrac{\text{lowest price offered}}{\text{price actually paid}}$	

Assume further that supplier A performed as follows during the past month. Five percent of its items were rejected for quality reasons; three unsatisfactory split shipments were received; and A's price was $100/unit, compared with the lowest offer of $90/unit. Table 20.1 summarizes the total performance evaluation calculation for supplier A.

This procedure can be used to evaluate any number of different suppliers whose performance is particularly important during a given operating period. The performance of competing suppliers can be compared quantitatively, and subsequent negotiating strategies developed accordingly. The user should always remember that valid performance comparisons of two or more suppliers require that the same factors, weights, and measurement formulas be used *consistently* for all suppliers.

In contrast to the categorical plan, which is largely subjective, the weighted point plan has the advantage of being somewhat more objective. The exercise of subjective judgment is constrained more tightly in the assignment of factor weights and the development of the factor measurement formulas. The plan is extremely flexible because it can accommodate any number of evaluation factors that are important in any specific case. Also, the plan can be used in conjunction with the categorical plan if buyers wish to include important subjective matters in the final evaluation of their suppliers.

Various research studies have noted, however, that a weighted point plan must be developed with care. The estimates of factor importance must be consistent from one situation to the next, and they must be consistent with the performance measurement formulas used because of the obvious interaction between them.

Table 20.1 Illustrative Application of the Weighted Point Plan

Supplier A Monthly Performance Evaluation			
Factor	Weight	Actual Performance	Performance Evaluation
Quality	50	5% rejects	$50 \times (1.00 - 0.05) = 47.50$
Service	25	3 failures	$25 \times [1.00 - (0.07 \times 3)] = 19.75$
Price	25	$100	$25 \times \dfrac{\$90}{\$100} = 22.50$
			Overall evaluation: = 89.75

Cost Ratio Plan

This plan evaluates supplier performance by using the tools of *standard cost* analysis that business people traditionally use to evaluate a wide variety of business operations. When using this plan, the buying firm identifies the *additional* costs it incurs when doing business with a given supplier; these are separated as costs associated with the quality, service, and price elements of supplier performance. Each of these costs is then converted to a "cost ratio", which expresses the additional cost as a percent of the buying firm's total dollar purchase from that supplier. These three individual cost ratios are then totaled, producing the supplier's overall additional cost ratio. For purposes of analysis, the supplier's price is then adjusted by applying its overall cost ratio. The adjusted price for each supplier is then compared with the adjusted price for other competitive suppliers in the final evaluation process.

For example, assume that for one supplier the quality cost ratio is 2%, the delivery cost ratio is 2%, the service cost ratio is –1%, and the price is $72.25. The sum of all cost ratios is 3%; hence, the adjusted price for this supplier is [72.25 + (0.03 × 72.25)] = $74.42. This is the price used for evaluation purposes vis-à-vis other suppliers.

Although the cost ratio plan is used by a number of large progressive firms, overall it is not widely used in industry. Operationally speaking, it is a complex plan. It requires a specially designed, company, wide, computerized cost accounting system to generate the precise cost data needed for effective operation. Consequently, the majority of supply management departments employing a quantitative type of evaluation rely on the simpler but effective weighted point plan—typically, modified specifically to meet their own unique circumstances. The use of activity-based cost accounting (ABC) systems is essential to provide data for cost ratio plans.[1]

For these reasons, the cost ratio plan is not discussed in detail in this book. Nevertheless, it is an excellent concept that has the ability to provide the most precise evaluation data of the three plans discussed. For firms using sophisticated information systems, the cost of designing and implementing the cost ratio plan typically is repaid many times by savings resulting from more precise analysis of supplier performance.

All three of the plans discussed—categorical, weighted point, and cost ratio—involve varying degrees of subjectivity and guesswork. The mathematical treatment of data in two of the plans often tends to obscure the fact that the results are no more accurate than the assumptions on which the quantitative data are based. In the final analysis, therefore, supplier evaluation must represent a combined appraisal of facts, quantitative computations, and value judgments. It simply cannot be achieved effectively by mechanical formulas alone.

Since the late 1980s, *scorecarding* has become a trendy, but confusing term to describe variations of these three methods. The confusion is due to misinterpretation of the balanced scorecard, a tool developed by Kaplan and Norton that focuses on four corporate goals: financial, customer, internal business processes, and learning and growth.

Cost-Based Supplier Performance Evaluation

A number of firms are experimenting with various types of cost-based evaluation plans, similar to the cost ratio plan. Such plans address the issue of how to measure overall supplier performance on a total cost basis. In addition to rationalizing lowest total cost performance suppliers, such plans demonstrate that supplier nonperformance costs can be measured. Recognizing the supplier as an integral member of the organization, competitive strategy requires the development of a system that provides supplier accountability and control while maintaining dependable, competitive suppliers. As organizations continue to secure longer-term supplier relationships, the ability to quantify performance becomes increasingly important. Companies can use the methodology as a contract monitoring tool when incorporated into long-term agreements. Again the use of ABC accounting systems greatly facilitates this kind of evaluation.

Motivation

Two common approaches are used to motivate suppliers to perform satisfactorily: punishment and reward. Many progressive supply managers use a combination of both approaches.

Punishment

Quite obviously, the greatest punishment for unsatisfactory performance (if the area of litigation and punitive damages is ignored) is *not* to award contracts for future requirements. This is a powerful motivator, especially during periods when supply professionals are reducing the number of suppliers with whom they do business.

Another method is to downgrade a supplier. Applied Materials, for example, grades suppliers into categories, the highest of which is preferred. A preferred supplier will have business increased over time and be included in Applied Material's strategic plans. A poorly performing supplier can be downgraded, thereby reducing future business opportunities.

A less drastic approach called the "bill back" is especially appropriate when dealing with a "collaborative" supplier or a defense contractor. Under the bill back, incremental costs resulting from quality problems or late deliveries are identified and then billed back to the appropriate supplier. Some progressive supply managers have increased the motivational effect of the bill back by sending the bill to the supplier's chief operating officer so that he or she is aware of problems within his or her organization.

Rewards

The biggest reward for satisfactory performance is follow-up business. As in the previous example, Applied Materials rewards superior performers with follow-up business, as well as greater opportunity for future business.

Additionally, as with raising children or dealing with "significant others," recognition also is a powerful stimulant to future successful performance. Several years ago, an Arizona supply manager divided her suppliers into three categories: outstanding, acceptable, and marginal. She wrote an appropriate letter to the CEO of each supplier firm. The results of her efforts were rewarding. The outstanding group performed even better! Most of the CEOs from the second and third groups requested meetings to discuss what they could do better to earn *an outstanding letter!*

A major appliance manufacturer publicly recognizes its most successful suppliers. Such suppliers are encouraged to share their recognition with their employees. The employees are encouraged to continue their efforts to improve quality and productivity. Many suppliers reward their outstanding employees with a trip to the buying firm to see how their products are used.

Each year, the appliance manufacturer selects its 100 "best" suppliers. This selection is based on a combination of service, responsiveness, value analysis suggestions, cost, and related factors. Each representative and CEO attends the Supplier Appreciation Group's Day. Over 50 senior managers from the buying firm also attend. Each of the 100 outstanding suppliers receives a plaque acknowledging its status and contribution. The buying firm publicizes this list in appliance and supply magazines, to the delight of those listed.

Assistance

Progressive firms have discovered that several types of assistance to suppliers pay big dividends.

Transformational Training

Many firms have learned the benefits of providing training to their suppliers in the approaches such as statistical process control (SPC), (JIT) manufacturing, and Six Sigma. Progressive firms recognize that their ability to procure quality products and services on-time requires suppliers with competence in these tools and philosophies.

Quality Audits and Supply System Reviews

As organizations realize their interdependence with their suppliers, they are becoming more proactive in ensuring that their suppliers' quality systems and procurement systems operate effectively. Such assurance is often provided by quality audits and supply system reviews given by the buying firm. Such reviews should not be focused on punishment, but on identifying opportunities for improvement.

A supplier's procurement system affects its quality, cost, technology, and dependability. In theory, competition rewards suppliers who have efficient procurement systems with *survival* and *profit,* and it penalizes suppliers with inefficient systems. But such theory may take years to show results. Further, as firms move from reliance on market competition to collaborative and alliance relationships, the implications of inefficient supplier procurement systems become even more frightening. The supply system review provides a framework that a buying firm may follow when reviewing and assisting its key suppliers to upgrade their supply systems. The review is conducted in a constructive, cooperative atmosphere.

Problem Solving

Most progressive firms provide technical and managerial assistance to their suppliers when quality and related problems are encountered. Progressive supply managers have replaced the attitude "It's their contract and it's their problem" with the knowledge that the buying firm's success is dependent on its suppliers' success. Many problem solving tools and techniques used in aiding suppliers are presented in Chapter 7 and 13. Honda has a rather elaborate trouble shooting team ready to visit a supplier to offer technical assistance.

Collaboration

Experience in recent years demonstrates that the most successful supplier management results are generated when the buyer and the supplier view their relationship as one of collaboration. Such relationships are based on mutual interdependency and respect. Collaborative relationships begin or are renewed with careful source selection during the product or service design and development process. At this point, the buying firm needs a dependable supplier to provide the required process, design, and technological input if a marketable, profitable product or a satisfactory service is to result. In turn, the supplier needs a responsible customer for its products and services. Supply professionals need the supplier as much as (or more than) the supplier needs them. This interdependence grows as a project moves from design into operations. Unexpected problems arise, which require a "We shall overcome" attitude by the partners. During production, the buyers and suppliers must mesh their schedules, requiring another phase of cooperation.

The *ultimate* in collaborative relationships is a virtual integration of buyer and supplier, wherein two independently owned entities integrate their energies for as long as the relationship benefits both parties. One- or two-page memoranda of agreement replace lengthy contracts, change orders, and other legalistic and defensive procedures. As discussed in Chapter 4, we reserve the term "alliances" for such relationships. Finally, many firms are asking suppliers to rate their (buying firm) purchasing practices. See Appendix C.

The efforts of Lockheed Martin Missiles and Fire Control (MFC), with major operations in Dallas and Orlando, to improve collaboration with their key suppliers are a good case history of achieving real buy-in from suppliers. They expanded on an already successful group of strategic performance management teams (SPMT). For 20 years, MFC used performance management teams (PMTs) to "power continuous improvement from within by using internal cross-functional teams applying tools such as Lean six sigma. Based on their experience from 126 PMT teams, MFC acquisition and management staff included key suppliers as partners on the SPMTs to improve quality, lower cost, increase productivity, and achieve better scheduling for both buyers and suppliers. In one example, an SPMT achieved $1 million in cost savings by a joint redesign effort that reduced raw materials, machining time, and freight expenses.[2] Both parties are saving money through these joint efforts, a form of concurrent engineering with early supplier involvement (ESI).

Managing the Relationship

As previously discussed, any supply management department normally will have a continuum of supplier relationships from arm's length through collaborative ones to strategic alliances. The latter two types of relationships are increasing in number. Developing and managing such relationships are both challenging and fulfilling.

Several actions must be taken to ensure the success of each supply alliance relationship. (Supply managers should select and tailor appropriate actions when planning the management of collaborative relationships.) For example:

▲ Ideally, an inter-firm cross-functional team should be established to develop and manage plans, facilitate integration, and develop and manage appropriate metrics.

▲ Appropriate cross-functional team members at both the buying and the selling firms should receive training on being constructive team players.

▲ An inter-firm team composed of representatives of both firms should be formed. Members should jointly receive training in cross-functional team skills.

▲ The two firms must develop an integrated communication system responsive to the needs of both parties in their area of cooperation.

▲ Plans to increase and measure trust between the two organizations should be developed and implemented.

▲ Arrangements for co-location of key technical personnel and for periodic visits to each other's facilities should be developed and implemented.

▲ Plans should be developed and implemented for training on issues, including the designing of variance out of products and processes, quality, supply management, value analysis and engineering, strategic cost analysis, activity-based cost management, etc.

▲ Measurable quantifiable objectives must be established in areas, including quality, cost, time, technology, etc.

▲ The results of such improvement efforts must be monitored and reported to appropriate management.

▲ Ethics should win over expediency.

▲ Inter-firm team members and others who are closely involved must recognize the need to change their orientation from adversarial to collaborative.

▲ Inter-firm team members should become champions who ensure that their organizations understand and support the alliance's goals.

▲ Transparency of cost and other critical data.

It is in the interest of both the buying and supplying firms for the buying firm's personnel to support the supplier's operations. For example, always pay the supplier in accordance with the contract terms. If the buying firm constantly engages in late payments, they will eventually pay the interest charges in the form of increased costs hidden in the pricing proposals. Remember, the supplier cannot and should not finance your working capital. IBM has had a long-standing policy of paying invoices on time and believes this has translated into lower pricing. There are cases of firms paying 45 days late and still taking the discount. Not only are these practices imprudent, they are also unethical and are good examples of "sharp practices." Constant delaying in paying might put the buying firm on C.O.D. status.

Summary

Without question, management of supply contracts is a critical and challenging activity. Perhaps the most challenging aspect is the evolution from managing or controlling the supplier to managing the relationship. New attitudes and skills are required. Supplier relationships require the same amount of attention as a good marriage, but the many benefits of successful relationships make the efforts worthwhile.

Appendix A: Supplier Reporting Requirements for Unique Major Projects

During the preaward conference, arrangements are made for the timely receipt of the following data, as appropriate:

▲ *A program organization chart*   For a large job, the supplier designates its program manager and shows the key members of the organization by name and function. The program manager's functional authority should be clearly defined.

▲ *Milestone plan* For a complex project, this plan identifies all major milestones on a time-phased basis, including those of the supplier's major subcontractors.

▲ *Funds commitment plan* (incentive and cost reimbursement contracts only) This plan shows estimated commitments on a dollar vs. month basis and on a cumulative dollar vs. month basis.

▲ *Labor commitment plan* This plan shows estimated labor loading on a labor-hour vs. labor-month basis.

▲ *Monthly progress information* This report should be submitted 10 days after the close of each month. The report should contain as a minimum:

▼ *A narrative summary of work accomplished* during the reporting period, including a technical progress update, a summary of work planned for the next reporting period, problems encountered or anticipated, corrective action taken or to be taken, and a summary of buyer-seller discussions.

▼ *A list of all action items,* if any, required of the buying firm during the forthcoming performance period.

▼ *An update of the milestone plan* showing actual progress against planned progress.

▼ *An update of the funds commitment plan* showing actual funds committed against the planned funds by time (incentive and cost reimbursement contracts only).

▼ *A report on any significant changes* in the supplier's program personnel or in the financial or general management structure, or any other factors that might affect performance.

▼ *A missed milestone notification and recovery plan.* The supplier should notify the supply manager by phone, fax, or e-mail within 24 hours after discovery of a missed major milestone or the discovery of an anticipated major milestone slip. The supplier should provide the supply manager with a missed milestone recovery plan within seven working days.

Such data can be costly to compile and should be required only when it has been determined that their cost and the cost associated with using them to manage an order will result in a net saving or the likely avoidance of a schedule slippage. The supply manager, on a case-by-case basis, must determine what level of detail is necessary.

Appendix B: How Critical Path Scheduling Works

Critical path scheduling begins with the identification and listing of all significant activities involved in the project to be planned and controlled. When the list of activities is complete, the sequential relationship between all activities is determined and shown graphically by constructing an activity network (see Figure 20.2). The network shows the time required

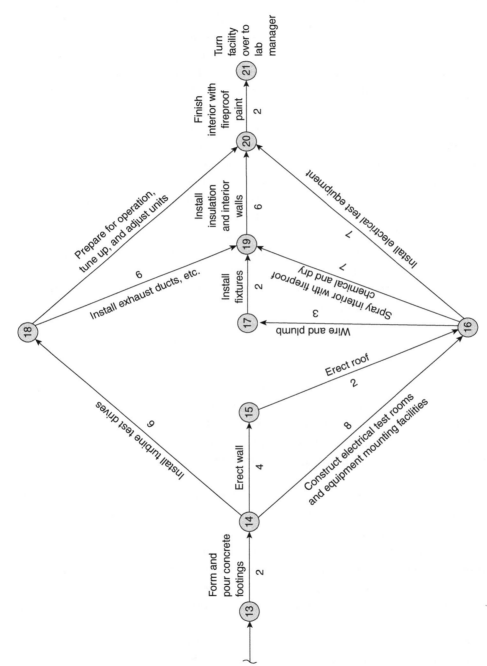

Figure 20.2 Simplified critical path network for subcontracted activities of a laboratory construction job.

611

to complete each activity, and it explicitly indicates the relationship of each activity to all other activities. Finally, it establishes the sequence in which activities should be scheduled for efficient completion of the total project. Some activities can be paralleled, allowing many different jobs to be carried on simultaneously; other activities must be placed in series to allow step-by-step completion of interrelated tasks. The complete interrelationship of all project activities is the important feature that distinguishes critical path analysis from Gantt and other bar chart planning techniques.

Construction of the network permits determination of a project's *critical path*—the sequentially linked chain (or chains) of activities requiring the most time for completion from start to finish of the project. Once the critical path is determined, the planner can identify precisely and completely the activities that require close control. He or she can also determine which activities permit the greatest latitude in scheduling and can, if necessary, most easily relinquish resources to more urgent jobs. An activity network also highlights the activities that can be expedited most effectively in case a stepped-up pace becomes necessary.

Figure 20.2 illustrates the mechanics of network development. This simplified example shows a partial network of the major subcontracted activities involved in the construction of a new laboratory. Engineering and purchasing activities prior to site preparation have been omitted for reasons of simplification. The example shows only the major activities to be completed after the site has been excavated and prepared for the concrete subcontractor.

1. Column 1 (Figure 20.3) lists the major activities to be completed.
2. Figure 20.2 shows the activities in network form. The required interdependencies and sequencing of operations have been determined and reviewed with the project engineer and the various subcontractors. Each arrow in the diagram represents an *activity* that will be conducted during a period of time. Each circle represents an *event* that will occur at a specific period of time (e.g., the start of an activity and the completion of an activity).
3. Careful estimates of the time required to complete each activity have been made by the respective contractors and are noted next to the appropriate arrows.[3]
4. The critical path (longest chain of activities) consists of activities 13-14-16-19-20-21; it will take 25 weeks to finish the project. The critical path is determined by totaling the time requirements for individual activities put together in every conceivable path from start to finish of the project. The critical path is the one requiring the longest time. Using the given time estimates, it is impossible to complete the project in less time.
5. At this point, the objective is to determine the amount of "slack time" (extra time, or leeway) existing in each of the non-critical paths. This is done first by developing

Activity (1)	Earliest (2)		Latest (3)		Slack (4)
	Start	Complete	Start	Complete	
13-14 Form and pour concrete footings	0	2	0	2	0
14-16 Construct electrical test rooms and equipment mounting facilities	2	10	2	10	0
16-19 Spray interior with fireproof chemical and let dry	10	17	10	17	0
19-20 Install insulatior and interior walls	17	23	17	23	0
20-21 Finish interior with fireproof paint	23	25	23	25	0
14-15 Erect walls	2	6	4	8	2
15-16 Erect roof	6	8	8	10	2
16-17 Install electrical wiring and plumbing	10	13	12	15	2
17-19 Install fixtures	13	15	15	17	2
14-18 Install turbine test drives	2	8	5	11	3
18-19 Install exhaust ducts. etc.	8	14	11	17	3
16-20 Install electrical test equipment	10	17	16	23	6
18-20 Prepare for operation, tune up, and adjust drive units	8	16	15	23	7

Figure 20.3   Data for Figure 20.2.

the figures in column 2 (the earliest start and completion dates for each activity) and then those in column 3 (the latest possible start and completion dates for each activity) of Figure 20.3.

The earliest start and completion date for each activity is determined by starting with the first activity and totaling the time requirements of the various paths through the network. If we say that the earliest start date for 13-14 (the concrete work) is today (the 0 point in time), the earliest completion date for 13-14 is the end of week 2 (0 + 2 = 2). Therefore, the earliest start date for both 14-15 and 14-16, which cannot begin before 13-14 is completed, is the end of week 2. The earliest completion date for 14-16 is the end of week 10 ($2 + 8 = 10$), and for 14-15 it is the end of week 6 ($2 + 4 = 6$). The earliest start date for 15-16 is the end of week 6, and the earliest completion date is the end of week 8 ($6 + 2 = 8$). What is the earliest start date for the electrical wiring and plumbing activity 16-17? Is it the end of week 8 or week 10? Clearly, it is the end of week 10 because the wiring cannot be started until the electrical test rooms and mounting facilities are completed. The analysis continues in this manner through the entire network until column 2 is complete. The latest possible start and completion date for each activity (column 3) is determined in exactly the reverse manner. Begin with the completion date for the final activity and work backward through the network, determining the latest completion and start dates that can be used for each activity without delaying completion of the project. All activities on the critical path will have identical "earliest" and "latest" dates. This is not true, however, for the non-critical path activities.

The latest completion date for the final painting activity (20-21) is the end of week 25, and the latest start date is the end of week 23 ($25 - 2 = 23$). Therefore, the latest completion date for 18-20 is the end of week 23, and the latest start date is the end of week 15. Notice, however, that the latest completion date for 14-18 is not the end of week 15. Activity 14-18 must be completed by the latest start date for activity 18-19, which is the end of week 11 ($23 - 6 - 6 = 11$). This procedure, then, is followed back through the network to complete column 3.

6. The purpose in developing columns 2 and 3 is to determine the amount of *slack time* in each non-critical, or slack, path (column 4). The slack time existing for a particular activity is simply the difference between the earliest start date and the latest start date (or between the earliest completion date and the latest completion date). This important factor represents leeway, which can be used in scheduling the slack path activity most efficiently in light of other demands for facilities and labor.[4]

The preceding discussion has dealt only with the rudiments of the critical path planning concept. In practice, performance is monitored, and progress data are periodically compared with the original plan. The technique is therefore an effective control device as well as an aid in making future planning decisions as changes in plans occur. The projects to which the technique is applied in practice are often made up of several thousand activities. After the initial network is constructed for such projects, a computer program such as Microsoft Project is typically used.

Its current popularity indicates that a large number of users find critical path analysis profitable primarily because it forces suppliers to do more planning than they otherwise would do. Numerous governmental and industrial buyers require subcontractors to submit a critical path network (for major events, or "milestones") with their bids on subcontract and construction jobs.

Appendix C: Supplier Questionnaire

The Supply Management and Internal Audit Departments at _____ have developed a short questionnaire to determine weaknesses and strengths of our procurement and payable functions. To ensure creditability, all suppliers being requested to complete this form were selected by Internal Audit. We would appreciate your taking a few minutes and completing the form. While you need not identify yourself, we would appreciate a generic description of the product or service you provide—see question #7. A self-addressed postage-paid envelope is enclosed for your convenience.

1. Do you find the supply managers you interact with at _____ to possess the necessary expertise to evaluate your products or services fairly?
___ Yes ___ No

2. Are you accorded a prompt and courteous interview in the Supply Management Department?
___ Yes ___ No

3. When requested to respond to written bids, is your firm allowed sufficient time?
___ Yes ___ No

4. On a scale of 1 to 5, 5 being the highest, how would you rate the ethics (honesty, willingness to listen to both sides of an issue, impartiality, etc.) of members of the Supply Management Department?
Circle one: 1 2 3 4 5

5. Are invoices paid on time? ___ Yes ___ No

6. Do you receive a prompt response when calling regarding a delinquent payment?
___ Yes ___ No

7. Please identify the product or service you provide—examples: laboratory equipment/supplies, agricultural-related products, computer equipment/supplies, etc.

_____

_____

_____

8. Are there any additional comments you have relative to the operation of the purchasing and accounts payable functions at _____ ?
Please return the completed questionnaire to: _____. We would appreciate a return date by _____. Your comments will be appreciated. Thank you!

Endnotes

1. Mary Lu Harding, "The ABC's of Activities and Drivers," *NAPM Insights*, November (1994): 6. For further information on ABC the OSD comptroller icenter is a comprehensive online site related to the U.S. Department of Defense budget process, financial management, and best practices.

2. Lisa Arnseth, "Engaging Suppliers to Excell," *Inside Supply Management*, September, (2010). See www.ism.ws and email to author@ism.ws.

3. The PERT technique is often used for jobs whose time requirements cannot be accurately estimated. For this reason, PERT requires three time estimates for each activity: (1) t_l = longest time required under the most difficult conditions (expect once in 100 times), (2) t_s = shortest time required under the best conditions (expect once in 100 times), (3) t_m = most likely time requirement. The expected time t_e, the figure actually used on the network, is a weighted average computed as

$$t_e = \frac{t_l + t_s + 4t_m}{6}$$

4. Column 4 indicates that both activities 14-15 and 15-16 have two weeks' slack. Notice that this is an either-or situation. Two weeksí slack cannot be utilized in both activities; two weeks represents the total combined slack for both activities.

Suggested Reading

Anderson, Ruth, and Maytee Aspuro. "Managing the Contract." *NAPM InfoEdge* 2(10)(1997).

Crowder, Mark A. "Project Organization and Management." *NAPM InfoEdge* 4(2)(1998).

Giannakis, Mihalis, and Simon R. Croom. "Toward the Development of a Supply Chain Management Paradigm: A Conceptual Framework." *Journal of Supply Chain Management*, Spring (2004): 27–37.

Greico, Peter Jr., Michael W. Gozzo, and Jerry W. Claunch. *Supplier Certification II: A Handbook for Achieving Excellence Through Continuous Improvement*, 3rd ed. Plantsville, CT: PT Publications, 1992.

Harding, Mary Lu. "The ABC's of Activities and Drivers." *NAPM Insights*, November (1994): 6.

Kaplan, Robert S., and David P. Norton. "Translating Strategy into Action: The Balanced Scorecard." Harvard Business School Press, Boston, MA, 1996.

Parker, Geoffrey G., and Edward G. Anderson Jr. "From Buyer to Integrator: The Transformation of the Supply Chain Manager in the Vertically Disintegrating Firm." *Production and Operations Management* 11(1)(2002): 75–91.

Rollman, Mary. "Mind the Gap: Collaborative Sourcing and Supplier Relationship Software." *Inside Supply Management*, November (2006): 12–13.

CHAPTER 21

Ethics and Social Responsibility

Ethics in the Supply Management Context

Corporate and academic interest in business ethics has blossomed in recent years. It is, however, well beyond the scope and purpose of this chapter to undertake a review of the extensive literature pertaining to business ethics. This is particularly true when one considers how ethics varies from one culture to another. Instead, most of this chapter focuses on ethical issues and guidelines frequently encountered in the context of supply management.

For the purposes of this chapter, our operational definition of ethics can be stated as the guidelines or rules of conduct by which we aim to live. Organizations, like individuals, have ethical standards and, frequently, ethics codes and policies. The ethical standards of an organization are judged by its actions and the actions of its employees, not by pious statements of intent put out in the company's name. The *character* of an organization is a matter of importance to its employees and managers, those who do business with it as customers and suppliers, and those who are considering joining it in those capacities.

Whether the market rewards good corporate behavior with respect to social responsibility and ethics is open to debate. Nevertheless, a review of the literature shows that increased social responsibility most likely will result in favorable financial performance. The review found 33 studies showing a positive relationship, 5 showing a negative relationship, and 14 showing no effect or inconclusive results.[1]

Procter & Gamble (P&G) understands the connection between ethical corporate behavior and financial performance. P&G takes the issue of ethics to a higher level by linking it to trust. In a visit to the University of San Diego, Stephen Rogers, former director of Technology Purchases at P&G, said the following:

We believe buyers and suppliers optimize the results of the working relationship when there is a foundation of trust. By treating suppliers honestly, ethically, and fairly, we do our part in building that foundation, just as the supplier must do its part. We honor the confidentiality of proprietary supplier information regarding technology, cost, or other sensitive information unless given written permission to share it. We do this not only because we believe it to be right, but also because it makes working with P&G and supplying our requirements attractive to current and future suppliers. We strive to give a clear understanding of the "rules of the game" to suppliers before and after commercial interactions.[2]

There is little doubt that in the minds of most people, including responsible businessmen and businesswomen, ethical conduct is a more substantive and relevant issue today than it was several decades ago. Unfortunately, however, this is not a unanimous view. One of the classic dilemmas facing management at all levels of any organization is the issue of "time focus": Do we focus on the short run or the long haul? There are still a fairly significant number of individuals who are willing to "operate on the margin," or sacrifice long-term values for short-term gains acquired through unethical activities. A survey of 4035 employees at all levels across a variety of industries[3] is quite revealing on this point:

> Ninety-seven percent of the employees surveyed said good ethics are good business. But responses to other questions indicated that many employees don't think that their companies agree. Two-thirds of the respondents said that ethical conduct isn't rewarded in American business. Eighty-two percent believe that man agers generally choose bigger profits over "doing what's right." One-fourth said that their companies ignore ethics to achieve business goals. One-third reported that their superiors had pressured them to violate company rules.[4]

This research may appear dated, yet a 2007 National Business Ethics Survey found that the "ethics and corporate social responsibility picture for American business really hasn't changed since the intense period of business scandals early in this decade."[5]

Professional Supply Management Ethics

We live and work in a highly competitive market economy with an emphasis on results. There is pressure on margins, pressure to compromise, pressure to succeed in an environment of both internal and external competition, and pressure resulting from government regulations. The *pressures* that the marketplace exerts indirectly on supply management departments and individual buyers make it essential that top management and supply management recognize and understand both the professional and the ethical standards required in the performance of their duties.

Principles and Standards of Ethical Supply Management Conduct[6]

The Institute for Supply Management, whose membership primarily consists of U.S. supply managers, has addressed the issue of ethics since its inception over 80 years ago. ISM published *Principles and Standards of Ethical Supply Management Conduct* in 1992 and updated the principles in 2008.[7] The principles given in the 2008 ethical standards are:

▲ *Integrity in your decisions and actions*
▲ *Value for your employer*
▲ *Loyalty to your profession*

From these principles are derived standards of supply management practice, as presented in the following sections, with minor editing.

1. Perceived Impropriety

Prevent the intent and appearance of unethical or compromising practice in relationships, actions, and communications.

The results of a perceived impropriety may become, over time, more disruptive or damaging than an actual transgression. It is essential that any activity or involvement between a supply management professional and active or potential suppliers that in any way diminishes, or even appears to diminish, open and fair treatment of suppliers be strictly avoided. Those who do not know us will judge us on appearances. We must consider this—and act accordingly. If a situation is perceived as real, it is in fact real in its consequences.

2. Conflict of Interest

Ensure that any personal, business, or other activity does not conflict with the lawful interests of your employer.

Supply management professionals have the right to engage in activities that are of a private nature outside of their employment. They must not, however, use their positions in any way to induce another person to provide any benefit to themselves or to persons with whom they have a family, business, personal, or financial relationship. Even though a conflict technically may not exist, supply management professionals must avoid the *appearance* of such a conflict. Whenever a potential conflict of interest arises, a supply management professional should notify his or her supervisor for guidance or resolution.

3. Issues of Influence

Avoid behaviors or actions that may negatively influence, or appear to influence, supply management decisions.

This principle concerns what commonly are known as "gratuities." Gratuities include any material goods or services offered with the intent of, or providing the potential for, influencing a buying decision. Unfortunately, gratuities may, from time to time, be offered inappropriately to a buyer or to other persons involved in supply management decisions (or members of their immediate families). Having any influence on the supply management process constitutes involvement. Those in a position to influence the supply management process must be dedicated to the best interests of their employer. It is essential to avoid any activity that may diminish, or even appear to diminish, the objectivity of the supply management decision-making process.

Gratuities may be offered in various forms. Common examples are money, credits, discounts, supplier contests, sales promotion items, product test samples, seasonal or personal gifts, food, drinks, household appliances and furnishings, clothing, loans of goods or money, tickets to sporting or other events, dinners, parties, transportation, vacations, cabins, travel and hotel expenses, and various forms of entertainment. Extreme caution must be used in evaluating the acceptance of any gratuities (even if of nominal value), and the frequency of such actions, to ensure that one is abiding by the spirit of these guidelines.

The following site has selected guidelines for dealing with gratuities: http://www.ism.ws/tools/content.cfm?ItemNumber=4740&navItemNumber=15959.

Business Meals

Occasionally, during the course of business, it may be appropriate to conduct business during meals.

- ▲ Such meals shall be for a specific business purpose.
- ▲ Frequent meals with the same supplier should be avoided.
- ▲ A supply management professional should be in a position to pay for meals as frequently as the supplier does. Supply management professionals are encouraged to budget for this business activity.

Personal Relationships

Personal relationships are an inherent aspect of supply management. Supply management professionals interact extensively with suppliers' representatives. Individuals in many other functional areas in both the buying and supplying organizations also interact extensively with one another. The development of personal relationships from such interactions is both expected and desirable because this can lead to relationships based on understanding and trust. It also must be recognized that the purchasing decision must not be influenced by anything other than what is in the best interest of the organization and that personal relationships that develop beyond what is necessary to ensure understanding and trust may be inappropriate. It is important, therefore, for supply management professionals to monitor closely the nature of relationships with suppliers' representatives to ensure that personal friendships do not develop that would result in decisions that are not in the organization's best interest.

International Practices

In some cultures, *business* gifts, meals, and entertainment are considered part of the development of the business relationship, as are close personal relationships. It is important, therefore, for supply management professionals to understand such practices and establish policies and procedures to deal effectively with suppliers from different cultures to ensure that they make supply management decisions that are in the best interest of the organization. This requires that suppliers be informed of the organization's policies with respect to business gifts, meals, entertainment, and the nature of personal relationships. It also requires that supply management professionals act courteously to suppliers' representatives who inadvertently may act in ways contrary to the organization's policies.

▲ In many foreign cultures, business frequently is conducted in the evenings and over weekends, which may be the only time key executives are available. In these circumstances, it is understood that the supply management professional would be expected to accept or provide for meals and entertainment when business matters are conducted. This is typically a more sensitive issue during the initial phase of the business relationship and may be tempered as the relationship progresses.

▲ Reciprocal giving of gifts of nominal value is often an acceptable part of the international buying and selling process. When you are confronted with this—when company policy does not exist—an appropriate guide would be to ensure that actions are in the best interest of your employer, *never for personal gain.*

▲ The definition of nominal value may be higher or lower than nominal value in your country as a result of custom, currency, and cost-of-living considerations and often is guided by the duration and scope of the relationship. A supply management professional must evaluate nominal value carefully in terms of what is reasonable and customary. When in doubt, consult company managers, professional colleagues, and your conscience.

Political Considerations

All organizations are subject to internal and external forces and pressures. Internal forces and pressures result from an organization's culture. External forces and pressures may include economic conditions, laws, regulations, public opinion, special interest groups, and political entities. The negative influence of internal and external forces and pressures on supply management can be minimized when the organization adopts practices that are based on ethical principles and standards.

Advertising

Care should be exercised in accepting promotional items or participating in activities that tend to promote one supplier over another or that could be perceived as preferential supplier advertising by a supply management professional.

Market Power

Supply management professionals must be aware of their organization's position (economic size, power, etc.) in the marketplace and ensure that that position is used within the scope of ethical behavior by the supply management professional and the organization.

Specifications and Standard

Supply management professionals must ensure that specifications and standards are written objectively to encourage appropriate competition.

4. Responsibilities to Your Employer

Uphold fiduciary and other responsibilities using reasonable care and granted authority to deliver value to your employer.

A supply management professional's foremost responsibility is to achieve the legitimate goals established by the employer. It is his or her duty to ensure that actions taken as an agent for the employer will be in the best interests of the employer, *to the exclusion of personal gain.* This requires application of sound judgment and consideration of both the legal and the ethical implications of one's actions.

5. Supplier and Customer Relationships

Promote positive supplier and customer relationships through courtesy and impartiality.

It is the responsibility of a supply management professional to promote mutually acceptable business relationships with all suppliers and customers. Affording all supplier representatives the same courtesy and impartiality in all phases of business transactions will enhance the reputation and good standing of the employer, the supply management profession, and the person involved. Rudeness, discourteousness, or disrespectful treatment of a supplier may result in barriers to free and open communications between buyer and seller and ultimately in a breakdown of the business relationship.

A supply management professional should extend the same fairness and impartiality to all legitimate business concerns. It is natural and even desirable to build long-term relationships with suppliers on the basis of a history of trust and respect, but those relationships should not prevent the establishment of similar working relationships with other suppliers.

6. Sustainability and Social Responsibility

Champion social responsibility and sustainability practices in supply management.

It is generally recognized that all business concerns, large or small, majority- or minority-owned, should be afforded an equal opportunity to compete. Within existing legal constraints, many government entities and corporations have developed specific procedures and policies designed to support and stimulate the growth of socially diverse firms.

7. Confidential and Proprietary Information

Protect confidential or proprietary information with due care and proper consideration of ethical and legal ramifications and governmental regulations.

Supply management professionals and others in positions that influence buying decisions deal with confidential or proprietary information of both the employer and the supplier. Proprietary and confidential information should be released to other parties (internal and external) only on a need-to-know basis. It is the responsibility of a supply management professional sharing confidential or proprietary information to ensure that the recipient understands his or her obligation to protect that information.

The ISM recommends several practices to ensure that confidential or proprietary information is managed properly. Those recommendations include the following:

1. Developing and communicating a policy covering proprietary and confidential information.
2. Requiring that confidential and proprietary information be identified as such, whether disclosed in tangible form or orally.
3. Using confidential information agreements.
4. Not accepting another's confidential or proprietary information unless there is a need for such information.
5. When dealing with any information, whether it is confidential or proprietary, exercising care in determining the effects of its use.

A supply management professional should avoid releasing information to other parties, including those within the supply management professional's own company, until assured that they understand and accept the responsibility for maintaining the confidentiality of the material. Extreme care and good judgement should be used if confidential information is communicated orally. All confidential or proprietary information should be shared only on a need-to-know basis.

Information of one supplier must never be shared with another supplier unless laws or government regulations require that such information be disclosed. If one is unclear about disclosure requirements, legal counsel should be consulted. When a supply management professional is privy to cost or profit data or other supplier information not generally known, it is his or her responsibility to maintain the confidentiality of that information.

Examples of information that may be considered confidential or proprietary are:

1. Pricing
2. Bid or quotation information
3. Cost sheets
4. Formulas and/or process information
5. Design information (drawings, blueprints, etc.)
6. Organizational plans, goals, and strategies
7. Financial information
8. Information that may influence stock prices
9. Profit information
10. Asset information
11. Wage and salary scales
12. Personal information about employees, officers, and directors
13. Supply sources or supplier information
14. Computer software programs

8. Reciprocity

Avoid improper reciprocal agreements.

Transactions that favor a specific customer as a supplier or influence a supplier to become a customer constitute a practice known as reciprocity, as does a specific commitment to buy in exchange for a specific commitment to sell. The true test for reciprocity, however, is in the motive, because the process may be less vague than a written or formal commitment. In any such transactions, reciprocity may be illegal if it tends to restrain competition or trade and if the transactions are coerced. In organization structures in which the supply management and marketing functions report directly to the same individual, the potential for reciprocity may be greater.

Supply management professionals must be especially careful when dealing with suppliers who are customers. Cross-dealings between suppliers and customers are not antitrust violations per se. Nevertheless, giving preference to a supplier who is also a customer should occur only when all other factors are equal. Dealing with a supplier who is also a customer may not constitute a problem if, in fact, that supplier is the best source. A company is engaging in reciprocity, however, when it deals with a supplier solely because of the customer relationship. A professional supply manager must be able to recognize reciprocity and its ethical and legal implications.

9. Applicable Laws, Regulations, and Trade Agreements

Know and obey the letter and spirit of laws, regulations, and trade agreements applicable to supply management.

Supply management professionals should pursue and maintain an understanding of the essential legal concepts governing their activities as agents of their employers. For example, key laws and regulations a supply management professional should be aware of when conducting business in the United States are:

▲ Uniform Commercial Code (UCC) and the Uniform Computer Information Transactions Act (UCITA)
▲ Uniform Electronic Transactions Act (UETA) and the Federal Electronic Signatures in Global and National Commerce Act (E-Sign)
▲ The Sherman Act
▲ The Clayton Act and the Robinson-Patman Act
▲ The Federal Trade Commission Act
▲ The Federal Acquisition Regulations (FAR) and Defense Federal Acquisition Regulation Supplement (DFARS)
▲ Environmental Protection Agency (EPA) laws
▲ Equal Employment Opportunity Commission (EEOC) laws
▲ Occupational Safety and Health Administration (OSHA) laws
▲ The Foreign Corrupt Practices Act
▲ The Public Company Accounting Reform and Investor Protection Act of 2002 (Sarbanes-Oxley Act of 2002)

Knowledge of international laws also may be relevant to a supply management professional. The ISM provides several recommendations to supply management professionals for understanding and complying with applicable laws:

▲ Supply management professionals should obtain training in the legal aspects of supply management to understand the laws that govern their conduct and so that they know when to seek legal counsel.
▲ Supply management professionals involved in a cross-border transaction must understand the laws that govern the transaction, including laws specific to the country within which they are doing business and laws of the country in which the seller resides, especially if they conflict with or supplement the laws of the country in which they are doing business.
▲ Interpretation of the laws should be left to legal counsel. It is often beneficial to involve legal counsel early in analysis and planning to identify and avoid potential legal pitfalls rather than involve legal counsel only after problems arise.
▲ Supply management professionals involved in government procurement must understand and apply laws that are specific to their particular agency.

10. Professional Competence

Develop skills, expand knowledge, and conduct business that demonstrates competence and promotes the supply management profession.

Professional competence is expected of supply management professionals by their employers, their supply management peers, others in their organizations, suppliers, and society at large.

A distinguishing characteristic of a profession is that its practitioners combine ethical standards with technical skills. Because of the impact that the conduct of supply management professionals has on the stature of the profession, it is important for all those in the profession to consider what is meant by professional competence and the way it is perceived by others.

Professional competence can be defined in many ways. Most definitions include the concept of mastery of a body of knowledge, continued efforts to increase one's ability and knowledge of the profession, communications skills, willingness to share knowledge with others, and conformance to the highest standards of ethical behavior.

The ISM recommends that supply management professionals:

- Ensure a basic understanding of all the requirements to be recognized as a competent supply management professional.
- Monitor trends and developments in the profession.
- Conduct a self-assessment of talents and skills.
- Establish a development plan to meet the needs of immediate and future employment.
- Seek out mentors and role models.
- Serve as a mentor.
- Earn and maintain the CPM or other credentials such as the certified professional in supply management (CPSM).
- Become actively involved in one or more professional associations.

The stature of the profession is enhanced through ethical actions and behavior of supply management professionals. When individuals combine in professional groups or associations, their actions and behavior become highly visible and enhance the stature of the profession. This has a direct impact on the profession and on the professional's organization, peers, and suppliers.

The ISM recommends that supply management professionals:

- Support professional development and interchange of ideas through membership in professional and service organizations.
- Actively seek and support change in ethical standards and practice when appropriate (e.g., changes in the environment or technology).
- Be aware that supply professionals are obligated to support only those actions and activities that uphold the highest ethical standards of the profession.
- Support the ethical principles and standards of each organization with which a supply management professional is affiliated.

▲ Encourage, support, and participate in ongoing ethical training and review within business and professional organizations.

▲ Strive to achieve acceptance of and adherence to these principles and standards by all those who influence the supply management process.

National and International Supply Management Conduct

A supply management professional should conduct supply management in accordance with national and international laws, customs, and practices; his or her organization's policies; and the ISM ethical principles and standards of conduct.

Legal systems vary throughout the world, as do business customs and practices. Supply management professionals therefore must be knowledgeable about these variations and the potential conflicts inherent in them when doing business across borders.

The ISM recommends that supply management professionals:

▲ Be especially sensitive to customs and cultural differences with respect to social and business behavior and issues of influence.

▲ Recognize that suppliers may not be familiar with laws, customs, and practices of various countries or with the company's policies. Consequently, it is important to ensure that appropriate information is communicated effectively to each supplier.

▲ Recognize which national laws may apply and which do not apply in other countries.

▲ Utilize organization management, legal counsel, and other available resources for guidance whenever there is uncertainty about which actions to take.

▲ Maintain an awareness of national and international standards (ISO 9000, ISO 14000, Ethical Trade Initiative, etc.).

Important Areas Requiring Amplification

Avoid Sharp Practices

In business, the term "sharp practices" has been used for years, and this topic is particularly relevant for supply management professionals. One of the standards in an earlier version of ISM's *Standards of Conduct* reads, "Avoid sharp practices." The term "sharp practices" typically is defined as evasion and indirect misrepresentation that is just short of actual fraud. These unscrupulous practices focus on short-term gains and ignore the long-term implications for a business relationship.

Some examples of sharp practices are:

▲ A supply manager talks in terms of large quantities to encourage a price quote on that basis. The forthcoming order, however, is substantially smaller than the amount on which the price was based. The smaller order does not legitimately deserve the low price thus developed.

▲ A large number of bids are solicited in hopes that the buyer will be able to take advantage of a quotation error.

▲ Bids are obtained from unqualified suppliers that the supply manager would not patronize in any case. Those bids then are played against the bids of responsible suppliers to gain a price or other advantage. The preparation of bids is a costly undertaking that deserves buyer sincerity in the solicitation stage.

▲ A supply manager who places in competition the prices of seconds, odd lots, or distressed merchandise misrepresents a market.

▲ An attempt is made to influence a seller by leaving copies of bids or other confidential correspondence where that supplier can see them.

▲ A concession may be forced by dealing only with "hungry" suppliers. The *current* philosophy is that a purchase order should create a mutual advantage with a price that is fair and reasonable.

▲ Obscure contract terms of benefit to the supply manager's firm are buried in the small type of contract articles.

▲ Taking advantage of a supplier who is short of cash and who may seek only to cover its out-of-pocket costs. Such a situation poses a dilemma because the supplier may be saved from borrowing at a disadvantage and may look upon such an order as a blessing![8]

Acurex Corporation adds the following items to the preceding list of practices to be avoided:[9]

▲ Allowing one or more suppliers to have information about their competitors' quotations and *allowing such suppliers to requite.*

▲ Making statements to an existing supplier that exaggerate the seriousness of a problem to obtain better prices or other concessions.

▲ Giving preferential treatment to suppliers that higher levels of the firm's own management prefer to recommend.

▲ Canceling a purchase order for parts already in process while also seeking to avoid cancellation charges.

▲ Getting together with other supply professionals to take united action against another group of people or a company.

▲ Lying to or grossly misleading a salesperson in a negotiation.

▲ Allowing a supplier to become dependent on the supply management organization for most of its business.

While many of these practices were once commonplace, Dr. Felch notes that they have been replaced by a philosophy that holds that mutual confidence and integrity are far more desirable ends than the short-term gains obtained through willful misrepresentation.

Competitive Bidding[10]

Supply management professionals respect and maintain the integrity of the competitive bidding process. They:

▲ Invite only firms to submit bids to which they are willing to award a contract.

▲ Normally award the contract to the lowest *responsive*, responsible bidder. If the buyer anticipates the possibility of awarding to someone other than the low bidder, he or she notifies prospective bidders that other factors will be considered. (Ideally, these factors will be listed.)

▲ Keep competitive price information confidential.

▲ Notify unsuccessful bidders promptly so that they can reallocate reserved production capacity.

▲ Treat all bidders alike. Clarifying information is given to *all* potential bidders.

▲ Do not accept bids after the announced bid closing date and time.

▲ Do not take advantage of apparent mistakes in the supplier's bid.

▲ If the buyer needs to reopen the bidding, send the request for new bids to all who initially submitted bids.

However, it *is* ethical for a buyer to work with the low bidder in an effort to identify possible areas of savings. Those identified savings allow the potential supplier to reduce its *costs* and its price.

Negotiation

Supply professionals maintain high ethical standards during all negotiated supply management activities.

▲ Competitors are informed of the factors that will be involved in source selection.

▲ All potential suppliers are given equal access to information and are afforded the same treatment.

▲ Supply professionals strive to negotiate terms that are fair to both parties.

▲ They do not take advantage of mistakes in the supplier's proposal.[11]

Samples

Many potential suppliers offer, even push, the acceptance of samples: "Just try it and see if it doesn't do a superior job for you." When a sample is accepted, supply professionals ensure that appropriate tests are conducted in a timely manner. The potential supplier then should be informed of the test results and suitability of the item in meeting the buyer's needs. In some instances, it may be wise to pay for the actual production cost of the sample.

Treating Salespeople with Respect

As was noted earlier, the use of courtesy and consideration by supply management personnel can influence the effectiveness of supplier relationships. The treatment of salespeople clearly produces overtones in the area of professional standards of conduct.[12]

▲ Salespeople should not be kept waiting for protracted periods; appointments should be kept meticulously. The power that is attached to the patronage position of a supply professional must not be abused so that the long-run interests of the buying company will be advanced.

▲ A mutually effective policy is for supply management personnel to see every salesperson on his or her first call. The appropriateness of follow-up visits should be determined by the potential strength of the buying firm's need for the supplying firm's product.

Substandard Materials and Services

When substandard materials or services are received, two proprieties should be observed:[13]

1. The supplier should be given prompt notice.
2. The appropriate supply manager should conduct negotiations for adjustments with the appropriate sales personnel in the supplier's organization.

Gifts and Gratuities

Nothing can undermine respect for the supply management profession more than improper action on the part of its members with regard to gifts, gratuities, favors, and so forth. People engaged in supply management should not accept from any supplier or prospective supplier any substantive gifts or favors. All members of the supply management system must decline to accept or must return any such items offered to them or members of their immediate families. The refusal of gifts and favors should be done discreetly and courteously. When the return of a gift is impractical for some reason, disposition should be made to a charitable institution, and the donor should be informed of the disposition.

Personal business transactions with suppliers or prospective suppliers should be avoided scrupulously. Offers of hospitality, business courtesies, or favors, no matter how innocent in appearance, can be a source of embarrassment to all parties concerned.

The buying firm's representatives should not allow themselves to become involved in situations in which unnecessary embarrassment results from refusal of hospitality or a business courtesy from suppliers. Generally, the best policy is to decline any sort of favor, hospitality, or entertainment to *ensure that all relationships are above reproach at all times.*

Clearly, all situations require the use of common sense and good judgment. For example, acceptance of company-provided luncheons during the course of a visit to a supplier's plant in a remote area is certainly reasonable. Similarly, the acceptance of supplier-provided automobile transportation on a temporary basis when other means are not readily available is a reasonable course of action for a supply professional to take.

Supply management personnel may ethically attend periodic meetings or dinners of trade associations, professional and technical societies, or other industrial organizations as the guest of a supplier when the meetings are of an educational and informative nature and attendance is considered to be in the professional interest of the buyer-seller relationship. Even so, the repeated appearance of an individual at such regularly scheduled meetings as the guest of the same company is the type of situation that should be avoided tactfully.

A simple casual lunch or cocktail with a supplier's representative typically is a normal expression of a friendly business relationship or frequently a timesaving expediency. It would be prudish to raise any serious questions on this score. The individual is in the best position to judge when this point has been passed. Any breach of ethics can be rationalized, but members of the supply management system can avoid embarrassment or possible unethical behavior by asking, "How would this look if reported in the company newsletter?" The desire to continue to talk shop or resolve a business issue accounts for most buyer-supplier lunches. Because the buying firm's prestige also is involved, there is good reason why an adequate expense account should be available to the supply professional. This permits him or her to pick up the bill for lunch. It is a small price to pay for maintaining a position free from any taint of obligation.

There's No Such Thing as a Free Lunch

Several years ago, Mike Darby of the ISM–Silicon Valley organization addressed the issue of free lunches. His comments were so appropriate that they are included here for the benefit of our readers.

The person who came up with this quote many years ago was probably a materials manager. There are many pros and cons about buyers going to lunch with their suppliers. I don't propose to advocate one or the other, but I would like to point out some ideas that may help you to make up your own mind.

First and foremost in my mind is *why* would a company want to spend some of its dollars that would normally flow directly to the bottom line to have me join them for lunch? I've asked this question of many suppliers. Some of the answers I've received include: "I have to take somebody to lunch every day, and it might as well be somebody I like"; "If I take you to lunch, then my lunch is free also"; "It will give us an opportunity to get to know each other better." These are all good answers, some of them more honest than I expected.

When I asked the same question of the buyers, the answer changed slightly. "We may need this supplier in the future, and I want to develop a close relationship with him"; "We have some serious problems to discuss"; "This will give me a good opportunity to negotiate a better deal with them."

Whatever the justification, the bottom line to me gets down to this. The supplier is willing to commit some of his profit dollars to this form of entertainment in *the belief that it will help him generate more profit in the future.* He may be in hope that the relationship he is building will help sustain him in your company through rough times, such as poor delivery performance or bad quality. He may be expecting to increase his prices, and a friend will never complain about a small change in price. He may ask you to lunch expecting that the slight social obligation he has just obtained from you may be paid back by getting another crack at a quote, first look at a new drawing, etc. Or he may join you for lunch hoping to obtain some information that will be useful to him in future negotiations.

All the above tend to add up to a one-sided deal, favoring the supplier and putting the buyer in the position of having to be very careful. Let's face it, the sales force of the supplier is being paid to perform a service to the supplier, and this is a good tool for him to use. He gets your undivided attention for a good hour or so, to use as he sees fit. Remember, it's impolite in our society to refuse to answer a question. A question as simple as "How's business?" provides the supplier with important information as to its market share within your company, business trends, and helps to set his expectations in future negotiations. It's also very difficult as a supply professional not to show a slight amount of favoritism toward the supplier that you had lunch with last Friday. Should he ask, "Where do I stand on that quote, and what do I need to do to get the job?" It gets difficult to be firm and say no to such a request.

Now before I get blasted out of the water for being one-sided, let me say that there are times when a business meal is appropriate. If you have a supplier in from out of town, and your business discussions extend through lunch or past business hours, a meal is probably appropriate. In this circumstance, I like to use the Host rule—If the meeting is in your territory, then you should be the host. If you are visiting the supplier's territory, then he can be the host. Make it fair and equitable, and the supplier loses any advantage he might have held from an obligation point of view. Remember, an obligation is in the mind of the person who received the favor. If, because of the manner in which you handle your conduct, there was no potential for a favor, then there can also be no obligation.

As a supply professional, picking up the check can quickly change a supplier's expectations, and as such is a very useful negotiation tactic. If he thought he had a contract in the bag, he suddenly begins to think that he may not be as secure as he thought. It also sends the message that you too are a professional and certainly not an easy target.[14]

Mature supply managers know that they are classified among the sales fraternity by the amount of entertainment they expect or will accept. Salespeople usually speak with real respect of the supply manager who pays his or her share of entertainment expenses. The supply management expense account is the most effective answer to this ethical problem.

Traditional Sales Techniques

Many supply management personnel feel that any form of gratuity constitutes a conflict with ethical standards. Others—in fact the majority, according to ISM studies—consider many of these items to be traditional sales tools. They therefore do not believe that such gratuities are offered with the expectation of favorable consideration.[15]

There are two common ways to control the acceptance of these kinds of gifts. The first is by placing a dollar limit on what can be accepted. In this case, a supply management department may have a stated policy of refusing any gratuity with a value in excess of, say, $10 or $15. Such policies provide a very simple, measurable guideline for how a buyer should decide about acceptance.

The other common policy is to forbid acceptance of any gratuity *the buying firm is not in a position to reciprocate*. Thus, if a firm's supply professionals accept sales promotion items such as pens or planning calendars, they should be in a position to reciprocate with similar items from their own firm.

Cultural Ramifications

Executives of many foreign suppliers expect that supply professionals will exchange gifts with them. That action is an accepted part of many foreign cultures. Some supply professionals encounter situations in which refusing a gift would interfere with the development of relations prerequisite to consummating a transaction. In such cases, supply professionals *report* the situation to their superiors and arrive at a solution that may include acceptance *and* a reciprocal gift.[16] If a supply manager hides his or her action, he or she automatically knows that the action is unethical.

Management Responsibilities

Written Standards

Management's first responsibility is to develop a set of written ethical and professional standards that are applicable to all members of the organization's supply management system. Supply management supervisors, buyers, expediters, design engineers, manufacturing engineers, quality assurance personnel, maintenance supervisors, receivers, and accounts payable personnel all must accept and observe those standards. The standards should address the topics discussed in this chapter. Research has shown clearly that *written* policies dealing with ethical issues have a strong positive influence on the behavior of a firm's supply management professionals.[17]

Ethics Training and Education

Professional supply managers, with the *assistance of top management* and their colleagues in other functional areas, must ensure that appropriate personnel receive periodic training or education with respect to the organization's ethical and professional standards. That training cannot address all issues. It can, however, increase the sensitivity of those who receive it. All members of the supply management system must respect their roles as agents of their employer and must represent the best interest of their organization. Subsequent to such training, many organizations require the attendees to sign a statement to the effect that they have taken the training and understand and will honor the standards.

Supply managers also should ensure that their personnel receive training in current thinking and techniques in the areas of requirements planning, source selection, pricing, cost analysis, negotiation, and supplier management, as well as ethical and professional standards.

The vast majority of supply managers are dedicated and conscientious people. Accordingly, it is almost shocking to see how little high-quality training many of these individuals have received. Supply management system reviews conducted by the authors indicate that a substantial number of buyers are being asked to perform tasks for which they have received little current training, including training in the area of ethical and professional conduct.

Departmental Environment

Department policy should make it clear that buying personnel engage in any unethical activity at the risk of losing their jobs. It is a generally accepted view that a small percentage of Americans are dishonest, that an equally small percentage are completely honest, and that most people are honest *or* dishonest, *depending on the circumstances*. Consequently, after the basic policy and training frameworks have been established, it appears that the surest way to encourage ethical conduct is to create a working environment in which unethical temptations seldom become realities.

The foundation for that environment consists of the people themselves. Management will be repaid many times for the effort put into thorough, careful investigation and selection of buying personnel. Habits and attitudes are "catching" in the close working environment of a supply department. If most of the personnel are basically honest, departmental management has the major part of the ethics battle won.

The age-old adage "monkey see, monkey do" is certainly applicable in the matter of ethical conduct. Departmental management and supervisory people must *live*, to the letter, the department's policies and ideals. Numerous studies have confirmed beyond doubt that the actions and attitudes of supervisors are the most influential single factor in determining the attitudes of a work group.

Miscellaneous Factors

Two concluding thoughts are worthy of consideration. First, some progressive organizations have established an internal or external ombudsman who can be contacted with impunity about ethical issues. Once accepted, the practice seems to work well.

Second, the president of Seldon Associates suggests greater utilization of post-purchase audits as desirable safeguards. He writes, "When every buyer knows that his or her purchases may be audited, there is a built-in safeguard tending to assure ethical purchasing."[18]

Dealing with Gray Areas

All supply professionals have ethical obligations to three groups of people: *employers, suppliers, and colleagues*:

▲ *Employer*   Guidance should focus on the characteristics of loyalty, analytical objectivity, and a drive to achieve results that are in the very best interest of the employing organization.

▲ *Suppliers*   The essence of the guiding spirit in dealing with the supplier community is honesty and fair play.

▲ *Colleagues*   All individuals engaged in supply management work are regarded by outside observers as members of an emerging profession. As such, they have an obligation to protect and enhance the reputation of that body of professionals.

When supply management professionals must take action in a "gray area" not clearly covered by policy, they may find guidance by seeking answers to the following questions:

1. Is this action acceptable to everyone in my organization?
2. Is the action compatible with the firm's responsibilities to its customers, suppliers, and stockholders?
3. What would happen if *all* buyers and salespeople behaved this way?
4. If I were in the other person's shoes, how would I feel about this action if it were directed toward me?
5. Would it be comfortable to have this act or action reported on a newscast to the general public?

The Four Way Test

The businessmen and businesswomen of the Rotary International organization provide a follow-up of four questions. They apply the Four Way Test to the things they think, say, and do:[19]

▲ Is it the TRUTH?
▲ Is it FAIR to all concerned?
▲ Will it build GOODWILL and BETTER FRIENDSHIPS?
▲ Will it be BENEFICIAL to all concerned?

In making the final decision, a few moments' thought may well be devoted to the following lines, entitled "What Makes a Profession."[20]

If there is such a thing as a profession as a concept distinct from a vocation, it must consist in the ideals that its members maintain the dignity of character that they bring to the performance of their duties, and the austerity of the self-imposed ethical standards. To constitute a true profession, there must be ethical tradition so potent as to bring into conformity members whose personal standards of conduct are at a lower level, and to have an elevating and ennobling effect on those members. A profession cannot be created by resolution, or become such overnight. It requires many years for its development, and they must be years of self-denial, years when success by base means is scorned, years when no results bring honor except those free from the taint of unworthy methods.

Social Responsibilities

Aside from proper, ethical, and legal conduct and, whenever possible, using qualified minority suppliers, supply chain managers have a major role in protecting the physical environment. In addition to issues previously discussed in this and other chapters, supply managers must understand statutes such as the Clean Air Act of 1970 (1990) and the Comprehensive Environmental Response Compensation Act (CLERCA, the Super Fund law). In addition, the current focus on green purchasing has serious implications for all members of the supply chain as we discussed in detail in Chapter 13.

For example, various state and federal agencies are mandating minimum mileage standards on vehicles to reduce carbon emissions as part of the effort to control global warming.[21] This mandate has triggered the need for new product designs in vehicles and, in particular, engines. Cross-functional design teams and early supplier involvement (ESI), as described in previous chapters, will be even more important to accomplish the proper response to these new regulations.

In addition, specifications and contract provisions must consider the mandate from the U.S. Environmental Protection Agency (EPA) to reduce waste, stimulate reuse, reallocate, and recycle (the 4 Rs).[22] Supply Chain Pollution Avoidance (SCPA) is a good term to remind supply chain managers always to think from Mother Earth to disposal or back to Mother Earth."[23,24]

The Institute for Supply Management (ISM) has defined Seven Principles of Social Responsibility as follows:[25]

▲ *Community initiatives*   This includes providing cash, equipment, people, and other resources to nonprofits and educational or other community groups.
▲ *Diversity in supply management*   This has been discussed in many parts of this text, in particular increased sourcing with minority and disadvantaged firms and individuals.
▲ *Environmental impact or change*
▲ *Ethics*
▲ *Financial responsibility*   This includes full and accurate disclosure beyond what is required by law, such as risk assessment.
▲ *Human rights*   Having respect for individual differences of race, religion, age, and sexual orientation.
▲ *Safety* for employers, customers, and suppliers. This obviously includes OSHA compliance but goes beyond the stated requirements.

Organizations should embody these principles in a written document distributed to all employees and suppliers. Studies have revealed that organizations that have high ethical and social standards tend to be more profitable in the long run as opposed to the Enrons of the world.

Summary

As you finish reading this chapter about ethics and social responsibilities, consider an event that occurred at a career planning workshop for students who soon would be entering the job market. The workshop facilitator instructed the participants to write their own obituaries. Morbid? Maybe. Useful? Absolutely!

The purpose of the exercise was to force the students to project to the end of their lives and then summarize how their lives had been lived. This exercise of coerced self-reflection was designed to examine each student's values. What kind of career unfolded? How did this person treat other people? Did this person touch the lives of others in such a way that he or she was missed?

All of us, of course, write our own obituaries each day of our lives by the way we choose to live our lives. If your obituary discussed your ethics, what would you want it to say? Finally, what did you do to help protect our physical and social environment?

Endnotes

1. Tony McAdams with contributing authors James Freeman and Laura P. Hartman, *Law, Business, and Society*, 6th ed. (New York: Irwin/McGraw-Hill, 2001) 95.
2. Stephen Rogers, presentation at the University of San Diego. Reprinted with the permission of Procter & Gamble Company, November 2001.
3. McAdams, *Law, Business, and Society*, 42.
4. Shaun O'Malley, "Ethical Cultures—Corporate and Personal," *Ethics Journal* Winter (1995): 9.
5. National Business Ethics Survey® Shows Observations of Unethical Behavior in American Business Back to Pre-Enron Levels," *Business Wire*, November 26, 2007. http://findarticles.com/p/articles/mi_mOEIN/is_2007_Nov_26/ai_n21121151.
6. Based on the content of the Institute for Supply Management's 2004 publication, *Principles and Standards of Ethical Supply Management Conduct*. http://www.ism.ws/tools/content.cfm?ItemNumber=4740&navItemNumber=15959.
7. *Principles and Standards of Ethical Supply Management Conduct: With Guidelines* (Tempe AZ: Institute for Supply Management [ISM], 2008).
8. Robert I. Felch, "Proprieties and Ethics in Purchasing Management," in *Guide to Purchasing* (Oradell, NJ: ISM, 1986) 4.7–3 and 4.7–4.
9. Richard E. Trevisan, "Developing a Statement of Ethics: A Case Study," *Journal of Purchasing and Materials Management* Fall (1986): 13.
10. Much of the discussion included in this section is based on material in ISM's *Guide to Purchasing*, Section 4.7: "Proprieties and Ethics in Purchasing Management," by Dr. Robert I. Felch. Specific footnotes are utilized where appropriate.
11. Felch, *Guide to Purchasing*, 4–5.
12. Ibid., 7.
13. Ibid.
14. *Pacific Purchaser* November–December (1988): 11.
15. Michael H. Thomas, "Know Where You Stand on Ethics," *Purchasing World* October (1984): 90.
16. Somerby Dowst, "Taking the Mystery Out of Conflict of Interest," *Purchasing*, September 11, 1986, 70A1.
17. G. B. Turner, G. S. Taylor, and M. F. Hartley, "Ethics Policies and Gratuity Acceptance by Purchasers," *International Journal of Purchasing and Materials Management*, Summer (1994): 46.
18. Doyle Seldon, "Ethics, an Additional Look," *Purchasing Management* December (1988): 41.
19. *Manual of Procedures* (Evanston, IL. Rotary International, 2007) 80.
20. *NAPM Standards of Conduct*, 1959.

21. Edgar E., Blanco, "Stay Ahead of the GHG Curve," *Inside Supply Management* April (2011): 32–33. Email author at author@ism.ws.

22. R. Jerry Baker, "The Environment: Playing Our Part," *NAPM Insights* April (1994): 2. See also Jeff Marcus, "Trends, Purchasing and the Environment," NAPM Insights, April 1994, 28.

23. David N. Burt and Michael F. Doyle, *The American Keiretsu* (Homewood, IL: Business One Irwin, 1993) 109.

24. Steve V. Walton, Robert B. Handfield, and Steven A. Melnyk, "The Green Supply Chain: Integrating Suppliers into Environmental Management Process," *International Journal of Purchasing and Materials Management* 34 (2) (Spring 1998): 2.

25. Lisa Cooling, "Social Responsibility in the Real Business World," *Inside Supply Management* March (2008): 26–29.

Suggested Reading

Carter, Craig R. *Ethical Issues in Global Buyer-Supplier Relationships.* Tempe, AZ: Center for Advanced Purchasing Studies, 1998.

Cooling, Lisa. "Spread the Carbon-Neutral News: James Martin Plays a Key Role as News Corporation Commits to Reducing its Environmental Impact throughout the World." *Inside Supply Management* October (2007): 30–32.

Cooper, Robert W., Garry L. Frank, and Robert A. Kemp. "The Ethical Environment Facing the Profession of Purchasing and Materials Management." *International Journal of Purchasing and Materials Management* 33 (2) (May 1997): 2.

Cooper, Robert W., Garry L. Frank, and Robert A. Kemp. "The Ethical Environment Facing the Profession of Purchasing and Materials Management." *International Journal of Purchasing and Materials Management* 33 (2) (Spring 1997): 2–11.

Hartley, M. F., G. B. Turner, and W. C. Ferguson. "Purchasing's Role in the Development of Corporate Ethics Statements." *Ethics Policy Statements for Purchasing, Supply, and Materials Management*, Tempe, AZ: National Association of Purchasing Management, 1995, 137–141.

Lindsay, R. M., L. M. Lindsay, and V. B. Irvine. "Instilling Ethical Behavior in Organizations: A Survey of Canadian Companies." *Journal of Business Ethics* 15(1996): 393–407.

Marchetti, Anne M. *Beyond Sarbanes-Oxley Compliance: Effective Enterprise Risk Management.* Hoboken, NJ: John Wiley & Sons, 2005.

Roberts, Julie S. "Navigating the Ethics of E-Commerce." *Purchasing Today* 12 (12) (December 2001): 28.

Robertson, D. C. and B. B. Schlegelmilch. "Corporate Institutionalization of Ethics in the United States and Great Britain." *Journal of Business Ethics* 12 (1993): 301–12.

Schildhouse, Jill. "Corporate Ethics: Taking the High Road." *Inside Supply Management* 16(3) (March 2005): 30.

Shaw, B. "Business Ethics Today: A Survey." *Journal of Business Ethics* 15(1996): 489–500.

Trevino, Linda K. and Katherine A. Nelson. *Managing Business Ethics: Straight Talk about How to Do It Right*, 4th ed. Hoboken, NJ: Wiley & Sons, 2007.

Turner, G. B., M. F., Hartley, and S. Taylor. "Ethics, Gratuities, and Professionalization of the Purchasing Function." *Journal of Business Ethics* 14(1995): 751–760.

Zsidisin, George A. and Sue P. Siferd. "Environmental Purchasing: A Framework for Theory Development." *European Jounal of Purchasing and Supply Management* July (2000): 61–73.

CHAPTER 22

Implementing World-Class Purchasing in the Supply Chain

This chapter begins with the details for designing an effective and efficient purchasing organization to carry out the plans, and appropriately ends with evaluating the department through a look at an annual report and risk management. The appendices contain many case histories of organizations using best-in-class methods and procedures.

There are several "cookbook" implementation programs such as the 26 Weeks program by Long and Meyer and a very sophisticated ERP-type model called SCOR by Bolstorff and Rosenbaum.[1] We still feel the internal supply team must execute the traditional managing process in order to decide what programs to employ.

The effective design and organization of a company's purchasing and supply management department can significantly impact the firm's ability to achieve its goals and objectives. Necessary activities in building and sustaining an effective department entail going back to the basic processes of planning, organizing, directing, motivating, controlling, and evaluating the resources and functions of the group in order to assure purchasing's contribution to the organizational goals. These interdependent activities lay the foundation for a successful, competitive supply chain. This chapter will provide a retro, fundamental look at how to implement these processes to build an effective purchasing/supply management department within your organization.

Designing an Effective Purchasing/Supply Management Department

As you worked your way through this book, you revisited the basic fundamentals and the latest purchasing methods. Now it is time to put them into action. The previous 21 chapters have provided the rationale and components of best-in-class purchasing activities and practices within the supply chain. They are the components of the philosophy,

The Progression to World-Class Supply Management[SM]

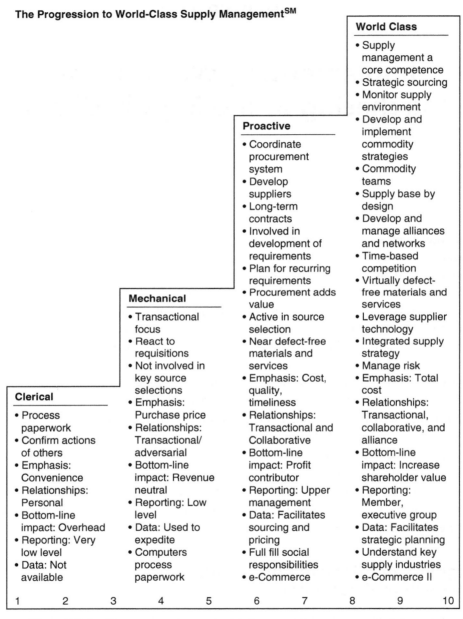

Figure 22.1 The progression to world-class supply management. Adapted from David N. Burt and Michael F. Doyle, *The American Keiretsu* (Homewood, IL: Business One–Irwin, 1993): 21.

the strategy, and the tactics to achieve the maximum purchasing value added within the supply chain. It is the implementation of world-class purchasing practices that results, in a value added network. Thus, value network management (VNM) is the product of an effective supply chain.

At this point, we suggest you reevaluate where your organization is in relation to the progression to world-class supply management step chart Figure 22.1 and consider the characteristics of a world class organization as presented in Figure 22.2. Where are you now after exploring the previously presented material? Another way of assessing the current status of your purchasing department is to look at the operations, practices, policies, and procedures that exist, commonly described as the purchasing system review.

An internal audit, complete with flow-charting of these operations, will give you a starting point when comparing your department to those described as world-class in the step chart. If you find yourself on a step lower than you would like, continue on to learn how you can move your organization to a world-class 10.

Planning makes the biggest contribution to a successful supply chain. It lays the foundation for all subsequent management activities and, therefore, must be done with great care. Whether they take advantage of it, the typical purchasing department has been provided a significant opportunity to leverage the supply chain for a competitive advantage in planning, primarily because it offers a chance to break with tradition. With the development of the supply chain concept, purchasing departments have moved from paper pushers to stewards of the bottom line. Not only should purchasing support organizational goals and strategies, it should contribute to the development of future organizational goals including involvement in corporate planning and budgeting.

The Purchasing Planning Function

Defining Planning

In our opinion, Peter Drucker, the father of modern management, still has the best definition of planning:

> Planning is a continuous process of making present entrepreneurial decisions systematically and with the best possible knowledge of their futurity, organizing systematically the effort needed to carry out these decisions, and measuring the results of these decisions against expectations through organized systematic feedback.[3]

Note the power of the key words: continuous process, entrepreneurial, knowledge of futurity, organizing the efforts to achieve the plan, and measuring results through a

1. The senior supply executive, often called the chief purchasing officer (CPO) is at an equal level with the production and marketing executives.
2. Purchasing has a major input into the organization's strategy, planning, and budgeting process.
3. Supply chain management (SCM) requires proactive vs. reactive thinking, policies, and procedures.
4. Cross functional teams for design and sourcing are necessary.
5. Partnerships and collaboration with suppliers is the basic philosophy. Collaboration stimulates innovation, and better designs, and reduces the lead time to market, which leads to higher share of market (SOM), and increased customer satisfaction—all major marketing goals. There is some evidence that 65 to 70% of product innovations come from suppliers.
6. A reduced but higher quality supplier base, with long-term contracts including meet competition requirements is required.
7. Early supplier involvement (ESI) in new product development programs and modifications is common.
8. Concurrent engineering vs. sequential—over the transom—nonintegrated steps and decisions is a major change.
9. Management has a viewpoint of total cost of ownership (TCO) vs. a focus on price. Remember, unit price is just one portion of total lifecycle costs.
10. Total quality management (TQM) is a prerequisite to world-class SCM. The philosophies and techniques of the gurus, that is, Deming, Crosby, Taguchi, Juran, etc. Six Sigma is the rather new training system to apply TQM, as popularized by Motorola in the 1980s, but GE has become the current disciple. The goal is to eliminate rework, rejects, returns—do it right the first time.
11. Do continuous flow, that is, Lean manufacturing.
12. Utilize demand management beyond naive forecasting including using advanced IT software such as MRPII, ERP, DRP (distribution resource planning), and SCOR.
13. Separate strategic from tactical purchasing. Even a one-person department can do this.
14. Outsource materials with low value added capacity, that is, MRO, office supplies, and other indirect commodity material and services such as security and janitorial. What is your core asset-operation? Put time, effort, and money on high value added activities. Purchasing now becomes a profit center vs. a cost center.
15. Eliminate the small order problem via consolidation, simplification, and buyer planner activities and procedures. For example, use long-term contracts with release authority given to planners.
16. Work backward from the ultimate customer and think the entire pipeline of materials in and products/services out. Burt calls this total value impact.
17. Total systems and interface vs. department self-interest, that is, think macro but take appropriate micro steps. Strive for the Gestalt, that is, $2 + 2 = 5$, the synergy.
18. The really new skill emphasis is on negotiation in order to get all the players in SCM on the same page, the same page of understanding with the contract, which must be managed.
19. In most cases, SCM begins with an audit, diagnostic, ABC analysis etc. to determine the current situation. We look for gaps on the road to SCM. Think of this stage as a radar fix to see where we are vs. where we want to go. This is major re-engineering and benchmarking.
20. It takes money to make money. SCM requires a fairly significant up-front investment in analysis, training, software, and education to get the desired ROI. This is a rather sophisticated collection of many policies, procedures, techniques, and tools. It requires an elaborate plan with input and cooperation from many different stakeholders. The conversation will take time and U.S. firms are notoriously impatient.
21. Finally, the supply chain exists whether you manage it or not, do not leave it to chance.

Figure 22.2 Characteristics of a world-class purchasing organization.[2]

systematic feedback. A budget is simply the dollar cost of the resources to be used in a plan, and a forecast is the predication of the results achieved from the execution of a plan. Far too many executives think a budget or a forecast is a plan and neglect attention to the strategy and tactics behind the numbers. No wonder so many plans fail.

Four Phases of Planning

There are four planning phases: the current situational analysis, the setting of objectives, the creation of new plans, and the implementation and monitoring phase.

The Current Situation Analysis Phase

This phase of planning takes a look at where a purchasing/supply management department currently stands in respect to existing functional and operational plans. The current situation analysis phase can be called the audit, the purchasing system review, or the diagnostic phase. To examine the present status of a department, look at the items listed in Figure 22.3 as a starting point. Be sure to include any others unique to your organization. Then look for gaps between corporate and functional objectives and results to date. For example, review the records regarding the ability of the supplier to meet delivery dates, quality standards, and costs. Assess whether there is a gap between goals and results. If a positive partnership relationship with suppliers is a goal, a confidential supplier survey must be conducted to obtain supplier input. Supplier councils, meetings, and performance reports help but there is no substitute for

---

Functional Analysis

▲ Do existing strategies support the corporate mission statement?

▲ Do existing strategies support the corporate and business strategies?

▲ Does a department manual contain documented policies on conduct of purchasing personnel, defining buyer-seller relationships, addressing ethical and social responsibilities of department and staff, operational issues, etc.?

▲ Are procedures in place to allow the department to perform the steps in the purchasing cycle, responsibilities of parties in the cycle, instructions on how to use forms, operational procedures, etc.?

▲ Does a supplier welcome manual exist?

▲ Consider how these manuals, policies, etc. have been working lately. Identify any weaknesses or needed revisions.

▲ Are resources being used effectively?

▲ Is departmental authority clearly defined? Does backdoor selling exist?

▲ Does the department structure support effective flow of material through the supply chain?

▲ Key supplier-commodity analysis: dollar volume and percentage of material purchased from individual suppliers. (Functionally decide acceptable amount; tactically put decision into place.)

▲ Internal departmental trend analysis regarding number of purchasing employees, number of purchase orders, requisitions, value per purchase order, total purchase dollar vs. total manufacturing cost, buyer training and education, ROI and ROA contributions, etc.

▲ Purchasing department budgets, especially the trend over several years. Are there sufficient travel funds for visits to suppliers and educational seminars?

▲ Development and effectiveness of cross-functional teams where appropriate.

Figure 22.3 The situational analysis: where you are at this time. (David N. Burt and Richard L. Pinkerton, *A Manager's Guide to Strategic Procurement*, New York: The American Management Association (AMACOM), 1996, 232. Used with permission from AMACOM, permission through Copyright Clearance Center, Inc.)

Tactical Analysis

▲ ABC inventory analysis, often called the spend analysis.

▲ Critical commodity list. What material can shut down the operation?

▲ Existing and potential sole source buy situations vs. single source by design, accident, or monopoly.

▲ Variance plus or minus from past objectives and goals.

　▼ Target price and contract terms.

　▼ On-time delivery rating.

　▼ Quality control rejection rate.

　▼ Supplier development ratings.

　▼ Inventory: average level, number of turns, safety stock, stock outs, excess, obsolescence, and procedures for disposal of materials including hazardous materials.

　▼ Stores: receiving and inspection, lost items, delays, damage, material handling, and flows.

　▼ Value analysis: engineering programs, other special projects.

　▼ Long-term contracting: blanket orders, system contracts, consignment-buying, formula pricing, etc.

　▼ Make or buy projects.

　▼ Traffic audit activities.

　▼ Surplus analysis: idle equipment, material reports.

　▼ External trend analysis regarding long-term movement by line item of price, lead-time, internal inventory levels, commodity availability, supplier availability, quality rates, etc.

　▼ Production control capacity for scheduling and past accuracy record.

　▼ Overall material and inventory savings.

▲ Long-term material availability—national and global. Is new technology tracking adequate?

▲ Special problems, such as price in effect at time of delivery, surcharges, sudden price increases.

▲ Paperless purchasing progress such as credit card, IT, and supplier stocking programs.

▲ Implementation of integrated purchasing systems with key suppliers.

▲ Cycle time reduction techniques such as flow tracking studies.

▲ Utilizing leading edge technology where appropriate.

▲ Other items unique to your organization.

Note: Many of the above items are also inputs to the annual materials report.

Figure 22.3 *(Continued)*

the confidential survey returned to a neutral staff such as the marketing research department. A sample survey, which also includes illustrative questions for internal organization personnel using the services of the purchasing department, is included as an appendix of Chapter 20. Many other items in the situation analysis are a matter of accurate record keeping and analysis if proper tracking systems are in place such as incoming defect and "late" reports. If you said in last year's plan that the department would set up a value analysis program, you either did it or failed—perhaps with some progress. The key is to be honest at this phase and to determine the reasons why objectives were achieved or neglected. The findings from this analysis should be the focus of any additions or changes to the existing purchasing plan.

Once you have compiled all the information about where you are, prepare the final written report, which compares status to objectives or the absence of objectives for particular items. Be sure to include your SWOT analysis in this situation analysis report.

This report, along with corporate level objectives, will serve as the basis for developing the purchasing/supply management plan.

The Objective Phase

Objectives are what an organization wants to accomplish. Objectives most often deal with change, goals are much more specific. It is the desire to improve or correct something that formulates an objective. Objectives tend to be open-ended, stated in relative, broad terms, and are often stated in the context of some relevant external environment. In February 2007, Susie Mesure reported in *The Independent Business News* that "Wal-Mart, the biggest retailer in the world, pledged to slash its carbon footprint by barring products that contribute to global warming from its shelves."[4] This statement represented a corporate-level strategy for Wal-Mart.

Every chapter in this book contains numerous objectives the authors believe produce "world-class procurement departments." They are the programs, policies, and procedures thought to be the most effective and efficient for present-day operations.

Again, these are the goals of value network management (VNM). Rather than repeat all these objectives, we leave it to the reader to select the most appropriate for his or her organization although we will repeat a few in this chapter for illustrations.

The term benchmarking is a form of objective setting against other organizations' metrics, described as the "best in their class" or some other industry top rating for a particular attribute. The danger in using benchmarks is that they may not be appropriate for a particular organization's mission or resources.

Development of Functional and Operational Plans and Goals

Objectives will remain aloof ideas of grandeur if goals are not developed to make them happen. Goals are time oriented, specific in nature, and internally focused. When contemplating their objective of carbon reduction, Wal-Mart determined that "the footprint of (their) global supply chain (was) many times larger than its operational footprint…"[5] It was apparent the goals to accomplish this objective had to involve the purchasing and supply management department's functional and operational plans. To help the corporation reach its corporate level objectives, purchasing formulated functional and operational goals to collaborate with suppliers and environmental experts to develop, measure, and assess carbon reductions. They committed to "to eliminate 20 million metric tons of greenhouse gas (GHG) emissions from its global supply chain by the end of 2015." The plan details a focusing on product categories with the highest embedded carbon; those with the greatest opportunity for reductions. The goal is to help suppliers reduce their energy use, costs, and carbon footprint.[6] This goal is time oriented, specific, and internal. To make this happen, Wal-Mart has structured these plans in a program format developed to support this corporate objective.

Figure 22.4 is an example of detailed operational planning. As the planning webs down the hierarchy to the day-to-day issues, the planning becomes more specific, short-term, and action oriented. The consolidation procedure is one of the most popular ways to achieve early major cost savings.

---

1. **Commodity Analysis**
   For all divisions, branches, subsidiaries, offices, plants, etc., obtain computer printouts and look for the same product codes and different descriptions, and the same description but different codes, including slight variations in sizes and other specifications.
2. **Total Last Year's Purchases**
   Review last year's purchases by product type and specification variations.
3. **Analyze the Volume by Supplier**
   Look for the same product purchases from four or more and in some cases two or three suppliers.
4. **Simplify and Standardize**
   If possible, change the specifications to reduce product variation and, if possible, also change from custom to commercial or standard shelf items. This part of the consolidation process is a type of value analysis procedure using a committee of users or cross-functional team with a purchasing manager or buyer as chair.
5. **Forecast the Future Requirements**
   Forecast at least one year; two or three if possible.
6. **Prepare Requests for Proposals for Major Consolidated Contracts**
   Make use of blanket orders, requirement contracts, or system contracts. Include estimated release schedules, clauses, price warning clauses, inventory control procedures, shipping, master catalogs, supplier stocking programs, and price discounts due to increased volume, etc. Reduce the number of suppliers to one prime and one backup, unless there are good reasons not to do so.
7. **The Distinction between Blanket Orders and Systems Contracts**
   A blanket order is a long-term contract for one class of product with one to three (as a rule) suppliers depending on volume. They are called many names including: international-national contracts, open-end orders, stockless purchasing, corporate-wide agreements, evergreen contracts, long-term contracts, multiyear contracts, and so on.
   A system contract is a consolidation agreement for an entire family of products such as office supplies, tools, forms, repair parts, and general MRO hardware items; often called stockless purchasing because the supplier usually owns the inventory until issued for use directly to the user (the supplier-operated tool room concept). Users have master catalogs and purchasing manages by periodic review. The systems contract is usually issued to one supplier and its use greatly reduces the "small order problem." Direct releases using computer data phone terminals, fax, or IT are optional features of many system contracts. The term stockless purchasing is not accurate as blanket orders and requirement contracts usually have the supplier holding large amounts of inventory. Consignment buys can also be a part of either a blanket order or system contract. Usually, the buying firm must purchase minimum amounts and be responsible for obsolete material held in inventory. The key to all long-term contracts is life cycle cost analysis.
8. **Negotiate the Contracts with as Few Suppliers as Possible**
   This requires more contract instructions regarding delivery, stocking programs, invoicing, and other such issues unique to this form of contracting, but overall costs should be drastically reduced. Don't forget to ask for the price discount. As a rule, this type of negotiation must be conducted in person. Many firms try for one prime and one backup supplier using commodity-sourcing teams to negotiate the contracts.
9. **Audit the Results**
   Review on a monthly basis for contract performance and issue corrective instructions to the supplier, users, or both if necessary. Remember, the buying company may be at fault because of unexpected demand, change orders, release error, etc. Watch for price creep. Prepare for the next negotiation cycle.

Figure 22.4 The consolidation procedure, a major planning tool–project. (David N. Burt and Richard L. Pinkerton, *Op. cit.*, 233. Used with permission from AMACOM, permission through Copyright Clearance Center, Inc.)

Linking Purchasing Objectives with Corporate Objectives

All plans start with the corporate mission, strategy, or charter. If your department is performing a task or function that does not relate back to the support of the corporate mission or strategy, resources are being wasted and the task should be discontinued. How can you be certain the activities being performed in your department echo the corporate goals and objectives? You must first understand your organization's mission and higher-level strategies. "At its very roots, corporate strategy addresses the long-term mission of an organization, including long-term survival."[7] Firms will generally use one or combinations of well-known models such as the Boston Consulting Group (BCG) growth-share matrix and GE's strategic business-planning grid. The popular strengths, weaknesses, opportunities, and threats (SWOT) analysis is an effective tool to use in developing planning documents. While these approaches and analysis methods are useful at the corporate and strategic business unit (SBU) level, they must be translated into more detailed operational plans at lower levels. The failure to move from the corporate mission to the selection of specific action steps at lower levels is a major reason for planning failure. We discuss transforming corporate objectives into purchasing strategies next.

Corporate strategies define how a company will compete, thereby setting the stage for the functional strategies and objectives of marketing, purchasing, and supply management to follow.

Strategy and objectives development takes place on four levels within the organization:

1. Corporate strategies—define the business of the organization.
2. Business level strategies—describe how the organization is going to compete.
3. Functional level strategies—define how each function will support the competitive strategy of the business.
4. Operational level strategies—define how goals will be met on a day-to-day basis.

If your department is represented at upper management level, you are likely to be involved in planning strategy at the corporate and business levels. At the very least, you must drive the functional and operational planning for your function (purchasing and supply management) and ensure that your plans fit with strategy at the higher levels. These strategies must be linked together in a hierarchical relationship. Figure 22.5 shows a manufacturer of children's shoes and what their strategies might be to support a mission of becoming the nation's most profitable producer of children's shoes. The figure proposes strategies from the manufacturing and purchasing perspective. Marketing would of course develop strategies and objectives unique to their function in support of the corporate mission.

The objectives on the lowest level of the pyramid, "lower production costs by 10%" and "searching out ISO certified suppliers," support the desire to "manufacture a high quality product at the lowest possible cost" and "searching out quality, low-cost suppliers,"

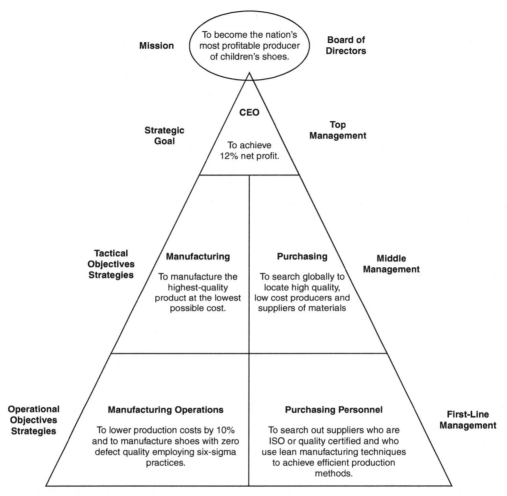

Figure 22.5 Hierarchy of unified goals and objectives. Adapted from a figure demonstrating the hierarchy of goal unification from Warren R. Plunkett, Raymond F. Attner, and Gemmy S. Allen, *Management: Meeting and Exceeding Customer Expectations*, 8th ed. (Mason, OH: Thomson South-Western, 2005): 65.

which will help "achieve a 12% net profit" that will move the company toward its mission of becoming "the nation's most profitable producer of children's plastic beach toys." Purchasing and supply management objectives must be designed to do the same; move the organization toward its mission.

The first step toward moving through this process is to identify or define the organization's mission. The strategic management process centers on the belief that a firm's mission can best be achieved through a systematic and comprehensive assessment of both its internal capabilities and its external environment. A SWOT analysis can stimulate the creative phase of planning and validate the integrity of the mission statement, as it exists. Once the mission statement is validated, findings from the SWOT analysis can trigger the development of lower level strategies. Opportunities and threats identified in the external environment can lead to development of long-term objectives and grand strategies, which will filter into operating strategies and objectives. For instance, it can identify weaknesses such as excessive number of purchase orders with particular suppliers, indicating a problem with paper control but also triggering an opportunity to negotiate long-term contracts based on electronic release procedures. An inordinate number of suppliers for the same material is a weakness that also presents an opportunity for consolidation. These findings can lead to written plans, strategies, policies, or procedures further down the hierarchy of planning for the organization.

Once the mission statement is reaffirmed, the task of developing strategy at the corporate, business, functional, and operational levels begins. As the formulation of the plan winds its way through the organization, the strategies become more specific, narrow, short-term, and action oriented. If the mission began with "We deliver high quality…" for instance, the plan might include a *corporate strategy* of vertical integration to control the quality, a *business strategy* of differentiation in an attempt to set its quality apart in the market, a *purchasing functional strategy* of buying only from ISO certified suppliers, and a *purchasing operational objective* of zero defect ratings on suppliers.

The Creative-New Action Plan Phase

This is the innovative step where we create new plans to either correct a problem and/or to achieve revised objectives.

Transforming Corporate Objectives into Purchasing Strategies and Objectives

Common objectives of all organizations consist of increasing revenues and reducing costs to increase profitability. Purchasing has a unique advantage over other functions when contributing to the achievement of these corporate objectives. With the role purchasing plays in the supply chain, it has a direct opportunity to leverage the supply chain to enhance the competitive position of the company through its supply base. "An effective purchasing strategy (therefore) is one that fits the needs of the business and strives for consistency between the internal capabilities and the competitive advantage being sought, as defined in the overall business strategy."[8] Purchasing must, therefore, align with corporate strategies by converting corporate objectives into purchasing strategies and goals. This must carry all of the way through to the commodity goals. Table 22.1 shows some samples of corporate objectives transforming to purchasing goals. These company-wide purchasing goals must then be translated into operational level goals.

Table 22.1 Sample Purchasing Goals in Support of Corporate Goals

Corporate Objective	Purchasing Goal
Sell off-site warehouse in 2 years	Move to JIT within 2 years; go to vendor managed inventory (VMI) for MRO items within 18 months; eliminate all excess and obsolete inventory within 2 years.
Reduce cycle time in new product development	Get purchasing involved early in design stage. Get suppliers involved early in design stage. Pre-qualify key suppliers.
Increase quality level of products	Require all major suppliers be ISO or Six Sigma certified. Work with suppliers to implement zero-defect policy.
Reduce dollar value of inventory	Develop systems contracts to manage office supplies and MRO items. Work with suppliers to bring inventory in on consignment through VMI.

Transforming Corporate Objectives into Functional and Operational Purchasing Plans

Planning is very creative work that requires proactive thinking. It requires input from all the players and is the primary responsibility of the procurement executive. The need for more planning and strategic activities is another reason why some large purchasing departments are separating the research and planning activities from the day-to-day buying tasks. Good managers have vision and know how to translate vision into operational plans with one eye on strategy for the long term and the other on tactics for the immediate operational period.

It is essential to remember that plans must be *sustainable*. They must have the resources to move from objectives to strategic and tactical actions.

As previously stated, strategic plans are long-term and involve considerable organizational resources to make them happen. Tactical plans address the more immediate needs of the organization using resources already available to the organization. Both functional and operational plans may be strategic or tactical and may include and be formatted as:

1. goals and objectives
2. programs, such as a quality program
3. standards for clarity and measurement
4. policies and procedures
5. budgets to plan and control expenditures and to provide resources for projects.
6. Detailed action steps such as who does what, when, and how.

Purchasing functional plans are designed to support the corporate plans. One of the key strategic objectives of purchasing is its role in the development of an effective supply chain.

A world-class supply chain adds value to the organization's ability to compete and maximize profits. This is the major goal of value network management (VNM). Purchasing works to develop and maintain supply sources and supplier relationships for goods and services in support of the organization's overall strategic goals. The following are samples of purchasing functional strategies that purchasing departments may use depending on the corporate strategy they are supporting.

- Early supplier involvement
- Single vs. multiple sourcing
- Share of supplier's capacity
- Local, national, and international sourcing
- Manufacturer vs. distributor
- Outsourcing
- Green supply management
- Purchasing risk management
- Minority- and women-owned business enterprises
- Ethical considerations
- Reciprocity
- Supply base optimization
- Long-term supplier relationships
- Long-term contracts

These strategies have been discussed at length in this book.

Purchasing operational plans are developed to carry out the functional plans and goals. They assure the continual flow of goods and services required for the day-to-day operations of the organization. They include everything from make vs. buy analyses, to policies defining the buyer-seller relationship, to establishing budgets to cover the costs of materials throughout the year, to determining a good inventory balance, to defining the conduct of purchasing personnel. Very often, operational plans take the form of goals, policies and procedures, and budgets. A good purchasing policy manual will address operational issues, buyer-seller relationships, social and ethical issues, and conduct of buying personnel. Procedural manuals will include things like the steps of the purchasing cycle, guidelines on contract development, and steps for the use of various forms, as well as various operational procedures.

Also as previously suggested, survey your internal customers or the users in operations-production, engineering, finance, marketing, quality control, and others about your efforts and results. The purchasing plan should address any complaints or service problems identified by the survey. For example, buyers can negotiate price cuts to such a degree they reduce quality or timely delivery. Such an objective could be in direct conflict with a corporate objective for high quality or leading edge technology. We must think long term and total cost of ownership (TCO).

The Implementation and Monitoring Phase

Development of Monitoring Process

After creating the final written plan, which contains many sub-plans and projects, it is important to keep your eye on how the plan rolls out. The control process has its beginnings in the planning process. See Figure 22.6 for an example of a planning project-control chart. Using the proper metrics for the deliverables is very important to measuring success. While we can track specific cost savings and other objectives data like the number of staff achieving certification, some measures are subjective. However, the use of surveys from suppliers, internal

Prepared by: _____ Date: _____ Approved by: _____ Date: _____ Date of preparation: _____

Start date: _____ Completion 10%[a]: _____ 50%-75%: _____ 100% Finish date: _____

Current Situation by Priority	Assumptions	Objectives, Creative Phases	Key Steps, Tasks	Budget	Control Responsibility	Scheduling Target Dates, Milestones	Variance Problems, Corrections
EXAMPLE 12 suppliers for industrial fasteners	1. Quantity need will continue for 2011–2012 based on past order activity. See attached and forecast 2. Cost analysis as attached indicates too many suppliers	1. Reduce administrative cost by 25% and prices by 20% 2. Negotiate blanket order or systems contract—B/O with two suppliers or one supplier on a system contract	1. Form commodity team 2. Coordinate with production 3. Analyze past orders 4. Standardize? 5. Supplier meetings 6. Supplier plant visits 7. Other users 8. Legal? 9. Contract terms 10. Negotiate	$5,000 for travel[b]	Dick Jones Sr., MRO Buyer	1. Form team 6/25/11 2. Coordinate with production 7/26/11 3. Supplier meetings 9/30/11 4. Supplier visits 9/30/11 5. Final negotiations 10/15/11 6. Effective start 11/1/11	1. Complete as planned as execution takes place. 2. Audit: Did the savings really occur? 3. Revise if necessary

Comments:

Signature of Project Leader

[a]The date the project is 10% completed, 20% completed, etc.
[b]In this actual case, the suppliers were all in the same industrial area, hence the small amount.

Figure 22.6 Procurement planning project chart (example is consolidation of industrial fasteners. (David N. Burt and Richard L. Pinkerton, *Op. cit.*, 237. Used with permission from AMACOM, permission through Copyright Clearance Center, Inc.)

customers (users), and supporting departments can provide quantitative responses. Supplier scorecards are another tool that can be useful in the monitoring process. It is important to note that this tool can be labor intensive if not supported by a software package. The metrics generated by your purchasing software can help you format a fairly objective performance measure. To ensure the best performance metrics for your organization, check to see that you have the systems and procedures in place to capture the necessary data to support it.

Finally, avoid the many planning mistakes many organizations have made. See Figure 22.7 for a list of planning hazards to sidestep in your planning.

---

1. Corporate planning has not been integrated into a firm's total management information system.
2. Lack of understanding of the different dimensions of planning vs. forecasting vs. budgeting, etc.
3. Management at different levels in the organization has not properly engaged in or contributed to planning activities. Failure to include purchasing is a major problem.
4. Planning vested solely in a planning department: the MBAs write it but nobody does it.
5. Many companies do not change plans as the situation or assumptions change.
6. Many companies fail to implement plans: nice studies, but no action.
7. Too much attempted at once.
8. Confusing financing and budgeting with planning and strategy.
9. Inaccurate IT inputs or GIGO (Garbage In, Garbage Out).
10. Failure to see the big picture: Hung up on details, computers, printouts, etc.
11. Overemotional commitment. Commitment to a pet project, supplier product, plan.
12. Lack of communication: fear, confusion, poor perception.
13. Past success egotism (Why change?).
14. What business are we really in (conglomerate confusion or single purpose myopia)?
15. Preoccupation with immediate ROI, ROA vs. long-term payoff.
16. We're too small to plan (but big enough to go bankrupt).
17. Our competition is going out of business (we hope).
18. What new technology?
19. Failure to translate strategy into tactics.
20. Insufficient resources for the plan. In particular, too few people, insufficient travel and training funds.
21. Bad timing.
22. Nonexistent, inadequate, or inaccurate records.
23. Loose planning structure and lack of marketing research and good forecasting methods.
24. Inability to accurately predict or naïve forecasting.
25. Failure to revise in light of new information: poor intelligence.
26. Simple mistakes can lead to big negative results.
27. Blind copying of someone else's plans, no creativity; misuse of benchmarking.
28. Poor memory, reinventing the wheel.
29. Failure to integrate plans to all levels.
30. Don't put it in writing, they might check up: few metrics.
31. Lack of top management interest and support.
32. Wrong assumptions.
33. We cannot think globally, the U.S. market is enough.

Figure 22.7 Planning hazards: mistakes and attitude problems that must be anticipated, avoided, or corrected. (David N. Burt and Richard L. Pinkerton, *Op. cit.*, 238. Used with permission from AMACOM, permission through Copyright Clearance Center, Inc.)

The Organizing Function

With the plans in place, it is time to organize the resources to support the plans. The organizational structure provides the framework of relationships between the different resources of the organization. It is the mechanism supporting the creation of products and services; in effect, supporting the achievement of the organization's mission.

As this section may be a review for many readers, keep in mind that it is also good study preparation for many of the certification programs.

The classic organization process has four primary functions:

▲ It clarifies the work environment by delineating tasks, responsibilities, and authority.
▲ It creates a coordinated environment by creating work units.
▲ It achieves the principle of unity of direction, defining who reports to whom.
▲ It establishes the chain of command defining an unbroken line of reporting.[9]

Management will improve the chances of creating an efficient work environment if they do an effective job on organizing their resources.

Most management textbooks recognize five basic steps in the organizing planning process. The purchasing and supply management executive must organize a department to carry out the plans and programs specific to their department. First, as we have previously stated, they must identify the corporate goals pertaining to their function in order to understand the more specific unit goals and objects they are responsible to achieve. Once they zero in on their responsibilities, they can align their resources to support the functional goals and objectives. The next step is to determine the activities necessary to reach organizational goals. These activities include things such as hiring, training, sourcing, ordering, contracting, expediting, and negotiating; all of the strategic and tactical functions of a purchasing and supply management department. This often involves determining the sequence and timing of necessary activities and tasks. The third step refines the activities by grouping them. The manager arranges the activities into logical groupings or classifications by functional similarity. In step four, the manager must assign these activities to the right individuals within the department and give them the level of authority required to carry out their activity. Without the necessary authority to complete the task, the individual will fail. The last step entails determining the vertical and horizontal operating relationships of the organization as a whole. This step results in the chain of command and establishes the hierarchy of decision making throughout the organization. It answers the age old questions "Who reports to whom?", "How many subordinates work for each manager?", and "How many steps are there to the final decision maker?"[10]

The Five Basic Steps to Organizing

Identify the Corporate Goals Pertaining to Purchasing

This first step in organizing has been discussed in detail in the section on planning, as planning and organizing are intimately related. It is the plan that gives authority to the organizational structure. As the goals of the organization filter through the hierarchy of planning down to the purchasing and supply management organization, the activities needed to achieve the goals become clear.

Determine Activities Necessary to Reach Organizational Goals

The next step in the organizing process is determining the activities necessary to achieve organizational goals. In many purchasing/supply management departments, these activities are identified as necessary to provide an uninterrupted flow of materials and services, or as profit-generating activities. The former are referred to as tactical or operational activities, the latter as strategic activities. Whenever possible, the organization will separate operational and strategic responsibilities formally in their organizational structures. This allows for a focus on the strategic sourcing process, which promises long-term increased profitable sales and cost savings, improved profitability, and competitive advantage. Some organizations group related activities into various projects

Grouping the Activities

The third step in the organizing process is to arrange the activities into logical groups that can be performed by individuals. Purchasing work divides into six distinct classifications, each of which encompasses a fairly wide range of these activities. These classifications— management, buying/supply management, contract and relationship management, strategic planning and research work, follow-up and expediting, and clerical activities—are discussed in great detail in Chapter 2.

Assigning of Activities to Individuals

As important as identifying the task is identifying the proper individual in your department to assign responsibility for it. In the section on staffing, we will discuss assessment and development of employees. It is critical to know the strengths and weaknesses of your staff in order to effectively delegate the activities within your department. The employees must be given the appropriate authority to accomplish the task.

Designing a Hierarchy of Relationships

This last step is most critical to purchasing's ability to impact the organizational goals; in particular, the goal of increased profitability. This step puts together all parts of the organization in a hierarchy that determines power based on position in this hierarchy. Because it lays out the chain of command, the question of where purchasing is placed in this puzzle and to whom it reports can effectively increase their influence on the executive level, or decrease it. As we stated earlier in the book, to attain commitment and respect from the top, purchasing and supply management must shift its focus from internal processes to big-picture issues such as determining and defining the requirements, contract negotiations, supplier relations, and strategic long-term goals of the organization.

Basic Concepts of Organizing

Authority

Authority is the formal and legitimate right of a manager to make decisions, give orders, and allocate resources.[11] All managers in an organization have authority in varying degrees based on their place on the organizational chart. That is why the position purchasing occupies on the hierarchy of organizational authority impacts their power and abilities to be effective.

The organizational chart delineates three types of authority demonstrating the relationship between individuals and departments. Solid lines on the chart designate the relationship between the subordinate and supervisor. This type of authority is referred to as line authority. Staff authority, designated by a dotted line, gives individuals or a department the authority to serve the organizational departments in an advisory capacity. Staff authority generally flows upward to the decision maker. For instance, the legal department would give advice to the upper level executives. A line of dashes flowing downward represents functional authority within the organization. Functional authority allows functional department managers to make decisions regarding activities specific to their department. Purchasing should have involvement in and authority over all purchases a company makes no matter which department is requesting the buy. "Back door selling" has been a major problem in many organizations. IT for example, should involve purchasing when buying new computers for the company. Once again, the position of purchasing on the organizational hierarchy will determine the breadth of purchasing's span of authority.

Delegation

Management texts define delegation as the downward transfer of formal authority from one person to another.[12] What this means for a purchasing/supply management department is the ability of the personnel to legally contract to buy for the organization. Through the process of delegation, purchasing and supply management personnel are given the power to manage their tasks and activities. Without this sharing of authority, the bonds of agency are not established. The buyer has no legal right to represent the company in contractual transactions. An adequate amount of authority must be transferred to the buyer to allow him or her to complete the task. For instance, a buyer might be granted a $50,000 purchasing limit. When spending above this amount, additional signatures would be required from someone with proper authority. Remember that with authority comes acceptance of responsibility and creation of accountability. Managers must monitor this process. Delegation does not relieve the manager of responsibility and accountability.

Centralization vs. decentralization

As introduced in Chapter 2, centralization and decentralization refer to the level of authority and decision-making within the organization. Philosophically an organization may believe it to be more effective to corral the authority within higher levels of the organization. Because this can result in an inflexible environment, companies competing in fast changing environments often choose to systematically delegate authority throughout the middle and lower levels of organizational management. However, the clear trend is toward centralization to achieve maximum leverage. In the appendix at the end of this chapter, you will find many examples of this as you read the section on the Richter Awards.

In terms of purchasing and supply management, these terms are concerned with the degree to which the authority of performing purchasing and supply management activities is consolidated at higher levels or delegated throughout the organization. Centralization of purchasing is concerned solely with the placement of purchasing authority. Centralization exists when the entire purchasing function is made the responsibility of a single person.

Organizing for Supply Chain Management

The impetus to capture control of materials as they move through the pipeline has resulted in a continuous effort to establish a structure that will help reduce inventories, improve quality, eliminate material stock-outs, reduce lead times, and improve overall communications throughout the process. There exists a variety of organizational structures that will move management of supply closer to these goals. The materials management structure is still used in organizations and allows for improved coordination and collaboration internally. It is the belief of the authors that, although effectively improving communication of materials personnel, it falls short of breaking down functional silos and integrating processes throughout the supply chain.

The broader concept of supply chain management is all encompassing with regard to the planning and management activities. It integrates supply and demand activities throughout the movement of materials in the pipeline.

Regardless of the structure of the overall materials management group within the organization, the purchasing department must consider whether to organize by product orientation or commodity orientation. Companies such as Saturn choose a cradle-to-grave approach to supporting a product line. This approach relies on use of cross-functional teams assigned to a product line for the life of the product. Other companies within the GM family organize by the various commodities required to support production. Commodity buying can be controlled solely by individuals within the procurement department, or managed by a cross-functional group consisting of representatives from purchasing, production and inventory control, engineering, and marketing. Chapter 2 in this book speaks specifically to these organizational structures.

In this global economy, organizational structures will vary depending on various factors that influence their development. One fact that will not vary, however, is the need for flexibility in the design. As plans change, the organizational structure must change with them if the company is to survive.

The Staffing Function

Staffing the purchasing/supply chain organization is a critical facet to making the supply chain work.[13] Selecting, hiring, and training new employees, as well as developing and advancing the skills of existing employees should be of high priority to management. Because of legal issues, staffing is one aspect of a manager's responsibilities that generally falls into the functional authority of the Human Resource departments in most companies. This is not to imply you have no part of it. On the contrary, the HR department will look to you to identify the skills and people needed to accomplish your organization's goals. Staffing links people and processes and the departmental organizational plan should detail the activities from which to develop the job descriptions and specifications to hire the staff.

An effective staff can ensure optimum performance in the supply chain. "A department's efficiency and, more importantly, its effectiveness depend on having the right people performing the right tasks. Purchasing managers must effectively forecast their personnel needs in terms of quantity and quality and must aggressively pursue purchasing professionals who will contribute to overall supply chain management goals and objectives."[14] However, the task of determining what skill sets are the right ones for achieving purchasing and supply management goals is a difficult task. The Center for Advanced Purchasing Studies (CAPS), affiliated with the Institute for Supply Management, published a study titled, "A Skills-Based Analysis of the World-Class Purchaser." This 55-page study details professional attributes as well as skills needed by purchasing personnel at all levels of the purchasing/supply management hierarchy.[15] In summary, the study identified eight key attributes for the world-class purchaser. Roberta Duffy, in her article "Skill Sets: Building for the Future,"[16] summarized them as follows:

▲ Interpersonal communication
▲ Team skills and facilitation
▲ Analytical problem solving
▲ Technical/industry-specific skills
▲ Computer literacy
▲ Negotiations
▲ Education and professionalism
▲ Continual learning

You will need to identify the set of skills needed by members of your department to achieve your organization's goals. Though many of these skills may be common to all supply chain management departments, some will be unique to your organization. If individuals in your organization do not possess all of the skills needed, you must develop a program to train and cross-train where necessary. A skills needs analysis can be developed for your department by job description. An employee's current performance levels can be evaluated by use of this analysis matrix and a training program, tailored to their individual needs, can be developed and incorporated into their individual goals and objectives. The organization must take responsibility for providing the resources for a training program. Education reimbursement programs can allow students to stay abreast of the current conventional wisdoms. Many training companies today offer programs both on-site and over the Internet. Not all programs need be formalized training programs. Informal training can be effective if structured properly. Many companies set up a mentoring program to build skills relating to company-specific processes. Whatever method you use, "getting the building blocks in place is the first step toward leveraging the effects for competitive advantage."[17]

Having evaluated the skill level of your personnel, you may discover a need for a massive transformation to reach world-class purchasing status. You must conduct a thorough review of the staff. At what educational skill level, experience, and attitude are the current staff? Is there excess capacity in terms of workload to allow additional special projects? We worry about these questions because our fellow supply colleagues in industry, you, the practitioner often tell us that they are overworked, that is, stressed out as even the giant organizations demanded more work from even less people during the great recession of 2008–2010. Remember, people are our greatest asset. This means that some of the current staff needs to be retrained, transferred, and sadly, some may have to be terminated. Many organizations who have dramatically improved their purchasing activities have had to add highly skilled and educated staff. David Nelson of Honda, John Deer, and Delphi fame always added significant numbers of very experienced and highly skilled personnel. Indeed, some employees will resist change and will need professional, remedial education. All the authors of this book have gone through the transformation process and have experienced the fact that not all of the people will get on the train at departure time.

It is important to remember that change requires a series of preparation steps. That is, it is better to have evolution vs. revolution. We need to take one step at a time to build the confidence of our team as they perform a progression of value-added results. Few organizations can cope with immediate, massive changes, that is, people simply cannot digest too much at one time. Finally, look for a champion within the current organization who is respected by his or her peers and will help sell the new programs and procedures. Champions must have vision, experience in the supply function, communicate well, have the ability to handle stress, be able to resolve conflict, and have the support of the top executive. Some organizations have found the need to hire consultants to plan and direct the implementation process and to train the personnel involved in the chance procedure.

The Directing Function

Directing has been defined as the use of communication and leadership to guide the performance of subordinates toward achievement of the organization's goals. This can best be achieved when the subordinates can see a common purpose in organizational goals and personal goals; that is, if they can connect personal gain such as a bonus to the achievement of the organization's goal of increased profits.

Leadership

Leadership is the ability to move others toward the organization's goals. Leadership styles vary depending on the amount of decision-making authority a leader is willing to give to his or her subordinates. These styles exist on a continuum ranging from total control of the decision-making process being held by the leader, to subordinates being empowered to make decisions. When the manager maintains control over decisions, he is said to lead with an autocratic style. Participative leadership styles allow for the input of subordinates in the decision-making process. Leaders who empower subordinates, permitting them to function without direct management involvement, are free-rein in their leadership style. Seasoned leaders have recognized that each style is of use and is dictated by circumstances of the situation.

As rigid, hierarchical structures are giving way to more fluid structures in an effort to stay competitive, the changing environment is creating more opportunities for more leaders. "Today's rapidly changing business climate requires purchasing and supply professionals to understand the changing face of leadership, and to attain the kinds of skills needed to be a leader who can 'get things done.'"[18] Marvin Bower, one of McKinsey & Company's founders, urges a change in how we run companies today. In his book *The Will to Lead*, Bower advocates replacing bosses with leaders and leadership teams placed strategically throughout the company. He identifies eight attributes he believes are essential to being an effective leader:

- ▲ Trustworthiness
- ▲ Truthfulness
- ▲ Fairness
- ▲ Unassuming
- ▲ Active listening
- ▲ Open-mindedness
- ▲ Initiative
- ▲ Flexibility and adaptability[19]

To these eight attributes we would add those of vision, charisma, and a willingness to listen. By providing vision, leaders inspire people to "be all they can be," charisma inspires the followers to take action, and good leaders are always good listeners.

Leadership and the Purchasing Profession

As liaisons to both internal customers and upstream suppliers, purchasing personnel must focus on leadership both internally and externally. As internal leaders, we must take an active role in developing supply chain strategies. Externally, we must provide guidance in developing our key suppliers, strengthening our supply chain, and seeking out opportunities to grow our business and sharpen our competitive edge. Many opportunities exist for the staff of a purchasing/supply management department to step up to the leadership plate. Pilot projects in new product development activities, work on commodity teams, supplier development initiatives, inventory reduction efforts, internal and external process audits, and VA and VE initiatives all present opportunities for purchasing personnel to take the lead.

Purchasing and supply professionals who can develop their leadership skills will lead the profession into a more strategic role for business. Vic Venettozzi of Schneider Electric North America states, "There has probably never been a more critical time for the purchasing and supply profession to be concerned with good leadership. Recognition of the supply chain, and the opportunities to create value in entirely new ways, will become the next engine for generating competitive advantage in all industries."[20]

The Controlling Function

Control is the process of establishing standards of performance, and evaluating actual performance against these standards. From a purchasing and supply management perspective, it is important to note that control does not implicitly imply centralization. The control process allows the department to identify problems early on in the process so that corrections can be made to bring the process back on track with less time and cost involved.

Controlling is the function that brings the management cycle full circle. It links all the preceding functions of organizing and staffing to the goals of planning. The planning process determines the goals and objectives that eventually become the foundation of the controls.

The control process consists of four steps: (1) establishing performance standards, (2) measuring performance, (3) comparing the measured performance to the establish standard, and (4) taking corrective action.

Performance standards (or metrics) are generally a quantitative or qualitative measuring device designed to monitor people, money, capital goods, or processes. These standards usually focus on measuring and monitoring productivity and quality. They are the spring-board for initiating corrective action, setting priorities, and evaluating progress. They should reflect customer requirements and expectations. Lastly, they must continually be re-examined to ensure that they are still required and operating effectively and efficiently. Some examples of measurement standards include standard costs of materials, measurement of price variances, level of expected quality, standards for on-time delivery, standard for quantity or dollar investment on inventory, and measurements established to measure supplier performance.

After standards are established, managers must measure actual performance to determine any variation from the standard. Generally, measurement tools result in objective results for gauging performance to plan. Managers may use reports, calibrated tools, rulers, gages, and even a "go/no go" rule for measurement. When the standard is such that it does not allow for quantitative assessments, a more objective way must be used. For instance, if a restaurant's partial measurement of employee performance includes ensuring employees wash their hands after using the restroom, managers may have to rely on observation to measure performance to standard. Because health officials' guidelines recommend a 20-second wash, observation may only result in an opinionated result for it is unlikely a manager could follow each employee through this process or measure the length of time spent.

When comparing the results of measurements to standards, we look to see if deviations exist; if so, are they within an acceptable range? If they are not, we will look for the cause of any deviations. By identifying a cause, we can develop a corrective action.

Corrective actions may be built into the control process in the planning stages. For instance, a policy or procedure may exist that prescribes the approved corrective action.

Some corrective actions are automatic as in the case of something controlled by a thermostat. However, automatic corrections can malfunction and be the cause of the deviation. There are situations when corrective actions call for exceptions to the prescribed modes of behavior. You may choose to accept a restocking charge from a key supplier although the action goes against your policy of not paying restocking charges.

During the planning process, managers should consider how they will control a particular plan. When the initial plans are being made, develop the controls at the same time. For a purchasing and supply management department, items to control are cost, time, quality, cycle times, supplier issues, employee issues and behavior, ethical issues, and so on.

Controls can be developed to prevent a problem, catch one as it is happening, or allow for an armchair quarterback review to determine what went wrong. As noted in Chapter 7, SPC (statistical process control) is a feedforward control system. As the machine is punching out product, data is being collected measuring the effectiveness of its performance. If a measurement is not to specification, the control system will alert the operator so that corrections can be made before the error moves outside of the acceptable tolerance of the drawing. When a buyer measures a supplier's on-time performance with every delivery made by the seller, the control process is considered a concurrent control. Monitoring is taking place as the process is happening allowing for the buyer to contact the supplier immediately to get delivery back on track before a line goes down. Most measuring methods are feedback control methods; focus will be called to the problem after the fact. Inventory reports may come out monthly. Excess and obsolete items may only be viewed quarterly. If this is acceptable, there is no need to change the process; however, use the results to help in future planning. Remember, control procedures must provide sufficient lead time in order to take any needed corrective action.

In establishing control processes to manage your department, it is not necessary to reinvent the wheel. There are many tools that already exist, often in electronic formats. Although we often see budgets as a planning tool, budgets are also at the heart of controlling the financial aspects of an organization. Budgets are used to show the sources and allocation of funds, allocate resources, control expenses, and measure performance. Materials are managed by computerized systems that control inventory levels by way of safety stock levels, planned ordering parameters, and economic order policies. Quality tools used for controlling include pareto charts, process flow charts, cause-and-effect diagrams, control charts, and others discussed in Chapter 7 of this book.

When developing controls, be sure to provide the resources to implement the control. If the process of monitoring the control is the responsibility of a staff person, he or she must be instructed as to how to carry out the process, and be empowered to make it work. Avoid over controlling, a common practice in many organizations who employ far too many review levels.

Evaluating the Purchasing/Supply Management Department

Evaluation and risk management are the final steps in the purchasing management cycle. The world-class purchasing department consistently reviews its purchasing practices to refine its activities in order to be more efficient and effective in its performance.

The Annual Report

A great tool for assessing your department's achievements and progress toward goals is an annual report. The creation of this report forces management to take a hard look at the department and evaluate any need to re-plan, reorganize, redirect, or make adjustments for changing conditions that have impacted the department.

We recommend the purchasing department prepare an annual materials report to document and communicate major activities. This report can be used in the situational analysis diagnostic phase of planning, as well as informing the appropriate individuals of purchasing activities. It is a good way to help sell the value of the purchasing function. This report should include information such as:

1. The year's spend by category and supplier (from ABC analysis)
2. Major contracts
3. Value engineering, value analysis projects
4. Total actual savings
5. Future cost reductions, that is, prevention of forecasted price-cost increases through various methods (relate to #3)
6. Quality ratings compared to goals
7. On-time delivery compared to goals
8. Price trends
9. Key supplier management: number of visits, supplier meetings, awards
10. Inventory levels, turns, stock-outs
11. Personnel profiles including certification progress
12. Training and education activities such as seminars and conference attendance, and contributions to professional societies
13. Updates to purchasing policy manuals, supplier welcome pamphlets
14. Green purchasing activities
15. Risk management actions
16. Description of major negotiations, reduction of POs
17. Analysis of industry trends affecting materials, forecasts of price, capacity, regulations, lead times, and other helpful data to internal users of materials
18. Forecasts of next year's major activities and projects

Again, this report should be sent to all parties interacting with the purchasing or supply department. Although the manager should omit confidential and sensitive data, the annual report is an excellent vehicle to educate all interested parties as to the procurement policies, activity, and contribution. Procurement managers must learn to sell their value added activities to the rest of the organization and avoid being a mystery unit in the backroom. In addition, distributing a purchasing newsletter on a monthly basis not only helps to inform all users regarding price trends, material availability, lead time requirements, and other pertinent news, it helps provide a record for the situation summary in the planning process and/or annual report.

Risk Management

We think it is appropriate to end the book with a discussion of purchasing's role in the organization's risk management, an activity that has become a popular management buzz term since Hurricane Katrina, the BP Gulf disasters, and the Japanese tsunami in March 2011. Boston-based ABERDEEN Group reported in a 2008 study, Supply Chain Risk Management; Building a Resilient Global Supply Chain; that 99% of study participants experienced at least one supply chain disaster over the previous year. These disruptions, including lack of supplier capacity, material shortages, and fuel prices, lead to financial losses of 58%, loss of market share for 30%, and brand reputation damage for 24%.[21]

The best way to reduce risk to an acceptable level is to follow the best practices described in the preceding chapters; you cannot insure against all risks. Astute managers will work to minimize risk, yet every decision, plan, and project has a certain probability of failure or performance lower than expected.

All operations need contingency plans and procedures. The March 2011 fire at MAGNA International's Howell, Michigan plant closed General Motors, Ford, Chrysler, Mazda, and Nissen plants for lack of dashboard and other interior trim parts.[22] This event begs the question of could the fire have been prevented or quickly controlled with better fire prevention tools or procedures, and were there any back-up plans to ensure supplier delivery?

A good example of purchasing crisis risk management is the action GM took following the Japanese earthquake and tsunami of 2011. General Motors identified 118 products that it needed to monitor for shortages caused by disruptions of Japanese parts suppliers. GM assembled hundreds of employees in a team, coordinated from three "crisis rooms" at its Vehicle Engineering Center in Warren, Michigan. They labeled the suppliers who were down in a red color code and flew 40 employees to Japan to assess the damage to suppliers and arrange for new sources in South Korea and other countries to supplement the lost capacity.[23]

New Focus on Risk Management

Consultant Steven C. Rogers, a 30-year veteran as Manager of Strategic Sourcing at Procter and Gamble, offers some insight. In his book, *The Supply-Based Advantage: How to Link Suppliers to your Organization's Corporate Strategy*, Chapter 12 is titled "Risk Management in the Supply Base: Insuring Against Damage, Loss, and Liability."[24] Roger lists five major categories of purchasing risk:

1. Legal   Such as anti-trust violations
2. Financial   Price volatility and supplier financial health
3. Operational   Late deliveries, quality problems, etc.
4. Regulatory   OSHA, EPA, trade agreements, customs
5. Business environment   Domestic and global economy, security issues, competitive forces, technology, etc.

He then outlines plans to manage these risks including avoiding decisions involving excessive risk, monitoring changes, developing contingency plans, collaborating with suppliers to share/spread the risk, and having a risk plan.

Included in any risk plan are purchasing policies and procedures that can help manage risks in areas such as back door selling, unauthorized personnel buying, or overextending purchasing buying authority. These policies should be included in the suppliers' welcome booklet and the purchasing policy manual. Smart organizations insist that only specific personnel operate as agents, authorizing them to sign purchase orders or contracts. They also ask suppliers to register with the purchasing department upon entering the building. They instruct their receiving department to refuse packages shipped without purchase order numbers and proper paperwork. And they make sure all involved personnel have copies of these policies, procedures, and other relevant guidelines.

What follows is a checklist of purchasing risk factors and our brief comments regarding minimization of the risks. We hope it provides the outline for your risk plan:

1. *Supplier Failure* Late delivery, rejects, poor service, excessive expediting, and other problems with suppliers can be avoided by careful sourcing and suppliers' certification methods and procedures as outlined in Chapter 13. Key risk factors include failure to monitor supplier contract compliance and failure to conduct regular site visits at key suppliers' facilities. Your key suppliers should have disaster plans that include contingency plans for major issues that can impact on-time delivery.
2. *Inventory Levels, Turns, Stockouts* Regular ABC analysis as described in Chapter 10 is a must. Closely manage the A items, that is, the high value materials and parts that can shut you down (it could be a custom-made 10-cent part). Special commodity analysts can play an important role.

3. *Single Sources* They obviously must be monitored on a periodic basis with enough lead-time to switch to the backup suppliers. Whenever possible, have a backup supplier. Many firms have a prime supplier for 80% of the buy and a backup supplier for the remaining 20% who can rapidly ramp up production when needed (a fire in the primary plant). Even the Japanese do this as part of their JIT systems. Some buyers with enough clout require their sole source suppliers to separate finished goods stock from WIP and store it in a fire-proof area.

4. *Currency-Price Volitility Risk* We have covered the currency risks in Chapter 14 including hedging techniques. Have appropriate contract language to handle fuel surcharges, price re-opener clauses, and proper indexes using total cost of ownership (TCO) and ABC accounting and the other methods in Chapters 15 and 16. Use VA-VE teams to help control costs.

Continuous training can reduce internal risk and good negotiations should provide better win-win contracts. It is worth repeating that the key to reducing external risk is proper sourcing. As we wrote in Chapter 13, this is the most important purchasing activity. The authors still find buyers who have never visited their key suppliers—that is very risky.

Some experts advocate simulating disaster interruption in order to test the various contingency plans.[25] This helps identify the high-risk suppliers and transportation facilities and the ability of alternate suppliers to increase production. However, the costs of having two warehouses, higher safety stocks, and backup distributor systems may be prohibitive.

Patrick Lynch advocates mitigation mapping to identify potential risk situations and estimate the probability of occurrence, duration, and disruption correction–contingency plans. These "failure modes," as Lynch describes the risk factors, include human errors, logistics delays, climate, quality control, regulatory changes, and supplier insolvency.[26]

5. *Ethical Risks* A strong, detailed code of conduct with strict enforcement can greatly reduce the risk of bad behavior including kick backs, stealing, fraud, and other illegal and unprofessional conduct. Organizations must prosecute the offenders, not simply fire them.

6. *Natural Disaster Risk* Buyers and key suppliers, especially those in high storm risk areas, such as Florida, the Gulf states, and parts of California, must have disaster reaction plans for hurricanes, floods, volcanoes, droughts, earthquakes, and other acts of God. Contingency plans must include the establishment of recovery teams to reduce panic, provide alternative sources of supply, plan alternative logistics, and network communications for sources of help including FEMA and city-state emergency agencies.[27] This assessment should be part of the sourcing procedure.

Finally, recognize that we cannot reduce risk to zero and the organization must avoid becoming paranoid to the point of inaction. We all know managers and even entire organizations that are so risk averse, they underachieve. These are the average performers who "play it safe." We certainly hope no organization hires a risk manager that adds to staff overhead with little value added. Only direct managers of specific functions can reduce risk by following the best practices in their specialties. This should provide plans and actions based on reasonable risk, but actions that also maximize value added operations.

Appendix A: The Institute for Supply Management Awards for Excellence in Supply Management

The ISM Awards for Excellence in Supply Management (formerly the R. Gene Richter Awards for Leadership and Innovation in Supply Management) are given to supply management departments within organizations that demonstrate leadership and innovation in supply management in at least one of five categories: Process, Organization/Structure, People, Technology, or Sustainability (added 2011). The first awards were given in 2006 to the following recipients.

Process

Johnson & Johnson, manufacturer of health care products, for unique and innovative partnership of the supply management organization and the law department for both in-house and outside counsel with savings/cost avoidance of $48.7 million in the first year (www.jnj.com).

KLA-Tencor, a leading supplier of process control and yield management solutions for the semiconductor industry. For moving from a transaction-based procurement system to a strategic organization with a cross-functional focus, including sourcing in the company's well-established product life cycle (PLC) management. Before a product can exit the investigation phase, a strategic sourcing input is required. Product cost reductions 5 to 10 times that of former projects were achieved (www.kla-tencor.com).

Technology

Fluor Hanford manages several major activities for the U.S. Department of Energy at the Hanford site in southern Washington State. For development of its Web-based material sourcing "Wizard" search application. The goal was to develop a system that would allow workers in the field to order needed routine supplies through an easy-to-use, one-stop shopping system while the supply staff concentrated on strategic issues. Savings include approximately $300,000 in 12 months from redundant buying, $2 million per year in reduced transaction costs, and reduced unit prices from 10 to 60% (www.Fluor.com/government).

Rockwell Collins, a leader in the design, production, and support of communication and aviation electronics for commercial and government customers. For taking its supply collins.com portal, a successful Web-based tool for suppliers, to the next level by adding applications that would automate three key purchases-to-pay processes, requests for quotes, quote receipts, total cost of ownership analysis, and automated purchase orders. The volume of transactions from quote release to final payment has increased from 46 to 70% (www.rockwell.com).

People

London-based BP, P.L.C., one of the world's largest energy companies, for developing eight core training modules that address most skill sets, including living strategy and program management, communication and engagement, internal performance, financial decision making, strategic sourcing, and supplier management. Compressed into six-month periods for 140 members of the global refining and marketing (downstream) and corporate and functions (indirect) procurement community. Nearly 50% of the program evaluations indicated a perceived significant improvement in the skills area (www.bp.com).

Organization/Structure

Daimler Chrysler, a major German automobile manufacturer, for consolidation of its global procurement. Its International Procurement Services (IPS) department focused on consolidation of global nonproduction material procurement responsibility in Germany and for the Chrysler Group in the United States and Canada. The results include an 80% increase in value added, increased internal customer satisfaction, and the implementation of one common information technology solution (www.daimlerchrysler.com).

2007 Award Winners

Process

Johnson & Johnson, a major health care product manufacturer, for launching a consulting sourcing initiative for North America. Using three teams, J&J increased the level of consulting procurement expertise and increased the value of and emphasis on master services agreements, including improved rating of consultant capabilities (www.jnj.com).

Technology

Alltel Communications, Inc., which operates the largest wireless network in the United States, for designing the Workflow Management Tool, which was created entirely in-house by the Procurement and Logistics Group (PLG) at Alltel Wireless. This system assigns works, tracks savings and project status, and provides visibility into the sourcing/savings funnel, allowing the organization to see "what tomorrow might look like" across the company. The benefits include project tracking and management, status updates, workflow management, single-source reporting for PLG savings and metrics reporting, visibility into all sourcing projects at a macro level, deal summaries and saving funnels by group, ability to forecast gross margin impacts, ability to recommend changes to the financial planning/budgeting process, and increased integrity in savings reporting (www.alltel.com).

People

BP, P.L.C. This London-based firm is one of the world's largest energy companies. For developing the rapid sourcing team, a program to develop and train new university recruits to improve the sourcing process, drive adoption of e-sourcing tools, and address low-value spend categories that often are overlooked. Results include savings of more than $550 million through the use of e-sourcing tools during 2005–2006 (www.bp.com).

2008 Award Winners

Process

Lockheed Martin (LM) Aeronautics Company, a leader in design, research and development, systems integration, and production and support of advanced military aircraft and related technologies. To prevent shortages of critical raw materials such as titanium, the LM Material Management and IS&T departments joined forces to launch a "preemptive strike" with their forecast raw material application (FORM), which generates a comprehensive, time-phased raw material forecast for the F-35 joint strike fighter program. This achieved a significant reduction in negotiated pricing for long-term raw material contracts and much better demand planning for raw materials for all program trading partners (www.lockheedmartin.com/aeronautics).

Pfizer, Inc., a global leader in pharmaceuticals. Three years ago, Pfizer World Wide Procurement was challenged to achieve $2 billion in savings in purchased goods and services by 2008. Using a new reverse auction system to source commodities considered core to the research and development organization saved significant dollars. Pfizer used cross-functional/cross-disciplined teams to develop a unique collaborative model between World Wide Procurement and the development of medical organizations. This model facilitated the prequalification of suppliers to participate in reverse auctions with major cost savings (www.pfizer.com).

Royal KPN telecom provides telephone, Internet, and television services to customers through its fixed network in the Netherlands and internationally. KPN had been experiencing an adverse relationship with its outsourced IT supplier, ATOS Origin. To restore trust and desired performance, KPN formed a steering committee composed of senior KPN and ATOS Origin Executives that eventually produced what they called the collaborative KPI program, which revised the original outsourcing contracts to eliminate unrealistic revenue guarantees for ATOS Origin and focused on cost reduction. Thus, the new collaboration efforts created a value-based relationship rather than one based on cost-driven goals. This new contract established an escrow account for revenues, a set of performance measures for the supplier, and a two-way set of KPIs containing performance targets of both supplier and buyer. Complaints of poor service from KPN's customers dropped from 15% to 5% monthly (www.kpn.com).

HP supplies printing, personal computing, software, services, and other IT products and services worldwide. HP has distinguished itself as a promoter of social and environmentally responsible (SER) programs within its operations, supply chain, and industry. Indeed, in 2002, HP was the first electronics company to publish a social and environmental responsibility supplier code of conduct. The company conducts regular supplier assessments and audits for all suppliers but in particular for its 16 direct material suppliers with more than 300 factories in countries identified by HP as "higher risk countries." Those suppliers listed in audits with major deficiencies in the SER program have from 30 to 180 days to address the particular issues. HP also has several "zero-tolerance" areas such as using underage child and forced labor, serious safety issues, and other violations of environmental laws. Special projects are now going on in China and Central Europe (www.hp.com).

Note: In 2008, there were four process, one technology, and no people or organization/structure awards. See the June 2008 issue of *Inside Supply Management* by Mary Siegfried.

What do all these winners have in common? In the opinion of the authors of this book, they are characterized by:

1. Vision for the future.
2. Proactive action.
3. A focus on supply as a value-adding member of the chain vs. a cost center.
4. Being creative: How can we do it better?
5. Use of cross-functional teams.
6. "Best-in-class" practices.
7. Use of the tools, methods, procedures, and philosophy contained in this and other leading texts, trade books, and ISM material.

Technology

Masco Corporation, one of the world's largest manufacturers of brand name consumer products for the home improvement and new construction markets. Masco Global

Purchasing and IT teams designed a supplier management tool (SMT) that allows Masco to leverage its spend, share supplier information, improve payment terms, track commodities, evaluate and rate suppliers, and track spending globally. This highly decentralized supply management organization uses SRM to unite disparate ERP systems into a robust spend analysis tool, integrating more than $10.8 billion in spend and 22 million transactions across 38 divisions representing 695 manufacturing and service locations. From October 2004 to September 2007, Masco divisions posted eight figures in savings, ranging from 4% to 67% (www.masco.com).

© Institute for Supply Management™. All rights reserved. Reprinted with permission from the publisher, the Institute for Supply Management™, Tempe, AZ; www.ism.

References: June 2006, June 2007, and June 2008 issues of *Inside Supply Management*, by Mary Siegfried, John Yuva, Lisa Cooling, and RaeAnn Slaybaugh.

2009 Award Winners

Organization/Structure Award Category

After merging with Verizon wireless in 2009, Alltel wire less is now the fifth largest wireless provider in the world, but is the largest network in the United States. Described as a "total supply chain transformation," Alltel created "a closed-loop system that combined and integrated all forward, reverse, and repair product flows into one organized system."[28] A good example of the ever increasing trend to centralized purchasing, Alltel, by centralizing in timed stages, brought purchasing, supply management, warehouse-distribution, and reverse logistics (repair, warranty, exchange process) under one supply management organization. The results include: three times more product available to retail, a 54% reduction in accessory cost enabling a 15% increase in profit margins, $500,000 weekly reductions in expenses, and superior best in industry inventory turns, less stockouts, and improved forecast accuracy[29] (www.alltel.com).

Process Award Category

Royal KPN Telecom provides telephone, Internet, and television services to customers in the Netherlands as well as voice, Internet, and data services to business customers. Royal KPN formed an energy management team (EMG). This cross-functional team studied consumption and ways to reduce energy use and that of its suppliers. EMG also sought to reduce KPN Telecom's carbon footprint, and tax rates, and improve forecasting. This team also entered the futures market with two-year intervals. Reduced risk and significant cost savings are impressive results achieved by EMG[30] (www.kpn.com).

Technology Category

Using what they call its "optimization tool", a variation of Web-based reverse auction, the United States Postal Service achieved savings of over $57 million over a three-year period ($15.2 million per sourcing event). USPS also claims improved collaboration with suppliers as the Web site offers training, case studies, and pre-bidding information.[31] The "optimization" of efforts focused on transportation sourcing (www.usps.com).

2010 Awards Winners

Leadership and Innovation Category

LG Electronics, is a global leader in consumer electronics with five business units: home entertainment (flat panel TVs, etc.), mobile sales, home appliances, air conditioning, and business solutions. By creating a global procurement strategy, a commodity council, and global commodity managers, LG Electronics went from decentralized to centralized procurement. This transformation resulted in savings of $3.2 billion for the period 2009–2012[32] (www.ige.com).

The 2011 Awards Winners

In the interest of space limitations, we will provide just a short summary of the winners' work.

Process Category

Pfizer Inc is the world's leading biopharmaceutical company. The Worldwide Professional Services (PS) team worked with Pfizer Knowledge Management Services (KMS)—consulting, legal, and financial services—to achieve cumulative savings of $370,000 during the target time period with annual savings of 14%. For example, the number of law firms used by Pfizer went from 85 to less than 20 and created a flat fee cost vs. billable hours model[33] (www.delphi.com).

Technology Category

Delphi is a collection of former GM part supplier plants and is now a leading global supplier of electronics and technology parts for the automobile industry. Delphi's cost management group within global supply management designed a user-friendly software desktop tool to measure total cost of ownership (TCO) with a focus on global logistics costs, that is, which is the better plant location? The cost management team worked closely with global logistics, manufacturing, engineering, and R&D[34] (www.delphi.com).

Process Category

L-3 Communications is a government prime contractor for a variety of electronic systems—electrical, electronic, and electromechanical components (EEE) for aircraft, homeland defense, and related industries. The Diminishing Manufacturing Sources and Material Shortages (DMSMS) team and the Counterfeit Parts Team (CPT) "developed a disciplined and comprehensive strategy to mitigate obsolescence and counterfeit risks." The savings include: "cost avoidance through early detection of potential obsolescence issues, a 53% reduction of component alerts (parts at risk) and savings from negotiating corporate pricing for third-party tools"[35] (www.I-3com.com).

Technology Category

U.S. State of Georgia, Department of Administrative Services, State Purchasing Division (SPD). SPD "created spend management analytics, a refreshable spend cube" to collect and analyze spend data from the states' more than 160 agencies, colleges, and universities in order to provide analysis of the total spend.[36]

This procedure, when used with "the e-procurement system, team Georgia marketplace, forms the spend under management (SUM) initiative, which is a process improvement effort to monitor and analyze state-wide buying activity" (www.doas.ga.gov).

Sustainability Category

IBM2 Integrated Supply Chain (ISC) includes procurement, logistics, engineering, manufacturing, customer fulfillment, and operations managing more than a U.S. $40 billion spend in 89 countries. Working with Corporate Environmental Affairs, ISC developed three initiatives.

1. Center of Excellence for Environmental Compliance/Social and Environmental Management System.
2. Supply Chain Social Responsibility Initiative
3. Green ISC Initiative"

IBM requires its first-tier suppliers to develop similar management review systems including public disclosure of results. IBM also requires the first-tier suppliers to communicate these requirements to their suppliers. In addition, all suppliers must comply with IBM's code of conduct, which includes working conditions, non–discrimination policies, environmental protection, and ethics. "IBM uses a third party to review supplier facilities and report their compliance with these principles."[37] Green ISC initiative is what one would expect, that is, recycled materials (in particular plastics), carbon disclosure, and energy conservation (www.ibm.com).

Appendix B: From Reactive to Proactive Procurement: A Case Study[38]

by Richard L. Pinkerton, Ph.D., C.P.M., Chair, Marketing and Logistics, Sid Craig School of Business, California State University, Fresno, California, and Kathleen A. Petitis, Purchasing Supervisor, ILC Technology, Inc., Sunnyvale, California

Abstract

This paper documents the efforts of a medium-sized hi-tech manufacturer of laser lamps and power sources to move from a reactive purchasing operation based on repetitive purchase orders with many suppliers to purchasing by long-term contracts with a few key suppliers based on partnerships. It includes the analysis of the costs of the old paper driven system and conversion to negotiation by cross-functional teams using a strategic- proactive approach. The sequence of events from evaluation of the old system to the training of personnel involved with material acquisition is covered including the first reports of initial annual dollar savings of approximately $260,000.

The Company

ILC Technology in Sunnyvale, CA is a firm of about $60 million annual sales, approximately 300 employees, and 3 subsidiaries: Precision Lamp in Cotati, CA, Converter Power, Inc. in Ipswich, MA, and Q-ARC, Ltd. in Cambridge, England. It manufactures and sells globally replaceable high-performance light source products for the medical, industrial, communication, aerospace, military, and entertainment industries. Total annual purchases are approximately $27 million under the control of a purchasing department. consisting of 1 supervisor, 3 buyers and 1 administrative assistant. The department reports to the Director of Materials and Logistics, who is also responsible for inventory-production control, stores, and quality assurance. ILC Technology, Inc. is traded on NASDAQ as ILCT.

The Situation Report

During the early part of 1994, the professor co-author of this paper (hereafter called the consultant) conducted 12 in-depth interviews with key managers, examined various computer printouts of purchasing activity, visited several key suppliers and, of course, held several discussions with the purchasing personnel. You will notice we use the planning term "situation analysis" rather than the traditional term, purchasing audit. The president and CEO of ILC wanted to reengineer the entire purchasing operation as opposed to the traditional implication of the purchasing audit, which focuses on compliance to existing policy. In other words, he wisely wanted a status report that would compare current practice with benchmark or state-of-the-art proactive purchasing practices.

After the president approved the situation report on July 29, 1994, he asked the then purchasing manager (no longer at ILC Technology) and the consultant to hold a general management meeting to present the findings of the situation report. After the introduction and endorsement by the president, the consultant listed the following major findings:

A. *Too many suppliers for the same material.* It was not unusual to have 5 to 8 suppliers per part as a result of lack of planning, multiple preference by engineers, back door buying by engineers, and a reliance on repetitive purchasing vs. long-term contracting. ILC had few, if any, real partnerships with key suppliers.

B. *Too many total suppliers.* During a 12-month period (1993–1994), ILC had a total of 360 direct material (material that becomes part of the finished product) and 2879 indirect suppliers for a total of 3239.

C. *Too many purchase requisitions and purchase orders.* As a consequence of an excessive number of suppliers and lack of long-term contracting, non value added paper costs were very high. For example, the supplier with the most purchases in dollar volume, a precision machine shop, required 442 purchase requisitions (PRs) and 442 purchase orders (POs) for a total of 884 transactions or 4 per day (based on 200 workdays per year) during a 1-year period. The controller estimated the processing cost per PO at $50, which meant ILC had spent $22,100 in one year just for the initial transaction cost for this one supplier. In addition, we actually counted a total of 5456 POs during a 1-year period for just direct material from all suppliers at a transaction cost of $272,800. This was the classic case of way too many purchasing cycles with far too many suppliers. The buyers in the purchasing department were largely relegated to paper pushers checking for proper product codes, tax accounting codes, and correcting mistakes.

D. *Unrealistic lead times given to customers.* Marketing and others in the engineering design departments often either ignored or failed to understand the elements of lead time. This often occurred for standard products but was especially abused when converting prototypes to production runs.

E. *Inaccurate forecasting and failure to follow forecasts lead to a failure to have a master production schedule.* This finding along with finding D caused serious delivery problems.

F. *Excessive engineering and other related changes created delivery, quality, and supplier performance problems.* Poor document control and failure to lock-in design prior to production resulted in many engineering changes, print revisions, changed delivery dates, and other corrections, which caused excessive rework, returns, and major confusion in house and with suppliers.

All of the above factors created a rush-rush atmosphere with frequent corrective paperwork plus a huge number of PRs and POs to document the changes. In addition, the many repeat small orders diluted our negotiation clout and resulted in high prices paid to suppliers who were never quite sure if they had future business with ILC.

The Corrective Action

After many revisions, a new procurement policy was issued in compliance with ISO 9000 standards with the following key provisions.

A. All purchasing will go through the purchasing department; no more back door buying by engineers.

B. Material needs will be consolidated based on a minimum of 1-year requirements.

C. Buying will be accomplished based on long-term contracts for at least 1 year with additional periods as the firm gained experience.

D. The supplier base for direct material will be reduced wherever possible, to one prime and one backup supplier on the approximate basis of 80% of the physical volume to the prime, 20% to the backup. All suppliers will be surveyed and certified by a documented procedure.

E. For indirect material such as MRO items, one supplier (whenever possible) will be selected to manage ILC inventory, including restocking on the basis of systems contracts for entire families of products such as office suppliers, hardware, tools, standard electronic components, and other such material commonly available from distributors.

F. Negotiation will be accomplished by cross-functional commodity and design teams and will include representation from purchasing, planning, production, engineering, quality assurance, and others as needed such as stores. All individuals will receive training in negotiation and team building. Department managers will form the teams with assistance from purchasing.

G. After the contract is issued, planner-schedulers in the manufacturing departments will order quantity releases direct to the supplier via fax and eventually by EDI or Internet. Purchasing will only be involved when there is a problem and there will be no PRs or POs for releases of direct production material and selected indirect material suppliers.

H. Purchasing credit cards will be used by selected personnel for "odds and ends" of small dollar value.

I. Marketing and production planning will establish a new forecasting system with accurate lead times. Change orders will be minimized.

J. We will treat our suppliers as "partners" with early supplier involvement at all stages via scheduled meetings.

K. Purchasing will move from only checking and preparing forms to professional sourcing investigation in order to be a major resource for the cross-functional teams.

L. Formal training sessions will be conducted to help install the new system.

The Results

As one might expect, initially the teams struggled but finally developed their negotiation and team skills. This process took approximately 18 months and one team had to be replaced with more effective members. There are four major teams: Quartz (raw material), equipment (machining-painting), Cermax (product line), and packaging with several smaller teams for MRO items. As of June 1996, the documented cost savings for 1997–1998 will be at least $260,000 plus the elimination of 1400 POs for an additional reduction of $70,000 in paperwork for just eight suppliers alone. We anticipate eventual yearly savings of $350,000 but documentation continues to identify all savings. At the time this paper was written, the reduction in the number of suppliers was being calculated but we are confident it will be at the 10% figure for the start-up period and should go to 25% in another year.

Because ILC engages in a great deal of R & D, special projects still create a lot of initial PRs and other paperwork. The purchasing department will have to devise a system to cope with this aspect of supply management. The entire company is reviewing and changing the forecasting-lead time procedure and it is too early to calculate the reduction in change orders. However, the new procurement system forced several management changes and brought attention to bear on the problem. We do have much better communications with our suppliers and have better pricing, delivery, and quality as reflected in the $260,000 direct savings. The purchasing department is very excited about their new role as a proactive force adding value rather than cost. The contract system and direct release policy has given the purchasing department the time they needed to manage materials rather than just purchase it.

Note: In the early 2000s, ILC was purchased by Perkin-Elmer Lamp Co. and ILC product lines were absorbed in the acquisition company. ILC no longer exists. The much improved purchasing operation helped negotiate the sale.

Reprinted with permission from the publisher, Institute for Supply Management, "From Reactive To Proactive Procurement: A Case Study" by R. Pinkerton and K. Pettis, 82nd Annual International Purchasing Conference Proceedings, 1997, pp. 169–172, Washington, DC. See www.ism.ws.

Appendix C: Raytheon Supply Base Optimization, April 2005

Introduction

This case study of the Raytheon Space and Airborne Systems Organization, Elsegundo, CA, is the product of a senior class project in the capstone course, MS-Supply Chain Management degree program, the University of San Diego. The students on the team were: Karen E. Grant, Dawn C. Garrett, Linda G. Keenan, Kelley A. O'Dell, and Shirley J. Patterson. It is important to note that these are adults in management positions at Raytheon. Their project director, James D. Reeds, is a consultant-educator with extensive procurement-supply chain experience. The following abstract gives the overall view of the project.

Abstract

How does a company optimize its supply base? Supply base optimization is a result of a well-executed supply base design tied to the strategic business plan. To achieve this, a firm needs to clearly define its business plan and strategic and core competencies. For everything that they decide to buy, Supply Chain in concert with Engineering, Business Development, Manufacturing, and Quality must characterize each commodity and define discreet strategies based on technology roadmaps. Establishing the strategies and objectives is the easy part! The challenge for most firms is execution. How is this work refreshed in the face of business strategy shifts, mergers, and acquisitions? How do we know that these strategies are contributing to the value of the business? The Supply Base Optimization project developed the process for defining, communicating, executing, and maintaining an optimal supply base in support of our overall strategic business environment.

The full project report includes the application of almost all the best practices listed in this book. It is for this reason that we primarily use their charts, diagrams, and figures to focus on the implementation of world-class supply chain management at Raytheon and also to avoid unnecessary duplication of the material covered in this book. By the way, this is a real case and the research was funded by Raytheon.

We will describe in some detail their work on industry benchmarking as this complements our material on planning. Benchmarking is part of the situation review, diagnostic phase of planning-project work. In our book, we started the planning with an internal review of a company's purchasing operation. The Raytheon team started with benchmarking comparable firms in their industry as to what are "the best practices." Then they compared them to the current practices in Raytheon. Both approaches are acceptable, and, indeed, both analyses can be done simultaneously.

The Raytheon team used the benchmarking definition coined by the American Productivity & Quality Center (APQC), "the process of identifying, sharing, and using best practices to improve processes."[39]

The team prescreened many candidates and then selected five to six industry leaders to attend a two-day benchmarking symposium. IBM, Rockwell Collins, Northrup Grumman, and Motorola attended, and each firm described their supply chain process followed by a full discussion led by a facilitator. In addition, two additional benchmarking events were held.

The team then compared Raytheon's current practice to the best in class practices and "closed the gaps" with the following recommendations:[40]

▲ "Preferred supplier selection is linked to documented commodity strategies.
▲ Key supplier selection is now linked to technology road maps.
▲ The supply base design is now aligned with the business strategy.
▲ Social responsibility is now incorporated into each commodity strategy.
▲ Strategic supplier development targets will be identified early.
▲ Employee skill set is being upgraded to effectively manage an integrated supply chain.
▲ Organizational structure is being revised to facilitate commodity management.

In closing these gaps, we shift dramatically from tactical to strategic supply management."[41]

We feel the Raytheon team's keys to supply base optimization generally support our same views as shown in Table 22.2.

The following material concerning "Sustaining Momentum," "Strategies for Implementing Change," and "Lessons Learned" are taken from the actual report. This is very valuable advice for any organization writing a master reengineering transformation plan.

Sustaining Momentum for Supply Base Optimization (SBO)

The SBO team identified several areas that need to be addressed to sustain the changes in the new process, and these are addressed as follows:

1. *Department Structure.* The SBO team is currently reviewing the number of resources, roles, and responsibilities required to sustain the new process. As this project moves from the pilot stage to full implementation, additional resources and work redesign will be required due to the high degree of complexity and integrated activities.

2. *Human Resource Systems.* In the new environment, the skill level of both commodity managers and buyers will need to change to sustain the momentum. They will need to increase their technical competency to lead the supplier selection process. Additionally, there will be a need for more experienced individuals in supplier relationship management and executive interface. The organization has begun to implement change in this

Table 22.2 Keys to Supply Base Optimization Derived from Industry Benchmarking

---

▲ Process must start with the customer and then work backward to define the supply chain strategy.

▲ Significant portion of product cost is driven by external sources.

▲ Strategies should be linked to business processes.

▲ Measure success through business performance improvement.

▲ Supply chain must own sourcing.

▲ Supply executive leader should report to the business president.

▲ Process improvements must be driven from top-down to be successful.

▲ Multifunctional senior leadership endorsement is a must.

▲ Tools are used to bring processes upstream into the design process to ensure critical early involvement with engineering.

▲ Strategic approach results in significant reductions in number of suppliers.

▲ Commodity strategy is prioritized around spend.

▲ Commodity teams are cross-functional.

▲ Focus on standardization to reduce complexity.

▲ Involvement of entire value chain including customer produces greatest success.

▲ Trust and collaboration with suppliers is a key enabler.

▲ Early involvement with the "right" suppliers is key to success.

▲ Process centralization adds to efficiency and focus.

▲ Staff for success.

▲ Move resources from tactical to strategic focus.

▲ Evaluate employee skill sets and train, develop, and make changes if needed.

▲ Create urgency through clear case for action.

Source: Contract Management, January (2005): 34. Reprinted with permission of the National Contract Management Association.

area through the hiring of experienced commodity managers who led the development of the pilot commodity strategies.

3. *Measurement System.* Both engineering and supply chain will need to incorporate the new process requirements into the performance development system at all levels of the organization. Engineering will need to be held accountable for percentage utilization of preferred suppliers in new designs. Supply chain will need to measure the percentage sourced to preferred suppliers as well as the number of commodity strategies completed. These measure will be monitored in monthly operations reviews and program design reviews. In addition, we will need to establish a team effectiveness measurement based on the new work redesign structure. Rewards and recognition need to reinforce

the behaviors required to sustain the new process. The team will use the advisory board to garner the political support required to implement these measurement and reward systems.

4. *Database.* Our project requires the implementation of an information system database as a repository and communication vehicle. A Web-enabled database is being developed to communicate and facilitate the use of preferred suppliers in the design and sourcing communities. It will also include key information on capabilities and performance data of the suppliers.

5. *Policies and Procedures.* With the three pilots we developed a set of standardized templates and process flows for the development of future commodity strategies. They were documented into formal work instructions and institutionalized within our SAS Integrated Product Development Process (IPDP).

Strategies for Implementing Change

The following steps were taken as effective components of our change management model:

1. *Diagnosis*
   - ▼ Spent a significant amount of time in the diagnostic phase of our project. This included multilevel interviews with key functional and program level stakeholders. Both the team members and external consultants conducted the interviews to offset any bias that the internal members might have.
   - ▼ Conducted internal benchmarking within Raytheon and a formal external benchmarking event was held in November 2003. Participants included IBM, Motorola, Rockwell Collins, Northrop Grumman, and Raytheon.
   - ▼ Performed an extensive literature review to validate our hypothesis and support our case for change.
   - ▼ Collected and reviewed historical data to characterize the current state. The past two years of spend data was broken out by commodity and used to facilitate the selection of the pilots.

2. *Motivating Change*
   - ▼ Facts and data were fed back to the advisory council and departments to motivate and create a readiness for change.

3. *Developing Political Support*
   - ▼ Established an internal network across Raytheon of resources that are engaged in supply base reduction initiatives. Weekly meetings were held to share best practices, successes, and failures.
   - ▼ Involved the functions impacted by the changes as part of the team. This includes programs, supply chain, engineering, manufacturing, and quality.

▼ Established an advisory council of vice presidents from Engineering, Supply Chain, Manufacturing, Quality, and Raytheon Six Sigma to support and allocate the resources required to successfully implement and sustain the changes.

▼ Secured the Vice President of Engineering as a co-sponsor of the team at an executive level.

4. *Managing the Transition*

▼ Applied project management methodologies including a detailed work breakdown structure and milestone plan. This aided in the integration of various levels of teams involved in the project, USD, local site and Raytheon enterprise.

▼ Attempted to develop the new processes internally but found that we lacked the skill sets and expertise to support the required changes. Consequently, the team failed to make progress in the early stages. To overcome this barrier, the team brought in external consultant resources that were able to bring the skill set and expertise required to facilitate the change process.

5. *Sustaining the Momentum*

▼ Initiated a supply chain recruiting effort to find the skilled resources with the level of experience and expertise required to sustain the changes. To date, three senior commodity management professionals have been hired by supply chain to support the new process.

Lessons Learned

The following areas were identified as lessons learned in that the implementation was not as effective as desired.

Motivating Change Communication upward to our advisory council was good; however, the communication flow downward had been lacking. These are the people that will be impacted by the implemented changes. Therefore, the team will spend a significant amount of time developing and implementing a detailed communication plan.

Creating a Vision. Although the team thought it had a clear vision, the local site team bogged down at times due to inadequate facilitation. Too much time and energy was spent on very tactical activities and not enough time on strategic aspects of the project. Adding additional direction and facilitation to the local site team from the USD (University of San Diego) team has helped resolve this problem.

Developing Political Support. Initially, the team focused on the changes within Raytheon. As this project affects our supply base, we had to ensure that our suppliers were aware of the project. This is seen as a key motivation for them to focus on world-class performance so that they are chosen as part of the optimized supply base. A communication plan is being developed to include suppliers in this process.

Managing the Transition. The team needed to draw better delineation between the different teams, USD, local site, and enterprise teams working on the project. Unclear boundaries caused some problems in certain situations.

Raytheon must focus on providing our customers with cost-competitive solutions. The Supply Base Optimization project will dramatically alter our sourcing processes and return substantial savings to the company and to our customers. These processes will help in our progression toward World-Class Supply Management[SM]. The only way that we were able to achieve our goals was to ensure that our change management plan addressed all of these issues. We had to both manage and lead the change. The team recognized the importance of the change management process and was committed to ensuring success.

Appendix D: At Rolls-Royce North America, an Organizational Shift Is the Driving Force Behind an Innovative New Journey to Engage Its Supply Chain[42]

For two years, Kathi D. Bndgewater, purchasing process and business development executive for Rolls-Royce North America, lived and breathed supplier relationship strategy. That's because Bridgewater and a team of nine other senior-level Rolls-Royce executives were on a journey to radically change the way the company works with its suppliers. The results of that journey not only brought about changes to the company's supplier strategy, it sparked an organizational transformation that continues to shape Rolls-Royce today.

The transformation began several years ago when the supply management organization took a close look at itself and its supply chain. The organization's focus at that time was on driving piece price down year after year. It realized, however, that such a focus was leading to adversarial behaviors in the supply chain, and failed to address the bigger issue of lower total costs. That initial examination prompted Rolls-Royce, which provides power systems and services for use on land, sea, and air, to hire an outside consulting firm to conduct a comprehensive supplier survey, Bridgewater recalls. When the "dismal survey results" came back in early 2004, it was clear to Rolls-Royce's top-level executives that significant improvements needed to be made. The U.K.-based company, with 38,000 employees in offices and manufacturing and service facilities in 50 countries, also realized that to make significant improvements, an organization-wide commitment to change was needed.

At that point, a working relationship team was formed with the goal of changing and redefining the manner in which Rolls-Royce engages its supply chain. Bridgewater was chosen as a team member, representing Rolls-Royce North America's manufacturing facility in Indianapolis. She emphasizes that the senior-level team was cross-regional, cross-functional, and cross-business because the challenge it was undertaking involved the entire company. "It was not a purchasing-related activity. This was about a transformation of how the company interfaces with our suppliers and the relationships we have with them," she says.

Team members signed a two-year commitment to the project, and it became their full-time job. For Bridgewater, that meant living in England for four months at the outset of the project, and then traveling back and forth regularly during the course of the two years.

She admits that the early days were tough, as the team worked to find common ground, common goals, and to identify its vision. Because Rolls-Royce is a company that is "trusted to deliver excellence and is very driven in that area," Bridgewater says the team had to step out of its comfort zone at times and remind itself that its goal was to improve relationships.

"It was not just about processes, and documenting and fixing processes," she notes. "It's about how we work with suppliers. We had to ask ourselves what a good relationship with a supplier looks like. Then we had to draw that as our vision."

A visual management room was set up, and the team used it throughout the two years as it identified goals and made its business case for change.

A Supplier Survey Reveals the Truth

Because the supplier survey results sparked the need to re-examine supplier relationships, Bridgewater says those results became the foundation from which the team worked to make improvements and changes. She admits that the results of the supplier surveys were not necessarily surprising, but they were sobering. "No one likes to hear that their scores are low, but those who worked in the supply chain area were aware of the issues brought out in the survey," she says. "By and large, I don't think there was a message in the survey that no one anticipated."

For example, the survey made it clear that Rolls-Royce "did not have one voice," Bridgewater says. One supplier company in the survey said that, on average, it interfaced with 44 different people at Rolls-Royce. Others pointed out that because of the focus on keeping piece price down, there seemed to be a lack of concern for suppliers' profitability.

Other key messages from the supplier survey were:

- A lack of trust
- Good performance did not translate into more business
- Functional silos, not integrated business
- A short-term approach
- Poor use of suppliers' capabilities

Deciding what issues to focus on was one of the team's first challenges. "The toughest part was boiling the issues down to a workable number of topics, say 10 or 12," Bridgewater notes. When the team identified specific areas that it wanted to address, such as organizational interface, earlier engagement of suppliers in projects, and technical interface with suppliers, then a team member became the "owner" of that workstream.

For example, the engineer on the team was responsible for the technical interface workstream. His responsibilities included taking the plans the team was working on back to the engineers to get their feedback and buy-in on proposals. Bridgewater says the team sought and received "constant feedback" from stakeholders to make sure it was on the right track.

Another key element in the process, Bridgewater emphasizes, was supplier participation. "We couldn't forget that this is all about suppliers. We had to test our thinking on the suppliers," she says. Supplier focus groups were formed within the first few months of the process, and were asked to participate in interactive workshops from time to time throughout the two years.

From the outset, the team understood that although it would disband, the process of improving supplier relationships would be continuous. The policies and plans developed by the team became part of Rolls-Royce's business plan. "It all fed back into the organization," she says. "For example, you can look at our deliverables and our business plan in the purchasing area and see a lot of the output from the team's activity feeding into the organization in terms of goals and objectives."

Some of the deliverables developed by the team are:

- ▲ A new supplier strategy process for developing, documenting, and sharing strategies with suppliers.
- ▲ A "Relationship Profiling Tool" that provides a framework for Rolls-Royce and a supplier to assess their strengths and weaknesses as a supply chain team.
- ▲ A program for training and "up-skilling" employees.
- ▲ A supply chain technical interface methodology to cover joint technology development and a new product introduction process to better use suppliers' capabilities.

Bridgewater cautions that organizations seeking to improve their own supplier strategies should not expect overnight results. "When you are talking about cultural transformation and managing relationships, you are on a long journey," she says. "It's about sticking with it and about being consistent."

The proof of success is in sustainability, Bridgewater notes, stressing that strategies should be dynamic and ongoing. Suppliers are telling Roll-Royce that they are impressed by the company's commitment to improving relationships and engaging them in the process, Bridgewater adds, although some are taking a wait-and-see attitude. A follow-up survey will be done, although the timing has not yet been decided.

Despite the challenges the team faced, Bridgewater says the journey was worthwhile because it demonstrated Rolls-Royce's commitment to its customers. It also was clear from the start that the company had to invest in a supplier relationship strategy if it wanted to survive and compete in today's business environment.

"The strategies and plans we developed are now part of the way we work, a continuation of the journey. In all organizations, from engineering to supply management, they (plans and strategies) are part of our core business plans. It is embedded in the way we move forward," she says.

Endnotes

1. C. Long and G. Meyer, *Sacred Cows Make the Best Barbecue: Supply Chain Management: A Revolutionary 26 Weeks Action Plan.* (Seal Beach, CA: Vision Press, 1998). Peter Bolstorff and Robert Rosenbaum, *Supply Chain Excellence: A Handbook for Dramatic Improvement using the SCOR Model,* 2nd ed. (New York: AMACOM [The American Management Association] 2007). See also the Supply Organization Appendix in Chapter 2. Another good program for a utility is by Dr. Emiko Banfield *Harnessing Value in the Supply Chain: Strategic Sourcing in Action,* (New York: John Wiley & Sons, 1999).

2. Richard L. Pinkerton, "Taking the Mystery Out of Supply Chain Management (SCM): The Vision and the Reality". An after-dinner address to the Purchasing Management Association of Cleveland, OH (PMAC), November 15, 2007.

3. Peter F. Drucker, "Long-Range Planning," *Management Science* 5 (1959: 240). Used with permission from AMACOM. We also recommend reading the classic work on planning, George A. Steiner, *Top Management Planning* (New York: Macmillan [Arkville Press Book], 1969). The first book devoted to procurement planning was David H. Farmer and Bernard Taylor, Eds., *Corporate Planning and Procurement* (New York: Wiley, a Halsted Press Book, 1975).

4. Susie Mesure, "Wal-Mart in pledge to slash its carbon footprint," *The Independent Business News,* February 2, 2007. http://www.independent.co.uk/news/business/news/walmart-in-pledge-to-slash-its-carbon-footprint-434761.html

5. Walmart's Corporate Home page, Press Room, "Walmart Announces Goal to Eliminate 20 Million Metric Tons of Greenhouse Gas Emissions from Global Supply Chain." Press Release dated February 25, 2010. http://walmartstores.com/pressroom/news/9668.aspx

6. Ibid.

7. Robert Monczka, Robert Trent, and Robert Handfield, *Purchasing & Supply Chain Management*, 3rd ed. (Mason, OH, Thomson South-Western, 2005) 160.

8. Monczka, Trent, and Handfield, op. cit., 171.

9. Warren R. Plunkett, Raymond F. Attner, and Gemmy S. Allen, *Management: Meeting and Exceeding Customer Expectations*, 8th ed. (Mason, OH: Thomson South-Western, 2005) 246.

10. Adapted Plunkett, et al., op. cit., 247–255.

11. Plunkett, et al., op. cit., 255.

12. Ibid. 260.

13. For a very good description of purchasing staffing, see Kenneth H. Killen and John W. Kamauff, *Managing Purchasing: Making the Supply Team Work*, (Burr Ridge, IL: Irwin Professional Publishing, 1995): 147–177.

14. Ibid., 147.

15. For a comprehensive study on skills likely needed by your departmental personnel, see Larry C. Giunipero, The CAPS Study: A Skills-Based Analysis of the World Class Purchaser, 2000, on the ISM Web site, www.ism.ws.

16. Roberta J. Duffy, "Skill Sets: Building for the Future," *Purchasing Today* March (2000): 44. See www.ism.ws.

17. Ibid.

18. Mary Siegfried Dozbaba, "Taking the Lead in Leadership Development," *Purchasing Today* February (2000): 46–56. See www.ism.ws.

19. Ibid., 46–47.

20. Ibid., 56.

21. Mary Siegfried, "The Resilient Supply Chain," *Inside Supply Management*, December (2008): 28–30. See www.ism.ws.

22. Robert Schoenberger, "Fire at Supplier Plant Forces Shutdown at Lordstown," *The Cleveland PD*, March 4, 2011, C1. Contact rschoenb@plaind.com.

23. Nick Bunkley (of The New York Times), "GM Actions After Quake Kept Assembly Lines Running," *The Cleveland PD*, May 13, 2011, C1-2.

24. Stephen C. Rogers, *The Supply-Based Advantage: How to Link Suppliers to your Organization's Corporate Strategy*, (New York: AMACOM, 2009) 272–299.

25. John Yuva, "Impact Ready: In the Eye of Natural Disasters," *Inside Supply Management* October-November (2010): 20–23.

26. Patrick Lynch, "Build a Better Disruption Plan," *Inside Supply Management*, December 2010–January 2011: 20–23. See www.ism.ws.

27. John Yuva, "Assess Your Vulnerability to Natural Disasters," *Inside Supply Management*, September (2010): 28–31. See www.ism.ws.

28. Lisa Arnseth, "Achieving Total Supply Chain Management," *Inside Supply Management*, April (2009): 28. Institute for Supply Management, Tempe, AZ. Reprinted with permission.

29. Ibid., 29.

30. John Yuva, "Cross Functional Synergy Drives Energy Management," *Inside Supply Management*, April (2009): 30–31. Institute for Supply Management, Tempe, AZ. Reprinted with permission.

31. Mary Siegfried, "Getting the Most From Optimization," *Inside Supply Management*, April (2009): 32–33. Institute for Supply Management, Tempe, AZ. Reprinted with permission.

32. Lisa Arnseth, "From Decentralization to Center-Led Supply Management," *Inside Supply Management*, April (2010): 20–23. Institute for Supply Management, Tempe, AZ. Reprinted with permission.

33. Mary Siegfried, "Ripe for Change," *Inside Supply Management*, April (2011): 22–23. Institute for Supply Management, Tempe, AZ. Reprinted with permission.

34. Mary Siegfried, "Precision Tool Tackles Complex Task," *Inside Supply Management*, April (2011): 24–25. Institute for Supply Management, Tempe, AZ. Reprinted with permission.

35. John Yuva, "Winning the War on Obsolescence and Counterfitting," *Inside Supply Management*, April (2011): 26–27. Institute for Supply Management, Tempe, AZ. Reprinted with permission.

36. Lisa Arnseth, "Smart Procurement on Georgia's Mind," *Inside Supply Management*, April (2011): 28–29. Institute for Supply Management, Tempe, AZ. Reprinted with permission.

37. Ibid., 31

38. Richard L. Pinkerton and Kathleen A. Pettis, "From Reactive to Proactive Procurement: A Case Study," *Proceedings*, 82nd Annual International Purchasing Conference, May 1997, Washington, D.C., 169–172. Institute for Supply Management, Tempe, AZ. Reprinted with permission.

39. Shirley Patterson, "Supply Base Optimization and Integrated Supply Chain Management," *Contract Management*, January (2005): 26–35. See also Michael J. Spendolini, *The Benchmarking Book*, New York: AMACOM, 1992, the classic work on the subject.

40. Ibid., 33.

41. Reprinted with permission of the National Contract Management Association.

42. Mary Siegfried, "Transforming Supplier Relationships: The Rolls Royce Case Study," *Inside Supply Management*, January (2008): 32–34. Reprinted with permission from the Institute for Supply Management.

Suggested Reading

Anklesaria, Jimmy. *The Supply Chain Cost Management: The Aim and Drive Process for Achieving Extraordinary Results*. New York: AMACOM, 2007.

Banfield, E. *Harnessing Value in the Supply Chain*. New York: John Wiley & Sons, 1999.

Bolles, Dennis, and Darrel Hubbard. *The Power of Enterprise-Wide Management*, 3rd ed. New York: AMACOM, 2004.

Cohen, Shoshanah, and Joseph Roussel. *Strategic Supply Chain Management: The 5 Disciplines for Top Performance*. New York: McGraw-Hill, 2005.

Dinsmore, Paul C. and Jeannette Cabanis-Brewin, *The AMA Handbook of Project Management*. New York: AMACOM, 2011.

Farmer, David H. and Taylor, Bernard, Eds., *Corporate Planning and Procurement*. Wiley, a Halsted Press Book, 1975. The first book devoted to purchasing-procurement planning.

Fogg, C. Davis, *Team-Based Strategic Planning: A Complete Guide to Structuring, Facilitating & Implementing the Process*, New York: AMACOM, 1994.

Giannakis, Mihalis, and Simon R. Croom. "Toward the Development of a Supply Chain Management Paradigm." *Journal of Supply Chain Management* Spring (2004) p: 27–37.

Goodman, Alvin S. and Makarand Hastak. *Infrastructure Planning Handbook: Planning, Engineering, and Economics*. Burr Ridge, IL: McGraw-Hill, 2009.

Harvard Business Review on Managing the Value Chain Boston. MA: Harvard Business School Press, 2000.

Harvard Business Review on Supply Chain Management. Boston, MA: Harvard Business School Press, 2006.

Hunt, Shelby D., and Donna F. Davis. "Grounding Supply Chain Management in Resource Advantage Theory." *Journal of Supply Chain Management* 44, (1) (January 2008): 10–21.

Jacobs, F. Robert, Willson Lee Berry, D. Clay Whybark, and Thomas E. Vollman. *Manufacturing, Planning and Control for Supply Chain Management*. New York: McGraw-Hill Professional Books, 2011.

Killen, Kenneth H. and John W. Kamauff, *Managing Purchasing: Making the Supply Team Work*, Vol. 2 of the NAPM Professional Development Series. Tempe, AZ: The National Association of Purchasing Management (now called ISM), and New York: Irwin Professional Publishing, 1995.

Long, C., and G. Meyer. *Sacred Cows Make the Best Barbecue: Supply Chain Management: A Revolutionary 26 Week Action Plan*. Seal Beach, CA: Vision Press, 1998.

Naper, Rod, Clint Sidle and Patrick Sanaghan. *High Impact Tools and Strategies for Strategic Planning: Creative Techniques for Facilitating Your Organizations Planning Process*. New York: McGraw-Hill Professional Books, 1997.

Parker, Devlon B., George A. Zsidisin, and Gary L.Ragatz. "Timing and Extent of Supplier Integration in New Product Development: A Contingency Approach." *Journal of Supply Chain Management* 44 (1) (January 2008): 71–83.

Siegfried, Mary. "Transforming Supplier Relationships: Supply Management Professional Profile: Rolls-Royce North America." *Inside Supply Management* 19(1) (January 2008): 32–34. See Appendix D in this chapter for a full version of this article.

Steiner, George A. *Top Management Planning*, New York: MacMillan (Arkville Press),1969. This is the classic work on planning.

Supply and Demand Chain Executive Magazine, news items, events. See www.sdcexe.com.

Supply Chain Management Review. See www.scmr.com.

The Center for Advanced Purchasing Studies (CAPS), See www.capsresearch.org, Tempe, AZ.

Thompson, Jr., Arthur A., John E. Gamble, and A.J. Strickland III. *Crafting & Executing Strategy: The Quest for Competitive Advantage. Concepts and Cases*, 17th ed. Burr Ridge, IL: Mc-Graw Hill, 2009.

Trent, Robert J. *Strategic Supply Management: Creating the Next Source of Competitive Advantage*. Fort Lauderdale, FL: J. Ross Publishing, 2007.

Wilson, J. Holton, and Barry Keating, John Galt Solutions, Inc. Business Forecasting with ForecastX Software & Student CD, 5th ed. New York: McGraw-Hill, 2007.

The Future of Supply Management[1]

From Tactical to Strategic

The future of supply management is at a crossroads. For those who see the future as a reflection of the past, supply management will look merely as a *tactical* method to obtain adequate flow of quality materials and services at the lowest cost.

However, for those who are more *strategic* in their vision, supply management will emerge as a major contributor to their organization's *competitive advantage*, especially in the following crucial areas:

▲ Rapid innovation
▲ Extraordinary quality
▲ Faster time to market
▲ Dramatic reductions in total cost of ownership
▲ Assurance of continuity of supply
▲ Shortened cycle time

Over the years, a number of companies have taken a strategic view of their supply base, seeing it as a vital asset in their efforts to improve corporate performance and gain a competitive advantage. Leading senior executives who have come to recognize that supply is responsible for the largest portion of their company's budget see supply from a strategic, competitive advantage perspective. Individual companies who have recognized the strategic nature of their supply base and sought to leverage it include such firms as Toyota, Honda, Procter & Gamble, IBM, Cisco Systems, Eli Lilly, and Rolls Royce.

However, this approach has not been broadly embraced across the supply management field. Organizations not embracing this role of providing competitive advantage run the risk of quickly losing their competitiveness and disappearing.

A Look into the Future: Three Powerful Elements of Competitive Advantage

While predicting the future is always chancy, we have enjoyed a pretty good track record over the past 50 years. Three concepts will enable the transformation of supply management to strategic status:

1. Innovation
2. Shifting from linear supply "chains" to "integrated value networks"
3. Trust between buyers and sellers

These powerful concepts are predicated on the supply managers understanding the nature of their company's strategy, what creates value for customers, the massive advantage trust can bring to both buyer and seller, and an ability to work across corporate and departmental boundaries. Each of these skills are, by definition, unique and hard for competitors to copy—key components of competitive advantage.

In the fall of 2010, Andy Grove, former chief executive of Intel, addressed an audience at Stanford University, saying that we need a renaissance of manufacturing in America. He pointed out that manufacturing pays some $22 per hour, approximately twice what is received in the services industries. And new manufacturing jobs will help our futile efforts to reduce our current intractable level of unemployment.

In order for manufacturing to achieve this impact, it must be combined with an approach to supply, which harnesses its potential strategic power, principally in the areas of innovation, cost (at the total cost of ownership level, not merely purchase price), time to market, a synergistic approach to R&D, continuity of supply, and leadership.

Thus, we are comfortable writing "For the well being of our nation and the survival and success of our firms, supply chain management *must* focus on its potential strategic contributions."

Innovation

To illustrate the power of innovation, we will use two examples from the automotive industry.

Toyota has demonstrated the critical importance of innovation to its success. Some 65% of Toyota's innovations come from its supply base. In Toyota's North America operations alone, this amounts to over 2 million innovations annually. (Much of this innovation is in process flow and design improvements.) This has led to an enormous competitive advantage. Toyota's U.S. suppliers have reported that their major innovations flow to Toyota because of the trusting relationships (compared with little innovation flow to the American auto manufacturers who treat their suppliers with disparagement).

During the 1980s Toyota and Honda began a revolution in the way suppliers were regarded—not as commodity vendors to beat up, but as valued suppliers of a strategic asset to provide competitive advantage. In the late 1980s, Honda reframed the strategic game by involving suppliers inside the product development process. More than 65% of the American-made Honda innovations were originated by suppliers; thus, their early involvement and creative talents could benefit Honda immensely.

In Honda's eyes, suppliers were not lowly vendors providing a commodity (as rival General Motors believed), but a valued partner in the game. Innovation flows (including process improvements) were just as important as reductions in cost. If suppliers suggested a cost improvement, they were rewarded:

> When we receive a suggestion from our suppliers, we split the savings 50/50. However, if a supplier is not making their profit numbers, we give them a larger percentage of the savings (in the short term), sometimes up to 100%. It helps them out.
>
> —Dave Nelson, Sr. V.P. of Procurement, Honda[2]

During the five-year period from 2004 to 2008, the U.S. auto manufacturers (Ford, GM, and Chrysler), who regularly abused their suppliers, collectively lost $100 billion, while the collaborative Japanese manufacturers (Toyota, Honda, and Nissan) were all profitable.

Value Networks and the World to Come

It is our prediction that supply network management (SNM) and its big brother, value network management (VNM), will be commonplace within the coming years—at least at firms that survive and thrive!

SNM and VNM address the fact that business in the modern world is conducted in a networked fashion. It is far easier to visualize chains, whether supply chains or value chains, but truth be known, we live and operate in a world of networks!

As advanced supply chains become more strategic and collaborative, their very nature changes, just as a worm goes through a stage of metamorphosis first as a caterpillar and lastly as a butterfly. So, too, with supply chains when they become more strategic and morph into value chains—from supplier to customer—and then again into value networks as more of the players become better connected, faster, and more integrated.

In the old supply chain model, the economic value is measured in terms of costs from the monetary perspective of "transactional exchange." In this mode of thinking, profit comes from one's power position in the industry, the intensity of rivalry of competitors, and the control of precious resources. Moreover, prosperity is derived from one's ability to compete against competitors, suppliers, and customers.[3]

When supply and value chains are synergistically linked through strategic alliance architecture, they begin the transformation into supply and value networks. Here, the economic value is measured in different terms, not as tactical, transactional exchange, but as "mutually strategic value creation," primarily based on collaborative use of the intellectual capital of new ideas and the recognition of the mutual needs of buyer and supply to win and remain competitively sustainable. This model reinforces the need to flow innovations through the value network to ensure continuing competitive advantage to both the network and the ultimate customer.

In the value network, the paradigm of economic value shifts dramatically. Profit now comes from speed of innovation and response, as well as coordination and leverage of members' strengths and value of their core competencies. Prosperity is derived from the network's ability to create mutual value for its customers, with its suppliers, and possibly join forces with competitors when needed to satisfy customers.

Networks are flexible virtual systems linked by communication systems and collaborative relationships. Within any network, many things are happening simultaneously. Cooperative relationships must be established with suppliers and their suppliers back to Mother Earth. VNM augments SNM to include the marketing and distribution systems, which bring the goods and services to the ultimate customer.

Networks are designed and managed to drive cost out while ensuring that one member does not benefit at the expense of another. World-class supply and value networks are highly adaptive. They focus on value and speed, they are innovative, and they are highly integrated. Quite obviously, trust is an essential precondition to successful supply networks and value networks.

The model in Figure E.1 depicts the evolution from "buying" (primarily cost-based) to "value creation" (primarily focused on creating competitive advantage). It also serves the important role of a benchmark, allowing senior managers and corporate directors to judge how their organizations are progressing.

Trust

The key building block for network management is finally being understood. Colleague Robert Porter Lynch and his co-author, Paul Lawrence, have cracked the trust code. Their article, "Leadership and the Structure of Trust,"[4] provides essential insight into the development and maintenance of trust and collaborative innovation. This insight will allow us to develop and nurture trust among the many players in a network. As Lawrence and Lynch write, "Trust enables everything to move faster, more effortlessly, and with less conflict."

The key principles of trust are:

▲ Fairness
▲ Transparency

World-Class Supply Management

Level of Evolution	1. Clerical	2. Transactional	3. Process Based	4. Strategic	5. Systemic
Type	Tactical Buying	Bid-Based Buying	Supply Chain Mgmt	Value Chain Mgmt	Value Network Mgmt
Value Driver	Timely Availability, Convenience	Lowest Purchase Price	Cost, Quality, Just In Time	Total Cost of Ownership, Top & Bottom Line, Integration	Total Value Impact Costs, Revenues, with Quality, Innovation, & Speed
Financial Impact	Overhead-Cost Center	Improve Bottom Line, No Consideration of Revenue Impacts	Improve Bottom Line, Some Consideration of Revenue Impacts	Revenue & Bottom Line Impacts, Increase Share Holder Value	Future & Present Revenue & Bottom Line impacts Supplier & Stakeholder Impact
Integration Level & Functional Elements	No Integration, Purchasing stands alone	Low Integration, Procurement & Logistics	Partial Integration, Procurement, Logistics, Operations, Engineering	Moderate Integration, Procurement, Logistics, Operations, Engineering, R&D	Full Integration, Procurement, Logistics, Operations, Engineering, R&D, Sales, Marketing, Service, Customer
Basis of Competitive Advantage	Do the Job	Leverage Size of Buyer	Through Put, Global Impact	Coordination, Interconnectedness	Hyper-Competition, Hybridization of Technology, Speed & Innovation, Customization
Performance Metrics	Timeliness & Efficiency	Low Component/Unit Cost, On Time Delivery	Coordination & Cost, Develop Suppliers	Speed, Effectiveness, Monitor Supply Environ't	Innovation, Synchronicity, Synergy, Monitor Customer Environment
Innovation Focused on	Not Considered	Squeeze the Vendor is first priority	Process Innovation, Develop Requirements, Near Defect Free Supply	Speed & Integration, Design Supply Base, Integrated Supply Strategy	New Processes, Systems Solutions, Leverage Supplier Technology
Time Focus	Isolated from Customer	Reactive to Customer	Responsive to Customer	Pro-Active to Customer	Pre-Active with Customer

*Copyright by David N. Burt and Robert Porter Lynch.
Evolved from *The American Keiretsu* by David N. Burt and Michael Doyle, Business One–Irwin, Homewood, IL, 1993, p. 21.

Figure E.1 Evolution of supply management portrays the Burt-Lynch supply chain evolution model.

▲ Reliability
▲ Integrity
▲ Security
▲ Honesty
▲ Accountability

What is important about trust is that Lawrence and Lynch's studies show that trust creates at least a 25% competitive advantage in terms of speed of transactions and the elimination of non-value added work.

Transformation

The road to world-class SNM and VNM is a tortuous one, filled with challenges. The preceding chapters have addressed the technical principles underlying world class. We now need to address the even more challenging human and strategic issues.

The following insight has been gained from a combined 100 years of our being buyers, directors of purchasing, chief procurement officers, consultants, and academics. The first major challenge is obtaining executive support.

Two approaches are especially attractive: (1) gaining support of the CEO and (2) gaining support of manufacturing and the firm's outward-facing activities. In order to gain CEO support, arrange for an hour-long meeting with the CEO. Share the list of questions with the CEO (see Figure E.2).

It is our experience that these questions will result in a request for an action plan, which will become the basis of the implementation initiative.

The second approach calls for a meeting with one or more directors of manufacturing and the following outward facing activities: new product development, marketing, and sales. Ask them the same questions. Then ask them to rate the firm's supply management system against the supply chain evolution model.

This process should lay the foundation for a recommendation to the CEO to undertake the necessary transformation. In a worst-case scenario, manufacturing and the outward looking division will work with you as teammates to implement as much change as possible. Both approaches will be more successful if a realistic estimated dollar impact on sales and savings is made for the next two years.

Transformation Prerequisites

During our combined 100 years of experience in procurement, we have been fascinated with the process of transformation—especially the transformation from reactive

Robert Porter Lynch works with many senior executives on alliances, innovation, and supply chain strategy. He has shared with us the questions he asks when first meeting with senior management.

Supply Chain: Questions for Senior Execs

1. What percentage of your corporate budget is spent on outside sources?
   ▼ Typical Answer: 40–70%
   ▼ Does this make your procurement function "strategic" to your company?
   ▼ Typical Answer: Yes, I guess so!
2. Do you treat your supply chain as an "Expense" or as a "Strategic Asset"?
   ▼ Do you see your major suppliers as strategic partners?
   ▼ In your supply chain, 80% of your purchases go to what percentage of your suppliers?
   ▼ Typical Answer: 3–8% of suppliers account for 80% of expenses. (These should be the initial target for strategic supply innovation.)
3. Do you think of your supply chain as a means of creating competitive advantage?
   ▼ How well is your supply chain strategy connected to your customer strategy?
   ▼ Typical Answer: No consideration has been given to creating competitive advantage—this is the opportunity for transformation.
4. Is innovation a competitive advantage in your industry?
   ▼ What percentage of your innovation streams come from your supply chain?
   ▼ Typical Answer: Less than 5%
   ▼ So, if 60% of total corporate expenses go to suppliers, and you get only 5% of your innovation streams from suppliers, is there something wrong with this picture?
5. What is the level of trust you have with your most strategic suppliers?
   ▼ What is the real cost of distrust?
   ▼ If you want strategic supplier alliances to produce innovation flows, then you must have a strategy, system, and method for establishing trust.

Figure E.2 Questions for senior executives.

purchasing to proactive procurement to world-class supply management, and in the future into VNM. We have had the good fortune to work with huge international corporations, governments, midsized businesses, and small firms. We cannot overemphasize the importance of four keys to successful transformation:

▲ Executive sponsorship
▲ The presence of a champion (this individual may be from supply, alliances, marketing, or elsewhere in the corporation)
▲ The involvement of dedicated, educated, and committed personnel
▲ The availability of an experienced outside advisor

And there you have it: our vision of the future and some thoughts on how to get there. Remember, your company's survival and success is counting on you. Good luck!

Endnotes

1. Appreciation is expressed to colleagues Robert Porter Lynch of the Warren Co. and Stephen C. Rogers, formerly with Procter and Gamble, for their invaluable assistance with this Epilogue.
2. Interview with Robert Porter Lynch, October 21, 1997.
3. This is the model made renowned by Michael Porter in his treatises on strategy.
4. Lawrence, Paul R., and Lynch, Robert Porter, "Leadership and the Structure of Trust," *European Business Review*, May-June 2011.

INDEX

Note: Page numbers followed by *f* denote figures; page numbers followed by *t* denote tables; page numbers followed by *n* denote notes.